Astrophysics and Space Science Proceedings

Chaos in Astronomy

Conference 2007

G. Contopoulos
Editor

Research Center for Astronomy, Academy of Athens, Athens, Greece

P.A. Patsis
Editor

Research Center for Astronomy, Academy of Athens, Athens, Greece

Springer

Editors

G. Contopoulos
P.A. Patsis

Research Center for Astronomy
Academy of Athens
Soranou Efessiou 4
11527 Athens
Greece

ISBN: 978-3-540-75825-9 e-ISBN: 978-3-540-75826-6
DOI: 10.1007/978-3-540-75826-6

Library of Congress Control Number: 2008930098

Printed on acid-free paper

9 8 7 6 5 4 3 2 1

springer.com

In memory of Prof. Nikos Voglis

Preface

The conference *Chaos in Astronomy* was held in Athens from 17 to 20 September 2007 and was dedicated to the memory of Nikos Voglis, who was director of the Research Center for Astronomy of the Academy of Athens until his death on 9 February 2007. It was attended by 73 registered participants coming from 18 different countries. A total of 40 oral papers were delivered including a conference summary. Furthermore 16 posters were presented. The conference was the main event in a series of talks, public lectures and discussion meetings about "Chaos in Astronomy" that have taken place at the Research Center for Astronomy. We underline three special talks that have been given by D. Kazanas (NASA-Goddard Space Flight Center), D. Lynden-Bell (Cambridge, UK) and C. Tsallis (Brazilian Academy of Science).

The main sponsor of the conference was the "Alexander S. Onassis" Foundation. In addition the conference has been supported by the General Secretariat for Research and Technology, the National Observatory of Athens, and the Hellenic Ministry of Culture. We are grateful to all sponsors for their generosity.

We also express our gratitude to the Academy of Athens and to the directorship of the Biomedical Research Foundation, where the conference took place.

The objective of the conference was to bring together experts from all branches of Astronomy that work on nonlinear phenomena and Chaos. The scientific discussions concentrated in applications of the Chaos Theory to Galactic Dynamics. We discussed the chaotic motion of stars in the spiral arms of barred spiral galaxies, the chaotic phenomena associated with the boxy/peanut morphology of the central parts of galactic disks, the chaotic motion of stars in the solar neighborhood, the classification of ordered and chaotic orbits in elliptical galaxies, and in general the role of nonlinear phenomena in the observed structure of galaxies. Besides Galactic Dynamics, there were many talks about nonlinear phenomena and chaos in Celestial Mechanics, accretion disks around Black Holes, in Solar Physics and Cosmology. The present volume offers the state of the art of the ongoing discussions in topics of the Chaos theory in all these fields.

The present proceeding contains 49 papers. All submitted papers were refereed and almost all were revised. The procedure followed was the same as when papers are submitted to a research journal. We thank all the participants for their contributions and the referees for their help.

Athens
August 2008

G. Contopoulos
P.A. Patsis

Contents

List of Contributors XV

Nikolaos Voglis (1948–2007)
G. Contopoulos XXI

Part I Galactic Dynamics

Ordered and Chaotic Orbits in Spiral Galaxies
G. Contopoulos 3

Order and Chaos in Spiral Galaxies Seen through their Morphology
P. Grosbøl 23

The Flow through the Arms of Normal and Barred-Spiral Galaxies
P.A. Patsis 33

Ansae in Barred Galaxies, Observations and Simulations
I. Martinez-Valpuesta 45

Orbital Structure in Barred Galaxies and the Role of Chaos
M. Harsoula, G. Contopoulos, and C. Kalapotharakos 53

Chaos in Galaxies
D. Pfenniger 63

Boxy/Peanut Bulges: Formation, Evolution and Properties
E. Athanassoula and I. Martinez-Valpuesta 77

Invariant Manifolds as Building Blocks for the Formation of Spiral Arms and Rings in Barred Galaxies
M. Romero-Gómez, E. Athanassoula, J.J. Masdemont and C. García-Gómez 85

Secular Instabilities of Stellar Systems: Slow Mode Approach
E.V. Polyachenko, V.L. Polyachenko, and I.G. Shukhman 93

Stellar Velocity Distribution in Galactic Disks
Ch. Theis and E. Vorobyov 105

Dynamical Study of 2D and 3D Barred Galaxy Models
T. Manos and E. Athanassoula 115

Chaos in the Mergers of Galaxies
P.O. Vandervoort 123

Hamiltonian Normal Forms and Galactic Potentials
G. Pucacco 137

Local Phase Space: Shaped by Chaos?
D. Chakrabarty 151

Barred Galaxies: An Observer's Perspective
Dimitri A. Gadotti 159

Invariant Manifolds and the Spiral Arms of Barred Galaxies
C. Efthymiopoulos, P. Tsoutsis, C. Kalapotharakos and G. Contopoulos 173

Collisional N-Body Simulations and Time-Dependent Orbital Complexity
N.T. Faber 183

Resonances in Galactic and Circumstellar Disks
A.C. Quillen 191

Regular and Chaotic Motion in Elliptical Galaxies
J.C. Muzzio 203

Orbital Distributions and Self-Consistency in Elliptical Galaxies
C. Kalapotharakos 215

The Connection Between Orbits and Isophotal Shape in Elliptical Galaxies
R. Jesseit, T. Naab, and A. Burkert 225

Gas Orbits in a Spiral Potential
G.C. Gómez and M.A. Martos 231

The Structure of the Phase Space in Galactic Potentials of Three Degrees of Freedom
M. Katsanikas, P.A. Patsis, and L. Zachilas 235

Regular and Chaotic Orbits in Narrow 2D Bar Models
D.E. Kaufmann and P.A. Patsis 241

The Coalescence of Invariant Manifolds in Barred-Spiral Galaxies
P. Tsoutsis and C. Efthymiopoulos 247

Part II Celestial Mechanics

Chaotic Dynamics in Planetary Systems
R. Dvorak 255

Routes to Chaos in Resonant Extrasolar Planetary Systems
J.D. Hadjidemetriou and G. Voyatzis 269

Prometheus and Pandora, the Champions of Dynamical Chaos
I.I. Shevchenko 285

Planets in Multiple Star Systems: A Symplectic Approach
P.E. Verrier and N.W. Evans 293

Stabilized Chaos in the Sitnikov Problem
A.R. Dzhanoev, A. Loskutov, J.E. Howard, and M.A.F. Sánju 301

Part III Fundamental Concepts and Methods

Nonextensive Statistical Mechanics – An Approach to Complexity
C. Tsallis 309

Absolute Versus Relative Motion in Mechanics
D. Lynden-Bell 319

Far Fields, from Electrodynamics to Gravitation, and the Dark Matter Problem
A. Carati, S.L. Cacciatori, and L. Galgani 325

Distribution Functions for Galaxies using Quadratic Programming
V. Dury, S. De Rijcke, V. Debattista, and H. Dejonghe 337

Chaos Analysis Using the Patterns Method
I. Sideris 347

On the Topology of Regions of 3-D Particle Motions in Annular Configurations of n Bodies with a Central Post-Newtonian Potential
T. Kalvouridis 357

The Average Power-Law Growth of Deviation Vector and Tsallis Entropy
G. Lukes-Gerakopoulos, N. Voglis, and C. Efthymiopoulos 363

Global Dynamics of Coupled Standard Maps
T. Manos, Ch. Skokos, and T. Bountis 367

Part IV Other Astronomical Systems

The Chaotic Light Curves of Accreting Black Holes
D. Kazanas 375

3D Accretion Discs Dynamics: Numerical Simulations
D.V. Bisikalo 387

Growth of Density Fluctuations at the Time of Leptogenesis
E.A. Paschos 399

Fully Developed Turbulence in Accretion Discs of Binary Stars: Turbulent Viscosity Coefficient and Power Spectrum
A.M. Fridman, D.V. Bisikalo, A.A. Boyarchuk, L. Pustil'nik, and Y.M. Torgashin 403

Solar and Stellar Active Regions
Loukas Vlahos 423

Chaos and Self-Organization in Solar Flares: A Critical Analysis of the Present Approach
L. Pustil'nik 435

Charged Particles' Acceleration through Reconnecting Current Sheets in Solar Flares
C. Gontikakis, C. Efthymiopoulos, and A. Anastasiadis 449

The Perturbed Photometric-Magnetic Dynamical Model for the Sunspot Evolution
G. Livadiotis and X. Moussas 455

The Dynamics of Non-Symmetrically Collapsing Stars
G.S. Bisnovatyi-Kogan and O.Y. Tsupko 461

Cosmic Order out of Primordial Chaos: A Tribute to Nikos Voglis
B. Jones and R. van de Weygaert 467

Cosmological Inflation: A Personal Perspective
D. Kazanas 485

List of Contributors

A. Anastasiadis
National Observatory of Athens
Institute for Space Applications
and Remote Sensing
15236 Penteli, Greece
anastasi@space.noa.gr

E. Athanassoula
Laboratoire d'Astrophysique de
Marseille
Observatoire Astronomique de
Marseille-Provence Pole de l'Etoile
Site de Chateau-Gombert
38, rue Frederic Joliot-Curie
13388 Marseille cedex 13
France
lia@oamp.fr

D.V. Bisikalo
Institute of Astronomy of the
Russian Academy of Sciences
Moscow, Russia
bisikalo@inasan.ru

G.S. Bisnovatyi-Kogan
Space Research Institute of
Russian Academy of Science
Profsoyuznaya 84/32
Moscow 117997
Russia
and
Joint Institute for Nuclear Research,
Dubna, Russia
and
Moscow Engineering Physics
Institute
Moscow, Russia
gkogan@iki.rssi.ru

T. Bountis
Department of Mathematics
Center for Research
and Applications of Nonlinear
Systems (CRANS)
University of Patras
26500 Patras, Greece
bountis@math.upatras.gr

A.A. Boyarchuk
Institute of Astronomy of the
Russian Academy of Sciences
Moscow, Russia

A. Burkert
Universitätssternwarte
München
Scheinerstrasse 1
81679 München, Germany

S.L. Cacciatori
Department of Physical and Mathematical Sciences
Insubria University
Via Valleggio 11
22100 Como, Italy
sergio.cacciatori@uninsubria.it

A. Carati
Department of Mathematics
Milan University
Via Saldini 50
20133 Milano, Italy
andrea.carati@unimi.it

Dalia Chakrabarty
School of Physics and Astronomy
University of Nottingham
Nottingham NG7 2RD, UK
dalia.chakrabarty@nottingham.ac.uk

George Contopoulos
Research Center for Astronomy
Academy of Athens
Soranou Efessiou 4
11527 Athens, Greece
gcontop@academyofathens.gr

V. Debattista
Center for Astrophysics
University of Central Lancashire
Preston PR1 2HE, UK
vpdebattista@uclan.ac.uk

H. Dejonghe
Sterrenkundig Observatorium
Universiteit Gent
Krijgslaan 281
9000 Gent, Belgium
Herwig.Dejonghe@UGent.be

S. De Rijcke
Sterrenkundig Observatorium
Universiteit Gent
Krijgslaan 281
9000 Gent, Belgium

V. Dury
Sterrenkundig Observatorium
Universiteit Gent
Krijgslaan 281
9000 Gent, Belgium

Rudolf Dvorak
Institute of Astronomy
University of Vienna
Türkenschanzstr. 17
1180 Vienna, Austria
dvorak@astro.univie.ac.at

Arsen R. Dzhanoev
Departamento de Fisica
Universidad Rey Juan Carlos
28933 Mostoles
Madrid, Spain
arsen.dzhanoev@urjc.es

C. Efthymiopoulos
Research Center for Astronomy
Academy of Athens
Soranou Efessiou 4
11527 Athens, Greece
cefthim@academyofathens.gr

N. Wyn Evans
Institute of Astronomy
University of Cambridge
Madingley Road
Cambridge CB3 OHA, UK
nwe@ast.cam.ac.uk

N.T. Faber
Observatoire Astronomique
Université de Strasbourg
and CNRS UMR 7550
11 rue de l'Université
67000 Strasbourg, France
nfaber@astromaster.u-strasbg.fr

A.M. Fridman
Institute of Astronomy of the Russian Academy of Sciences
Moscow, Russia
afridman@post.tau.ac.il

Dimitri A. Gadotti
Max Planck Institute for Astrophysics
Karl-Schwarzschild-Strasse 1
85741 Garching bei München
Germany
dimitri@mpa-garching.mpg.de

L. Galgani
Department of Mathematics
Milan University
Via Saldini 50
20133 Milano, Italy
luigi.galgani@unimi.it

C. García-Gómez
D.E.I.M., Universitat Rovira i Virgili
Campus Sescelades
Avd. dels Països Catalans 26
43007 Tarragona, Spain
carlos.garciag@urv.cat

Gilberto C. Gómez
Centro de Radioastronomía y Astrofísica
Universidad Nacional Autónoma de México
Apartado Postal 3-72 (Xangari)
Morelia
Mich. 58089, México
g.gomez@astrosmo.unam.mx

C. Gontikakis
Research Center for Astronomy
Academy of Athens
Soranou Efessiou 4
11527 Athens, Greece
cgontik@academyofathens.gr

Preben Grosbøl
European Southern Observatory
Karl-Schwarzschild-Strasse 2
85748 Garching, Germany
pgrosbol@eso.org

John D. Hadjidemetriou
Department of Physics
University of Thessaloniki
Thessaloniki, Greece
hadjidem@auth.gr

M. Harsoula
Research Center for Astronomy
Academy of Athens
Soranou Efessiou 4
11527 Athens, Greece
mharsoul@academyofathens.gr

James E. Howard
Laboratory for Atmospheric and Space Physics and Center for Integrated Plasma Studies University of Colorado
Boulder, CO 80309, USA

R. Jesseit
Universitätssternwarte München
Scheinerstrasse 1
81679 München, Germany
jesseit@usm.lmu.de

Bernard Jones
Kapteyn Astronomical Institute
University of Groningen
P.O. Box 800
9700 AV Groningen
The Netherlands
bernard@astrag.demon.co.uk

Constantinos Kalapotharakos
Research Center for Astronomy
Academy of Athens
Soranou Efessiou 4
11527 Athens, Greece
ckalapot@phys.uoa.gr

Tilemahos Kalvouridis
Department of Mechanics
National Technical University of Athens
Athens, Greece
tkalvouridis@gmail.com

M. Katsanikas
Research Center for Astronomy
Academy of Athens
Soranou Efessiou 4
11527 Athens, Greece
mkatsan@yahoo.gr

David E. Kaufmann
Department of Space Operations
Southwest Research Institute
Suite 300
1050 Walnut Street
Boulder, CO 80302-5142, USA
kaufmann@boulder.swri.edu

D. Kazanas
Astrophysics Science Division
Code 663, NASA/GSFC
Greenbelt, MD 20771, USA
Demos.Kazanas@nasa.gov

George Livadiotis
Department of Astrophysics
Astronomy and Mechanics
National University of Athens
Panepistimiopolis
15784 Zografos, Athens, Greece
glivad@phys.uoa.gr

Alexander Loskutov
Moscow State University
119992 Moscow, Russia

G. Lukes-Gerakopoulos
Research Center for Astronomy
Academy of Athens
Soranou Efessiou 4
11527 Athens, Greece
gglukes@phys.uoa.gr

Donald Lynden-Bell
The Institute of Astronomy
Madingley Road
Cambridge CB3 OHA, UK
and
Clare College
Cambridge CB3 OHA, UK
dlb@ast.cam.ac.uk

T. Manos
Department of Mathematics
Center for Research and Applications
of Nonlinear Systems (CRANS)
University of Patras
26500 Patras, Greece
thanosm@master.math.upatras.gr

Inma Martinez-Valpuesta
Instituto de Astrofísica de Canarias
C/Vía Láctea
38200 La Laguna, Tenerife, Spain
imv@iac.es

Marco A. Martos
Instituto de Astronomía –
Universidad Nacional Autónoma
de México
Apartado Postal 70-264
Ciudad Universitaria, D.F. 04510
México
marco@astroscu.unam.mx

J.J. Masdemont
I.E.E.C & Department of
Mathematica Aplicada I
Universitat Politècnica de
Catalunya, Diagonal 647
08028 Barcelona, Spain
josep@barquins.upc.edu

Xenophon Moussas
Department of Astrophysics
Astronomy and Mechanics
National University of Athens
Panepistimiopolis
15784 Zografos, Athens, Greece
xmousas@phys.uoa.gr

Juan C. Muzzio
Facultad de Ciencias Astronómicas y
Geofísicas de la Universidad Nacional
de La Plata and
Instituto de Astrofísica de La
Plata (CCT–CONICET La Plata
and UNLP)
La Plata, Argentina
jcmuzzio@fcaglp.fcaglp.unlp.edu.ar

T. Naab
Universitätssternwarte München
Scheinerstrasse 1
81679 München, Germany

Emmanuel A. Paschos
Institute of Physics
University of Dortmund
44221 Dortmund, Germany
paschos.e@gmail.com

P.A. Patsis
Research Center for Astronomy
Academy of Athens
Soranou Efessiou 4
11527 Athens, Greece
patsis@academyofathens.gr

Daniel Pfenniger
Geneva Observatory
University of Geneva
1290 Sauverny, Switzerland
daniel.pfenniger@obs.unige.ch

E.V. Polyachenko
Institute of Astronomy
Russian Academy of Sciences
48 Pyatnitskya St.
Moscow 119017, Russia
epolyach@inasan.ru

V.L. Polyachenko
Institute of Astronomy
Russian Academy of Sciences
48 Pyatnitskya St.
Moscow 119017, Russia

Giuseppe Pucacco
Dipartimento di Fisica – Università di Roma "Tor Vergata"
Via de lla Ricerca Scientifica
00133 Rome, Italy
pucacco@roma2.infn.it

L. Pustil'nik
Israel Space Weather and Cosmic Ray Center
Tel Aviv University, Israel
levpust@post.tau.ac.il

Alice C. Quillen
Department of Physics and Astronomy
University of Rochester
Rochester, NY, USA
aquillen@pas.rochester.edu

M. Romero-Gómez
Laboratoire d'Astrophysique de Marseille
Observatoire Astronomique de Marseille-Provence Pole de l'Etoile
Site de Chateau-Gombert
38, rue Frederic Joliot-Curie
13388 Marseille cedex 13
France
merce.romerogomez@oamp.fr

Miguel A.F. Sánju
Departamento de Fisica
Universidad Rey Juan Carlos
28933 Mostoles
Madrid, Spain

Ivan I. Shevchenko
Pulkovo Observatory of the Russian Academy of Sciences
Pulkovskoje ave. 65/1
St. Petersburg 196140, Russia
iis@gao.spb.ru

I.G. Shukhman
Institute of Solar-Terrestrial Physics
Russian Academy of Sciences
Siberian Branch
P.O. Box 291
Irkutsk 664033, Russia
shukhman@iszf.irk.ru

Ioannis Sideris
Institute for Theoretical Physics
University of Zürich
Winterthurerstrasse 190
8057 Zürich, Switzerland
sideris@physik.unizh.ch

Ch. Skokos
Astronomie et Systèmes
Dynamiques, IMCCE
Observatoire de Paris
77 Av. Denfert–Rochereau
75014 Paris, France
hskokos@imcce.fr

Christian Theis
Institute of Astronomy
University of Vienna
Türkenschanzstrasse 17
1180 Vienna, Austria
theis@astro.univie.ac.at

Yu.M. Torgashin
Institute of Astronomy
of the Russian Academy
of Sciences
Moscow, Russia

Constantino Tsallis
Centro Brasileiro de Pesquisas
Físicas
Rua Xavier Sigaud 150
22290-180 Rio de Janeiro, Brazil
tsallis@santafe.edu

P. Tsoutsis
Research Center for Astronomy
Academy of Athens
Soranou Efessiou 4
11527 Athens, Greece
ptsoutsi@phys.uoa.gr

O. Yu. Tsupko
Space Research Institute of
Russian Academy of Science
Profsoyuznaya 84/32
Moscow 117997, Russia
tsupko@iki.rssi.ru
and
Moscow Engineering Physics
Institute, Moscow, Russia

Peter O. Vandervoort
Department of Astronomy
and Astrophysics
The University of Chicago
5640 S. Ellis Avenue
Chicago, IL 60637-1433, USA
voort@oddjob.uchicago.edu

Patricia E. Verrier
Institute of Astronomy
University of Cambridge
Madingley Road
Cambridge CB3 OHA, UK
pverrier@ast.cam.ac.uk

Loukas Vlahos
Department of Physics
University of Thessaloniki
54124 Thessaloniki, Greece
vlahos@astro.auth.gr

Eduard Vorobyov
University of Western Ontario
London, ON, N6A 3K7, Canada
vorobyov@astro.uwo.ca

George Voyatzis
Department of Physics
University of Thessaloniki
Thessaloniki, Greece
voyatzis@auth.gr

Rien van de Weygaert
Kapteyn Astronomical Institute
University of Groningen
P.O. Box 800
9700 AV Groningen
The Netherlands
weygaert@astro.rug.nl

L. Zachilas
Department of Economic Studies
University of Thessaly
Volos, Greece
zachilas@uth.gr

Nikolaos Voglis (1948–2007)

G. Contopoulos

I welcome all the participants of this symposium on "Chaos in Astronomy", in memory of the late director of the Center for Astronomy of the Academy of Athens, Dr. Nikos Voglis.

Nikos Voglis was the main organizer of the previous symposium on "Galaxies and Chaos" in 2002, and of the present symposium until his death in February 2007.

His death was shocking, as it was quite unexpected. The previous evening I had presented at the Academy of Athens a 10-year summary of the activities of our Center and Nikos Voglis role was emphasized. The next morning he came to his office as usual and had an important collaboration with his students. But when he returned to his office he had an unexpected cardiac arrest and despite the doctors' efforts he never recovered.

The shock for all of us was great because Voglis was the guiding spirit of our Center. He devoted many hours every day not only to his own scientific work, but also in helping his students and colleagues. He had supervised the work of 4 PhD students and 6 Masters' students, and several more people profited from his advice, as he taught two postgraduate courses at the University of Athens.

Voglis was born in 1948. He was my student at the University of Thessaloniki and graduated in 1976. Then he was appointed teacher at a high school, but later (1981) he accepted my invitation to come to our Department of Astronomy as an assistant. After that time he progressed fast. He got his PhD in Astronomy in 1984, and he was appointed lecturer in 1986, assistant professor in 1990 and associate professor in 1997 at the Department of Physics of the University of Athens.

In 2000 he was elected researcher A at the Center for Astronomy of the Academy of Athens (equivalent to full professor) and in 2001 he became director of the Center until his death.

He published about 100 scientific papers and he was editor of a proceedings volume.

Voglis' main contribution in Astronomy was in the field of Galactic Dynamics. He developed new methods in this field, that led to significant new results.

His method of distinguishing between Order and Chaos in dynamical systems, was based on his "stretching number", i.e. the short time deviations of nearby orbits. This led to the dynamical spectra of stellar systems that opened new areas in the theory of chaos with applications to galaxies and more general dynamical systems.

Another import contribution was the discovery of the role of distant galaxies in providing angular momentum to a particular galaxy. He could explain the counterotating populations observed in many galaxies due to tidal effects.

He worked also on the formation of galaxies, by the collapse of protogalaxies and on the role of the gas in collapsing systems.

He explored systems that evolve but still retain a memory of the initial conditions. He emphasized the role of chaotic orbits in establishing self-consistent models of galaxies.

He studied thoroughly the role of a central black hole in the center of a galaxy. He noticed that the addition of a black hole changes many orbits form ordered to chaotic. However, later on, a new equilibrium is reached, where many chaotic orbits become again ordered, but of different type. Then, he explained how this evolutionary effect does not contradict the principle of entropy increase.

He worked further on chaos in Relativity and Cosmology and on various mathematical problems. In particular, he studied the role of solitons and breathers in galactic dynamics. He established a connection between the third integral that is derived from ordinary differential equations and various partial differential equations that generate solitons and breathers.

Finally, his recent work was devoted to the role of chaos in the spiral arms of strong barred galaxies. He noticed that in such systems the spiral arms outside the bar are composed mainly of chaotic orbits. The spirals follow the asymptotic curves from the unstable periodic orbits close to the ends of the bar. Up to now, the theory of density waves emphasized the spiral arms in normal galaxies, that are composed of ordered orbits. In this case, Chaos is unimportant. However, in strong bars Chaos is dominant and the formation of spiral arms follows a different mechanism. This opens a new dimension in Galactic Dynamics.

Thus, it is appropriate that the present symposium on "Chaos in Astronomy" should be devoted to the memory of Nikos Voglis.

The conference photo with the participants at the main entrance of the Academy Conference Center.

Nikos Voglis at the “bernard60” conference (Valencia, June 2006). Picture taken by Phil Palmer.

Part I

Galactic Dynamics

Ordered and Chaotic Orbits in Spiral Galaxies

G. Contopoulos

Research Center for Astronomy, Academy of Athens,
Soranou Efessiou 4, GR-11527, Athens, Greece
gcontop@academyofathens.gr

Summary. The spiral arms in normal spirals produce perturbations of the order of 2–10% of the axisymmetric background. Chaos is in general quite limited. Thus, the spiral arms are composed mainly of regular orbits. On the other hand, barred galaxies produce large perturbations of the order of 100% and generate a large degree of chaos. Chaos is most important around and outside corotation and also in an outer envelope of the bar. The bar consists mainly of ordered orbits, while the outer spiral arms consist mainly of chaotic orbits. The structure of the spiral arms depends on the unstable asymptotic curves from the unstable periodic orbits around the Lagrangian points L_1, L_2 at the end of the bar. Stars, starting near these Lagrangian points, move close to the unstable asymptotic manifolds. Orbits, starting close to other unstable periodic orbits inside and outside corotation, follow similar manifolds. Various recent developments on this topic are discussed.

1 Introduction

The density wave theory of spiral structure started with the work of B. Lindblad [23, 24]. However, his work did not draw the attention it deserved for two reasons: (a) Its style was rather difficult to follow, and (b) he believed that the spiral arms are leading. Thus, although the density wave theory is valid both for trailing and leading spirals, he applied this theory mainly to leading spirals.[1] But leading spirals were, later, in general excluded by observations.

The revival of the density wave theory was due to Lin and Shu [21]. This was a linear theory of spiral density waves, that explained the observed spiral arms.[2]

[1] However in his late years he turned his interest to trailing spirals and developed a circulation theory of trailing spiral structure [25, 26].

[2] I met C. C. Lin at MIT in 1963 and he told me that he wanted to develop a theory of spiral density waves. I pointed out to him that B. Lindblad already worked on this problem. We went together to the library and borrowed a number of volumes of the Stockholms Obserbatoriums Annaler. But when I met again Dr. Lin the

G. Contopoulos, P.A. Patsis, *Chaos in Astronomy.* Astrophysics and Space Science Proceedings, doi: 10.1007/978-3-540-75826-6,

A similar, but independent, work was done around the same time by Kalnajs [17, 18]. Much further work was done later by Lin [20], Lin et al. [22], Shu [37], Toomre [38, 39], Lynden-Bell and Kalnajs [27], Bertin et al. [2] and others.

My own work on the theory of density waves dealt with resonances. The usual way to formulate the density wave theory is by developing the potential V, the density σ, and the distribution function f (density in phase space) in powers of the amplitude of the spiral perturbation, e.g.

$$f = f_0 + f_1 + f_2 + \cdots , \tag{1}$$

where f_0 represents the axisymmetric distribution, etc. This series is a particular form of the third integral of motion. In fact, I had already developed with L. Woltjer such a theory for a very rough model of spiral arms [13].

The linear theory deals with the first order term f_1. However, such a theory fails near the main resonances of a galaxy, namely the Lindblad resonances and corotation, because in these cases f_1 contains small divisors that make $|f_1|$ larger than $|f_0|$. But in resonant cases we can find other forms of the third integral [3–5]. There is a different form of the third integral for each resonance, but all forms are different from the original form (1).

The form (1) (away from the main resonances) together with the resonant forms explain the main characteristics of the spiral galaxies whenever the perturbation is not large. However, in cases of large perturbations this theory is not applicable. In fact, in such cases, the various resonances interact with each other and we have a large degree of chaos.

Because of that we must differentiate between normal galaxies which have relatively small perturbations and only a small degree of chaos and barred galaxies that have very large perturbations and a large degree of chaos.

2 Normal Spirals

In normal galaxies the amplitude of the perturbation (in the radial force) is in general of the order of 2–10% of the axisymmetric background. The Sc, Sb spirals are of order up to 10% (open spirals), while the Sa spirals are of order 2% (tight spirals). In normal cases, chaos is insignificant and most orbits are regular. Nevertheless non-linear effects are important for amplitudes of the order of 10%. In fact, it was shown by Contopoulos [9] that while the stable periodic orbits between the inner Lindblad resonance (ILR) and the 4/1 resonance are oriented in such a way, as to support the spirals, the periodic orbits beyond the 4/1 resonance do not support the spirals (Fig. 1).

next day, he told me that he had difficulties in understanding Lindblad's papers and that he decided to do the theory independently, from scratch. In fact, his theory was quite successful and it was widely used in the years to come.

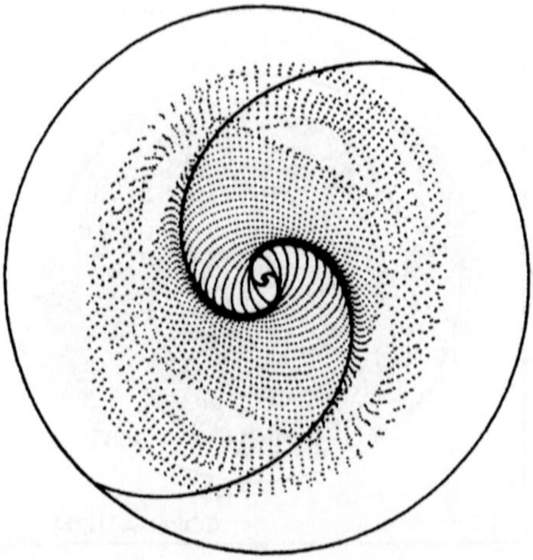

Fig. 1. Stable periodic orbits in a model of the spiral galaxy NGC5247. The orbits support the spiral up to the 4/1 resonance. The *circle* represents corotation (Contopoulos [9])

Non periodic orbits follow the stable periodic orbits. Thus, relatively strong self-consistent spirals terminate near the 4/1 resonance.

Efforts to explain the strong spiral arms as terminating near corotation, or near the outer Lindblad resonance (OLR) [12, 31] have failed. Nevertheless, weak extensions of the spirals extend beyond the 4/1 resonance up to corotation [30].

The deviations of the orbits beyond the 4/1 resonance are due to nonlinear effects [12]. Only in weak spirals the nonlinear effects are small and the spirals can extend all the way to corotation.

The response of gas to self-consistent stellar spirals follows the stars up to the 4/1 resonance but beyond that we have in general multiple weak arms extending up to corotation (Fig. 2) [31, 32].

3 Barred Spirals

The perturbations due to bars are in general very strong, of the order of 100% of the axisymmetric background, therefore chaos is expected to be important. Chaos is most important near and beyond the end of the bar at corotation.

There are two reasons that limit the extent of bars up to corotation: (1) The forms of the regular orbits support the bar inside corotation, but they tend to destroy any extension of the bar outside corotation [7]. (2) Near corotation, most orbits are chaotic [8].

The fact that the main bars terminate near corotation is now generally accepted [14, 34, 36]. (However, there may be also short bars inside the inner ILR, if the ILR is double [14].)

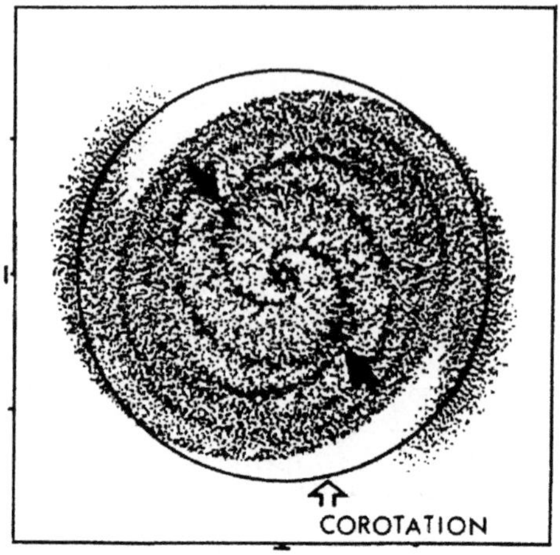

Fig. 2. The response of a gaseous disk to a stellar spiral terminating at the 4/1 resonance (*arrows*). The response is along the imposed spirals up to the 4/1 resonance. Beyond that there is a four armed extension up to corotation (*circle*) (Patsis et al. [31])

Beyond the end of the main bar we have spirals. We can consider spirals rotating with the pattern velocity of the bar, which emanate from the ends of the bar [19], or spirals rotating with a different pattern velocity [35].

A surprising result of the early studies of orbits near corotation was that chaotic orbits are partly supporting the spiral arms [19]. Such orbits, shown in Fig. 3, support both the spiral arms outside corotation and the outer regions of the bar.

A more detailed study of the chaotic orbits near and outside corotation was made by Voglis and his associates [42, 43]. His work dealt with N-body systems rotating rather fast. In such cases corotation is inside the main body of the galaxy. The first important result of these studies of Voglis was that the spiral arms are composed mainly of chaotic orbits (Fig. 4). Chaotic orbits dominate also an outer envelope of the bar, while ordered orbits form the central body of the bar.

Ordered orbits appear also in the outermost parts of the galaxy, beyond the $-4/1$ resonance. However, the density of stars in these regions is quite small, thus the number of stars in ordered orbits is small there.

The proportion of chaotic orbits in N-body models of rotating galaxies is twice the corresponding proportion in non-rotating systems (elliptical galaxies) (60% vs. 30%). Furthermore, the Lyapunov characteristic numbers in rotating systems are by one order of magnitude larger than in non-rotating systems [42]. This increase is due to the overlap of resonances near corotation [11]. In fact, near corotation there are infinite resonances between the angular velocity in the rotating frame (Ω-Ω_s) and the epicyclic frequency κ. These resonances interact if the amplitude of the bar is not very small. The

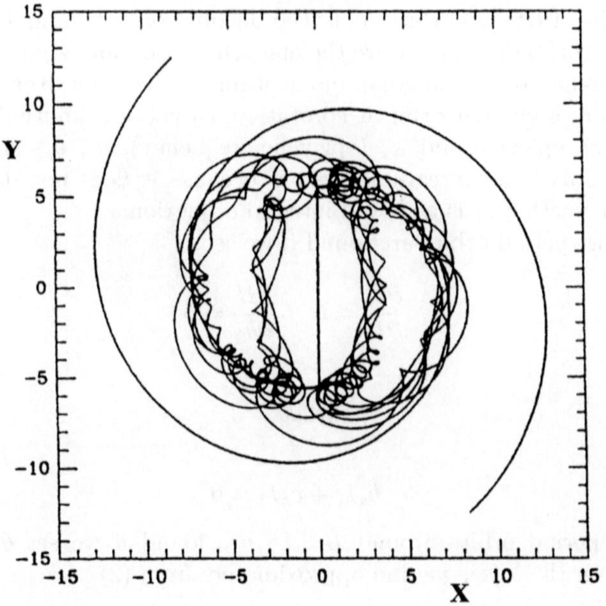

Fig. 3. A chaotic orbit supporting the outer parts of the bar inside corotation and partly supporting the bar outside corotation (Kaufmann and Contopoulos [19])

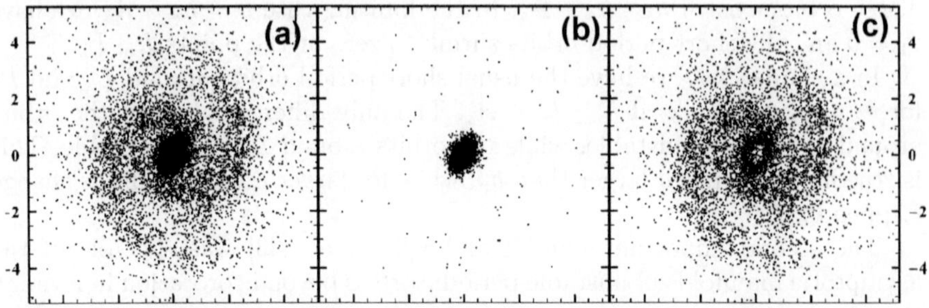

Fig. 4. The particles forming an N-body barred galaxy (**a**), are ordered inside the bar (**b**) and chaotic along the spiral arms and along the envelope of the bar (**c**) (Voglis et al. [42])

interaction increases as the amplitude of the perturbation increases. In non-rotating systems there is no corotation and chaos is due to different types of resonances.

If the rotation is faster (larger Ω_s), corotation moves closer to the center of the galaxy and the various resonances affect a larger number of stellar orbits.

The periodic orbits can be found theoretically in a simple case of a barred galaxy with Hamiltonian in action-angle variables (I_i, θ_i), $(i = 1,\ 2)$:

$$H \equiv h_s + \kappa_s I_1 + a_s I_1^2 + 2b_s I_1 I_2 + c_s I_2^2 + A_s \cos(2\theta_2) = h \qquad (2)$$

(Contopoulos 1978 [6]). Here, h is the Jacobi constant and h_s the Jacobi constant at corotation. I_1, θ_1 are the epicyclic action and angle, $I_2 = J_0 - J_s$ is the difference of the angular momentum of the orbit from the angular momentum of a circular orbit at corotation, θ_2 is the azimuthal angle of the center of the epicycle and κ_s (epicyclic frequency), a_s, b_s, c_s and A_s are constants. This is an integrable system because it does not depend on θ_1, therefore the action I_1 is a second integral of motion.

The short period orbits are found if we set

$$\frac{\partial H}{\partial I_2} = 0, \quad \frac{\partial H}{\partial \theta_2} = 0. \tag{3}$$

Therefore,

$$\sin 2\theta_2 = 0 \tag{4}$$

and

$$b_s I_1 + c_s I_2 = 0. \tag{5}$$

The short period orbits around L_1, L_2 are found if we set $\theta_2 = \pi/2$, or $\theta_2 = 3\pi/2$. In this case, we find approximately from (2)

$$I_1 = \frac{1}{\kappa_s}\left(h - h_s + A_s\right). \tag{6}$$

These orbits exist if $h \geq h_s - A_s$. In the limiting case $h = h_s - A_s$, we have $I_1 = 0$ and the short period orbits shrink to zero at the points L_1, L_2.

In a similar way we have the usual short period orbits around L_4 and L_5 for $\theta_2 = 0$, or $\theta_2 = \pi$ if $h \geq h_s + A_s$. The only difference is that the orbits around L_1, L_2 are unstable, while the orbits around L_4, L_5 are stable. This is true for h not much larger than h_s, while for larger h the stabilities change (Appendix A).

The second important remark of Voglis et al. [43] was the role of the asymptotic manifolds of unstable periodic orbits beyond corotation in forming the spiral arms.

At the unstable Lagrangian points L_1, L_2 in corotation start the two families of unstable short period orbits and they become larger as the Jacobi constant h increases.

For every value of h ($\geq h_s$), there are two such unstable orbits that generate two stable and unstable manifolds in phase space. Orbits near these manifolds are chaotic. The particles on the unstable manifolds move either outwards, forming the outer spiral arms, or inwards, forming the chaotic envelope of the bar (Fig. 5).

In order to visualize the behaviour of these orbits Voglis et al. [43] used a special surface of section on the configuration plane (r, θ), that is defined by the apocentra $\dot{r} = 0$ of the orbits. Two unstable asymptotic curves from the apocentra P_{L_1}, P_{L_2} of the unstable periodic orbits around L_1, L_2 form spirals extending outwards in a trailing way (along U and U′ in Fig. 5). The orbits of

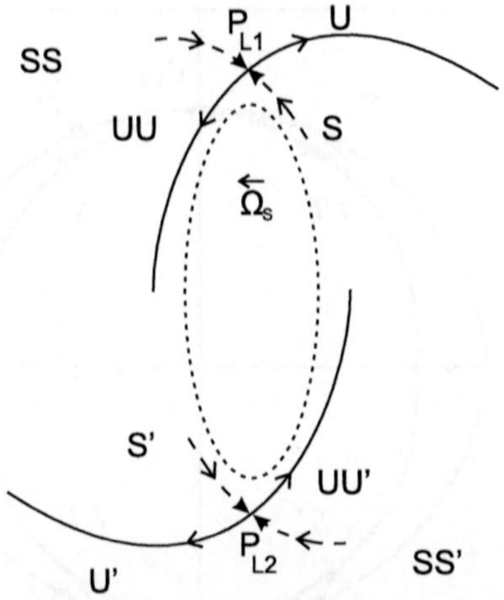

Fig. 5. Orbits starting close to the unstable periodic orbits P_{L_1}, P_{L_2} move towards P_{L_1}, P_{L_2} along the stable asymptotic curves S, SS and S′, SS′ and away from P_{L_1}, P_{L_2} along the unstable asymptotic curves U, UU and U′, UU′

stars starting close to L_1, L_2 have their apocentra near these spirals, therefore the stars on these orbits stay for a relatively long time close to these lines forming a spiral pattern. Of course the individual stars stay only temporarily near their apocentra, therefore the spiral arms are waves and not material arms. The other two unstable asymptotic curves are along the outer envelope of the bar (UU and UU′ in Fig. 5).

If the bar perturbation is small the unstable asymptotic curves are nearly separatrices joining P_{L_1} with P_{L_2}. In Fig. 6 we see the unstable asymptotic curve U from P_{L_1}, that reaches P_{L_2}, passing outside the Lagrangian point L_4, while the curve UU reaches P_{L_2} passing inside the Lagrangian point L_5. These asymptotic curves are similar to the ones from the unstable Lagrangian points L_1 and L_2, that appear if $h = h(L_1)$ (Jacobi constant at L_1). Weak perturbations can explain the double rings observed in certain galaxies [33].

When the curve U from P_{L_1} approaches P_{L_2}, it oscillates across the curve SS′ from P_{L_2}. These oscillations become larger as the curve U comes closer to P_{L_2} and they approach asymptotically the curves U′ and UU′ from P_{L_2}. In fact, these oscillations are the main characteristic of chaos. The curve UU′ performs similar oscillations across the curve S′ from P_{L_2}, when it approaches P_{L_2}. The overall picture is isomorphic to a pendulum-like phase portrait from $P_{L_1}$and P_{L_2} if we use the action-angle variables I_2, θ_2 (Fig. 7). If the perturbation is not small, the unstable asymptotic curves are not very close to separatrices but deviate considerably from them forming large oscillations.

The motions of particles along the asymptotic curves can be found by calculating the consequents (images of points on the I_2, θ_2 plane) of initial points close to P_{L_1}. The theoretical motions can be described approximately as a flow given by a Sine-Gordon partial differential equation [41, 43]. This

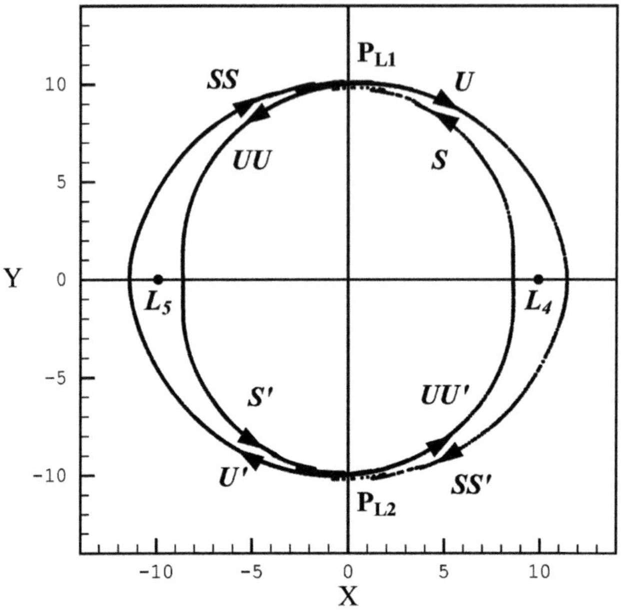

Fig. 6. Asymptotic curves from the unstable periodic orbits P_{L_1}, P_{L_2}. The asymptotic curve U from P_{L_1} approaches P_{L_2} close (but not exactly along) the curve SS′ and makes large oscillations close to U′ and UU′ approaching again P_{L_1}

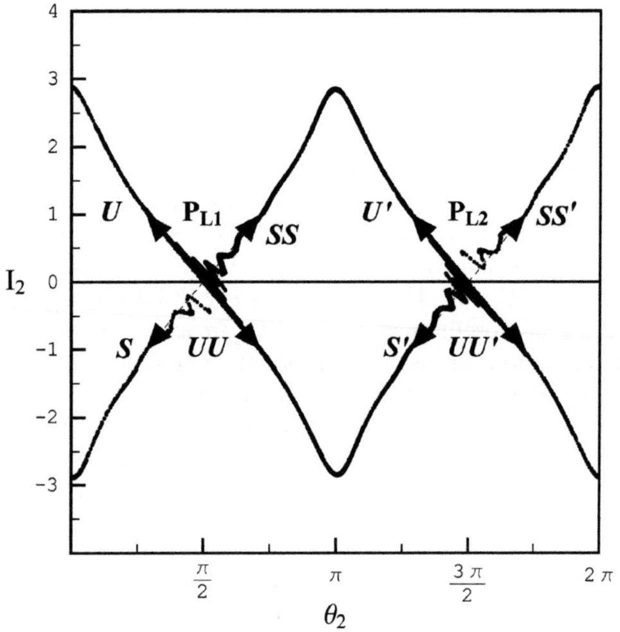

Fig. 7. The asymptotic curves U, UU from P_{L_1} and U′, UU′ from P_{L_2}, in the action-angle variables i_2, θ_2

equation secures that the phases along the asymptotic curve remain correlated. Namely, the angles θ of successive apocentra of nearby orbits are along similar sequences, despite the fact that the orbits are chaotic.

The Sine-Gordon equation is an integrable PDE and its main solutions are called kinks and antikinks. In the galactic case, these solutions are exact in the limit of separatrices that appear for vanishing perturbations. However, it is remarkable that these solutions are approximately correct even for relatively large perturbations. Thus, the phase correlations of the apocentra of the orbits persist even in strong bars. These phase correlations can be seen along the spiral arms composed of numerically computed apocentra of orbits.

In the case of large perturbations the oscillations of the asymptotic curves become larger and they cover a relatively large part of the galaxy. However, the overall figure is clearly spiral (Fig. 8).

Orbits starting close to the asymptotic curve S approach the point P_{L_1} (Fig. 5) and deviate either close the curve U outwards, along the spiral, or close to the curve UU inwards, along the envelope of the bar. Orbits deviating along U start by making a number of loops close to the unstable periodic orbit P_{L_1} (Fig. 9a,b). Further on they make large loops and after they approach P_{L_2}, they deviate either along U′ (Fig. 9a), or they go close to UU′ (Fig. 9b).

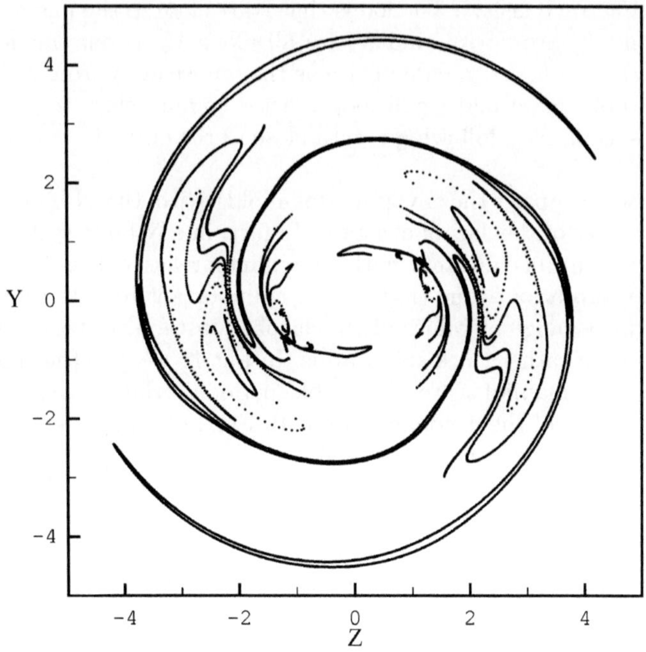

Fig. 8. The invariant manifolds U and U′ from P_{L_1} and P_{L_2} respectively (Voglis et al. [43])

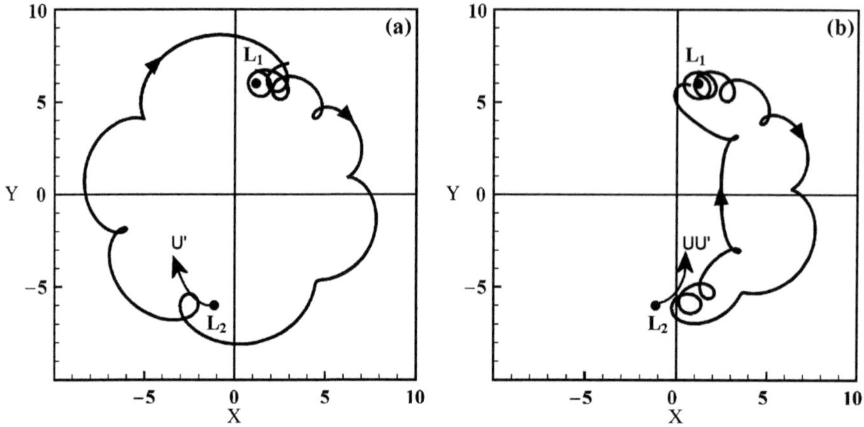

Fig. 9. Orbits starting close to the unstable periodic orbit P_{L_1} and deviating close to U. After they approach the region of P_{L_2} they deviate either along the direction of U′ (**a**), or along the direction of UU′ (**b**)

Note that Fig. 9a,b correspond to orbits with a large bar perturbation (the maximum density on the bar in this model is of the order of 100% of the axisymmetric density). The loops that start close to the unstable periodic orbit around L_1 are not of similar size all along U, as one might expect to see. Instead, they first decrease in size as they move away from P_{L_1}, but later they form both large and small loops. After coming close to P_{L_2} the orbit approaches again P_{L_1}, following a path close to the curve U or UU′ inside the point L_4.

When we compare the invariant manifolds with the N-body spirals we use many values of the Jacobi constant. In fact, in a N-body system, the stars have different Jacobi constants, with a maximum relative density $\Delta n/(n^*\Delta h)$, around the corotation value (Fig. 10). It is important to note that the spiral manifolds for a range of values of the Jacobi constant are very close to each other, and form rather thick spiral arms. In Fig. 11 we give the superposition of the unstable manifolds from unstable periodic orbits corresponding to a number of values of the Jacobi constant. In Fig. 12 we give the projections of N-bodies on the disk plane.

Chaotic orbits starting close to L_1 come again close to L_1 after supporting both the spiral and the envelope of the bar. Thus, they provide material for a new bunch of particles, starting again near P_{L_1}, that support the spiral. As a consequence, the spirals are quite long-lived, although they may become weak for certain intervals of time. Voglis et al. [43] noticed that good spirals may continue for about a Hubble time. It is remarkable that the spiral arms last for so long, because in the end, the chaotic motions become so complicated that the spiral arms have to disappear.

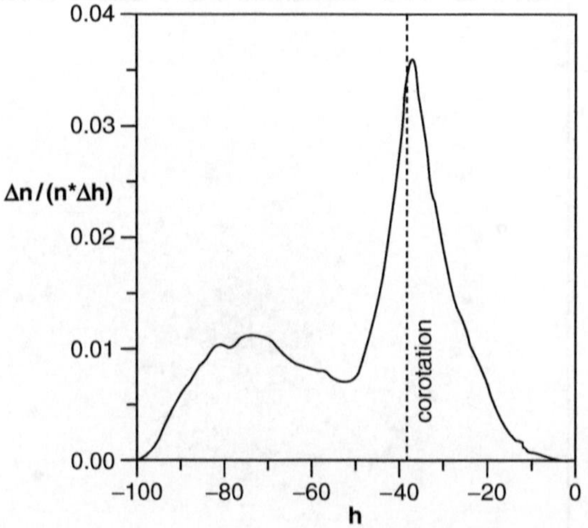

Fig. 10. The distribution of stars of various Jacobi constants h in a model barred galaxy (Voglis et al. [44])

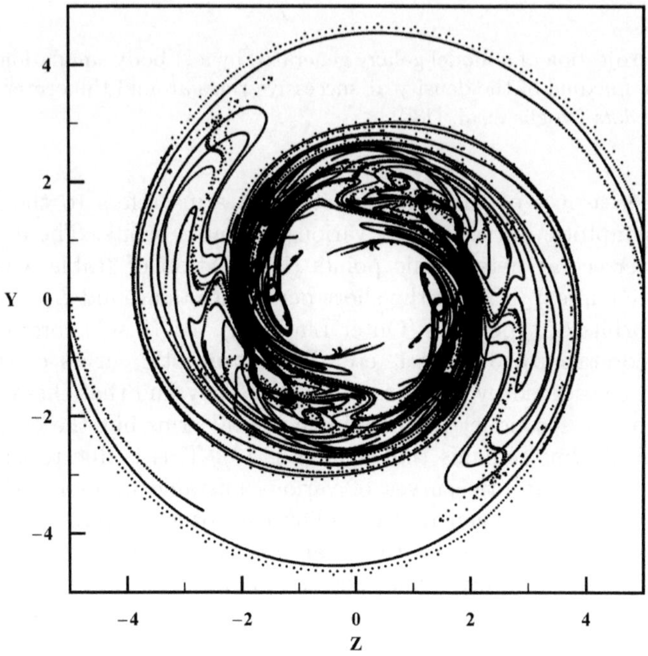

Fig. 11. Superposition of the unstable manifolds from the unstable orbits P_{L_1} and P_{L_2} with Jacobi constants $E_J = -900,000$, $-1,060,000$ and $-1,116,000$. For comparison, the Jacobi constant at L_1 is $E_J = -1,133,000$

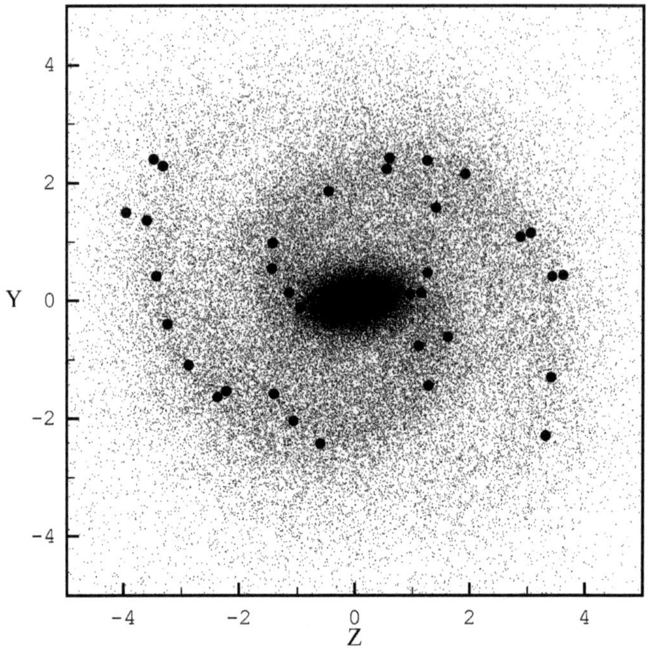

Fig. 12. Projection of a model galaxy generated by a N-body simulation on the disk plane. The maxima of the density at successive rings around the center are marked with *thick dots* (Voglis et al. [43])

Another aspect of the problem of spiral arms refers to the interaction of the asymptotic curves of the various unstable orbits. These asymptotic curves intersect at heteroclinic points (intersection of stable with unstable asymptotic curves). Besides the short period orbits around L_1, L_2, there are unstable orbits of type -2:1 (Outer Lindblad), 4:1 (inside corotation), -4:1 (outside corotation), 3:1, -1:1, etc. The asymptotic curves of these orbits have been considered by Tsoutsis et al. [40]. They find that these asymptotic curves produce secondary features in the spiral arms but the overall picture of two spiral arms remains the same (Fig. 13). This is due to the fact that the unstable asymptotic curves of various unstable periodic orbits cannot intersect. Therefore, in general they follow each other despite some small local deviations. As a consequence, the totality of the asymptotic curves of orbits close to corotation simply forms thick spiral arms.

The thickness of the system along the third dimension does not produce any appreciable effect on the form of the spirals.

Finally, one may consider the question of different pattern speeds between the bar and the spiral. In the course of time the material around L_1 and L_2 is temporarily reduced, as most stars move away along the spirals (Fig. 14). Thus, although the angular velocity of the spiral manifold is the same as that

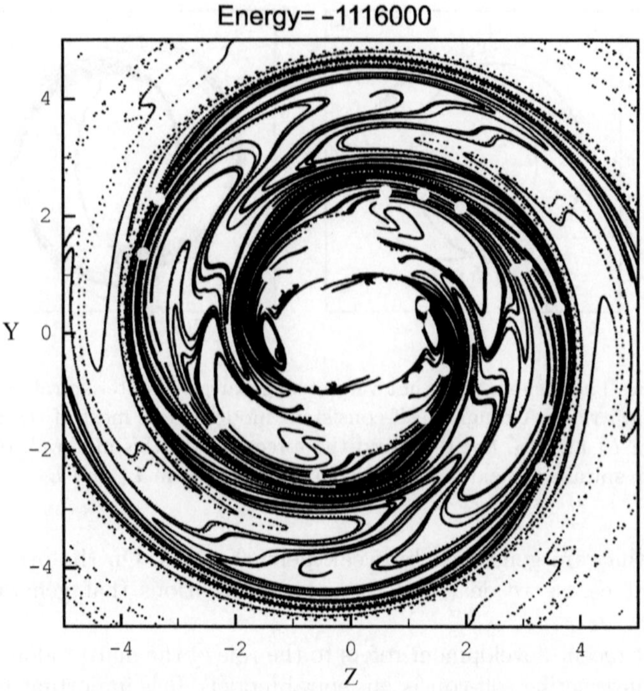

Fig. 13. The invariant manifolds of seven different families: P_{L_1}, 4:1, 3:1, −4:1, −1:1, and period doubling families from the families −1:1 and −2:1. The Jacobi constant is $E_J = -1.116 \times 10^6$. The *dots* represent the maxima of density of Fig. 12

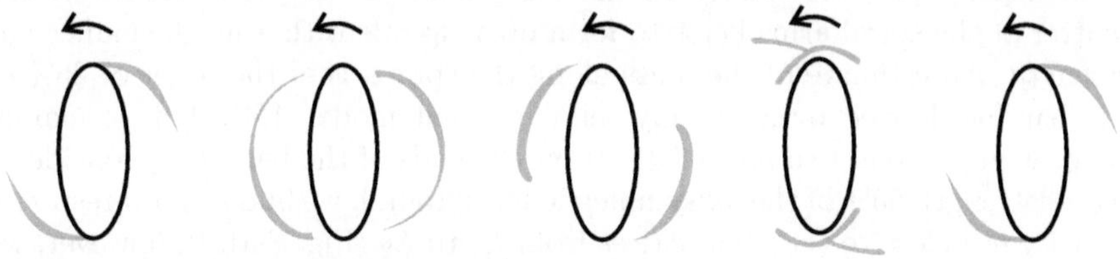

Fig. 14. The removal of the material, forming the spiral arms from the end of the bar, produces an apparently different pattern velocity of the spiral arms. The subsequent replenishment of this material close to the ends of the bar allows the formation of new spiral arms

of the bar, the observed angular velocity of the maxima of the density may be different. This is a kinematic effect that lasts until the material from L_1 reaches L_2. However, as new material arrives close to L_1, L_2, one should find a recurrence of the original pattern.

This effect is rather different from the usual approach to the problem of two different pattern velocities [35], which is based on different eigenvalues of the linear self-consistency problem. The various eigenvalues refer to ordered

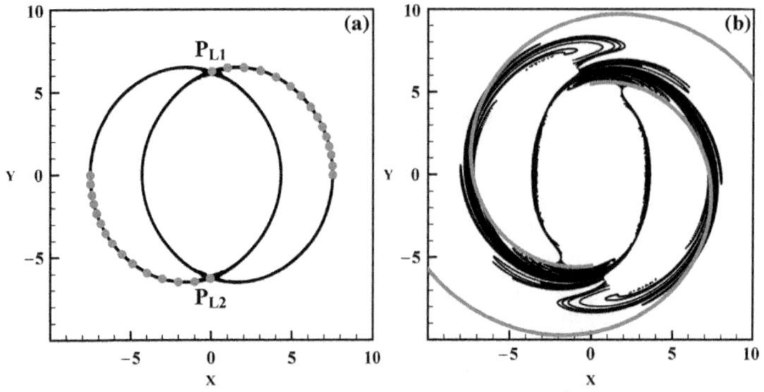

Fig. 15. (**a**) The response of a not very strong bar gives outer spiral arms that are almost separatrices. A roughly self-consistent model places most of the material up to 90° from L_1 and L_2. (**b**) If we add the forces due to this material, the response is two thick spiral arms extending up to about 180° form L_1 and L_2

motions, that are generated by weak perturbations. On the other hand, in the present cases, we deal with strong perturbations that generate chaotic motions.

A most recent development refers to the role of the matter along the spiral arms in constructing self-consistent spiral models. It is important to take this mass into account, because otherwise if we consider only the effect of the bar, in order to find an open spiral response, we would require a perturbation larger than 100% of the axisymmetric background, and this means that the total density perpendicularly to the bar should be negative. However, the matter of the spiral arms helps to form open spirals with a much smaller bar strength. An estimate of the mass along the spirals is of the order of 25% of the bar (as derived by a roughly self-consistent model [19]). This is mainly concentrated along two arcs of 90° from the ends of the bar. If we consider a bar of strength 50% of the axisymmetric background, we find a spiral response that is very close to the separatrices from L_1 to L_2 (Fig. 15a). If, however, we take into account also the forces due to the mass along the response spiral arms (dots in Fig. 15a) [19], we find a new response model (Fig. 15b) that contains thick open spirals along a length almost 180°. A detailed discussion of this effect is given by Efthymiopoulos et al. [15].

4 Velocity Fields

The velocity field of a normal spiral in the rotating frame is almost circular. As the spiral arms are inside corotation, the stars reach the spiral arms and pass through them. Then, their velocities turn slightly inwards, thus allowing the stars to stay longer close to the spiral arms. This effect is much more pronounced in the gas that forms shocks along the spiral arms. This type

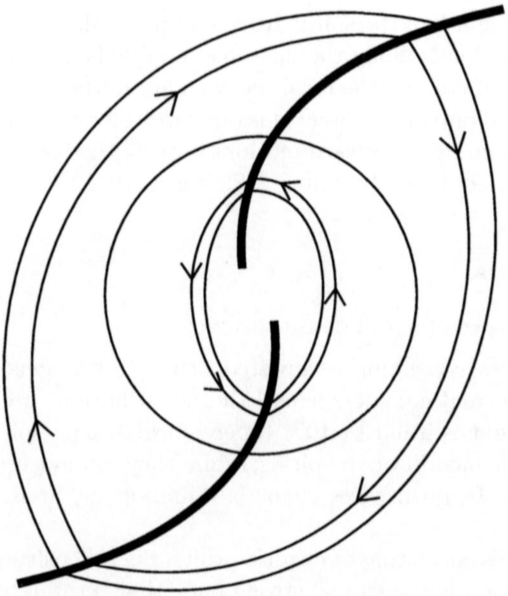

Fig. 16. The flow close to the spiral arms well inside and outside corotation

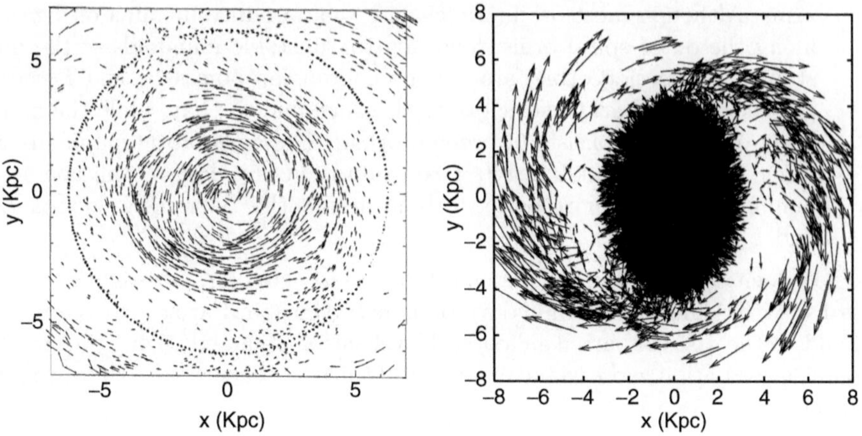

Fig. 17. The velocity field of stars (**a**) inside corotation (*circle*) in a model of the galaxy NGC 1566, and (**b**) outside corotation in a strong bar model

of velocity field is basic in the linear density wave theory (Fig. 16). Such a pattern of velocities is observed in model spiral galaxies (Fig. 17a, [29]).

The spiral arms well outside corotation (beyond the outer −4/1 resonance), where again the orbits are regular, define a velocity field which is almost circular with outward deviations close to spiral arms, but these deviations occur before the stars reach the spiral arms (Fig. 16).

On the other hand, in the spiral arms of barred galaxies outside corotation, the dominant motion is along the unstable manifolds, mainly along the spiral arms outwards. Of course, the local motions are initially along epicycles, but the overall motion outside corotation is outwards along the unstable manifolds (Fig. 5). These dominant outward motions outside the bar have been observed in a model of a strong bar by Patsis [28] (Fig. 17b).

5 Conclusions

There are two types of spiral density waves:

1. Density waves consisting of ordered orbits. Such density waves appear mainly in normal galaxies, where the perturbations are relatively weak, from 2% (tight spirals) to 10% (open spirals). The spirals extend up to the 4/1 resonance in open spirals, while they extend up to corotation in tight spirals. In both cases chaos is unimportant in shaping the spirals arms.
2. Density waves consisting of chaotic orbits. Such density waves appear beyond corotation in the case of strong bars, that terminate near corotation. In such cases the perturbations are of order 100% of the axisymmetric background and chaos is important near corotation and along the spiral arms. Although most orbits are chaotic they produce maxima of density along the outer spiral arms, following the unstable manifolds of the unstable short period orbits around the Lagrangian points L_1 and L_2 near the ends of the bar. Chaotic orbits also surround the bar, while the main body of the bar consists of ordered orbits. The asymptotic manifolds of other unstable periodic orbits near corotation are close to the unstable manifolds of the short period orbits around the Lagrangian points L_1 and L_2.

The velocities of stars in regular spirals are nearly circular, crossing the spiral arms, with somewhat abrupt deviations near the spiral arms. On the other hand, in the cases of barred galaxies, the velocities are mainly along the spirals outside corotation, close to the manifolds of the unstable periods orbits around L_1 and L_2.

Appendix: The Families of Short Period Orbits

There are two couples of families of short period orbits, one around L_4 and L_5 and another one around L_1 and L_2. The first families (SPO) are stable close to L_4 and L_5 and the other families (SL) are unstable around L_1 and L_2. As an example we consider a barred galaxy consisting of an isochrone background potential

$$V_0 = -\frac{1}{1+\sqrt{r^2+1}}$$

and a bar perturbation

$$V_b = \varepsilon r^{1/2} (16 - r) \cos 2\theta$$

(Barbanis and Woltjer [1]) (valid up to $r = 16$) and $V_b = 0$ for $r > 16$.

In the present case we take $\varepsilon = 0.0001$ (intermediate bar) and a pattern velocity $\Omega_b = 0.05$ (relatively fast bar). The Lagrangian points L_4, L_5 are at distances $r = 6.6308$ with Jacobi constant $h(L_4) = -0.1823$, while L_1, L_2 are at distances $r = 6.6533$ with Jacobi constant $h(L_1) = -0.1871$. The family SPO (Fig. 18) is stable from L_4 to the points A_0 ($h = -0.1824$), where we have a bifurcation of a bridge of double period orbits [10], and then unstable up to the points B_0 ($h = -0.1810$), where we have a bifurcation of another bridge of orbits of double period. Beyond that point, the SPO is stable up to the points S ($h = -0.1480$), where we have a bifurcating family of asymmetric orbits

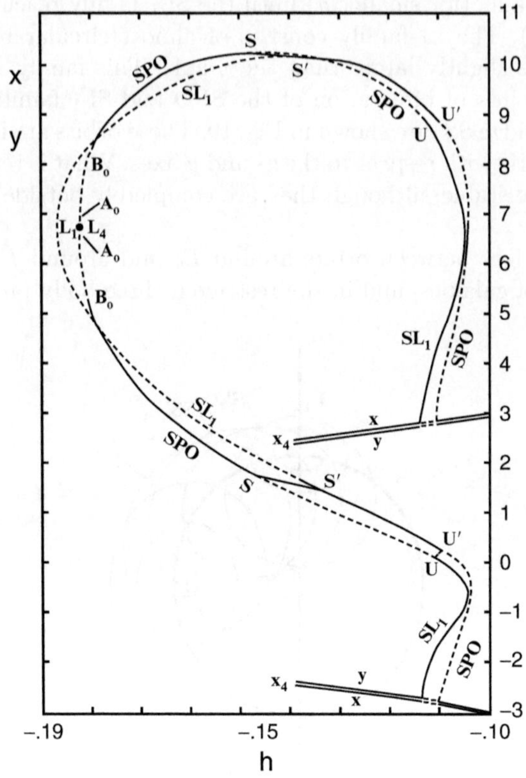

Fig. 18. The characteristics of the short period families SPO and S_{L_1} (x vs. the Jacobi constant h) (—–) stable and (----) unstable parts. The bridges forming the long period family start along the SPO at the points A_0, B_0, and further points beyond B_0. Two bridges of asymmetric orbits connect the SPO and SL_1 families, one stable (SS′) and the other unstable (UU′). (These bridges are only schematic)

(SS′ with respect to the x- and y-axes that joins the SPO family with the family SL_1, starting at L_1. Between B_0 and S there are infinite bifurcations of bridges of long period orbits of periods 3, 4, ... The orbits of these bridges are symmetric with respect to the x-axis [10].

Beyond the points S, the SPO family is unstable up to the points U ($h = -0.1102$), where we have another bifurcation of asymmetric orbits UU′, joining the SPO family with the SL_1 family. Then SPO is again stable up to a maximum $h = -0.1046$. Beyond this maximum, the value of h decreases along the SPO family until this family joins the x_4 family of retrograde orbits around the origin. Before reaching the maximum h, the orbits of the SPO family cross the origin $x = 0$ (Fig. 18) and from then on they can be considered as retrograde orbits around the origin.

The family SL_1 is unstable from L_1 up to the points S′ ($h = -0.1350$), where we have the bifurcation SS′, then stable up to the points U′, where we have the bifurcation U′U, unstable up to a maximum ($h = -0.1043$), and beyond that stable (for smaller h) until the SL_1 family reaches the x_4 family (on the y-axis). The x_4 family consists of almost circular retrograde orbits with the x-axis slightly larger than the y-axis. This family is stable, except between the points of bifurcation of the SPO and SL_1 families. Some stable orbits of the bridge SS′ are shown in Fig. 19. These orbits are large in size, and quite asymmetric with respect to the x- and y-axes. What is remarkable is that these orbits are stable, although they are completely outside the Lagrangian points.

Similar bridges between orbits around L_4 and around L_1 were found in other models of galaxies, and in the restricted three-body problem [16].

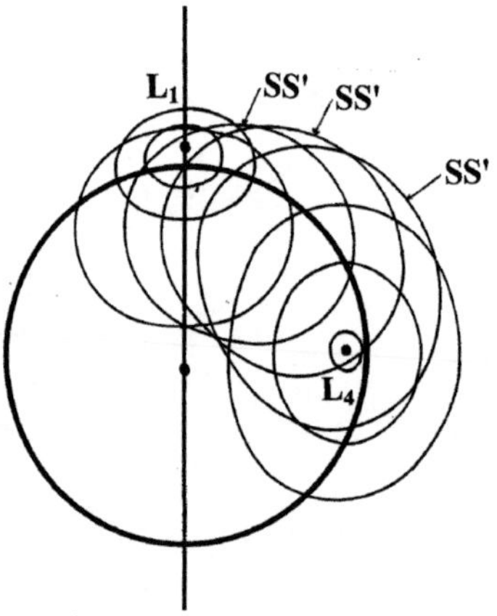

Fig. 19. Three orbits of the SPO family around L_4, which are symmetric with respect to the x-axis, three orbits of the SL_1 family around L_1, which are symmetric with respect to the y-axis, and three asymmetric orbits of the SS′ branch that joins the SPO family with the SL1 family. There are also an SPO family around L_5, an SL_2 family around L_2 and three more SS′ families symmetric to the above SS′ family with respect to the center and the x- and y-axes

References

1. Barbanis, B. and Woltjer, L. (1967), *Astrophys. J.* **150**, 461.
2. Bertin, G., Lin, C.C., Lowe, S.A. and Thurstans, R.P. (1989), *Astrophys. J.* **338**, 78, 104.
3. Contopoulos, G. (1970), *Astrophys. J.* **160**, 113.
4. Contopoulos, G. (1973), *Astrophys. J.* **181**, 657.
5. Contopoulos, G. (1975), *Astrophys. J.* **201**, 566.
6. Contopoulos, G. (1978), *Astron. Astrophys.* **64**, 323.
7. Contopoulos, G. (1980), *Astron. Astrophys.* **81**, 198.
8. Contopoulos, G. (1983), *Astron. Astrophys.* **117**, 89.
9. Contopoulos, G. (1985), *Comments Astrophys.* **11**, 1.
10. Contopoulos, G. (1988), *Cel. Mech. Dyn. Astron.* **43**, 147.
11. Contopoulos, G. (2002), Order and Chaos in Dynamical Astronomy, Springer, New York.
12. Contopoulos, G. and Grosbol, P. (1986), *Astron. Astrophys.* **155**, 11.
13. Contopoulos, G. and Woltjer, L. (1964), *Astrophys. J.* **140**, 1106.
14. Elmegreen, E.G. and Elmegreen, D.M. (1985), *Astrophys. J.* **288**, 438.
15. Tsoutsis, P., Kalapotharakos, C., Efthymiopoulos, C. and Contopoulos, G., 2008 (submitted).
16. Henrard, J. and Navarro, J.F. (2004), *Cel. Mech. Dyn. Astron.* **89**, 285.
17. Kalnajs, A. (1965), Thesis, Harvard University.
18. Kalnajs, A. (1971), *Astrophys. J.* **166**, 275.
19. Kaufmann, D.E. and Contopoulos, G. (1996), *Astron. Astrophys.* **309**, 381.
20. Lin, C.C. (1966), *J. SIAM Appl. Math.* **14**, 876.
21. Lin, C.C. and Shu, F.H. (1964), *Astrophys. J.* **140**, 646.
22. Lin, C.C., Yuan, C. and Shu, F.H. (1969), *Astrophys. J.* **155**, 721.
23. Lindblad, B. (1940), *Astrophys. J.* **92**, 1.
24. Lindblad, B. (1955), *Stockholm Ann.* **1**, No. 6.
25. Lindblad, B. (1963), *Stockholm Ann.* **22**, No. 5.
26. Lindblad, B. (1964), *Astrophys. Norvegica* **9**, No. 12, 103.
27. Lynden-Bell, D. and Kalnajs, A. (1972), *Mon. Not. R. Astron. Soc.* **157**, 1.
28. Patsis, P.A. (2006), *Mon. Not. R. Astron. Soc.* **369**, L56.
29. Patsis, P.A. (2007), (this volume).
30. Patsis, P.A., Contopoulos, G. and Grosbol, P. (1991), *Astron. Astrophys.* **243**, 373.
31. Patsis, P.A., Hiotelis, N., Contopoulos, G. and Grosbol, P. (1994), *Astron. Astrophys.* **286**, 46.
32. Patsis, P.A., Grosbol, P. and Hiotelis, N. (1997), *Astron. Astrophys.* **323**, 762.
33. Romero-Gómez, M., Masdemont, J.J., Athanassoula, E. and García-Gómez, C. (2006) *Astron. Astrophys.* **453**, 39.
34. Sanders, R.H. and Tubbs, A.D. (1980), *Astrophys. J.* **235**, 803.

35. Sellwood, J.A. and Sparke, L.S. (1988), *Mon. Not. R. Astron. Soc.* **231**, 25p.
36. Sellwood, J.A. and Wilkinson, A. (1993), *Rep. Prog. Phys.* **56**, 173.
37. Shu, F.H. (1970), *Astrophys. J.* **160**, 89 and 99.
38. Toomre, A. (1969), *Astrophys. J.* **158**, 899.
39. Toomre, A. (1977), *Ann. Rev. Astron. Astrophys.* **15**, 437.
40. Tsoutsis, P., Efthymiopoulos, C. and Voglis, N. (2007), *Mon. Not. R. Astron. Soc.* **387**, 1264.
41. Voglis, N. (2003), *Mon. Not. R. Astron. Soc.* **344**, 575.
42. Voglis, N., Stavropoulos, I. and Kalapotharakos, C. (2006), *Mon. Not. R. Astron. Soc.* **372**, 901.
43. Voglis, N., Tsoutsis, P. and Efthymiopoulos, C. (2006), *Mon. Not. R. Astron. Soc.* **373**, 280.
44. Voglis, N., Harsoula, M. and Contopoulos, G. (2007), *Mon. Not. R. Astron. Soc.* **381**, 757.

Order and Chaos in Spiral Galaxies Seen through their Morphology

P. Grosbøl

European Southern Observatory, Karl-Schwarzschild Strasse 2,
D-85748 Garching, Germany
pgrosbol@eso.org

Summary. It is clear from dynamical considerations and N-body models that both order and chaos are important for spiral galaxies. To study the relative importance of order and chaos in real galaxies, one needs accurate kinematic information for the stellar population. Unfortunately, such data are still very difficult to obtain outside the very central parts of spiral galaxies due to the relative low surface brightness of their disks. As a second option, one can consider morphological features in the disks (e.g. bars, and spiral arms) and interpret them as indicators of the underlying dynamics.

The presence of large-scale, well-define structures (such as bars and grand-design spirals) in disks of many spiral galaxies suggests that collective phenomena like density waves supported by regular orbits are important. The alignment of both young objects and dust lanes along spiral arms indicates a regular gas flow which would be shared by newly formed stars. At the end of bars, where one expects their co-rotation, a more diffuse region is often seen, possibly caused by chaotic stars. A few percent of barred galaxies, mainly early-type spirals, show broad, tight spirals with a morphology very close to that seen in N-body models where the arms are dominated by stars on chaotic orbits.

1 Introduction

At first sight, spiral galaxies with their thin disks and spiral arms give an impression of being dominated by ordered motions. However, orbital analysis and dynamical models suggest that chaotic motions also play a significant role in many spiral galaxies (see, e.g. [3, 4, 6, 28]). A comparison between observed spiral galaxies and theoretical models may contribute to the understanding of the relative importance of these different types of motion. Unfortunately, high-resolution kinematic data, required for a detailed study, are normally not avaliable for the major parts of the disk in such galaxies limiting the comparison to their morphology.

G. Contopoulos, P.A. Patsis, *Chaos in Astronomy.* Astrophysics and Space Science Proceedings, doi: 10.1007/978-3-540-75826-6,

The current paper reviews the morphology of spiral galaxies and discusses possible relations to order and chaos in them, mainly using near-infrared (NIR) data for which stellar population effects and attenuation by dust are less severe than in visual bands.

2 Observations vs. Models

Observations of galaxies provide a view of their current morphology but say little on previous or future evolution. Thus, it is not possible to use the morphology directly to determine to what degree regular or chaotic motions are dominant in any specific region. However, a comparison between observations and dynamic models may suggest the frequency of morphological features which in models can be identified as clear indicators of their dynamical structure.

Dynamic models of galaxies can crudely be divided into two groups namely response models which study the orbital structure of test partials in a given potential [1], and N-body models which compute the self-consistent evolution of a large set of stars [13]. Whereas response models are essential for the understanding of the dynamics of a given system (e.g. the degree of order and chaos in different regions), they are normally applied to conservative systems and cannot track the behavior of time dependent potentials. By identifying the main families of orbits in a potential and populating them according to a given distribution function, one can estimate the response phase density [22]. This can be compared to the potential imposed, to check if the model is self-consistent, and to observed galaxies. On the other hand, N-body models have the advantage of being intrinsically self-consistent but may suffer of limited resolution and numerical approximations.

A comparison between real and model galaxies meets several basic problems. Whereas models deal with the total mass distribution, one can only observe the 2-dimensional sky projection of the luminous components in galaxies. Beside assumptions on a dark matter halo, also the mass-to-light (M/L) ratio for the luminous matter must be estimated. This ratio may change significantly within a galaxy which has ongoing star formation and be affected by attenuation of dust. Although such variations are expected to be much smaller in the NIR bands than at visual wavelengths, population effects can be significant even in the K-band at 2 μm where one may see bright young stellar clusters along spiral arms [11].

In the Cold-Dark-Matter cosmology, galaxies assemble over a significant period of time through accretion and mergers. The historic star formation rate shows a peak around $z \approx 1$ with a steep decline to the present epoch [16]. This suggests that most galaxies have experienced major mergers, possibly as late as 8 Gyr ago, and may well have accreted minor systems up to the present epoch. Thus, it may not always be correct to regard galaxies as isolated systems even in the case of field galaxies with no obvious companions.

The outer parts of disk galaxies may not have had time to reach a relaxed state since the last external perturbation due to the long dynamic time scale in these regions. A detailed comparison with models may only be fruitful in the more central parts where time scales are shorter and the potential well is deeper, limiting the effects of external perturbations.

3 Special Regions and Features

Spiral galaxies can be divided into several regions with different morphological and kinematical properties. Their central part often has a strong intensity peak surrounded by a bulge region. In many spiral, a bar (possibly with nested bars and spirals) is present in the inner parts with spiral arms occupying the regions outside. Both bars and spirals are assumed to be associated to density waves each of which rotating with a fixed pattern speed Ω_p. From dynamic considerations, one expects that chaos is likely in resonance regions where the difference between angular speed of disk material and density wave is equal to a rational number times the local epicyclic frequence κ (i.e. $n \times (\Omega(\mathrm{r}) - \Omega_\mathrm{p}) = k \times \kappa$). The strongest effects are seen for small even n values and $k = 1$ such as the Lindblad resonances (ILR and OLR) with $n{:}k = \pm 2{:}1$ and co-rotation (CR) for $\Omega = \Omega_\mathrm{p}$. Due to crowding of resonances close to CR, chaos is more likely in this region. Perturbations stronger than 5% of the radial force introduce non-linear dynamical effects [9] and chaos is likely to be more important.

Although Ω_p has been measured directly for relative few galaxies (see, e.g. [17]) using the method of Tremaine and Weinberg [27], the location of resonances are suggested by morphological features. The end of the main bar is expected to be just inside CR [2] while the main stellar spiral pattern is likely to start beyond ILR or CR. Strong symmetric spiral arms may not extend beyond the 4:1 resonance [5] whereas weak, outer spirals may reach CR or even OLR. Thus, radial regions where the morphology of azimuthal perturbations in the disk change significantly (e.g. from bar to spiral, or shape of spiral) are likely to be associated to a resonance.

In barred spirals, the transition region between bar and spiral is of special interest since it is close to CR of the bar and therefore a likely location of chaos especially for strong bars. Two cases may be important namely one where bar and spiral share the same Ω_p, and the other where they have different Ω_p [24, 26] but are coupled through their major resonances, e.g. the bar CR could coincide with the spiral ILR.

In the former case, it is expected that the spiral emerge from the end of the bar. The majority of orbits in this regions are chaotic but can still support the spiral structure [14]. It is also possible that the spiral just outside the bar consists of chaotic particles which escape from the bar [18, 28] leading to a relative short, tightly wound spiral pattern.

The orientation of the spiral relative to the bar would be random if they rotate with different speed. Although this should be observable, it is not always clear when the inner part of the spiral is tight or joins a ring structure at the end of bar [24]. Two different Ω_ps were measured in NGC 6946 by Fathi et al. [8] using kinematic data.

The spiral structure can be approximated by logarithmic spiral patterns each occupying a radial region (typically between major resonances such as ILR, 4:1, CR, or OLR). Typically, grand-design spirals have a strong 2-armed, symmetric spiral in their inner disk while it often becomes multi-armed in the outer parts. Another significant class of disk galaxies have less regular spiral structures with weaker, multiple arms. On optical images, this appearance may be caused by attenuation by dust which can hide weak spiral perturbations in the disk [11]. Thus, it is preferable to use NIR images for analysis of spiral structure since absorption by dust is much less significant than in optical bands. Resonance may be indicated by sudden changes in the spiral pattern such as its pitch angle, amplitude, or number of arms.

4 Interpretation of Morphology

Before one can study specific morphological features in a galaxy and compare them with dynamic models, one needs to consider the convolution process which converts a mass density to an observed surface brightness. This involves projection effects, integration along the line of sight, resolution issues, estimates of M/L ratios, and attenuation by dust.

The two main visual components of spiral galaxies are an ellipsoidal, central bulge and a flattened disk in which bar and spiral structures are observed. To analyze perturbations in disk component, it is necessary to remove the bulge and determine the sky-projection parameters of the disk. The bulge is standardly assumed to be a nearly spherical, homogeneous component with a unique radial luminosity profile such as a $\mathrm{r}^{1/n}$ profile [25] with an index n in the range of 1–2. The de-composition is normally base on surface brightness distribution since a physical definition of the bulge is non-trivial. Several techniques may be used to determine to sky-projection, all with their own advantages and biases (see, e.g. [12]). Errors in projection parameters or bulge removal will introduce residuals in the disk (e.g. artificial bar structures) and may bias the analysis.

Population effects and attenuation by dust make it difficult to compare images taken in visual bands with models especially for late-type spirals with significant star formation. This is illustrated in Fig. 1 which shows B- and K-band images of NGC 5085. It is more difficult to trace the spiral arms on the B-band image due to the heavy dust absorption in them and their inner parts appear very different due to the strong dust lanes which reach all the way into the bulge region. The K-band shows the distribution of the old stellar disk population much better than visual bands and is better suited for estimating

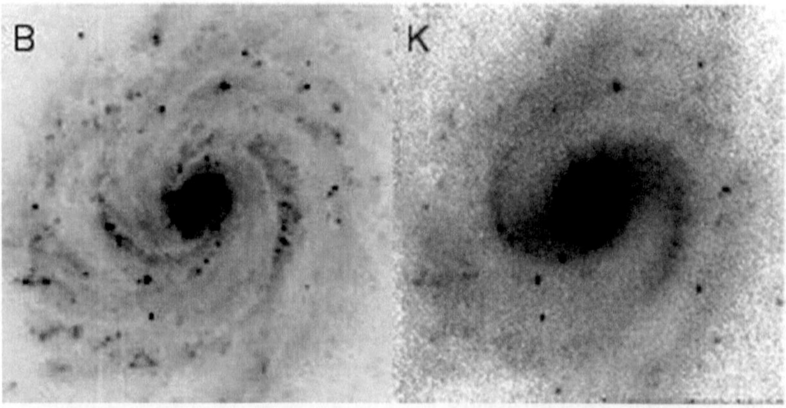

Fig. 1. Direct B- and K-band images of the spiral galaxy NGC 5085

the mass distribution of luminous matter. The appearance of spirals in the NIR is smoother and bulges are relative larger. Whereas one still can use the general Hubble classification, spirals in NIR are typically classified one class earlier than on optical images and the frequence of bars is higher [7].

In order to identify possible regions with regular or chaotic orbits, one option is to look for sharp or more diffuse features. A population of stars with low dispersion velocities and on regular orbits would be able to maintain their structure over a long period of time. They also may support collective phenomena like density waves [15]. In general, one would look for structures which have a well defined, regular, azimuthal profile which correlate over some radial interval. Star formation and attenuation by dust can make it difficult to identify the variation of the stellar mass density in the disk especially on visual images. Even in the NIR, population changes may play a significant role. Sharp features can also be produce by other mechanisms such as mergers and close encounters with other galaxies [23]. Such structures are transient and often less symmetric.

From a pure morphological point of view, it is virtually impossible to distinguish a region with chaotic stars from one populated with stars on quasi-periodic orbits with significant velocity dispersion as chaotic stars can remain in a relative small region for a long time before they escape. Detailed analysis of their kinematics would probably be able to reveal the difference but is not feasible with current instruments.

5 Pitfalls and Other Issues

Trying to interpret images of spiral galaxies, it is important to bear in mind limitations and possible problems. The idealistic model of disk galaxies consisting of a spherical bulge and a flat exponential disk is seldom correct.

The actual shape of the bulge cannot be determined uniquely from photometric data alone and may be oblate or triaxial. Since weak oval distortions in the disk may be present in the central regions, a proper modeling of the bulge is very delicate. Both neglecting to correct for the bulge or doing it wrongly will typically introduce spurious bar-like intensity variations in the inner disk. Such pseudo bars can often be recognized by their orientation which normally is along or perpendicular to the major axis of the galaxy.

The inner disk may be planner, due to its short dynamic time scale, but could have a significant thickness. For galaxies with substantial amounts of dust, the location of dust lanes could produce projection effects at wavelengths where absorption is important. The outer disk has often warps (e.g. due to external perturbations) which will distort the view and lead to misinterpretations of the outer spiral structure.

Dust in late-type spirals may lead to both absorption and reflection of light [19]. The effects depend on the detailed distribution of dust and stars, and cannot easily be removed by corrections based on broad band colors. Although star formation may take place over the major part of the disk, many young objects (e.g. HII-regions and OB associations) are concentrated in arm regions. They often dominate the visual impression of spiral structure on images taken in optical bands where both population effects and attenuation by dust are strong. A comparison with theoretical models is therefore better done at NIR (e.g. K-band) where these problems are reduced. However, it is often misleading to assume a constant M/L ratio in the disk of a galaxy. Young red-super-giant stars can contribute significant flux in NIR and cause significant changes of the M/L ratio between arm and inter-arm regions.

Detailed analysis of perturbations in the disks of spiral galaxies is frequently done by Fourier Transform techniques. Since spiral arms are well approximated with logarithmic spiral, this is often done using 2D transforms on $\theta - \ln(r)$ maps of the disk. Divergences from this shape will introduce errors. Significant care must be exercised in selecting appropriate radial regions (i.e. typically between major resonances) to avoid introduction of alias signals. The usage of digital filters can reduce such problems but it is still important to limit such analysis to structures which have the same origin.

6 Case Studies: Bars and Spirals

This section illustrate morphological features in spiral galaxies which may be associated to either regular or chaotic motions. Most of the images used are taken in the NIR K-band to minimize the effects caused by dust and variation in stellar populations.

The models of Voglis et al. [28] suggest that spirals in galaxies with massive bars can be dominated by stars on chaotic orbits. Such spiral arms are relative broad and tight. It is also possible to form short spiral arms from stars escaping through the end of the bar [18]. The stars flow along the arms

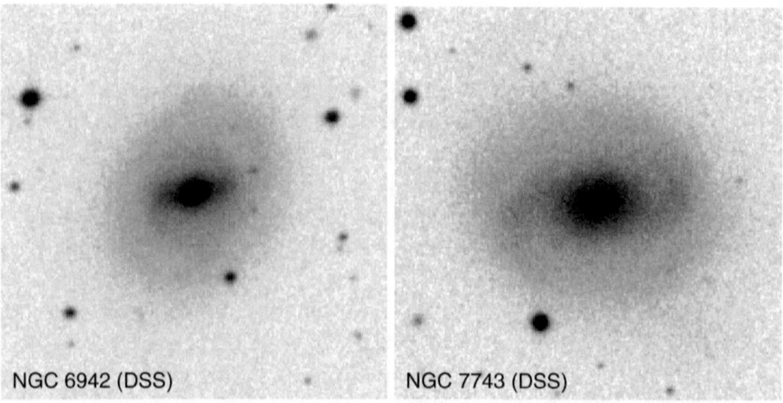

Fig. 2. Optical, direct images of NGC 6942 and NGC 7743 from the Digital Sky Survey

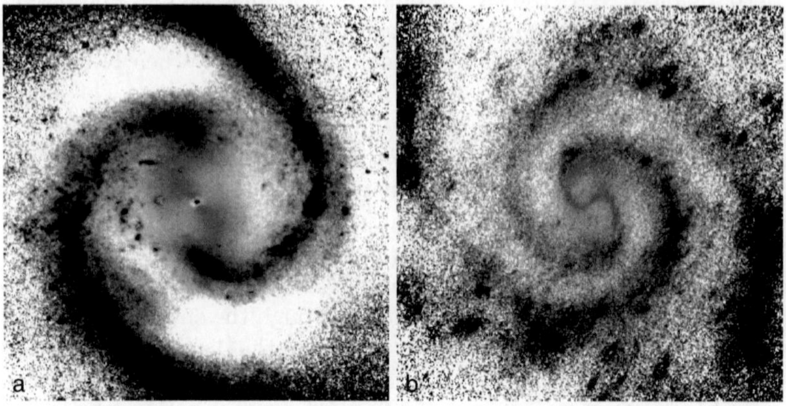

Fig. 3. Relative, K-band intensity variation in the disks of (**a**) NGC 1566 and (**b**) NGC 6118. The bulge components was subtracted and the K intensity divided by the azimuthal average. *Black* corresponds to 30% light excess while *white* to 30% deficiency

which originate at the end of the bar and are very tight. Optical images of two early-type spirals, NGC 6942 and NGC 7743, are shown in Fig. 2. Both galaxies show broad, tight spiral arms which resemble the ones seen in the N-body models with chaotic motions. In the Revised Shapley–Ames Catalog of Bright Galaxies [21], one finds a few percent of barred galaxies with this morphology mainly among early-type spirals. This supports the existence of spirals supported by chaotic motions but not as a main mechanism.

The transition region between bar and spiral is expected to contain chaos and is illustrated by NGC 1566 and NGC 6118 as seen in Fig. 3 where the relative azimuthal K-band intensity variation is displayed. In both cases, the bulge component was removed and the disks de-projected to a face-on view, to provide a clear picture of the perturbations in the inner part of their disks. There is a clear misalignment (i.e. $\approx 30°$) between the end of the bar and the start of the spiral in NGC 1566 which suggests a different pattern speed of the two components [24]. The amplitude of the two-armed perturbation decreases slightly in this region which could imply an increase of chaotic motions. There is very intense star formation in the spiral arms just outside the bar. This could be induced by a more turbulent (chaotic) gas flow in this region due to interactions between the two patterns.

The other example, NGC 6118, shows a very regular and well defined perturbation in its central parts. A small bar appears connected to the inner spiral without any significant change in amplitude. Narrowness and symmetry of these features suggest more regular motions.

Grand-design spirals have long, symmetric arms in their inner parts as shown in Fig. 4 where K-band images of the ordinary spiral NGC 2997 and the barred galaxy NGC 1365 are presented. The long, regular and often open patterns indicate the presence of a density wave supported by regular orbits. Many such galaxies have bright knots along their arms in their K-band images which have been identified as very young stellar clusters [10]. The alignment of these clusters and dust lanes along the arms suggest a regular gas flow, as proposed by Roberts [20], and therefore also of the flow of the newly formed stars. Although chaotic motions still may be present at the end of the bars, grand-design spirals seem to be dominated by regular motions.

In some case, the spiral arms seem not to open as logarithmic spirals but turn more inwards as in NGC 4535 and NGC 4548 (see Fig. 5). This could point to stars with chaotic orbits escaping through the end end of the bar but could also be projection effects due to a warp in the disk. Bright clusters and dust lanes in NGC 4535 indicate a regular flow while the break between bar and spiral in NGC 4548 makes a significant flow of stars from the bar to the spiral unlikely.

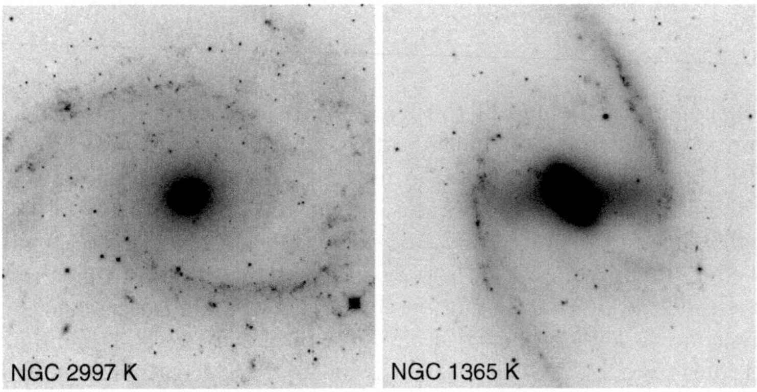

Fig. 4. Direct K-band images of the grand-design spiral galaxies NGC 2997 and NGC 1365

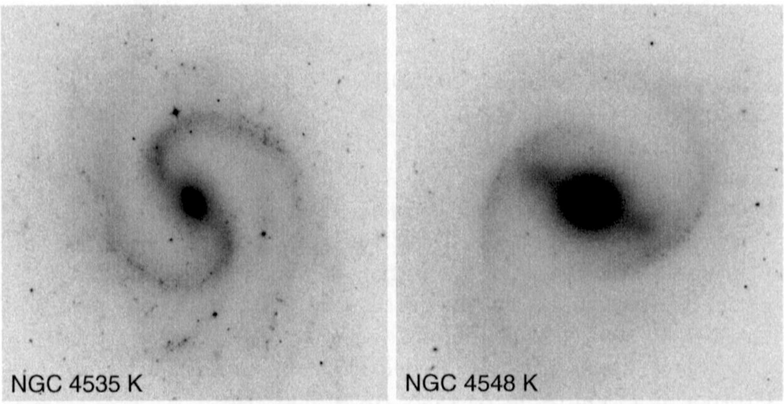

Fig. 5. K-band maps of the spiral galaxies NGC 4535 and NGC 4548

7 Conclusions

One cannot expect to get strong, obvious evidence for chaos in spiral galaxies from studies of their morphology. However, a comparison with dynamic models may indicate what regions may be affected by chaotic motions and the frequency of such features.

Smooth, large-scale structures in grand-design spiral galaxies (e.g. bars and main spiral arms) are likely to be dominated by regular orbits which can support collective phenomena such as density waves. Long, narrow dust lanes and young stellar clusters aligned with spiral arms indicate a regular gas flow. Although the gas does not directly show the stellar orbital structure, young stars are formed from the gas and will, on average, share its general kinematics. This does not exclude that some stars in these regions have chaotic motions with very long escape times.

It may be possible to identify major resonance regions in the disks of spiral galaxies as places where the nature of azimuthal perturbations change significantly. This includes the interface between bar and spiral, and the end of the main symmetric spiral pattern. Based on dynamic arguments, chaos is likely to be present in these regions especially in the case of galaxies with strong perturbations. In galaxies with several pattern speeds (e.g. for bar and spiral), stronger effects would be expected.

The morphology of several early-type spiral galaxies is very similar to that of models by Voglis et al. [28] where the spiral arms are dominated by chaotic stars. This suggests that such models may have a counterpart in real galaxies although they only constitute a few percent of the total population of barred spiral galaxies.

References

1. Contopoulos, G. 1965, AJ, 70, 526
2. Contopoulos, G. 1980, A&A, 81, 198
3. Contopoulos, G. 1983, A&A, 117, 89
4. Contopoulos, G. 1995, Ann. NY Acad. Sci., 751, 112
5. Contopoulos, G. & Grosbøl, P. 1986, A&A, 155, 11
6. Contopoulos, G., Varvoglis, V., & Barbanis, B. 1987, A&A, 172, 55
7. Eskridge, P. B., Frogel, J. A., Pogge, R. W., et al. 2000, AJ, 119, 536
8. Fathi, K., Toonen, S., Falcón Barroso, J., et al. 2007, ApJ, 667, L137
9. Grosbøl, P. 1993, PASP, 105, 651
10. Grosbøl, P., Dottori, H., & Gredel, R. 2006, A&A, 453, L25
11. Grosbøl, P. & Patsis, P. A. 1998, A&A, 336, 840
12. Grosbøl, P., Patsis, P. A., & Pompei, E. 2004, A&A, 423, 849
13. Hohl, F. 1971, ApJ, 168, 343
14. Kaufmann, D. E. & Contopoulos, G. 1996, A&A, 309, 381
15. Lin, C. C. & Shu, F. H. 1964, ApJ, 140, 646
16. Madau, P., Pozzetti, L., & Dickinson, M. 1998, ApJ, 498, 106
17. Merrifield, M. R. & Kuijken, K. 1995, MNRAS, 274, 933
18. Patsis, P. A. 2006, MNRAS, 369, L56
19. Pierini, D., Gordon, K. D., Witt, A. N., & Madsen, G. J. 2004, ApJ, 617, 1022
20. Roberts, W. W. 1969, ApJ, 158, 123
21. Sandage, A. & Tammann, G. A. 1981, A Revised Shapley–Ames Catalog of Bright Galaxies, Carnegie Inst. of Wash. Publ. No. 635 (Washington: Carnegie Inst.)
22. Schwarzschild, M. 1979, ApJ, 232, 236
23. Schweizer, F. & Seitzer, P. 1988, ApJ, 328, 88
24. Sellwood, J. A. & Sparke, L. S. 1988, MNRAS, 231, 25
25. Sérsic, J. L. 1968, Atlas de galaxias australes (Cordoba: Obs. Astron. de Cordoba)
26. Tagger, M., Sygnet, J. F., Athanassoula, E., & Pellet, R. 1987, ApJ, 318, L43
27. Tremaine, S. & Weinberg, M. D. 1984, ApJ, 282, L5
28. Voglis, N., Stavropoulos, I., & Kalapotharakos, C. 2006, MNRAS, 372, 901

The Flow through the Arms of Normal and Barred-Spiral Galaxies

P.A. Patsis

Research Center for Astronomy, Academy of Athens,
Soranou Efessiou 4, GR-11527, Athens, Greece
patsis@academyofathens.gr

Summary. We examine the flow at the arms of spiral galaxies from a qualitative point of view. In particular we investigate the properties in cases of ordered and chaotic stellar flows. Ordered flows are associated with normal (non-barred) spiral galaxies, while chaotic arms appear in barred-spiral systems. The two cases are characterized by different forcing at corotation. The chaotic orbits that reinforce the spiral arms traverse also the bar region and are associated with the outer boxy isophotes surrounding early type bars. Chaotic spiral arms either stop at an azimuth smaller than $\pi/2$, or present clear gaps at this angle.

1 Spiral Flows and Orbital Theory

Most theories about the dynamics of the spiral arms of disk galaxies are decoupled from the presence, and subsequently the dynamics, of a bar. The flow of stars through the spiral arms is in general believed to be determined by a so called "precessing-ellipses" pattern. Sketches that show this pattern can be found, e.g. in [7, 9] and in several textbooks. Expressed by means of stable periodic orbits this stellar flow is very illustratively described by Contopoulos and Grosbøl [2] (hereafter CG86, their Fig. 9). The bisymmetric grand design spiral structure stops at the point, where the assumed potential stops providing the system with elliptical stable periodic orbits that have the appropriate orientation with respect to the spiral arms. The orientation of the elliptical orbits changes as their Jacobi constant (E_J) varies, so that their apocentra are along the spiral potential minima. Beyond the 4:1 resonance region the apocentra are not found anymore to be at the potential minima (CG86). So, beyond that point, there is no appropriate backbone to support two open, strong, symmetric, spiral arms.

In barred-spiral galaxies the situation is more complicated. One possibility is that bar and spirals rotate with different pattern speeds, as is proposed in [18]. Then, the flow at the spiral arms can be considered to be independent of the presence of a bar, at least well beyond the corotation region of the

G. Contopoulos, P.A. Patsis, *Chaos in Astronomy*. Astrophysics and Space Science Proceedings, doi: 10.1007/978-3-540-75826-6,

bar, close to its end. At this region, in many cases, exists already the inner part of the spiral, which rotates with a different pattern speed than the bar. As Sellwood and Sparke [18] propose, we may have a resonance coupling (the corotation of the bar overlapping with the inner Lindblad, or another resonance, of the spiral). The detailed orbital dynamics in such a system have not been studied yet. We mention though the study of Boonyasait et al. [1], where orbits have been calculated at the arms of the barred-spiral galaxy NGC 3359. In the potential estimated from nearinfrared observations for this galaxy, we found an optimal value for the pattern speed of the spiral. For this value, there is an almost perfect agreement between (stellar and gaseous) response models and observations, as regards the spiral structure (Fig. 1). The model predicts a kind of "precessing-ellipses flow". However, with this pattern speed we cannot obtain simultaneously a bar supported by x_1 orbits, ending close to its corotation. We have to assume another pattern speed for the bar if

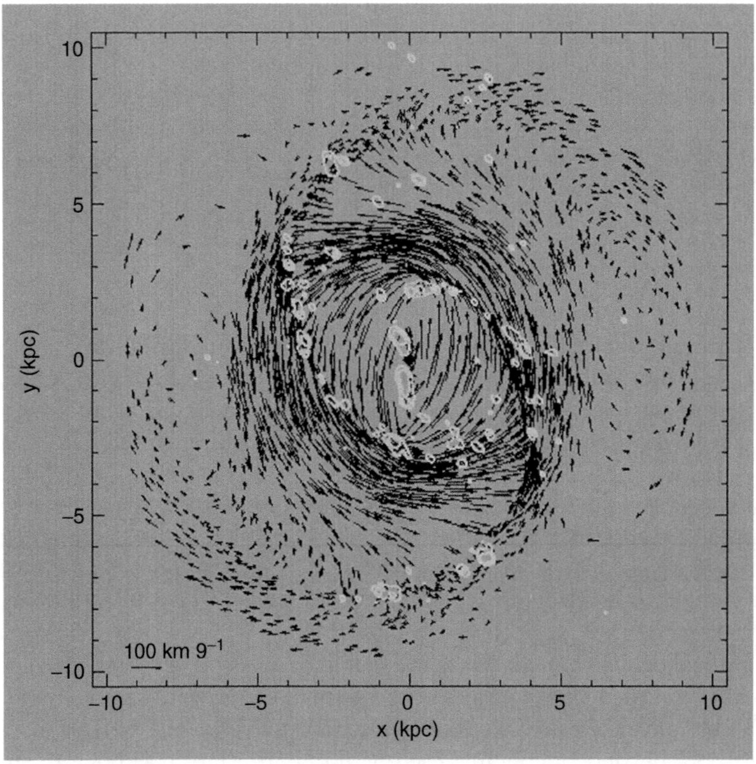

Fig. 1. The velocity field of an SPH gaseous model by [1] for the barred-spiral galaxy NGC 3359. A "precessing-ellipses flow" can model the spirals of this galaxy, assuming corotation beyond the end of the observed arms. We observe the agreement between the flow of the model (*black arrows*) and the overplotted Hα (*white*) contours, located on the arms of the galaxy (taken from observations)

we want to give it these features and match better the observations. This is a strong indication that bar and spiral arms rotate with different pattern speeds. In cases like the NGC 3359 models the "precessing-ellipses flow" is applicable also in barred-spiral systems. The reader is referred to [1] for details on the backbone of stable periodic orbits associated with the flow we present in Fig. 1.

The orbital structure of the arms in barred-spiral systems with one pattern speed, has been mainly investigated by Kaufmann and Contopoulos [8] (hereafter KC96) in rigid potentials of this type. In those models the bar ends close and before its corotation, thus the spiral arms have to extend beyond the corotation of the system. KC96 found that stable periodic orbits support the observed spiral arms between the −4:1 and −2:1 resonances. Thus, at the outer part of the spirals the flow can be described again with a kind of distorted "precessing-ellipses" pattern (see their Fig. 19). However, for the explanation of the inner part of the arms, close to the end of the bar, they invoked chaotic orbits. It was the first time that *chaotic* and not regular orbits have been associated with the support of a morphological feature in disk galaxies, i.e. the inner part of the spiral arms.

Also, Patsis, Athanassoula and Quillen [11, 12] using the potential estimated by Quillen, Frogel and Gonzales [14], have found that the shape of the outer boxy isophotes in the bar of NGC 4314, are due to *chaotic orbits.* These orbits support the boxiness of the bar more efficiently than the stable, rectangular-like, regular orbits, which occupy a tiny area of the phase-space. Recently Patsis [10] (hereafter P06) has shown that the orbits of the particles located on the spiral arms in a response model of the same potential (i.e. the one obtained in [14]) are chaotic as well. In the P06 model the spiral arms are just due to chaotic orbits. There is good agreement between the morphology of the response model and the nearinfrared image of the galaxy. In both model and galaxy the arms are emerging out of the end of the bar, are faint, and do not extend azimuthally over an angle larger than $\pi/2$. The NGC 4314 morphology is typical for early type bars, and this indicates that chaotic orbits could explain the spiral structure in this galactic type. Families of stable periodic orbits, other than those proposed by KC96 have been found in P06, and could potentially support spiral arms away from the ends of the bar in similar potentials, if populated.

It has to be emphasized that a large percentage of the orbits supporting the spirals in the response models in P06 have a strong "4:1" character (Fig. 2). This means that these orbits simultaneously enhance the spirals and present a morphology resembling the morphology of the simple periodic orbits encountered at the 4:1 resonance region, i.e. they support rectangular-like (Fig. 2h,i) or rhomboidal (Fig. 2e) structures. There are also intermediate types (Fig. 2c,d,f,g). All these are orbits of test particles of the response model for NGC 4314 given in Fig. 2b, that matches the morphology of the galaxy (Fig. 2a).

Comparing the P06 *chaotic* orbits with those of KC96, we realize that the former are of the general kind presented in Fig. 21a of the latter work,

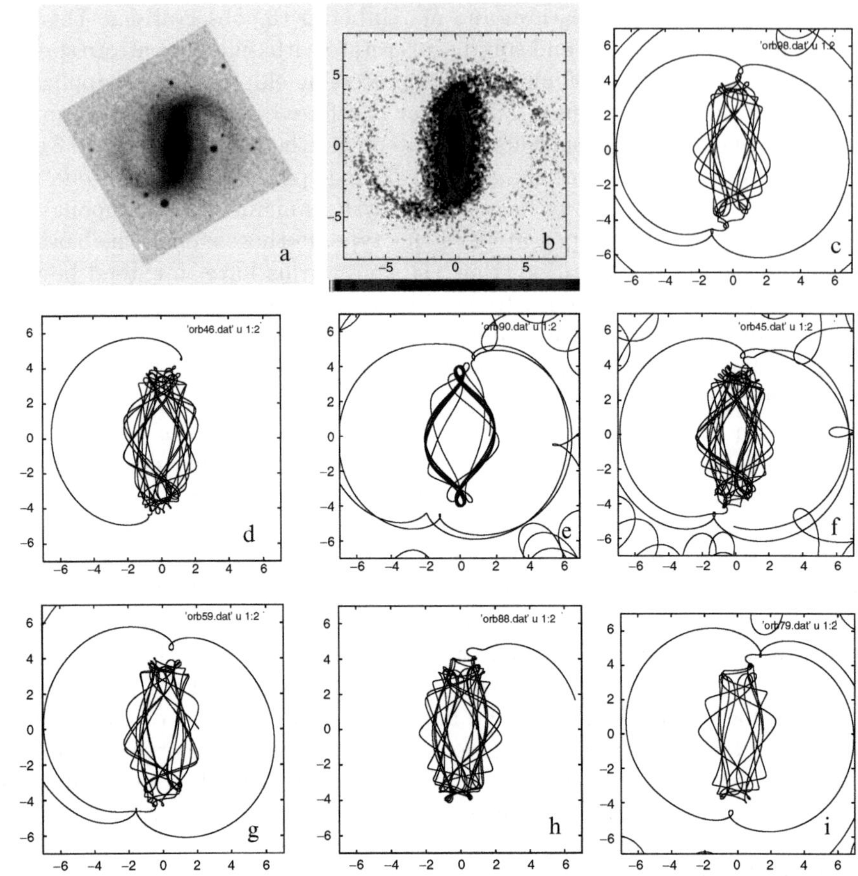

Fig. 2. (**a**) The DSS image of NGC 4314. (**b**) A stellar response model from those discussed in [10]. (**c**) to (**i**) Chaotic orbits demonstrating a 4:1 resonance region morphology and simultaneously supporting the spirals that emerge out of the ends of the bar. The orbits are integrated for about 10 orbital periods of the bar

that means that the orbits of the particles on the arms traverse also the area of the bar. We note that the loops at the arms in the P06 models are large, so the trajectories of the test particles remain almost parallel to the response spiral arms. Also the orbits in P06 open, visiting areas at large galactocentric distances. This is obviously the reason why these orbits support a spiral pattern that does not complete an angle larger than $\pi/2$. In other models, where for a large fraction of the orbits, the test particles return close to the end of the bar for several revolutions, rings could be formed as suggested by Romero-Gomez et al. [15]. In a subsequent paper Romero-Gomez et al. [16] investigated the parameter space in analytic barred potentials and correlated different morphologies of barred spiral systems with the variation of different parameters in their models. It seems that all these models describe a common type of orbital behavior close to the ends of the bar, which despite the fact that refers to chaotic motion, supports observed morphological features such as spiral arms or rings.

In N-body models now, orbits that traverse both bar and outer disk have been found in [17] and [13], who estimated that 30% of the particles in their simulation were on such orbits, which comprised a "hot" orbital population. According to response and orbital models, these orbits are the best candidates to support a spiral structure beyond the end of the bars. Recently, Voglis et al. [20] presented N-body models with a well developed long-lived barred-spiral structure, where they find that the arms in their full extent are due to chaotic orbits. The spiral arms in their models extend azimuthally over an angle of about π and do not need help by regular orbits of the $-2{:}1$ family in a part of them (as in the orbital models in KC96). Furthermore in [21] it is argued that chaotic orbits associated with the invariant manifolds of the short period family around L_1 or L_2, support the spiral structure found in [20] by spending a large part of their radial period at the location of the arms. This way the wavy nature of the arms is retained, while a mechanism totally different from the "precessing ellipses flow" is acting.

The present study tries to associate morphological and kinematical features with the one or the other type of flow. In Sect. 2 we present the features that are related with the "precessing-ellipses flow", using response models with a spiral perturbation of the CG86 type. In Sect. 3 we do the same for the case of chaotic spirals in the Quillen et al. potential for NGC 4314, and in Sect. 4 we summarize our conclusions.

2 The Flow in Normal Spirals

A typical response of a 2D disk of particles moving in an axisymmetric background potential Φ_0 to an imposed bisymmetric logarithmic spiral perturbation Φ_s, added to Φ_0, is as the one given in Fig. 3. We have constructed a large number of response models differing in their initial conditions and the strength of the spiral perturbation. The angular velocity of the spiral (Ω_s) is the critical parameter. The overall response is similar in models with the same Ω_s. In the particular model we present in Fig. 3, the background potential is of the form

$$\Phi_0(r) = -v_{max}^2 (f_b \exp(-\epsilon_b r) - [\mathrm{lnr} + \mathrm{E_1}(\epsilon_\mathrm{d}\mathrm{r})]) \tag{1}$$

and the perturbing potential is

$$\Phi_s(r,\theta) = Amp\, r \exp(-\epsilon_\mathrm{s}\mathrm{r}) \cos\left(\frac{2\,\mathrm{lnr}}{\mathrm{tani_0}} - 2\theta\right), \tag{2}$$

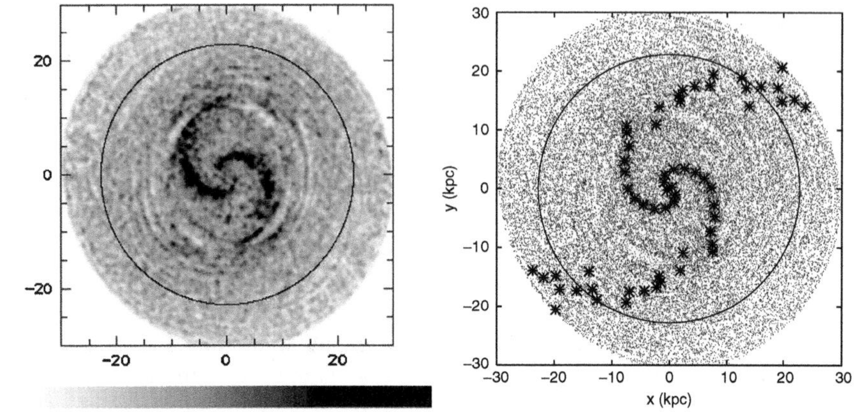

Fig. 3. The response model for a normal spiral. *Left*: an image representation of the response snapshot. *Darker* are denser regions. The strong, symmetric spiral ends at the 4:1 resonance. The *circle* indicates corotation. *Right*: the m = 2 phases overplotted on the same snapshot

where Amp is given by

$$Amp = A \, \frac{1}{2} \left(1 + \tanh[\kappa_1 (r - r_1)]\right) \, \frac{1}{2} \left(1 + \tanh[\kappa_2 (r - r_2)]\right) . \qquad (3)$$

The values of the parameters used are: $v_{max} = 200\,\mathrm{km\,s^{-1}}$, $f_b = 0.03$, $\epsilon_b = 0.694\,\mathrm{kpc^{-1}}$, $\epsilon_d = 0.1\,\mathrm{kpc^{-1}}$, $A = 155\,\mathrm{km^2\,s^{-2}\,kpc^{-1}}$, $\epsilon_s = 0.1\,\mathrm{kpc^{-1}}$, $r_1 = 1.5\,\mathrm{kpc}$, $\kappa_1 = 1\,\mathrm{kpc^{-1}}$, $r_2 = 45\,\mathrm{kpc}$, $\kappa_1 = 10\,\mathrm{kpc^{-1}}$, and $\mathrm{i_0} = -30°$. $\mathrm{E_1(x)}$ is the exponential integral, i.e.

$$\mathrm{E_1}(x) = \int_x^{\infty} \frac{\exp(-t)}{t} dt.$$

The spiral rotates with pattern speed $\Omega_s = 8.32\,\mathrm{km\,s^{-1}\,kpc^{-1}}$, and in the snapshot we have chosen for presentation it has completed more than 22 revolutions. Hereafter, we will call the above model with the particular set of parameters, "model M1". At time $t = 0$, 80,000 particles have been distributed randomly on a disk of radius $\mathrm{r} = 30\,\mathrm{kpc}$ so that we start with a homogeneous particle distribution. The particles are initially put on circular motion in the axisymmetric potential Φ_0. Added to this are peculiar radial and tangential velocities randomly chosen from Gaussian distributions, so that Toomre's $Q = 1$. This is just for obtaining a reasonable set of initial conditions, since the model is not self-gravitating.

The maximum strength of the imposed perturbation $F_{max}(r)$, expressed in terms of perturbing over axisymmetric forces $(\max|\nabla\Phi_s|/(d\Phi_0/dr))$ is $F_{max}(r) \approx 0.08$ at the end of the strong spirals at $r = 12.2\,\mathrm{kpc}$. At corotation ($r = 22.2\,\mathrm{kpc}$) $F_{max}(r) \approx 0.04$ and it continues decreasing outwards.

Equations of motion are derived from the Hamiltonian

$$H \equiv \frac{1}{2}\left(\dot{x}^2 + \dot{y}^2\right) + \Phi(x,y) - \frac{1}{2}\Omega_s^2(x^2+y^2) = E_J, \tag{4}$$

where (x, y) are the coordinates in a Cartesian frame of reference corotating with the spiral with angular velocity Ω_s. $\Phi(x,y) = \Phi_0 + \Phi_s$ is the total potential in Cartesian coordinates, E_J is the numerical value of the Jacobian integral and dots denote time derivatives.

By Fourier analyzing the surface density $\Sigma(r,\theta)$ of the snapshot in Fig. 3 we obtain the radial variation of the amplitudes of the various Fourier components. The variation of the amplitudes of the m = 2 and m = 4 components of the surface density are given in Fig. 4. The amplitudes $A_{2,4}(r)$ are normalized over the mean surface density $A_0(r)$ at each radius. Lower dispersion of velocities in the initial conditions result in stronger amplitudes in the response models under the same imposed potential. Characteristic for all spiral response models we tried is the abrupt increase of the m = 4 amplitude at the 4:1 resonance (around 12 kpc). On a disk with an initially homogeneous distribution of particles, one can observe at this region four crescent-shaped areas of low surface density just beyond the end of the strong spirals. Initially homogeneous disks allow us to study possible morphologies even beyond the end of the spirals. Contrarily the outer parts of initially exponential disks are depopulated after a few spiral revolutions.

As we see in Fig. 4 the m = 2 amplitude reaches a maximum $A_2(r)/A_0(r) \approx 0.38$, while at the end of the spirals it is of the order of 0.2. Despite the fact that no outer cut-off is introduced in the imposed potential (r_2 is larger than the maximum radius of the modeled disk) the m = 2 amplitudes

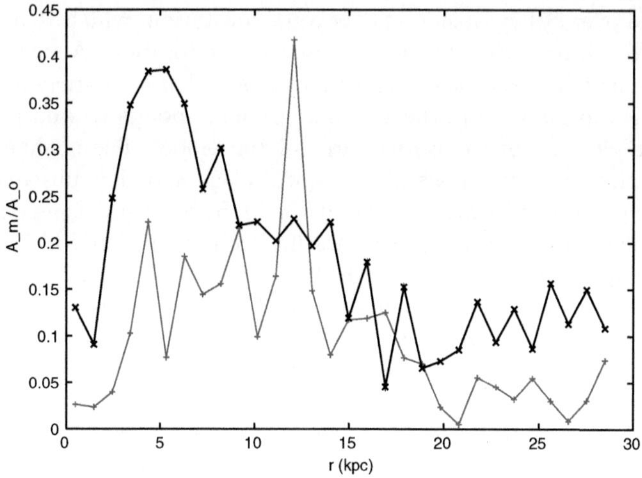

Fig. 4. The m = 2 (*black*) and the m = 4 (*grey*) amplitudes as a function of radius, for the snapshot presented in Fig. 3

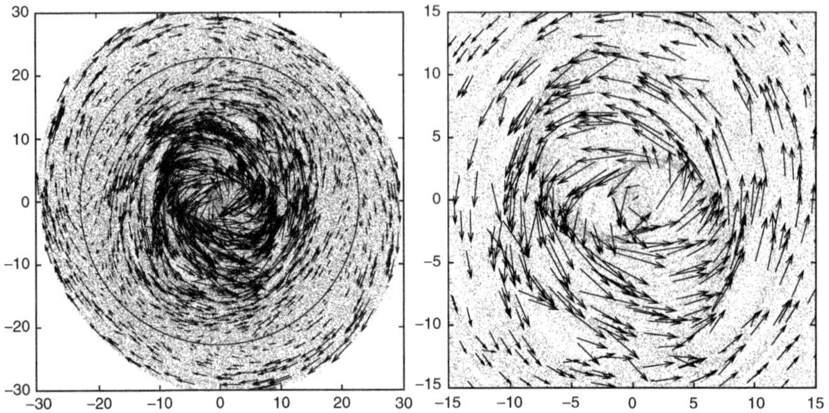

Fig. 5. Velocities in the corotating with the spiral frame of reference. *Left*: the *circle* denotes corotation. We observe the change in the sense of rotation inside and outside corotation. *Right*: zoom in the region where the strong symmetric spiral exists. We observe a typical case of the "precessing-ellipses flow"

decrease beyond the 4:1 region to the order of 0.1–0.15. These are values similar to those found from the K band surface density analysis in [6]. The scattered points of the phases at the outer regions of the disk indicate that a weak m = 2 component continues existing in larger distances, even beyond corotation (Fig. 3 right).

In our model the flow of the particles through the spiral arms is a typical "precessing-ellipses flow". The velocity field in the corotating with the pattern frame of reference, is given in Fig. 5. Velocities are indicated with arrows, scaled so that they are discernible in the figure. In Fig. 5 (left) we can see the overall flow inside and beyond corotation, which is indicated by a circle. We observe the change in the sense of rotation. At the corotation region the magnitudes of the velocities decrease characteristicly and we have anticyclonic motion around the stable Lagrangian points L_4 and L_5. Figure 5 (right) is a close-up of the central area of the model. The change in the direction of the velocity arrows at the spiral arms and the subsequent longer stay of the particles at the arm than at the interarm region is associated with the existence of the spiral structure itself. The flow depicted in Fig. 5 can be directly compared with the flow in Kalnajs [7] (his Fig. 3b).

A way of studying the kinematics of disk galaxies, is by means of the residual velocity field. The residual velocity field is obtained by subtracting from the original one the circular motion due to the axisymmetric component Φ_0. Residual velocity fields have been used by Fridman et al. [3, 4] and recently by Vorobyov [22], in order to show the existence of cyclonic and anticyclonic motion on the galactic disks. For the purpose of the present study it would be interesting to know whether or not the motion in vortices takes place beyond the end of the spiral structure or in the middle of the arms in the case of

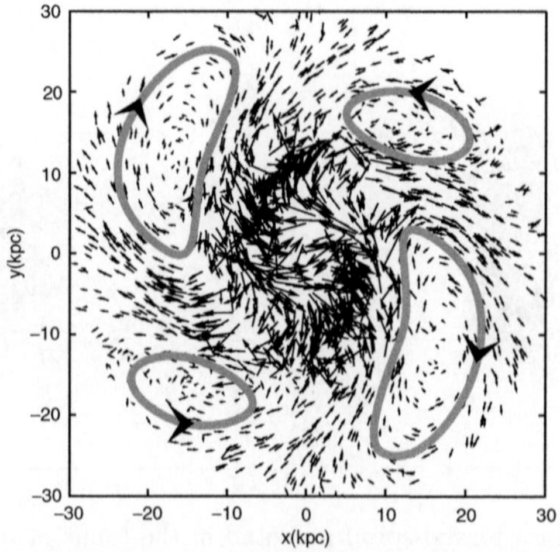

Fig. 6. The residual velocity field reveals the presence of cyclonic and anticyclonic motion beyond the end of the bisymmetric spiral structure. The *grey curves* indicate schematically the directions on the vortices. Spirals rotate counterclockwise

normal spirals. This would determine whether or not the "precessing-ellipses flow" characterizes the arms over their whole extent, or if we also have a flow determined by the presence of vortices along part of them. The corresponding to Fig. 5 residual velocity field is given in Fig. 6. We have drawn by hand four thick grey curves in order to underline the cyclonic and anticyclonic patterns, that can be observed at the corotation region. Anticyclones are around $L_{4,5}$ and cyclones around $L_{1,2}$. In both our model M1 and Vorobyov's model the amplitude of the spiral perturbation decreases outwards. Corotation radius is at 22.2 kpc, while the end of the symmetric spiral arms is at $r = 12.2$ kpc. This stellar residual velocity field is similar with the gaseous "Model 2" in [22] and places the cyclonic and anticyclonic patterns beyond the end of the response spiral structure. Thus, the flow up to the end of the arms is purely of the "precessing-ellipses" kind. We draw similar conclusions if we consider the response of SPH gaseous models to normal spiral potentials. The flow resembles the one suggested by Visser in [19] for M81.

A feature that can be assigned to ordered flows, is the alignment of bright knots along spiral arms in distances as long as 4 kpc ([5], see also Grosbøl in this volume). This characterizes K-band images of non- (or weakly) barred-spiral galaxies. These spirals are characterized also by continuity in their morphology. We do not observe gaps.

3 Barred Galaxies with Spirals

Spirals are encountered also in barred galaxies. They appear beyond the end of the bar. In the present section we consider systems where bar and spirals rotate with the same pattern speed. As mentioned in Sect. 1, the stellar flow at the

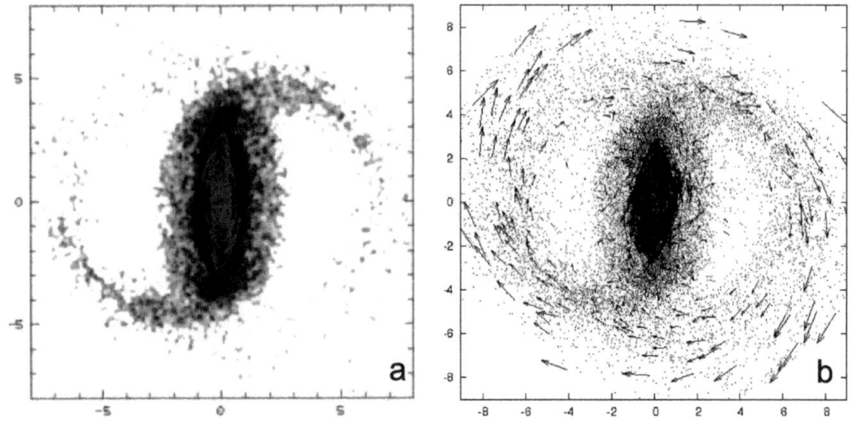

Fig. 7. The flow in a barred-spiral potential in the frame of reference co-rotating with the pattern. Bar and spirals share the same pattern speed. (**a**) An image representation of the modeled snapshot. (**b**) *Arrows* that indicate the flow of the test particles. The particles on the spiral arcs that are attached to the ends of the bar are in chaotic motion

spiral arms of such a system has been presented in P06 and is associated with chaotic motion in the corotation region and specifically with the motion along the manifolds at the unstable Lagrangian points L_1 and L_2, or other unstable periodic orbits at and beyond the 4:1 resonance. Figure 7a is a snapshot from a response model using the imposed potential

$$\Phi(r,\theta) = \Phi_0(r) + \sum_{m>0} \Phi_{mc}(r)\cos(m\theta) + \Phi_{ms}(r)\sin(m\theta), \tag{5}$$

where $m = 2,4,6$ [14].

The amplitudes of the components in (5) are written in the form $\sum_{n=0}^{8} \alpha_n r^n$. The values of the coefficients α_n in $(\mathrm{km\,s^{-1}})^2$ can be found in Table 1 in [14]. This potential is directly estimated from K-band observations of the early type barred-spiral galaxy NGC 4314. Calculations have been done in the rotating with $\Omega_b = 38.23\,\mathrm{km\,s^{-1}\,kpc^{-1}}$ system.

The velocity field of this snapshot is given in Fig. 7b. The chaotic orbits of the particles at the arms have been presented in P06. In this model there is no "precessing-ellipses flow" supporting the spiral structure. The particles that reinforce the arms visit both the bar as well as the disk area beyond the end of the bar. Most of their trajectories close to the bar resemble quasiperiodic orbits at the 4:1 resonance region (see Fig. 2). The rest of the formalism is as in Sect. 2 for the normal spiral galaxies.

As mentioned in Sect. 1, the arms now do not complete an angle larger than $\pi/2$ in azimuth. A continuation of the spiral structure is observed in some models at larger angles. In response models these continuations can be either due to material trapped around families of stable periodic orbits (KC96,

P06) or due to chaotic orbits. In the latter case it is observed a gap between the outer part of the spirals and the one attached to the end of the bar. In N-body models [20] however, there are no major discontinuities on the arms until azimuths larger than π.

A basic element that differentiates the flow in the barred-spiral case from the normal spiral one, is the strength of the perturbation. In model "M1" the relative m = 2 perturbation at the end of the spirals, in the surface density, is of the order of 0.2 (normalized over the amplitude of the 0-th order term – Fig. 4), falling to 0.1–0.15 at corotation. The corresponding m = 2 strength at the end of the bar (close to corotation) in the K surface brightness of the galaxy NGC 4314 is 0.8–0.9, as measured in [14].

4 Conclusions

Below we enumerate our conclusions:

1. In response models we find two kinds of flow that characterize the stellar spiral arms. The "precessing-ellipses flow", which is associated with the arms of normal (non-barred) spiral galaxies, and the chaotic stellar flow, *along* the arms, in models of barred-spiral galaxies.
2. In barred-spiral systems with two pattern speeds, the flow on the arms can be, to a large extent, as in normal spirals.
3. The m = 2 amplitude beyond the end of the spirals, near corotation, is of the order of 0.1–0.2 of the axisymmetric term in models for normal spiral galaxies in agreement with the analysis of galaxies with a similar morphology. Contrarily in barred-spiral systems with chaotic spiral arms, at the end of the bar it is larger than 0.8. This indicates that the strength of the perturbation at the corotation region is associated with the presence or not of chaotic spiral arms beyond that resonance.
4. The orbits of particles on chaotic arms traverse also the area of the bar. Most trajectories, during the time they spend close to the bar, resemble the morphologies of quasi-periodic orbits trapped close to stable periodic orbits at the 4:1 resonance region. Thus they can reinforce the boxiness of the outer isophotes of the bar. This also suggests that the chaotic orbits that support the spirals are not associated just with the unstable Lagrangian points, but with other unstable periodic orbits as well.
5. There are morphological characteristics, that can help in distinguishing "ordered" from "chaotic" spirals. The models show that if a "precessing ellipses flow" is present, the resulting spirals are continuous without gaps, all the way out to their end. This is consistent with the nearinfrared images of *normal* spirals. Spirals that emerge out of the ends of a bar do not complete an angle larger than $\pi/2$. Our models show that chaotic spirals either end at that azimuth or that they are characterized by clear large gaps at that azimuth. Again these are features frequently observed in the near infrared images of *barred* spiral galaxies.

References

1. V. Boonyasait, P.A Patsis, S. Gottesman: In *Nonlinear Dynamics in Astronomy and Physics* ed by S.T. GottesmanD, J-R. Buchler, M.E. Mahon (N. Y. Ac. Sci., 2005) pp 203–224
2. G. Contopoulos, P. Grosbøl: Astron. Astrophys. **155**, 11 (1986) (CG86)
3. A. Fridman, O.V. Khoruzhii, V.V Lyakhovich et al.: Astroph. Sp. Sc **252**, 115 (1997)
4. A. Fridman, O.V. Khoruzhii, V.V Lyakhovich et al.: Astron. Astrophys. **371**, 538 (2001)
5. P. Grosbøl, H. Dottori, R. Gredel: Astron. Astrophys. **453**, L25 (2006)
6. P. Grosbøl, P.A. Patsis: Astron. Astrophys. **336**, 840 (1998)
7. A. Kalnajs A: PASA **2**, 174 (1973)
8. D.E. Kaufmann, G. Contopoulos: Astron. Astrophys. **309**, 381 (1996) (KC96) 381 (KC96)
9. S. Onodera, J. Koda, Y. Sofue, K. Kotaro: PASJ **56**, 439 (2004)
10. P.A. Patsis: MNRAS **369**, L56 (2006) (P06)
11. P.A. Patsis, E. Athanassoula, A.C Quillen: In *Spiral Galaxies in the Near-IR*, ed by D. Minitti, H-W. Rix (Springer, Berlin Heidelberg, 1996) pp 246–247
12. P.A. Patsis, E. Athanassoula, A.C Quillen: ApJ **483**, 731 (1997)
13. D. Pfenniger, D. Friedli: Astron. Astrophys. **150**, 112 (1991)
14. A.C. Quillen, J. Frogel, R. Gonzales: ApJ **437**, 162 (1994)
15. M. Romero-Gómez, J.J. Masdemont, E. Athanassoula: Astron. Astrophys. **453**, 39 (2006)
16. M. Romero-Gómez, E. Athanassoula, J.J. Masdemont: Astron. Astrophys. **472**, 63 (2007)
17. L.S. Sparke, J.A. Sellwood: MNRAS **225**, 653 (1987)
18. J.A. Sellwood, L.S. Sparke: MNRAS **231**, 25 (1988)
19. H.C.D. Visser: Astron. Astrophys. **88**, 159 (1980)
20. N. Voglis, I. Stavropoulos, C. Kalapotharakos: MNRAS **372**, 901 (2006)
21. N. Voglis, P. Tsoutsis, E. Efthymiopoulos: MNRAS **373**, 280 (2006)
22. E.I. Vorobyov: MNRAS **370**, 1046 (2006)

Ansae in Barred Galaxies, Observations and Simulations

I. Martinez-Valpuesta

Instituto de Astrofísica de Canarias, C/Vía Láctea E-38200,
La Laguna, Tenerife, Spain
imv@iac.es

Summary. Some barred galaxies show two symmetric enhancements at the ends of the stellar bar, called *ansae*, or the "handles" of the bar. The existence of ansae was noticed some decades ago, and the discussion about their origin is still open. Based on a quantitative analysis of the occurrence of ansae in barred galaxies, and making use of *The de Vaucouleurs Atlas of Galaxies* by Buta and co-workers we found that ∼40% of SB0s show ansae in their bars, thus confirming that ansae are common features in barred lenticulars. The frequency of ansae decreases dramatically with later types, and hardly any ansae are found in galaxies of type SBb or later. In this paper we study the dynamics of ansae regions, by means of their orbital structure, concluding that the orbits giving shape to ansae are part of the bar orbital families. We are able to identify particles trapped in this orbits and their origin. These particles have been trapped by the bar. We conclude that bars presenting ansae could be those which are currently growing.

1 Introduction

Disk galaxies very often present bars. Old studies based on optical images established that about 1/3 of all disk galaxies are strongly barred, and additionally 1/3 are moderately barred (e.g., [21, 22]). In the last decade, using near-infrared (NIR) imaging the frequency of barred galaxies rises to close to 80% of local disk galaxies [11, 12, 14–16]. Recently, using images obtained with the *Hubble Space Telescope*, it has been determined that the fraction of strong bars remains almost unchanged from intermediate redshifts, of $z \sim 1.2$, to the present day (e.g., [10, 13, 24, 26]). This is in agreement with numerical simulations, which show that bars are indeed long-lived phenomena (e.g., [3, 8, 9, 17, 19, 23, 25]).

To a first approximation, bars can be considered as ellipsoidal features in disk galaxies. Following this, the isodensity contours of bars have been traditionally fitted with ellipses. Later on, and to account for the boxiness of observed stellar bars, a method of generalised ellipses was introduced by [4]. This method can be applied to both observed bars and numerical simulations

G. Contopoulos, P.A. Patsis, *Chaos in Astronomy.* Astrophysics and Space Science Proceedings, doi: 10.1007/978-3-540-75826-6,

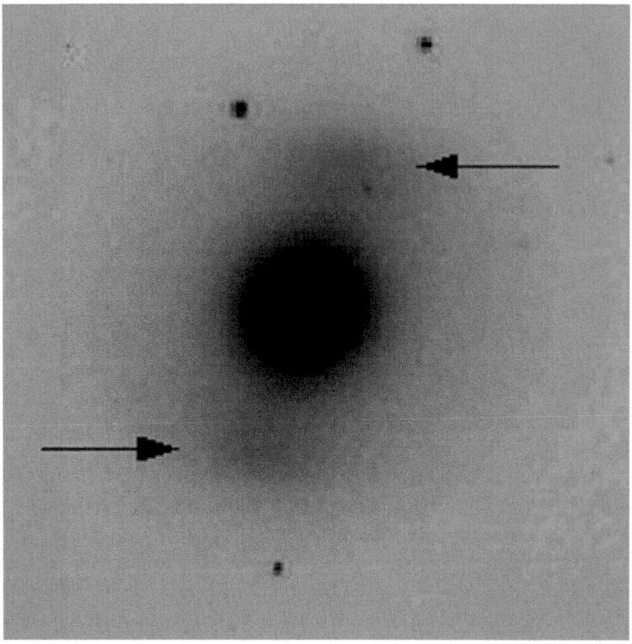

Fig. 1. *R* band image of the barred lenticular (type (R)SB(r)0$^+$) NGC 2859, with the locations of the ansae indicated with arrows. The images were obtained in service time on the night of 2007 February ninth with the ALFOSC camera on the 2.5 m Nordic Optical Telescope. Images are about 80 arcsec on the side, N is up and E to the left

of bars. Today the existence of different shapes of observed and simulated bars is widely accepted. Stellar bars can go from mild oval distortions to highly elongated and rectangular bars. The former coincide generally with weak bars, and the latter, with generally strong bars. In this paper, we are interested in the outer parts of the bar, specifically in the ansae, which appear in this region.

Ansae can be described as a pair of "handle-shaped" density enhancements at the ends of the bar (Fig. 1), sometimes also mentioned as "condensations" in earlier papers (e.g., Danby 1965). Ansae have been known in the literature for decades, and they appear to be rather frequent in early-type barred galaxies (e.g., NGC 4262, NGC 2859 and NGC 2950; [1, 21]). Recently, a statistical study has been published on the prevalence of ansae [18]. In this paper it was found that ~40% of SB0 present ansa bars. Ansae were discussed by [7] in the context of an "outflow" from the bar into the associated spiral arms. In *N*-body numerical simulations, ansae are seen on both ends of a bar as characteristic density enhancements in face-on or edge-on disks [2, 17].

The dynamical significance of ansae remains unclear. Are they regions of trapped disk or bar particles? Do they reflect any underlying dynamics? Do they appear in a particular stage of the bar evolution? We will try to answer these questions in this paper, by means of numerical simulations and orbital analysis.

2 Observations

The first step when talking about ansae is to give no just an "optical" definition of ansae but also a more mathematical one. Then, based on observations, a more correct definition could be the local maxima before the end of the bar located along the major axis. As an example we show in Fig. 2 two surface density cuts. One perpendicular, and one parallel, to the major axis of the bar in the observed galaxy NGC 7098.

Details about observations of ansae can be found in [18]. In this paper the authors assessed the frequency of occurrence of ansae in barred galaxies of various morphological types. Although such ansae have been described in the literature for decades, there was no yet information at all on how common they are. The analysis is based on classifications given by [6] for a sample of 26 early-type barred galaxies, and in the *The de Vaucouleurs Atlas of Galaxies* [5] (hereafter "the Atlas") for a much larger sample of 267 barred galaxies, of all types. The main conclusion from this paper is that ansae never occur

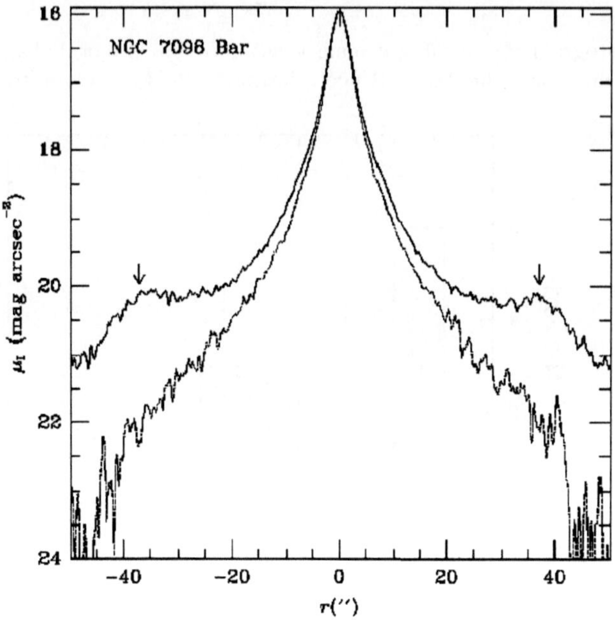

Fig. 2. Surface density cuts of parallel (*dark solid line*) and perpendicular (*thin solid line*) to the bar major axis of the observed galaxy NGC 7098 (provided by Ron Buta)

in late-type galaxies (types SBc or later) and only very rarely in intermediate (SBab-SBbc) types, but that they are very common in early-type barred galaxies. The highest fraction of ansae is found in strongly barred lenticulars, with ∼40% of SB0 galaxies showing ansae (Fig. 3). The overall fraction among SB0 and SBa galaxies is around 1/3. It is also found that the median bar strength, as based on literature measurements of the gravitational bar torque (Q_b), is significantly higher among barred galaxies with ansae than among those without (Fig. 4). We have made a histogram where the ansa and

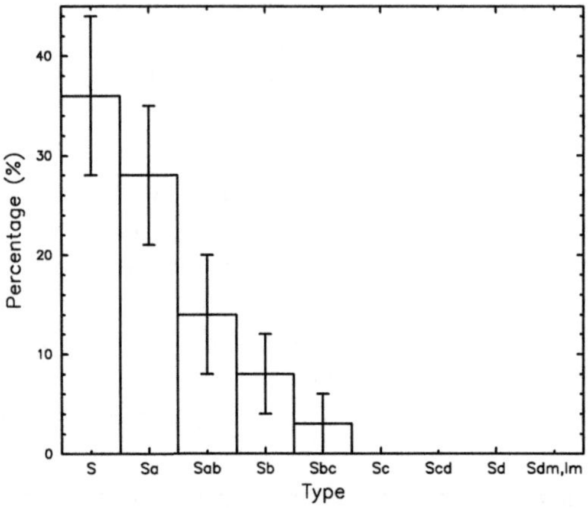

Fig. 3. Histogram of percentages of ansae as a function of morphological type, based on the results from "the Atlas". Uncertainties are Poisson errors (from [18])

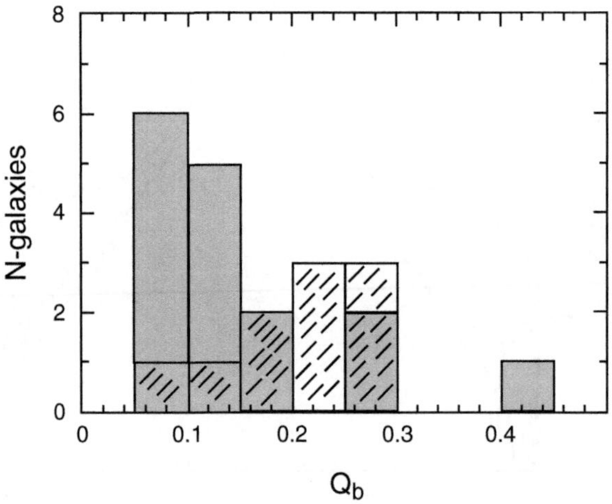

Fig. 4. *Grey areas* correspond to those barred galaxies without ansae, and *dashed areas* correspond to barred galaxies with ansae. Q_b values are obtained from the literature, mainly from [6]

non-ansa bars are shown as two different distributions. Grey areas correspond to those barred galaxies without ansae, and dashed areas correspond to barred galaxies with ansae. We have also performed the K-S test, which results in a p-value of 0.05. This is just about significant, although with the small numbers involved and the large spread in Q_b values this is obviously a statistical result rather than a prescription.

3 Simulations of Barred Galaxies and Ansae

The evolution of barred galaxies in numerical simulations is governed by dynamical and secular processes during which the bar experiences structural changes as well as changes in its kinematics. Previous numerical works have shown clearly that an evolving bar changes its structure over the timescales of secular evolution, but the details of this growth remain obscure. We will try to relate this growth to the ansae.

In [2], the appearance of ansae was related to the initial conditions of the models (e.g., the halo-to-disk mass ratio). In a later paper, [3], it was found that ansae can be seen in models of type "MH", with a centrally concentrated halo. In these models the bars grow stronger. In general, in simulated galaxies ansae appear after some Gyrs of evolution.

3.1 Ansae: The Particle Trapping Signature

In order to have a more complete study of ansae we use a single simulation with very good particle resolution (see [17] for details). The ansae can be detected in this simulation shortly after 4 Gyr. In order to understand the particle population of ansae, we select particles within a spherical volume located in the ansae region of the bar. Only a relatively small fraction of their orbits, ∼18%, are trapped by the bar, i.e., obey $\frac{\Omega-\Omega_b}{\kappa}=\frac{1}{2}$, as the spectral analysis has shown. However, this fraction has increased dramatically to ∼50% after about four bar rotations, as we have verified by repeating our analysis at a later stage of the simulation.

We followed the live trajectories of particles and find that the trapping occurs predominantly from the mildly oval and/or 3:1 orbits. The particles get mostly trapped by the x_1 family and by associated multi-periodic orbits, librating around them (Fig. 5). The orbits appear resonant not only with the x_1 (i.e., 2:1) but also with the $n{:}m = 2p{:}p$ multi-periodic families, where $p = 2$, 3, etc. We have verified this by using snapshot Poincare diagrams (not shown in here).

Patsis [20] noticed that ansae may appear by populating quasi-periodic orbits around x_1 stable periodic orbits of elliptical morphology (without loops at its apocentra). However, to show that a similar phenomenon occurs in non-analytical, grainy potentials, in the same way, is not a trivial extension of Patsis' work. The main issue here is to demonstrate that the islands, corresponding to the multi-periodic orbits in Poincare diagrams, will be as

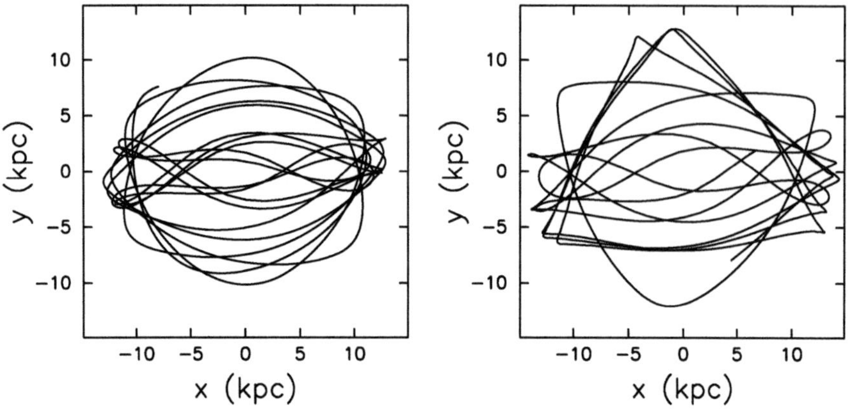

Fig. 5. Trajectories of two ansae particles before and after their trapping by the bar, in the rotating bar frame. The disk is face-on, the bar major axis is horizontal and the particles stream counter-clockwise. The particle orbits have been projected onto the xy-plane and the orbits have been integrated between 7.05 and 11.75 Gyr. The particles are captured at the bar region by a mildly oval stable periodic orbit (*left frame*) and a 3:1 stable periodic orbit (*right frame*)

efficient in trapping a substantial population of orbits under these conditions as the idealised case of the test particles in an analytic potential. A more detailed study is out of the scope of this paper.

It is crucial that the nearly abrupt switch from the disk particles to the bar particles happens in the region corresponding to ansae. Furthermore, the particles are trapped by the bar when they are in the leading part of the ansae with respect to the bar rotation. In fact, the ansae in our simulations appear to be, sometimes, asymmetric with respect to the bar major axis, normally skewed in the leading direction. This provides a clear hint that the orbit capture by the bar could be due to the action of its gravitational torques. Hence the corresponding density increase in the ansae region will reflect the general slowdown of a particle close to the apocentre of its elongated orbit.

4 Summary and Future Work

The important implication of the above analysis is that the ansae serve as a signature of the secular bar growth. They appear transient, if this growth is uneven, but are stable otherwise. As such, the ansae observed in early-type barred galaxies can be associated with the later phase of bar evolution, the phase when stellar bars exchange the angular momentum with the surrounding dark matter halos. Additional work is required to show beyond doubt that the absence of ansae is related to the lack of growth in the physical size of the bar.

A more detailed study will be performed making use of more numerical simulations with different initial conditions. We hope to clarify in more depth, where and how the ansae can exist and survive. And also a more detailed orbital analysis will be performed.

Acknowledgements

I would like to thank the collaborators of this project: Johan Knapen and Ron Buta, for their observational expertise, and Isaac Shlosman and E. Athanassoula for their theoretical and dynamical knowledge. The author is also very grateful to the organisers of this conference for providing the participants with a great time for discussions. The author thanks the Gruber Foundation Fellowship for partial economical support.

References

1. E. Athanassoula: Phys. Rep., **114**, 319 (1984)
2. E. Athanassoula: New Horizons of Computational Science, (Eds.) T. Ebisuzaki, J. Makino (Dordrecht: Kluwer, 2001), pp 69
3. E. Athanassoula, A. Misiriotis: MNRAS, **330**, 35 (2002)
4. E. Athanassoula, S. Morin, H. Wozniak, M. J. Pierce, et al: MNRAS, **245**, 130 (1990)
5. R. Buta, H. Corwin, S. Odewahn: The de Vaucouleurs Atlas of Galaxies (Cambridge: Cambridge University Press, 2007)
6. R. Buta, E. Laurikainen, H. Salo, D. L. Block, J. H. Knapen: AJ, **132**, 1859 (2006)
7. J. M. A. Danby: AJ, **70**, 501, 1965
8. V. P. Debattista, J. A. Sellwood: ApJ, **493**, L5 (1998)
9. V. P. Debattista, J. A. Sellwood: ApJ, **543**, 704 (2000)
10. B. G. Elmegreen, D. M. Elmegreen, A. C. Hirst: ApJ, **612**, 191 (2004)
11. P. B. Eskridge, J. A. Frogel, R. W. Pogee, A. C. Quillen et al.: AJ, **119**, 536 (2000)
12. P. Grosbøl, E. Pompei, P. A. Patsis: ASP Conf. Ser. 275: Disks of Galaxies: Kinematics, Dynamics and Perturbations, 305 (2002)
13. S. Jogee, F. D. Barazza, H. W. Rix, I. Shlosman et al.: ApJ, **615**, L105 (2004)
14. J. H. Knapen, I. Shlosman, R. F. Peletier: ApJ, **529**, 93 (2000)
15. I. Marinova, S. Jogee: ApJ, **659**, 1176 (2007)
16. K. Menéndez-Delmestre, K. Sheth, E. Schinnerer, T. H. Jarrett, N. Z. Scoville: ApJ, **657**, 790 (2007)
17. I. Martinez-Valpuesta, I. Shlosman, C. Heller: ApJ, **637**, 214 (2006)
18. I. Martinez-Valpuesta, J. H. Knapen, R. Buta: AJ, **134**, 1863 (2007)
19. J. K. O'Neill, J. Dubinski: MNRAS, **346**, 251 (2003)

20. P. A. Patsis: MNRAS, **358**, 305 (2005)
21. A. Sandage: The Hubble Atlas of Galaxies (Washington, DC: Carnegie Institute, 1961)
22. J. A. Sellwood, A. Wilkinson: Rep. Prog. Phys., **56**, 173 (1993)
23. J. Shen, J. A. Sellwood: ApJ, **604**, 614 (2004)
24. K Sheth, M. W. Regan, N. Z. Scoville, L. E. Strubbe: ApJ, **592**, L13 (2003)
25. O. Valenzuela, A. Klypin: MNRAS, **345**, 406 (2003)
26. X. Z. Zheng, F. Hammer, H. Flores, F. Assémat, A. Rawat: A&A, **435**, 507 (2005)

Orbital Structure in Barred Galaxies and the Role of Chaos

M. Harsoula, G. Contopoulos, and C. Kalapotharakos

Research Center for Astronomy, Academy of Athens, Soranou Efesiou 4, GR-11527, Athens, Greece
mharsoul@academyofathens.gr, gcontop@academyofathens.gr, ckalapot@phys.uoa.gr

Summary. We study the orbital structure of a self-consistent N-body configuration of a barred galaxy, having a value of spin parameter near the one of our Galaxy. We find that 60% of the orbits are chaotic. We examine the phase space using 2-D projections of the surfaces of section for test particles as well as for real N-body particles. The real particles are not uniformly distributed in the whole phase space but they avoid orbits that do not support the bar. We use frequency analysis for the regular as well as for the chaotic orbits to classify certain types of orbits. We find the main resonant orbits and their statistical weight in supporting the shape of the bar emphasizing the role of weakly chaotic orbits. Finally we show that the effectiveness of Arnold diffusion is restricted.

1 Initial Conditions and Characteristics of the Bar

We consider the orbital structure in a bar generated by using two N-body codes: a tree code developed by Hernquist [6] and a conservative technique code designed by Allen et al. [1]. We use initially "quiet initial conditions" to simulate a non rotating elliptical galaxy and we evolve the system using the tree code until the violent relaxation phase is finished. (The details of this model can be found in [13].) Then we use a conservative technique code (hereafter c-t) for 150 half-mass crossing times. The total number of particles is about 1.3×10^5.

Following [7] we find the monopole, quadrupole and triaxial terms of the potential (120 terms) that is used for orbits' calculations. For more details see [19].

In order to create a system rotating around its smallest axis (i.e. the z-axis) we simply rotate the projections of the velocity vectors of all the particles on the X–Y plane to become perpendicular to their position vectors with a counterclockwise sense of rotation, and thus we give the maximum possible rotation on this plane. Then we let the system evolve using the c-t code again, for another 250 dynamical times and calculate the "spin parameter" λ (see [11]):

G. Contopoulos, P.A. Patsis, *Chaos in Astronomy.* Astrophysics and Space Science Proceedings, doi: 10.1007/978-3-540-75826-6,

$$\lambda = \frac{J|E|^{1/2}}{GM^{5/2}}, \tag{1}$$

where G is the gravitational constant. The angular momentum J, the energy E (in the inertial frame), and the mass M are measured as cumulative quantities along cylinders on the X–Y plane, having the Z-axis as their main axis.

The unit of the radius (runit) that corresponds to the limiting radius of the whole bound system is approximately 140 kpc in real units and the unit of the time (tunit) corresponds to the half mass crossing time of the system (hereafter hmct). A Hubble time is ≈400 hmct, thus 1 hmct ≈ 32 Myrs.

At a time $t = 250$ tunits the value of λ is close to $\lambda = 0.22$ over most of the extent of the system. This value corresponds to the maximum value of λ for a rotationally supported galaxy and is close to the one of our Galaxy (see [5]).

In Fig. 1a we plot a snapshot of the isodensities on the X–Y plane at 12 half mass crossing times (hmct) after the beginning of the c-t run. We notice a trailing spiral structure, as the galaxy is rotating as a whole counterclockwise. This spiral structure travels outwards before it vanishes later on. Therefore angular momentum is transferred to the material outwards. This is a well known mechanism that makes the bar grow stronger and slow down [10]. This transference is due to the interaction of the bar perturbation of the gravitational field with stars moving in resonant orbits, particularly near the ILR and corotation. N-body simulations have confirmed this evolution scenario of bar-like galaxies (e.g. [2, 3]). The spiral structure in our experiment vanishes after about 20 hmct. However in experiments with larger pattern speeds these spiral structures can survive even for times comparable to a Hubble time [18].

The angular velocity Ω_p of the bar at the end of the c-t run is $\Omega_{\rm p} \approx 21^\circ\,\mathrm{hmct}^{-1} = 0.37\,\mathrm{rad\,hmct}^{-1}$.

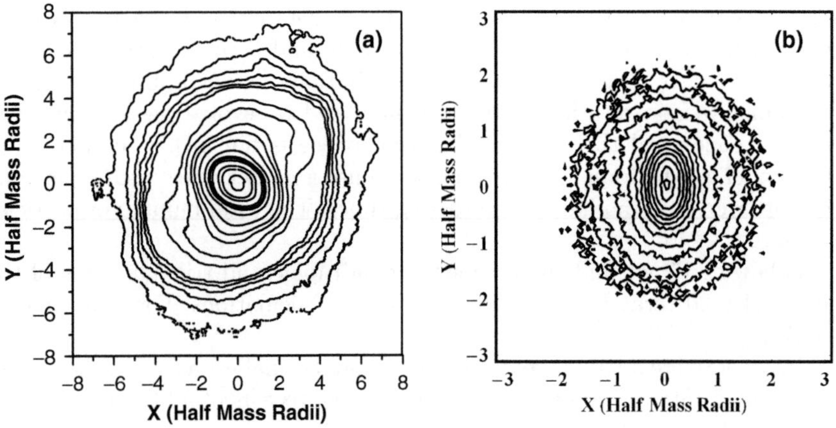

Fig. 1. (**a**) A snapshot of the isodensity contours at time = 12 hmct. (**b**) Isodensity contours at the end of the N-body run

In Fig. 1b we plot the isodensity contours on the X–Y plane at the end of the N-body run that corresponds approximately to a Hubble time. It is obvious that no spiral structure has survived.

The maximum value of the bar perturbation is close to 80% and is located a little inside the half mass radius of the system, while the perturbation drops to ≈25% near the end of the bar. This perturbation is considered large and is often observed in real barred galaxies. See for example [9], where the m = 2 and m = 4 components are calculated for a sample of 112 spiral galaxies.

2 Classification in Ordered and Chaotic Orbits

In order to classify the orbits in ordered and chaotic ones we use the combination of two methods, first a specific form of the Lyapunov Characteristic Number (LCN) method that is called the Specific Finite Time Lyapunov Characteristic Number (SFTLCN) [17] and then the Smaller ALignment Index, (SALI), or simply Alignment Index (AI) [12, 17]. For more details see [19].

The Specific Finite Time Lyapunov Characteristic Number ($SFTLCN$), or simply L_j for every orbit which is given by the formula:

$$L_j(T_{rj}, t_j) = \frac{T_{rj}}{t_j} \sum_{i=1}^{N_j} a_{ij}, \tag{2}$$

where t_j is the integration time, N_j is the number of time steps ($\Delta t = t_j/N_j$), T_{rj} is the average radial period of the orbit of the particle j and a_{ij} is the stretching number [14] at the time step i ($i = 1, \ldots, N_j$).

In the second method (using AI), we use the properties of the time evolution of the deviation vectors [15, 16].

In particular, we consider the time evolution of two arbitrary different initial deviation vectors ξ_{j1} and ξ_{j2} of the same orbit. If the orbit is chaotic, then the smallest of the two norms $d_{j-}(t) = |\boldsymbol{\xi}_{j1}(t) - \boldsymbol{\xi}_{j2}(t)|$ and $d_{j+}(t) = |\boldsymbol{\xi}_{j1}(t) + \boldsymbol{\xi}_{j2}(t)|$ is called Alignment Index (AI) and tends exponentially to zero. However, in our calculations we impose a cutoff limit, e.g. $AI = 10^{-10}$ to save integration time. On the other hand, if the orbit is ordered, then d_{j-} or d_{j+} oscillate around a roughly constant mean value. This value is usually close to unity and in any case it is not less than 10^{-3}. Thus, by following the evolution of the smaller of the two indices d_{j-}, d_{j+}, we can distinguish between regular and chaotic orbits.

Comparing these two methods we find that 60% of the orbits are chaotic which is almost twice the percentage calculated in the non rotating system, representing an elliptical galaxy (see for example [17]).

3 Orbital Structure of the Barred Galaxy

3.1 Surfaces of Section

In order to study the orbital structure of our system we use two powerful tools: (a) surfaces of section and (b) frequency analysis [8] of the orbits. For this purpose we use the values of the coefficients of the analytical expansion of a frozen potential at a snapshot corresponding to 300 hmct and construct surfaces of section (SOS) for different energies E_j in the rotating frame (Jacobi constants):

$$E_j = \frac{1}{2}(v_x^2 + v_y^2 + v_z^2) + V(x, y, z) - \frac{1}{2}\Omega_p^2 R_{xy}^2, \tag{3}$$

where V(x,y,z) is the frozen potential of both the axisymmetric part and the bar perturbation, given by the N-body run, v_x, v_y, v_z are the velocities in the rotating frame of reference and Ω_p is the angular velocity of the bar rotating as a solid body and calculated numerically.

In our 3-D model the SOS are 4D and we plot their projections on the $x, \dot{x}$ plane, for $y = 0$, $\dot{y} > 0$ having initially $z = 0$ and $\dot{z} = 0$. All sections are made in the rotating frame of reference.

In Fig. 2 the surface of section for test particles is plotted for a value of the Jacobi constant $E_j = -42$ (in our units which are normalized to -100 at the potential well). Figure 2a presents the region outside corotation. Some small islands of stability exist corresponding to the OLR and −1:1 resonances. For the area inside corotation (Fig. 2b) the islands of stability correspond to the 3:1, 4:1 and 5:2 resonances, as well as to the "x1" type orbits with loops at the edges and the retrograde "x4" type orbits.

The same surface of section for the real N-body particles is shown in Fig. 3. For this purpose we have taken the initial conditions of particles having the

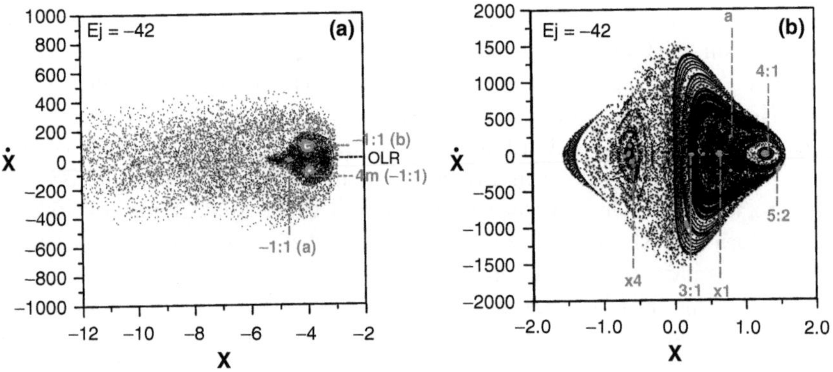

Fig. 2. The surface of section of test particles with a value of the Jacobi constant equal to −42: (**a**) for the region outside corotation and (**b**) for the region inside corotation

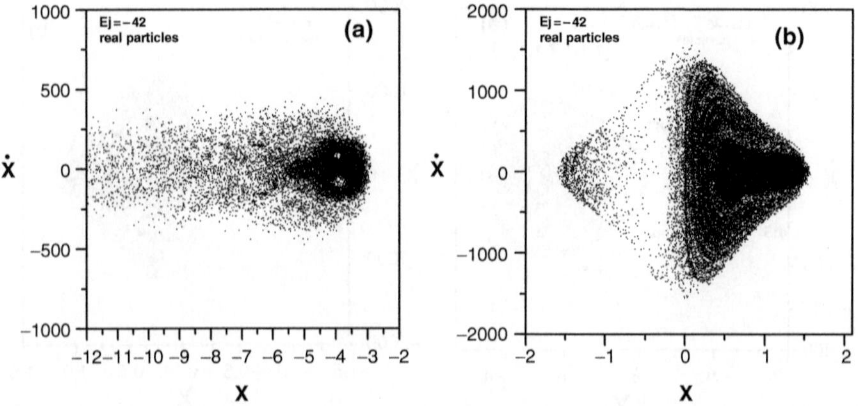

Fig. 3. Same as in Fig 2 but for the real N-body particles: (**a**) for the region outside corotation and (**b**) for the region inside corotation

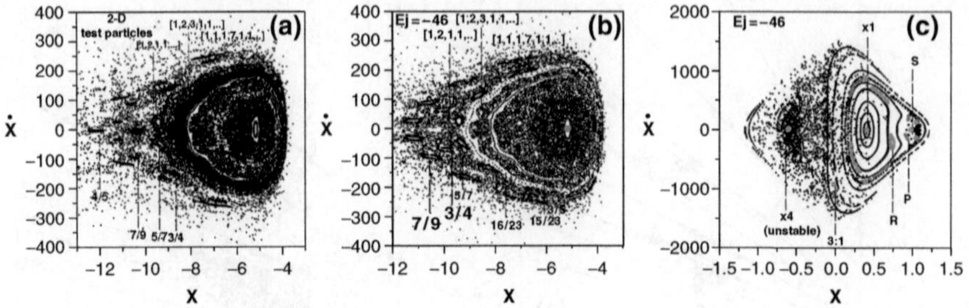

Fig. 4. SOS for $E_j = -46$ for test particles: (**a**) for the 2-D approximation of the region outside corotation (**b**) for the projection of the region outside corotation and (**c**) for the region inside corotation

same Jacobi constant with a certain tolerance and we have integrated them in a fixed potential for 100 iterations. It is remarkable in Fig. 3b that areas that correspond to the main resonances supporting the bar are populated while the area around the "x4" orbit is depopulated.

In Fig. 4 the SOS is plotted for test particles, with $E_J = -46$ for the region outside corotation (Fig. 4b) and for the region inside corotation (Fig. 4c). Figure 4a shows the 2-D approximation of the SOS where the forces on the z-axis are exactly zero. In this figure we mark some important tori. By comparing Figs. 4a and 4b we see only minor differences.

In Fig. 5 the projection of SOS is plotted for real N-body particles for the same Jacobi constant as in Fig. 4, for the region outside corotation (a) and for the region inside corotation (b). Here again it is obvious that for the region outside corotation, the real orbits (that are chaotic in their majority) are restricted in areas that are bounded by the same tori or cantori, as in Fig. 4, while there are areas with almost no N-body particles at all. Therefore tori

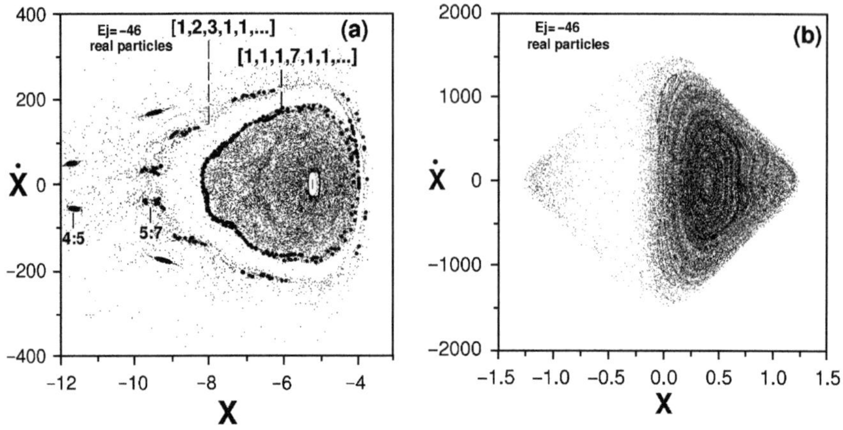

Fig. 5. Same as in Fig. 4b,c but for the real N-body particles

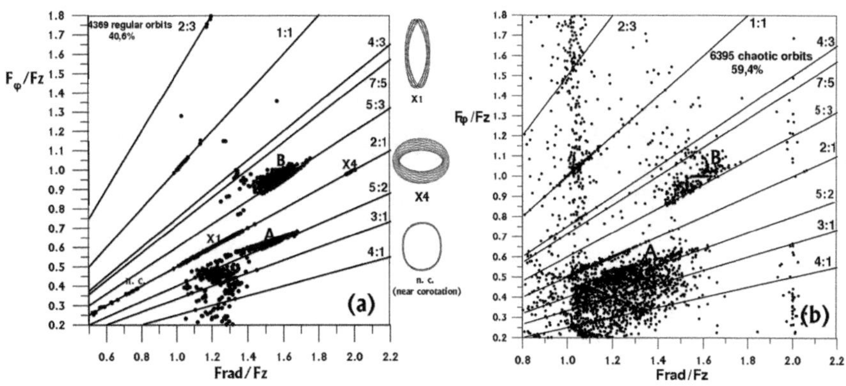

Fig. 6. Frequency analysis for (**a**) the regular orbits and (**b**) the chaotic orbits

and cantori on the projection of the SOS, play an important role in retaining chaotic orbits projected on the 2-D space of the galaxy, and the effectiveness of Arnold diffusion through the third dimension is limited (for a definition of Arnold diffusion see [4]).

3.2 Frequency Analysis

In Fig. 6 we plot the main frequencies of all the regular orbits (a) and all the chaotic orbits (b) in polar coordinates. We have used time series that correspond to 100 radial periods for the regular orbits and to 300 radial periods for the chaotic ones. The abscissa corresponds to the radial frequency F_{rad} and the ordinate corresponds to the angular frequency F_{ϕ}. Both frequencies are regularized by the frequency in the third dimension (F_z). In Fig. 6a we observe groups of orbits concentrated along lines of specific resonances and others that are more scattered in between the lines of resonances. We also see some groups of orbits that are concentrated in rather small areas, like Group A and Group B. Group A contains orbits near the limit of the bar and boxy in shape. On the other hand Group B orbits are more spherical. Figure 6b gives the frequency analysis of the chaotic orbits. Although the distribution is more scattered we can still identify concentrations around resonances. This is a consequence of the fact that most of these orbits are only weakly chaotic.

The spatial distribution of the chaotic orbits is in general more spherical than the one of the regular orbits, i.e. the shape of the bar is supported mostly by the regular orbits [18]. However there is a layer of weakly chaotic orbits that belong mostly to the outer part of the bar and support the shape of the bar.

In Fig. 7 we give the distribution of the orbits at different resonances in the inertial frame of reference where Ω_p is the pattern speed of the bar, Ω is the mean angular velocity of each orbit in the inertial frame of reference and κ is the corresponding radial frequency. Figure 7a corresponds to the regular orbits and it is obvious that the most important family is the "x1" type family also named ILR or 2:1 family. The second important family is the group A family (5:2 resonance) and a small peak on the right of A corresponds to the group B family. On the other hand the distribution of chaotic orbits (Fig. 7b) presents a greater number of families with most important the family near corotation. Other important families are the ILR (2:1) as well as the OLR (−2:1). There is also a peak that corresponds to groups A and B. It is remarkable that the distribution of the chaotic orbits presents peaks around the 4:1 and 3:1 resonances and some smaller peaks around other resonances, while these families are not present in the distribution of the regular orbits.

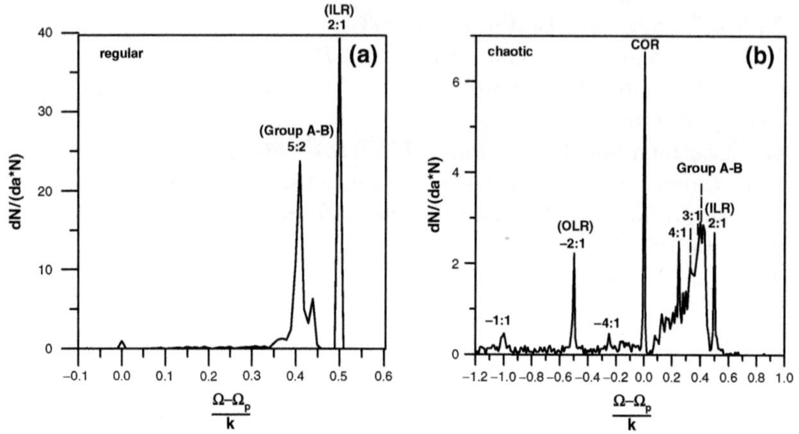

Fig. 7. The distribution of the resonant orbits in the inertial frame of reference for (**a**) regular and (**b**) chaotic orbits

4 Conclusions

In our model that simulates a rotating barred galaxy having a maximum bar perturbation ≈80%, chaos is very appreciable, i.e. about 60% of the total bound matter. However chaos is in general weak and only a small fraction of the total mass is able to develop chaotic diffusion in a Hubble time. The projections on a 2-D SOS show important differences between test particles and real N-body particles. Real particles avoid areas which correspond to orbits that do not support the bar. We have detected two mechanisms that force chaotic orbits to stay bound inside specific areas and make them dynamically important for the system. The first one, is the existence of cantori, that put a partial barrier on the 2-D SOS and reduce the effectiveness of Arnold diffusion. The second one is the stickiness effect of a considerable fraction of weakly chaotic orbits near resonances that support the bar. The latter mechanism makes the role of the chaotic layer near the limits of the bar important in supporting its shape. Frequency analysis has identified clearly the most important resonances: Ordered orbits are mostly located in the 2:1 (x1 type) and the so called group A. Chaotic orbits present similarities with the ordered (only more diffused), because the chaos is weak. The maximum of their distribution is located around corotation. Other important resonances are the OLR (−2:1) outside corotation, but also ILR (2:1), 4:1 and 3:1 near the end of the bar. These resonances together with group A support the shape of the bar.

References

1. A.J. Allen, P.L. Palmer, and J. Papaloizou: MNRAS, **242**, 576 (1990)
2. E. Athanassoula: in "Galaxies and Chaos", eds. G. Contopoulos and N. Voglis, Springer, Berlin, 313 (2003)
3. D. Ceverino and A. Klypin: MNRAS, **379**, 1155 (2007)
4. G. Contopoulos: "Order and Chaos in Dynamical Astronomy", Springer, Berlin (2002)
5. G. Efstathiou and B.J.T. Jones: MNRAS, **186**, 133 (1979)
6. L. Hernquist: Comput. Phys. Commun., **48**, 443 (1988)
7. C. Kalapotharakos, N. Voglis, and G. Contopoulos: Astron. Astrophys., **428**, 905 (2004)
8. J. Laskar: Icarus, **88**, 266 (1990)
9. E. Laurikainen, H. Salo, R. Buta, and S. Vasylyev: MNRAS, **355**, 1251 (2004)
10. D. Lynden-Bell and A.J. Kalnajs: MNRAS, **157**, 1 (1972)
11. P.J.E. Peebles: Astrophys. J., **155**, 393 (1969)
12. Ch. Skokos: J. Phys. A: Math. Gen., **34**, 10029 (2001)
13. N. Voglis: MNRAS, **267**, 379 (1994)
14. N. Voglis and G. Contopoulos: J. Phys. A, **27**, 4899 (1994)

15. N. Voglis, G. Contopoulos, and C. Efthymiopoulos: Phys. Rev. E, **57**, 372 (1998)
16. N. Voglis, G. Contopoulos, and C. Efthymiopoulos: CeMDA, **73**, 221 (1999)
17. N. Voglis, C. Kalapotharakos, and I. Stavropoulos: MNRAS, **337**, 619 (2002)
18. N. Voglis, I. Stavropoulos, and C. Kalapotharakos: MNRAS **372**, 901 (2006)
19. N. Voglis, M. Harsoula, and G. Contopoulos: MNRAS **381**, 757 (2007)

Chaos in Galaxies

D. Pfenniger

Geneva Observatory, University of Geneva, 1290 Sauverny, Switzerland
daniel.pfenniger@obs.unige.ch

Summary. After general considerations about limits of theories and models, where small changes may imply large effects, we discuss three cases in galactic astrophysics illustrating how galactic dynamics models may become insufficient when previously neglected effects are taken into account:

1. Like in 3D hydrodynamics, the non-linearity of the Poisson–Boltzmann system may imply dissipation through the growth of discontinuous solutions.
2. The relationship between the microscopic exponential sensitivity of N-body systems and the stability of mean field galaxy models.
3. The role of quantum physics in the dynamics of structure formation, considering that cosmological neutrinos are massive and semi-degenerate fermions.

1 Introduction

1.1 Chaos in Science

Chaos plays a key role in many sciences, but particularly in galactic astrophysics where it appears under different aspects in various galaxy models. Before discussing chaos in galaxies, some space will be spent discussing chaos in science.

The notion of chaos has deeply modified the scientist view of Nature, but also of the scientific process itself. But what makes chaos so special? In sciences chaos expresses commonly two important properties of *dynamical models* of natural phenomena:

1. *The sensitivity to initial conditions.* Initially close solutions of a dynamical model separate in average at least exponentially with time, $\langle \|\Delta x(t)\| \rangle > \exp(t/\tau)$, such that after a few characteristic Liapunov time τ the small scale effects neglected either in the model, or in the numerical model of the mathematical model, become macroscopic. This restricts a deterministic use of chaotic models over a model dependent finite time, and means that in chaotic systems Laplace's determinism is only applicable approximately over a limited time.

G. Contopoulos, P.A. Patsis, *Chaos in Astronomy.* Astrophysics and Space Science Proceedings, doi: 10.1007/978-3-540-75826-6,

2. *The rapid mixing of the model solutions.* An exponential divergence of nearby solutions is not sufficient to produce chaos in the technical sense. Indeed, if a uniformly diverging flow preserves the neighbourhood of solutions, a smooth transformation of coordinates allows to remove the formal divergence back: the regularized model is then no longer diverging. Therefore the divergence effect may be model dependent. To be truly chaotic, a system should remain chaotic even after smooth transformations of coordinates. It should *mix* the solutions, such that after a few Liapunov times not only initially close solutions become widely separated by the exponential divergence, but some initially distant solutions come close together at some time. With time, most of solutions become at least once arbitrarily close to each other, which expresses the mixing property of chaotic systems. Mixing occurs frequently in systems with bounded phase space, while in systems with unbounded phase space escaping solutions, i.e., solutions going to infinity, do not return often in almost all phase space regions.

Chaos is therefore important because it touches to an essential aspect of the scientific activity: to represent faithfully natural phenomena with formal models. At the heart of the scientific process is the determination of the domain of applicability and the limits of models and theories.

Actually we can enlarge the notion of sensitivity of solutions to initial conditions to the sensitivity of solutions to functionally close dynamical systems. The former case is a particular case of the latter one when the initial displacement has been produced by an initial impulse perturbation. Therefore in chaotic systems not only possible perturbations of the initial conditions lead to unpredictability, but also close but distinct functional approximations adopted in the model may lead to very different solutions after a finite time.

Therefore when considering chaos in natural systems, it is not only important to discuss how the ignorance of the real initial conditions influences the model predictions, but also how the functional "distance" from the real system to the modelled one participates to unpredictability. This point is actually central to physical modelling, because a good model should be robust to perturbations of the model functional form, since a model is always an approximation of the real system.

1.2 Epistemological Digression, Theories vs. Models

Until recent times theories and models could be viewed as fundamentally different: physical theories, like quantum physics or general relativity, have been thought to be more fundamental than models, because applicable to many more different cases. For a long time, since about the rise of classical mechanics in the seventeenth and eighteenth centuries, theories were even taken as absolutely exact. Isaac Newton's "laws" were widely considered as absolute at the same level as a God given, revealed truth.

In the meanwhile, classical mechanics was found to be only an approximation of Nature, and had to be corrected by relativistic and quantum effects. Further, the new physical theories also could not claim to be exact, because quantum physics and general relativity have remained incompatible since then, the theory of quantum gravity is still a work in progress. The so-called fundamental physical theories, like the elementary particle Standard Model, are today known to be incomplete, and must all be considered as approximations of Nature.

Models, like a galaxy or star model, concern often more specific phenomena than theories. Models are simplified formal approximations of particular phenomena, while theories aimed in the past to be exact isomorphisms with reality, but aren't. In models, scientists deliberately simplify reality in order to keep essential features and discard inessential ones. Doing so allows to describe in a formal way, with usual mathematics but more and more with computer programs, the gist of the phenomena. For example, a planet or a star may be approximated first by a perfect sphere, which allows to concentrate the discussion on its most essential features that we want to understand, and leave out the inessential aspherical features, like rotation induced flattening, or mountains. This simplification or reality has the important virtue to reduce the complexity of phenomena to a level compatible with the finite capacity of human brains to grasp complex systems. In the end, the subjective feeling of understanding a phenomenon comes from a simplified, i.e., inexact, description that captures essential features and discards inessential ones. This essential-inessential separation is often not unique, contains subjective assumptions, and of course is constrained by our finite brain capacity, which is not uniquely specified among human beings.

A minority of models and theories are successful, and often only for a limited time. The scientific process, like biological evolution, is selective. In this view, the understanding of Nature by human beings looks no longer as much miraculous as Albert Einstein thought.[1] All theories are derived by human beings from the search of a formal simplified description of Nature that is both as faithful as possible, but also that is adapted to the brain of at least some other human beings. Furthermore, accepted theories are repeatedly checked, scrutinized for correctness by the scientific community, so in the end only effective theories survive.

At this stage it is worth to put some attention to mathematics, which is used as the most solid formal language in physical models and theories. Mathematics is not a frozen field and its view by mathematicians has deeply changed during the twentieth century. Since about the Greek school of mathematics around −300 BC, and particularly since Euclid, mathematics and logic have been based on a limited set of axioms from which all the theorems were expected to follow with rigorous demonstrations. But during the twentieth

[1] "The eternal mystery of the world is its comprehensibility... The fact that it is comprehensible is a miracle." Albert Einstein.

century, famous mathematicians like Kurt Gödel, Alan Turing, and Gregory Chaitin showed however that such an axiomatic mathematics would always be incomplete (see [1] for an introduction to these problems and for references). Not by a little part, since actually most theorems ("mathematical facts") escape demonstration when starting from a limited set of axioms. Since physical theories try to build an isomorphism between Nature and a subset of mathematics, it follows that no "theory of everything" (TOE) can summarize with provable derived theorems the implicit complexity of its content. Similar ideas have been expressed by Stephen Hawking.[2] Therefore even if a theory would be exactly true, most of the consequences of it would escape an axiomatic description. Only left is the possibility to *compute* by brute force methods the consequences of theories. This is why the exponentially growing capacity of computers plays an increasingly important role in science, because computers allow to explore better and better the vast domain of theories that were out of reach by older methods using mainly theorems (like analytical methods). The best illustration of the power of the computer approach vs. the traditional analytic approach is the very notion of deterministic chaos that arose from computer models in the 1960s [6, 7].

2 Chaos in Galaxies

Let us now apply these above general ideas to galaxies. These particular structures in the Universe offer an excellent case illustrating the scientific method at work applied to complex phenomena where chaos takes an important part. Galaxies indeed contain a large variety of physical problems at widely different scales, most of them chaotic at some level, which have been more or less successfully described by distinct models. No galaxy model may claim to describe fully and exactly the galaxies, but the rich variety of models provide complementary descriptions of galaxies that overall improve our understanding of these objects. With these models, all deliberately simplified for making a part of galaxies understandable, we can much better think about galaxies than without.

Let us mention some types of chaos in galaxies existing at very different levels.

2.1 Chaos Linked to Newtonian Dynamics

Before and even after the discovery of interstellar matter and dark matter (>1930), galaxies have been described by purely stellar dynamical models of point masses following classical mechanics, the N-body model. Such a simplified description already demands substantial efforts to extract useful information helping us to understand how galaxies work. For a long time the mean

[2] `http://www.damtp.cam.ac.uk/strings02/dirac/hawking/`.

field approximations, models of the N-body model, were studied, with further time-independence and spherical or axisymmetric symmetry assumptions:

1. *Orbit description.* Assuming a fixed, smooth mean field potential allows to decouple the motion of point masses. At this stage one considers all the possible individual orbits in the galactic potential. As soon as the potential departs from a spherical shape, or differs from some particular functional form like the Stäckel potential family (for example due to the presence of a rotating bar), orbital chaos appears and occupies large regions of phase space associated to resonances. The neglected granularity of the mass become then relevant on chaotic orbits, for which the relaxation process is much faster than the one estimated in integrable potentials [12].

   Over several decades George Contopoulos and collaborators (e.g., [2, 3, 11, 16, 17]) have explored by numerical means the orbital complexity in the phase space of galactic potentials, illustrating well how numerical "brute force" orbit integration did allow to make understandable a part of the complexity of the mean field model of galaxies. In contrast, traditional analytical approaches have been much less efficient in extracting the information implicitly contained in the classical mechanics models of galaxies.
2. *Phase space fluid description.* Another approach is to approximate by infinity the large but finite number of stars or weakly interacting dark matter particles in a galaxy. In addition, these collisionless particles are supposed to be smoothly distributed in phase space, and to make a continuous and differentiable flow of matter in the 6-dimensional phase space of space and velocity coordinates. The differentiability of such a collisionless flow is a strong hypothesis that is by far not justified, neither by the observations of the stars in the solar neighbourhood, where stellar streams abound, nor by the usual arguments used in collisional flows in which smoothness is expected to arise through the microscopic chaos resulting from the frequent particle collisions (the relaxation time is short), erasing quickly irregularities and decorrelating particles.

   Further, one should also keep in mind that the number of stars in galaxies ($<10^{12}$) is actually not very large for a smooth fluid description in a 6-dimensional phase space. If one would represent phase space with cells each containing, say, at least 100 particles in order to have reasonably smooth average quantities between contiguous cells, one would obtain a number of bins per coordinate of only $(10^{12}/10^2)^{1/6} \sim 50$. Dark matter particles are more numerous but are not expected to contribute much density in the optical part of galaxies like the Milky Way, therefore the graininess of the mass distribution in galaxies is certainly already a difference with the smooth model that must be taken into account.
3. *N-body description.* The full N-body model of a galaxy is a much more faithful model than the two previous ones, at least when one wants to

describe a galaxy with about 10^{12} point masses. In a not too distant future, 5–10 years, it is likely that it will be possible to integrate such as number of particles with the forces calculated by tree approximation techniques. Therefore the N-body model has certainly a bright future and will make the collisionless smooth flow model less relevant in galactic dynamics, and even less in star cluster dynamics. However the N-body model is also limited:

a) The Miller's exponential instability [9] of individual particle trajectories and of the whole N-body system constrains to interpret N-body simulations in a statistical way. An ensemble of close but otherwise uncorrelated initial conditions produces an ensemble of round-off error dependent evolutions, which may contain statistical useful information about typical representative evolutions.

b) Another important limitation of the N-body approach when the bodies are supposed to represent real stars is that one often neglects the internal evolution of stars. At formation, most of the mass transformed into stars consists of small mass stars that eject a large fraction of their mass in the red giant phase several Gyr later. This ejected gas from planetary nebulae mixes mass, orbital momentum and kinetic energy in the interstellar gas, a highly energy dissipative process. Galactic models that ignore this dissipative aspect obviously miss an important part of reality over long time scales, especially in 5–12 Gyr old systems like elliptical galaxies that have been considered for a long time as prototypical systems for using pure dissipationless N-body dynamics. In fact ellipticals, like spiral galaxies but for different physical reasons, should be seen as substantially dissipative systems when described over several Gyr.

2.2 The Complex Physics of Baryons

Besides stars, galaxies contain gas in sometimes large amount, even exceeding the stellar content. Some very gas rich galaxies like Blue Compact Dwarfs galaxies are misnamed: they are called dwarf only because the visible stellar mass is tiny, but when the gas content revealed by HI emission is considered, they appear just as massive as normal galaxies, except for the fact that gas has not yet turned into stars. So for galaxies with even lesser amount of gas, stellar dynamical models are more or less rapidly invalidated by the rest of the physics that baryons can be subject to.

The interstellar gas physics is very complex and far from being under control. Typical interstellar gas is multiphased, has supersonic turbulence, and density and temperature contrasts covering several orders of magnitudes, a very chaotic state that defies description with the present physics tools. For example, thermodynamics supposes for its use that a local thermal equilibrium can be established, provided a local mechanical equilibrium has been reached. But supersonic turbulence means precisely that strong pressure gradients are ubiquitous, out of mechanical equilibrium regions frequent. Despite such incoherences, thermodynamical quantities like temperature are used in models and observations due to a lack of better theoretical tools about supersonic compressible turbulence.

What is apparent from simple order of magnitude estimates is that dissipative effects and the exchange of energies between the stars and the gas is not negligible for the whole galaxy equilibrium over Gyr timescales. For example the power radiated by the stars at their different stages of evolution, known to be partly recycled by the dust in the infrared, or known to feed a part of the turbulence in the interstellar medium, is comparable to the power necessary to change the whole galaxy shape against its own gravity [13]. Therefore, the galaxy global parameters and shapes can be expected to depend also on its internal dissipative micro-physics, and not only on the initial conditions at earlier epochs, or external effects like accretion.

2.3 The Dynamics of Non-Baryonic Matter

Solid cosmological and particle physics arguments exist for the existence of large amount of non-baryonic matter. 1 s after the Big Bang a number of neutrinos comparable to the photon number must have been produced mostly from electron–positron annihilations. The involved physics is the well known, far from exotic MeV nuclear physics. The discovery of neutrino oscillations between the e-, μ- and τ-neutrinos was a proof of their positive mass, and solved the 40 year old solar neutrino deficit. With the present constraints about the neutrino mass ($\sim$0.01–0.1 eV), this average leptonic density turns out to be comparable to the average identified baryon density [14]. The neutrino case shows that a particle predicted by Pauli in the 1930s for resolving an apparent violation of energy conservation during the β decay demanded huge efforts to arrive to the present solid conclusion that indeed much matter is in non-baryonic form. The gained knowledge about neutrinos suddenly doubles the amount of identified matter, which is this time leptonic.

With similar arguments, many other particle candidates (axions, neutralinos, super-symmetric particles, ...) have been proposed, often with strong theoretical motivations based on symmetries and conservation laws. For example, axions are invoked for explaining the zero neutron electric dipole moment, an empirical fact that escapes predictions of the Standard Model. Therefore, in view of all the oddities remaining to be explained in elementary particle physics, it is natural to expect a rich variety of different dark matter components that remain to be identified.

The consequence for galaxies is that each kind of matter can imprint different effects during structure and galaxy formations. Some phases of structure formation, like during the formation of Zel'dovich's pancakes, are highly sensitive to the neglected physics. Yet a high fraction of matter is expected to participate at least once, even over a brief time interval, to a sheet-like singularity where the outcome of such highly non-linear singularities is known to be very sensitive to the exact physics of the participating matter, so also from non-baryonic particles. The often adopted collisionless property of cold dark matter is just an assumption that may be acceptable in present day galaxy models, but may turn out to be invalid during perturbation sensitive events like pancake or filamentary collapses.

3 Cases of Sensitive Dependence in Galaxy Models

In the following we will concentrate on illustrative cases of sensitivity to perturbations of galactic models, where slight changes in the model may turn out to lead to radically different conclusions. We will discuss the perfectly smooth fluid phase space description of the star ensemble and of collisionless matter used in the collisionless Boltzmann equation, where the collisionless limit may turn out to lead to severe approximations. The discrete point mass models used in N-body simulations is also limited by its strongly chaotic character. Finally the role of quantum mechanics at extra-galactic scales related with cosmological neutrinos and possibly other relic dark matter particles will be argued to be not so negligible as usually assumed.

3.1 Collisionless Chaos

The main equation of collisionless galactic dynamics is the collisionless Boltzmann equation. Suppose that we describe the mass density at the instant t in space $x \in \mathbb{R}^3$ and velocity space $v \in \mathbb{R}^3$, i.e., $\{x, v\} \in \mathbb{R}^6$ by a density distribution $f(t, x, v) \in \mathbb{R}$. The projection of f onto the x-space

$$\rho(x,t) = \int d^3v \, f(x,v,t), \tag{1}$$

provides the usual mass density $\rho(x,t)$. This projection is well defined even when f is not differentiable. Poisson's equation gives us a constraint on the gravitational acceleration g induced by the mass density ρ,

$$\nabla \cdot g(x,t) = -4\pi G \rho(x,t) = -4\pi G \int d^3 \, f(x,v,t) \ . \tag{2}$$

Finally, the collisionless Boltzmann equation

$$\partial_t f + v \cdot \partial_x f + g \cdot \partial_v f = 0, \tag{3}$$

tells us that the mass flow in $\mathbb{R}^6$ is conserved, the characteristics curves of this equation are the trajectories of particles in the acceleration field g. The above three equations form a system of non-linear integro-differential equations, similar to Euler's equation in $\mathbb{R}^3$ for incompressible fluids.

Many efforts have been dedicated by mathematicians to understand the Euler or Navier–Stokes equations with more rigorous tools that commonly used in physics. This has been useful to understand much more general facts about non-linear partial differential equations, such as the limits of their applicability.

The simplest case taken from the Navier–Stokes equation but still preserving its non-linear character is the 1-dimensional Burger's equation,

$$\partial_t u + u \partial_x u = \mu \partial_x^2 u \ , \tag{4}$$

which describes a constant density advection flow at velocity $u(x,t)$ along the direction x, with an optional viscosity term proportional to a parameter μ on the right-hand side. When $\mu = 0$ the system corresponds to the energy conserving Euler equation. For simple initial conditions, say $u(x,0) = -\sin(x)$, the solution becomes multi-valued after a finite time when $\mu = 0$, and develops a shock, a discontinuity when $\mu > 0$. This shows a prototypical behaviour of non-linear partial differential equations: they tend to break the initial assumptions about the solution after a finite time by violating the assumption that the solution remains continuous and differentiable everywhere. We have here an example of sensitive dependence on the functional form of the flow model. If viscosity is zero the flow develops multi-valued velocities, while if viscosity is small but positive the flow remains single-valued but becomes discontinuous, and most of the energy is dissipated in the shocks.

An old but relevant result by Onsager [10] about 3-dimensional turbulence is that energy dissipation in Navier–Stokes fluids does not vanish to zero when viscosity tends toward zero.[3] As viscosity tends toward zero, Navier–Stokes' *equation* tends well toward Euler's equation, but the *solutions* don't. The most astonishing fact verified in experiments of developed turbulence is that the energy dissipation tends toward a positive constant independently of the value of viscosity. As in Burger's equation entropy increasing discontinuous solutions (so-called "weak" solutions) are physically the relevant ones.

Therefore it may be physically misleading to consider Euler's equation and its energy conservation law, for application on systems where viscosity is small, because even a small viscosity term becomes essential when the flow becomes discontinuous, highly turbulent, which is precisely the rule in low viscosity fluids.

There is no ground to believe that the growth of discontinuities is restricted to non-linear hydrodynamics. On the contrary, the growth of shocks and discontinuous solutions occur frequently in other non-linear partial differential equations. One should expect shocks in more complex, higher dimensional non-linear systems like the Poisson–Boltzmann system, in which the collisional or diffusive term is small, but is never exactly zero. In cases of strong phase space "turbulence" during collapses and violent events we can expect that the small residual collisionality has its diffusive effects strongly amplified. High phase space density gradients and multi-streams are susceptible to develop fast from the actual particle noise.

---

[3] An extensive review about Onsager's work on turbulence is given in [4].

Also we should remind that differentiability in usual fluids is often a valid assumption because the microscopic molecular chaos does erase the growth of correlations faster than the flow develops gradients of density or velocity. Except for very particular cases, laminar flows do require some positive viscosity, low Reynold's number, to stay laminar. Otherwise low viscosity fluids become spontaneously turbulent, i.e., develop discontinuous, singular flows.

Since precisely collisionless flows in galactic dynamics lack of a strong microscopic collisional chaos that would justify the usual smoothness assumption of distribution functions, we should rather expect irregular, non-smooth distributions ("weak" solutions) as a rule in galaxies. Actually, the local distribution function of stars in the solar neighbourhood is highly structured with several star streams, and is far from resembling a Maxwellian distribution. What we see is at best a partly relaxed distribution.

To explain the smoothness in the galaxy distribution functions, the violent relaxation concept [8] has been proposed. It was initially attributed to the time-dependence of the gravitational potential, but today we view it rather as resulting from the highly sensitive to perturbations, chaotic stages of galaxy evolution where microscopic perturbations become fast macroscopically relevant. Contrary to a still popular opinion, this is not directly the time-dependence of the potential that leads to an enhanced relaxation, but the highly chaotic, sensitive stage of the system. Counter-examples demonstrating that time-dependence does not necessarily relax a collisionless flow are analytical time-dependent periodic solutions of the Poisson–Boltzmann system (e.g., [15]).

3.2 N-Body Chaos

Miller (1964) [9] discovered numerically the exponential divergence of particular gravitational N-body systems, and noticed that the divergence of close systems occurs not because of transient close 2-body encounters, but constantly by the $N(N-1)$ particle interactions. Gurzadyan and Savvidy (1986) [5] showed also with Riemannian geometry that the N-body problem is indeed generally chaotic in simple particle distributions. However these studies should not be seen as definitive, it is indeed not difficult to invent particular counter-examples of as weakly unstable as wished configurations of N-body systems, such as widely separated pairs of pairs, etc., of binaries. By natural selection we do observe in the sky the least unstable multiple star systems, often arranged hierarchically. The solar system is also an example where the Liapunov time is much longer than its dynamical time.

An interesting problem is to specify the relationship between Miller's type of chaos seen at the microscopic level, and the global stability of a stellar system. In usual gases, the molecular very rapid chaos is the key property that guarantees that the system seen at macroscopic scales can be modelled with the quasi-deterministic rules of thermodynamics. To be effective thermodynamics requires a fast relaxation of molecules, in other words, a strong molecular chaos. Is it similar in gravitational systems? From numerical experiments, systems like hot spherical models of many equal mass stars are examples where indeed a kind of statistical robust state appears to be reached over time-scales longer than the crossing time. For these systems Miller's microscopic chaos could be actually favouring a global statistical quasi-equilibrium.

Other gravitational systems, such as disks, or systems with strongly anisotropic velocity dispersions, can present macroscopic instabilities leading to evolution. It is presently unclear whether Miller's microscopic chaos is related in any way with large scale instabilities.

3.3 Sensitivity to Quantum Physics

The relict cosmological neutrinos are fermions and have been produced ~1–2 s after the Big Bang from electron–positron annihilations at a redshift of $z \approx 10^{10}$. At this epoch all the particles were strongly relativistic and close to thermal equilibrium. This means that the distribution of neutrinos had to be very close to a Fermi–Dirac distribution with a negligible chemical potential. In other words, the natural creation state of such relativistic particles is always semi-degenerate [18], which means that quantum mechanics plays a significant role from the start. The phase space density of neutrinos is therefore sufficiently high for quantum effects to be important at the macroscopic level. As the Universe expands the neutrino phase space density is little modified, since neutrinos have only two weak possibilities of interaction with matter, the weak force and the gravitational force, and their number cannot decrease through decay since they are supposed to be stable.

For a rest mass range such as 0.01–0.1 eV at the present time the relict neutrinos should be non-relativistic and much colder than the often quoted temperature of 1.9 K, which would apply if they were massless or still relativistic [14]. Their speed is estimated low enough (~1,000 km s^{-1}) to be trapped at least in galaxy clusters. So if neutrinos are massive and participate to structure formation, clearly they are able to perturb the outcome of pancake collapses since their total mass is comparable to the one of baryons, and they are still today semi-degenerate fermions, so their physics differs from classical ballistic particles. Structure formation is a highly chaotic process where most of the matter goes through Zel'dovich's pancakes, so one can expect that any perturbation to a purely classical model of structure formation, like when including the fermionic properties of neutrinos, may significantly change the results.

But here we have the non-trivial challenge to merge two widely different descriptions of Nature. Sometimes quantum physics is said to be "holistic" because it is a non-local theory. A "particle" does not necessarily represent a localized mass, but can be a plane-wave. Pauli's principle does not require to localize particles, and the wave-function in Schrödinger equation may extend over all space. In contrast, the notion of localized point mass is central in classical mechanics.

In the recent years these questions have considerably advanced. The decoherence theory [19–22] allows to specify when a finite system, supposed to be isolated, is well represented by quantum physics, and when classical physics is the best description: when the neglected but always existing perturbation rate of the outer world to *a given model* stays indeed negligible, quantum coherence build-up wins and quantum mechanics stays a faithful representation of the system. But when the outer world perturbation rate is faster and destroys quantum coherence, classical physics emerges as a good description of the system. A satisfactory aspect of the decoherence theory is that a conscious observer is no longer required. A measurement corresponds to coherence destruction by external perturbations to the system. Such considerations are central for building quantum computers, where coherence preservation is crucial, and clearly unrelated with the presence of a conscious observer.

So in order to describe a particle as a localized mass distribution with classical mechanics, as implicit in all cosmological simulations up to now, one needs to check that decoherence is effective. Otherwise particles can not be localized and must be described with quantum physics as non-local ensembles.

But since neutrinos have extremely low probability to interact with other particles by the weak interaction, the only remaining possibility is gravitational interaction. As long as the Universe is homogeneous, gravitational interaction cancels and is negligible. Only when structures form ($z \ll 100$) matter inhomogeneities perturb neutrinos by gravitation.

The entanglement time-scale τ_E of identical neutrinos is estimated by a classical collision time where the cross section diameter is given by the de Broglie wavelength $\lambda_{\rm dB} = h/m_\nu v$ (see [14]),

$$\tau_E \approx (n_\nu \lambda_{\rm dB}^2 v_\nu)^{-1} \approx 1.3 \times 10^{-8} (1+z)^{-2} \left(\frac{n_\nu}{56}\right)^{-1} \left(\frac{m_\nu}{1\rm eV}\right)^{-1} \rm s, \tag{5}$$

where n_ν is the neutrino density, m_ν the neutrino mass, and z the redshift. After a multiple of τ_E large numbers of neutrinos are entangled just because they are identical fermions, not because of their weak interaction.

The decoherence theory [19–22] gives an estimate of the decoherence time τ_D if we know the characteristic relaxation time τ_R over which the external world perturbs the system of size Δx:

$$\frac{\tau_D}{\tau_R} = \left(\frac{\lambda_{\rm dB}}{\Delta x}\right)^2 . \tag{6}$$

For ensembles of neutrinos above a given size Δx the decoherence time τ_D will be shorter than the entanglement time τ_E, and above such a scale the neutrino ensemble may be considered as localized and included in a classical description, like hydrodynamics.

In [14] we estimated that at the present epoch the fastest relaxing mass condensations for neutrinos are the galaxies, not the stars or the galaxy clusters. With this estimate of the "relaxation" time τ_R produced by the external world, we found that when considering scales much larger than $\sim 10^{13}$ cm, the ensemble of at least 10^{40} neutrinos is perturbed at a rate fast enough by the galaxy gravitational interactions to be considered as a classical fluid with a Fermi–Dirac equation of state. Thus a simple model of cosmological neutrinos above solar system scales is to describe them as a Fermi–Dirac fluid, like what is done in stellar models of white dwarfs and neutron stars. This is a very different physics than the collisionless classical mechanics representation that up to now has been used in cosmological simulations including neutrinos, but closer to the adhesion model sometimes used in pancake models. Of course the consequences for structure formation models may be drastic because fluids tend to develop shocks, contrary to collisionless flows, and shocks imply entropy production and dissipation.

What has been said about neutrinos may be applicable to other dark matter particles like axions or neutralinos, with possible complications for bosons. Much depends on their effective rest mass, number density, and if they are sufficiently non-relativistic to participate to structure formation.

4 Conclusions

The sensitivity to perturbations is a characteristics of chaos. Various aspects of galaxies can be represented by different models that may contain sensitive parts subject to limitations on the scope of applicability.

Here are a few points to consider for future galaxy models:

- The assumed smoothness of distribution functions for collisionless systems is not grounded on theoretical arguments or observational evidences. The effects of irregular, non-smooth distributions on the dynamics of a model have been little investigated. The results of incompressible fluid turbulence could be useful to extend to collisionless phase space flows.
- Therefore the collisionless Boltzmann differential equation is a decreasingly attractive model in galactic dynamics in regard of the fast growing capabilities of the much less constrained N-body techniques.
- The dissipative baryonic physics in galaxies coming from the gas but also from the stellar evolution is important over Gyr timescale, not only for spirals, but also for ellipticals due to the stellar important mass loss in the red giant phase.
- Quantum physics in the semi-degenerate sea of cosmological neutrinos produces a kind of collisional relaxation due to decoherence and the fermionic nature of such particles. These neutrinos and possibly other dark matter particles should not be modelled as collisionless, but as a collisional fluid with the proper quantum statistics equation of state. This point is crucial during pancake shocks.

Acknowledgements

I am grateful to the organisers of this stimulating interdisciplinary conference for the invitation. This work has been supported by the Swiss National Science Foundation.

References

1. Chaitin, G.J. 1982, Int. J. Theor. Phys. 21, 941 (`http://cs.umaine.edu/~chaitin/georgia.html`)
2. Contopoulos, G. 1980, Astron. Astrophys. 81, 198
3. Contopoulos, G., Patsis, P.A. 2006, Mon. Not. R. Astron. Soc. 369, 1039
4. Eyink, G.L., Sreenivasan, K.R. 2006, Rev. Mod. Phys. 78, 87
5. Gurzadyan, V.G., Savvidy, G.K. 1986, Astron. Astrophys. 160, 203
6. Hénon, M., Heiles, C. 1964, Astron. J. 69, 73
7. Lorenz, E.N. 1963, J. Atmos. Sci. 20, 130
8. Lynden-Bell, D. 1967, Mon. Not. R. Astron. Soc. 136, 101
9. Miller, R.H. 1964, Astrophys. J. 140, 250
10. Onsager, L. 1949, Nuovo Cimento, Suppl. 6, 279
11. Patsis, P.A., Contopoulos, G., Grosbøl, P. 1991, Astron. Astrophys. 243, 373–380
12. Pfenniger, D. 1986, Astron. Astrophys. 165, 74
13. Pfenniger, D. 1991, in "Dynamics of Disc Galaxies", B. Sundelius (ed). Göteborg, Göteborg University Press, 389
14. Pfenniger, D., Muccione, V. 2006, Astron. Astrophys. 456, 45
15. Sridhar, S. 1989, Mon. Not. R. Astron. Soc.238, 1159
16. Voglis, N., Contopoulos, G., Efthymiopoulos, C. 1999, Cel. Mech. Dyn. Astron. 73, 211
17. Voglis, N., Harsoula, M., Contopoulos, G. 2007, Mon. Not. R. Astron. Soc. 381, 757
18. Weinberg S. 1962, Phys. Rev. 128, 1457
19. Zeh, H.D. 1970, Found. Phys. 1, 69
20. Zeh, H.D. 2005, Séminaire Poincaré 2, 1 (`quant-ph/0512078`)
21. Zurek, W.H. 2002, Los Alamos Science 27, 2
22. Zurek, W.H. 2003, Rev. Mod. Phys. 75, 715

Boxy/Peanut Bulges: Formation, Evolution and Properties

E. Athanassoula[1] and I. Martinez-Valpuesta[1,2]

[1] LAM, OAMP, 38, rue Frederic Joliot-Curie, 13388 Marseille cedex 13, France
lia@oamp.fr

[2] Instituto de Astrofísica de Canarias, C/Vía Láctea E-38200, La Laguna, Tenerife, Spain
imv@iac.es

Summary. We discuss the formation and evolution of boxy/peanut bulges (B/Ps) and present new simulations results. Orbital structure studies show that B/Ps are *parts* of bars seen edge-on, they have their origin in vertical instabilities of the disc material and they are somewhat shorter in extent than bars. When the bar forms it is vertically thin, but after a time of the order of a Gyr it experiences a vertical instability and buckles. At that time the strength of the bar decreases, its inner part becomes thicker, so that, seen edge-on, it acquires a peanut or boxy shape. A second buckling episode is seen in simulations with strong bars, accompanied by a further thickening of the B/P and a weakening of the bar. Quantitatively, this evolution depends considerably on the properties of the halo and particularly on the extent of its core. This influences the amount of angular momentum exchanged within the galaxy, emitted by near-resonant material in the bar region and absorbed by near-resonant material in the halo and in the outer disc. Haloes with small cores generally harbour stronger bars and B/Ps and they often witness double buckling.

1 Introduction

Disc galaxies viewed edge-on often show in their central parts a characteristic thickening, which has the shape of a box, of a peanut, or of an 'X' (e.g. [17, 18]). Since these structures swell out of the disc plane, they are called bulges, and more specifically, boxy bulges, or peanut bulges, or 'X' shaped bulges, or, for short, B/Ps or B/P bulges. Yet their properties, as well as their formation and evolution are very different from those of classical bulges [4]. In fact, evidence from many studies has shown that they are just *parts* of bars seen edge-on. Here, we will briefly review relevant results about their orbital structure (Sect. 2) and describe their formation and evolution as witnessed in

G. Contopoulos, P.A. Patsis, *Chaos in Astronomy*. Astrophysics and Space Science Proceedings, doi: 10.1007/978-3-540-75826-6,

N-body simulations (Sect. 3). Based on the inner halo structure, we distinguish different types of models and discuss the role of the halo on the growth of the bar and of the B/P.

2 Orbital Structure

In order to understand the dynamics of any structure, it is essential to study first the periodic orbits that form its backbone. For two-dimensional bars, the backbone is the well studied family of x_1 orbits [8, 14]. These are elongated along the bar and have an axial ratio that varies both as a function of their Jacobi constant and of the properties of the bar potential used [1]. When stable, they trap around them other orbits. The problem is more complex in three dimensions, since a vertical instability of parts of this family [11] introduces a number of other families extending vertically well outside the disc, such as the x_1v_1, x_1v_2, x_1v_3, etc., [22, 23, 25, 26]. These families, together with the main x_1 family from which they bifurcate, are known as the x_1 tree. [25]. They are linked to the n:1 vertical resonances and extend well outside the disc equatorial plane. As shown in [22], some of them are very good building blocks for the formation of B/Ps, because their orbits are stable and have the right shape and extent. Studies of these orbits reproduced many of the B/P properties and helped explaining crucial aspects of B/P formation and evolution. For example, an analysis of the orbital families that constitute B/Ps predicts that they should be shorter than bars. This is indeed verified both in N-body simulations and in real galaxies [4, 7, 18]. Furthermore, such studies predict that stronger bars should correspond to stronger B/Ps, and this also is verified both in N-body simulations ([6] and Athanassoula and Martinez-Valpuesta, in preparation) and in real galaxies [18].

3 Formation and Evolution of Boxes and Peanuts

The formation of boxy/peanut bulges has been witnessed in a large number of simulations ([3, 4, 9, 12, 13, 15, 16, 19–21, 24], etc.). These have many aspects in common, but also many differences. We will describe and illustrate here two different characteristic types of evolution, corresponding to two simulation types which in [9] have been labelled MH and MD, respectively.

In both cases the simulation starts with an exponential disc of unit mass and unit scale-length. It is immersed in a live halo and the halo-to-disc mass ratio is 5. In MH type models the halo has a small core, smaller or of the order of the disc scale-length, i.e. it is centrally concentrated. Thus, in the disc region, the halo contribution to the circular velocity curve is of the same order as that of the disc. On the contrary, in MD type models the halo has a very big core, much larger than the disc scale length, so that in the inner parts the disc dominates.

The evolution of the bar in these two types of models is quite different [9]. In both cases the bar grows by exchanging angular momentum with the outer disc and with the halo. Angular momentum is emitted by near-resonant material in the bar region and is absorbed by near-resonant material in the outer disc and in the halo [2]. The halo density at the locations of the resonances is much larger in MH types than in MD types. Thus, provided the resonances are responsive, there will be more angular momentum exchanged within the galaxy in MH types than in MD types. This leads to strong bars in MH types and much weaker ones in MD types [3, 9].

The time evolution of characteristic properties of the bar and the B/P is shown in Figs. 1, 2 and 3 for two MH and one MD type simulations, respectively. From top to bottom, the panels give the buckling strength, the B/P strength and the bar strength. The time is given in Gyrs, using the calibration proposed in [9]. Figure 1 shows that the initially unbarred disc forms a bar roughly between times 3 and 6 Gyrs (lower panel). We define as bar formation time the time at which the bar-growth is maximum (i.e. when the slope of the bar strength as a function of time is maximum) and indicate it by the first vertical line in Fig. 1. The bar strength reaches a maximum at a time noted by the second vertical line, and then decreases considerably over ~0.5 Gyr. The time at which the bar amplitude decrease is maximum is given by the third vertical line. Subsequently, the bar strength reaches a minimum, at a time shown by the fourth vertical line, and then starts increasing again at a rate much slower than that during bar formation.

The upper panel shows the buckling strength, i.e. the vertical asymmetry, as a function of time. Before the bar forms the disc is vertically symmetric, with the first indications of asymmetry occurring after bar formation. The asymmetry grows very abruptly to a strong, clear peak and then drops equally abruptly. The time of the buckling (dashed vertical line) is very clearly defined as the maximum of this curve. The middle panel shows the strength of the B/P, i.e. its maximum vertical extent, again as a function of time. This quantity grows abruptly after the bar has reached its maximum amplitude and during the time of the buckling. This abrupt growth is followed by a much slower increase over a longer period of time.

Taken together, the three panels of Fig. 1 show that the bar forms vertically thin, and only after it has reached a maximum strength does the buckling phase occur. During the buckling time the bar strength decreases significantly, while the B/P strength increases. The time interval during which B/P formation, or buckling occur is rather short, of the order of a Gyr, and it is followed by a longer stretch of time during which the bar and B/P evolve much slower.

Figure 2 shows results for another MH simulation. The first part of the evolution is very similar to that of the previous example, except that the time for bar formation is shorter and the time during which the bar amplitude decreases is somewhat longer. This example, however, has a very interesting feature: it has a second, weaker buckling episode shortly after 8 Gyrs. This

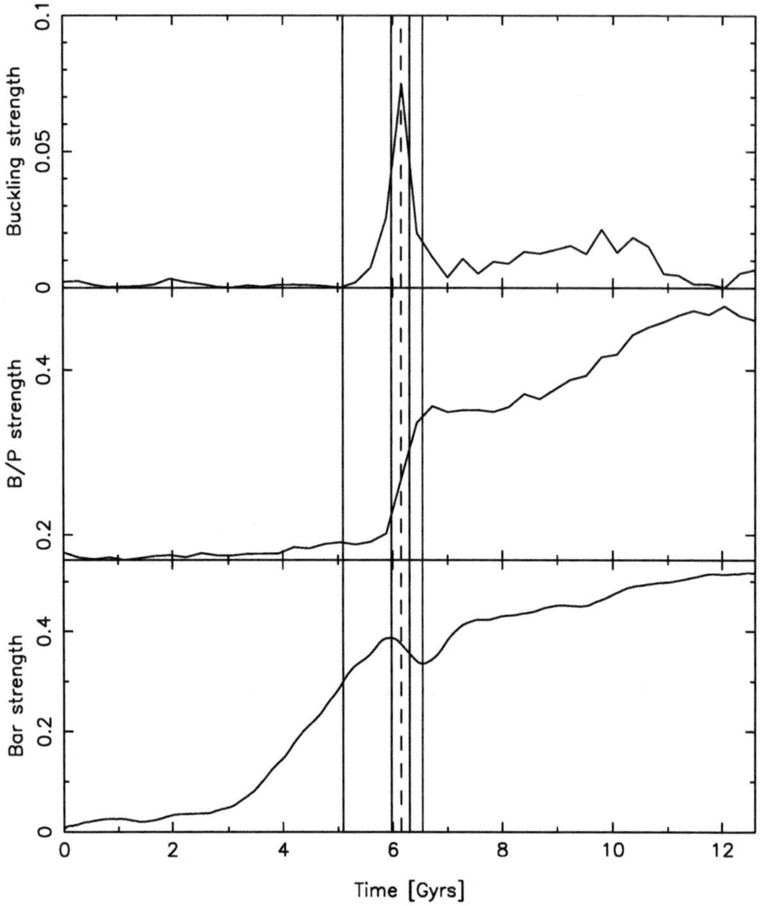

Fig. 1. Example of the time evolution of three B/P-, or bar-related quantities for an MH type model. These are the buckling strength (i.e. the vertical asymmetry; *upper panel*), the B/P strength (i.e. its maximum vertical extent; *middle panel*) and the bar strength (*lower panel*). The *vertical lines* mark characteristic times linked to bar formation and evolution. From left to right, these are the bar formation time, the maximum amplitude time, the bar decay time and the bar minimum amplitude time (see text). The *vertical dashed line* marks the time of the buckling

occurs very often in simulations developing strong bars and was discussed first in [5] and [20]. It is seen clearly in all three panels and has characteristics similar to those of the first buckling. Namely, there is an asymmetry, with a clear maximum of the buckling strength, a sharp increase of the B/P strength and a flattening of the bar strength. In many other examples with two buckling episodes, instead of a flattening there is a decrease of the bar amplitude. Thus, the main differences between the two buckling episodes are only that the

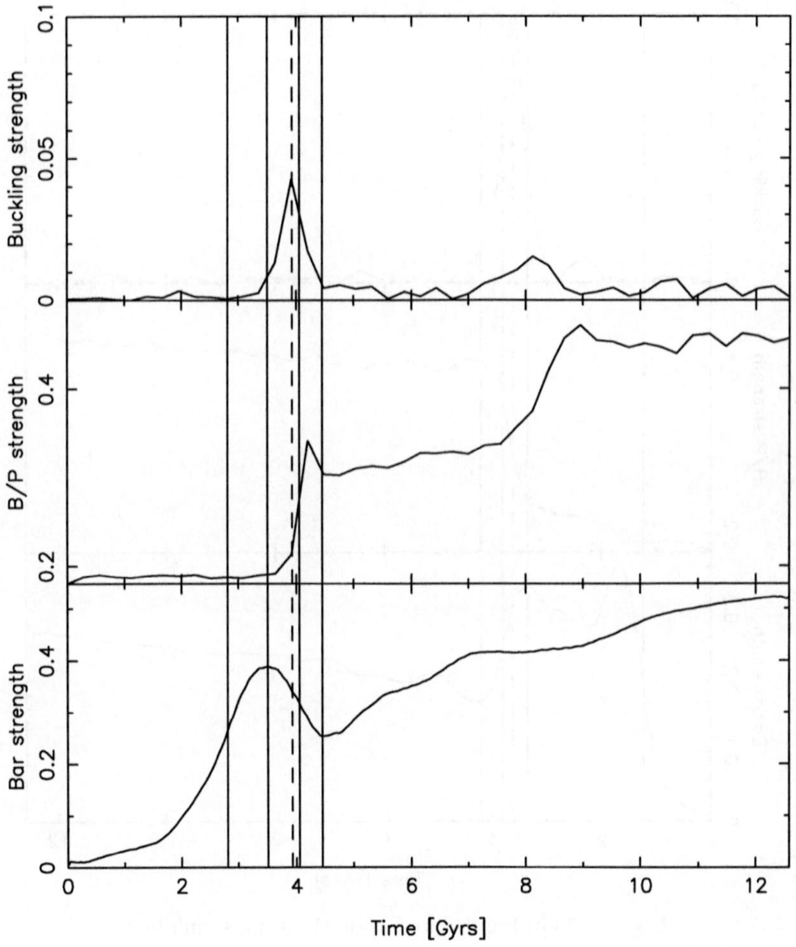

Fig. 2. As in Fig. 1, but for an MH-type simulation with two buckling episodes

buckling peak is broader and less high and that the drop in the bar strength is less strong. The time between the two bucklings, in this example, is about 4 Gyrs.

The evolution of the same quantities for the MD simulation is given in Fig. 3. Many crucial differences are immediately clear. The bar forms much faster than for the MH type models, in less than 2 Gyrs. This is in good agreement with results of previous simulations [2, 10]. There is a first maximum of the bar amplitude before 2 Gyrs, which, however, is not associated with a maximum of the asymmetry (buckling), nor with a sharp increase of the B/P strength. It thus can not be linked to the B/P formation, and we checked this further by viewing the evolution of the simulation by eye. This revealed that in the initial stages of the simulation a very long and thin bar forms, due to

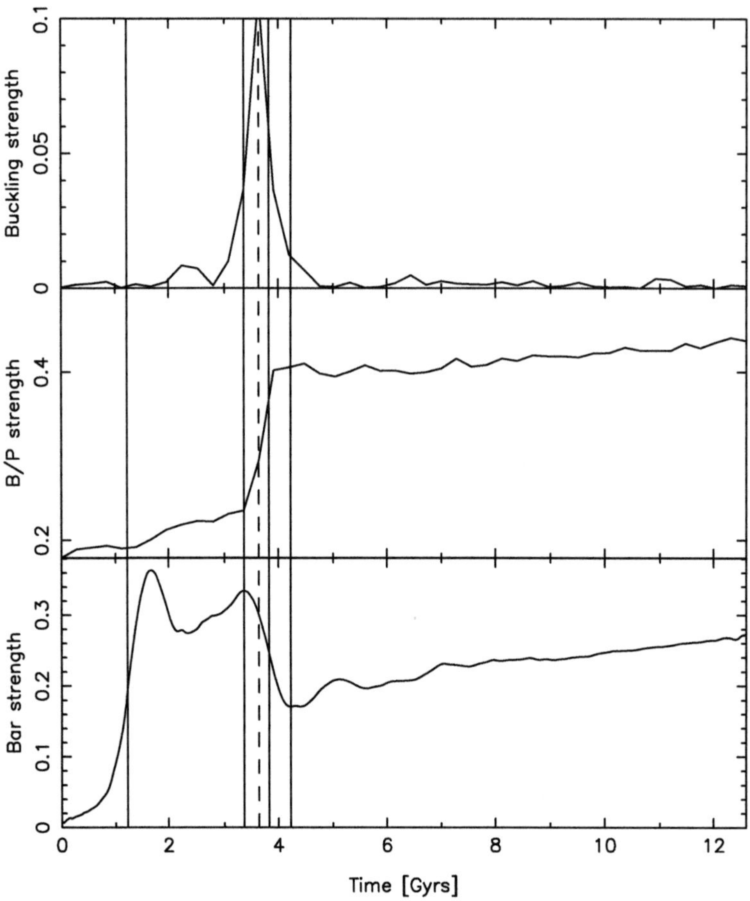

Fig. 3. As in Fig. 1, but for an MD-type simulation

the strong instability in the disc-dominated inner region. This bar drives a strong, two-armed spiral which heats the disc and thus lowers its own amplitude. Furthermore, the bar is so thin that it must render many of the orbits that support it chaotic [8], so that they can not support it further. Thus, the bar strength should decrease spectacularly and this is indeed witnessed in Fig. 3. Subsequently, the bar amplitude increases with time until the formation of the B/P, which, as in the previous examples, is clearly seen as a maximum of the buckling strength, a sharp increase of the B/P strength and a sharp decrease of the bar strength. The time between the bar growth and its decay is much longer than in the previous examples. In this specific case it is about 2.5 Gyrs, but in other cases it can be considerably longer.

It is thus clear that the halo properties, and in particular the size of its core, influence strongly the time evolution of the bar and of the B/P.

Acknowledgements

We thank A. Bosma for interesting and fruitful discussions. This work has been partially supported by grant ANR-06-BLAN-0172 and by the Grüber foundation.

References

1. Athanassoula, E. 1992, MNRAS, 259, 328
2. Athanassoula, E. 2002, ApJL, 569, 83
3. Athanassoula, E. 2003, MNRAS, 341, 1179
4. Athanassoula, E. 2005, MNRAS, 358, 1477
5. Athanassoula, E. 2005, in *Planetary Nebulae as Astronomical Tools*, eds. R. Szczerba, G. Stasińska, and S. K. Górny, AIP Conf. Proc. 804, Melville, New York, 333
6. Athanassoula, E. 2006, in *Mapping the Galaxy and Nearby Galaxies*, eds. K. Wada and F. Combes, Springer, Berlin, 47
7. Athanassoula, E., Beaton, R. L. 2006, MNRAS, 370, 1499
8. Athanassoula, E., Bienaymé, O., Martinet, L., Pfenniger, D. 1983, A&A, 127, 349
9. Athanassoula, E., Misiriotis, A. 2002, MNRAS, 330, 35 (AM02)
10. Athanassoula, E., Sellwood, J. A. 1986, MNRAS, 221, 213
11. Binney, J. 1978, MNRAS, 183, 501
12. Combes, F., Sanders, R. H. 1981, A&A, 96, 164
13. Combes, F., Debbasch, F., Friedli, D., Pfenniger, D. 1990, A&A, 233, 82
14. Contopoulos, G., Papayannopoulos, T., A&A, 92, 33
15. Debattista, V. P., Carollo, M., Mayer, L., Moore, B. 2004, ApJ, 604, L93
16. Debattista, V. P., Carollo, M., Mayer, L. Moore, B., Wadsley, J., Quinn, T. 2006, ApJ, 645, 209
17. Lütticke, R., Dettmar, R.-J., Pohlen, M. 2000, A&AS, 145, 405
18. Lütticke, R., Dettmar, R.-J., Pohlen, M. 2000, A&A, 362, 435
19. Martinez-Valpuesta, I., Shlosman, I. 2004, ApJ, 613, 29
20. Martinez-Valpuesta, I., Shlosman, I., Heller, C. 2006, ApJ, 637, 214
21. O'Neill, J. K., Dubinski, J. 2003, MNRAS, 346, 251
22. Patsis, P., Skokos, Ch., Athanassoula, E. 2002, MNRAS, 337, 578
23. Pfenniger, D. 1984, A&A, 134, 373
24. Raha, N., Sellwood, J. A., James, R. A., Kahn, F. D. 1991, Nature, 352, 411
25. Skokos, H., Patsis, P., Athanassoula, E. 2002, MNRAS, 333, 847
26. Skokos, H., Patsis, P., Athanassoula, E. 2002, MNRAS, 333, 861

Invariant Manifolds as Building Blocks for the Formation of Spiral Arms and Rings in Barred Galaxies

M. Romero-Gómez[1], E. Athanassoula[1], and J.J. Masdemont[2], and C. García-Gómez[3]

[1] LAM, OAMP, 38, rue Frederic Joliot-Curie, 13388 Marseille cedex 13, France
merce.romerogomez@oamp.fr; lia@oamp.fr
[2] I.E.E.C & Department of Mathematica Aplicada I, Universitat Politècnica de Catalunya, Diagonal 647, 08028 Barcelona, Spain
josep@barquins.upc.edu
[3] D.E.I.M., Universitat Rovira i Virgili, Campus Sescelades, Avd. dels Països Catalans 26, 43007 Tarragona, Spain
carlos.garciag@urv.cat

Summary. We propose a theory to explain the formation of spiral arms and of all types of outer rings in barred galaxies, extending and applying the technique used in celestial mechanics to compute transfer orbits. Thus, our theory is based on the chaotic orbital motion driven by the invariant manifolds associated to the periodic orbits around the hyperbolic equilibrium points. In particular, spiral arms and outer rings are related to the presence of heteroclinic or homoclinic orbits. Thus, R_1 rings are associated to the presence of heteroclinic orbits, while R_1R_2 rings are associated to the presence of homoclinic orbits. Spiral arms and R_2 rings, however, appear when there exist neither heteroclinic nor homoclinic orbits. We examine the parameter space of three realistic, yet simple, barred galaxy models and discuss the formation of the different morphologies according to the properties of the galaxy model. The different morphologies arise from differences in the dynamical parameters of the galaxy.

1 Introduction

Bars are a very common feature of disc galaxies. In a sample of 186 spirals, 56% of the galaxies in the near infrared are strongly barred, while an additional 16% are weakly barred [9]. A large fraction of barred galaxies show two clearly defined spiral arms [8], departing from the end of the bar. This is the case for instance in NGC 1300, NGC 1365 and NGC 7552. Spiral arms are believed to be density waves in a disc galaxy ([13], and [1] for a review). in [22], Toomre found that the spiral waves propagate towards the principal Lindblad resonances of the galaxy, where they damp down, and thus concludes that

G. Contopoulos, P.A. Patsis, *Chaos in Astronomy.* Astrophysics and Space Science Proceedings, doi: 10.1007/978-3-540-75826-6,

long-lived spirals need some replenishment. Danby argued that orbits in the gravitational potential of a bar play an important role in the formation of arms [6] and Kaufmann and Contopoulos argue that, in self-consistent models for three real barred spiral galaxies, spiral arms are supported also by chaotic orbits [11].

The origin of rings has been studied by Schwarz who calculated the response of a gaseous disc galaxy to a bar perturbation [18–20]. He proposed that ring-like patterns are associated to the principal orbital resonances, namely ILR (Inner Lindblad Resonance), CR (Corotation Resonance), and OLR (Outer Lindblad Resonance). There are different types of outer rings. Buta classified them according to the relative orientation of the ring and bar major axes [4]. If these two are perpendicular, the outer ring is classified as R_1. If they are parallel, the outer ring is classified as R_2. Finally, if both types of rings are present in the galaxy, the outer ring is classified as R_1R_2.

In Romero-Gómez et al. [16, 17], we note that spiral arms and rings emanate from the ends of the bar and we propose that rings and spiral arms are the result of the orbital motion driven by the invariant manifolds associated to the Lyapunov periodic orbits around the unstable equilibrium points. In Romero-Gómez et al. [16], we fix a barred galaxy potential and we study the dynamics around the unstable equilibrium points. We give a detailed definition of the invariant manifolds associated to a Lyapunov periodic orbit. For the model considered, the invariant manifolds delineate well the loci of an rR_1 ring structure, i.e. a structure with an inner ring (r) and an outer ring of the type R_1. In Romero-Gómez et al. [17], we construct families of models based on simple, yet realistic, barred galaxy potentials. In each family, we vary one of the free parameters of the potential and keep the remaining fixed. For each model, we numerically compute the orbital structure associated to the invariant manifolds. In this way, we are able to study the influence of each model parameter on the global morphologies delineated by the invariant manifolds.

Voglis, Stavropoulos and Kalapotharakos study the chaotic motion present in self-consistent models of both rotating and non-rotating galaxies, concluding that rotating models are characterised by larger fractions of mass in chaotic motion [23]. Patsis argues that the spiral arms of NGC 4314 are due to chaotic orbits and, to show it, he computes families of orbits with initial conditions near the unstable equilibrium points [15]. Voglis, Tsoutsis and Efthymiopoulos [24] reproduce a spiral pattern found in a self-consistent simulation using the apocentric invariant manifolds of the short-period family of unstable periodic orbits. They give the angular position of the apocentres, which is where they state the stars spend a large part of their radial period, as a soliton-type solution of the sine-Gordon equation.

In Sect. 2, we first present the galactic models used in the computations and the equations of motion. In Sect. 3, we give a brief description of the dynamics around the unstable equilibrium points and the role the invariant manifolds play in the transfer of matter around the galaxy. In Sect. 4, we show the different morphologies that result from the computations.

2 Description of the Model and Equations of Motion

In this section, we first describe the bar models used in the computations by giving the density distributions, or the potentials used. We then write the equation of motion and we define the effective potential and Jacobi constant.

2.1 Description of the Model

We use three different models, all three consisting of the superposition of an axisymmetric component and another bar-like. Our first model is that of Athanassoula [2]. The axisymmetric component is composed of a disc, modelled as a Kuzmin–Toomre disc [12, 21] of surface density $\sigma(r)$:

$$\sigma(r) = \frac{V_d^2}{2\pi r_d}\left(1+\frac{r^2}{r_d^2}\right)^{-3/2}, \tag{1}$$

and a spheroid modelled by a density distribution of the form $\rho(r)$:

$$\rho(r) = \rho_b\left(1+\frac{r^2}{r_b^2}\right)^{-3/2}. \tag{2}$$

The parameters V_d and r_d set the scales of the disc velocities and radii, respectively, and ρ_b and r_b determine the concentration and scale-length of the spheroid.

Our bar potential is described by a Ferrers ellipsoid [10] whose density distribution is

$$\begin{cases} \rho_0(1-m^2)^n & m \le 1 \\ 0 & m \ge 1, \end{cases} \tag{3}$$

where $m^2 = x^2/a^2 + y^2/b^2$. The values of a and b determine the shape of the bar, a being the length of the semi-major axis, which is placed along the x coordinate axis, and b being the length of the semi-minor axis. The parameter n measures the degree of concentration of the bar and ρ_0 represents the bar central density.

We also use two further ad-hoc bar potentials, namely a Dehnen's bar type, Φ_1, (Dehnen [7]):

$$\Phi_1(r,\theta) = -\frac{1}{2}\epsilon v_0^2 \cos(2\theta) \begin{cases} 2-\left(\dfrac{r}{\alpha}\right)^n & r \le \alpha, \\ \left(\dfrac{\alpha}{r}\right)^n & r \ge \alpha, \end{cases} \tag{4}$$

and a Barbanis–Woltjer (BW) bar type, Φ_2, (Barbanis and Woltjer [3]):

$$\Phi_2(r,\theta) = \hat{\epsilon}\sqrt{r}(r_1 - r)\cos(2\theta). \tag{5}$$

The parameter α is a characteristic length scale of the Dehnen's type bar potential, and v_0 is a characteristic circular velocity. The parameter ϵ is related to the bar strength. The parameter r_1 is a characteristic scale length of the BW bar potential and the parameter $\hat{\epsilon}$ is related to the bar strength.

2.2 Equations of Motion

In order to compute the equations of motion, we take into account that the bar component rotates anti-clockwise with angular velocity $\mathbf{\Omega_p} = \Omega_p \mathbf{z}$, where Ω_p is a constant pattern speed.[1] The equations of motion in this potential in a frame rotating with angular speed $\mathbf{\Omega_p}$ in vector form are

$$\ddot{\mathbf{r}} = -\nabla \mathbf{\Phi} - 2(\mathbf{\Omega_p} \times \dot{\mathbf{r}}) - \mathbf{\Omega_p} \times (\mathbf{\Omega_p} \times \mathbf{r}), \quad (6)$$

where the terms $-2\mathbf{\Omega_p} \times \dot{\mathbf{r}}$ and $-\mathbf{\Omega_p} \times (\mathbf{\Omega_p} \times \mathbf{r})$ represent the Coriolis and the centrifugal forces, respectively, and $\mathbf{r}$ is the position vector. Defining an effective potential:

$$\Phi_{\rm eff} = \Phi - \frac{1}{2}\Omega_p^2\,(x^2 + y^2), \quad (7)$$

(6) becomes $\ddot{\mathbf{r}} = -\nabla \mathbf{\Phi}_{\rm eff} - 2(\mathbf{\Omega_p} \times \dot{\mathbf{r}})$, and the Jacobi constant is $E_J = \frac{1}{2}|\ \dot{\mathbf{r}}\ |^2 + \Phi_{\rm eff}$, which, being constant in time, can be considered as the energy in the rotating frame.

3 Dynamics Around L_1 and L_2

For our calculations we place ourselves in a frame of reference corotating with the bar and place the bar major axis along the x axis. In this rotating frame we have five equilibrium points, which, due to the similarity with the Restricted Three Body Problem, are called Lagrangian points. The points located on the origin of coordinates, namely L_3, and along the y axis, namely L_4 and L_5, are linearly stable. The ones located symmetrically along the x axis, namely L_1 and L_2, are linearly unstable. Around the equilibrium points there exist families of periodic orbits, e.g. around the central equilibrium point the well-known x_1 family of periodic orbits [5] that is responsible for the bar structure.

The dynamics around the unstable equilibrium points is described in detail in [16]; here we give only a brief summary. Around each unstable equilibrium point there also exists a family of periodic orbits, known as the family of Lyapunov orbits [14]. For a given energy level, two stable and two unstable sets of asymptotic orbits emanate from the periodic orbit, known as the stable and the unstable invariant manifolds, respectively. We denote by $W^s_{\gamma_i}$ the stable invariant manifold associated to the periodic orbit γ around the Lagrangian

[1] Bold letters denote vector notation. The vector $\mathbf{z}$ is a unit vector.

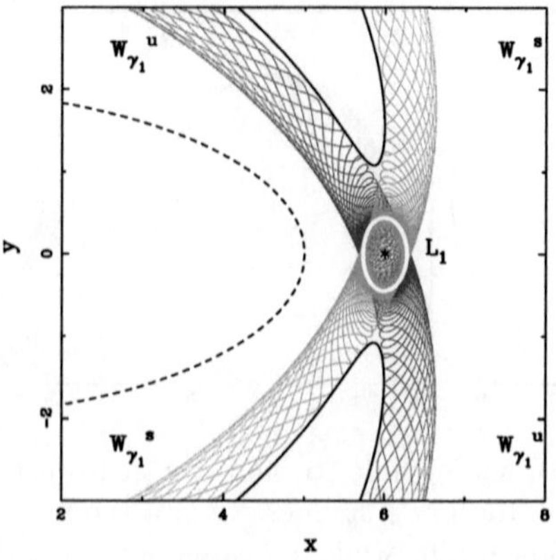

Fig. 1. Unstable (in *dark grey*), $W^u_{\gamma_1}$, and stable (in *light grey*), $W^s_{\gamma_1}$, invariant manifolds associated to the periodic orbit around L_1 (in *white solid line*). In *black solid lines*, we plot the zero velocity curves for this energy level and the *dashed curve* shows the outline of the bar

point L_i, $i = 1, 2$. The stable invariant manifold is the set of orbits that tends to the periodic orbit asymptotically. Similarly, we denote by $W^u_{\gamma_i}$ the unstable invariant manifold associated to the periodic orbit γ around the Lagrangian point L_i, $i = 1, 2$. The unstable invariant manifold is the set of orbits that departs asymptotically from the periodic orbit (i.e. orbits that tend to the Lyapunov orbits when the time tends to minus infinity) (Fig. 1). Since the invariant manifolds extend well beyond the neighbourhood of the equilibrium points, they can be responsible for global structures.

In [17], we give a detailed description of the role invariant manifolds play in global structures and, in particular, in the transfer of matter. Simply speaking, the transfer of matter is characterised by the presence of homoclinic, heteroclinic, and transit orbits. Homoclinic orbits correspond to asymptotic trajectories ψ such that $\psi \in W^u_{\gamma_i} \cap W^s_{\gamma_i}$, $i = 1, 2$. That is, they are asymptotic orbits that depart from the unstable Lyapunov periodic orbit γ around L_i and return asymptotically to it (Fig. 2a). Heteroclinic orbits are asymptotic trajectories ψ' such that $\psi' \in W^u_{\gamma_i} \cap W^s_{\gamma_j}$, $i, j = 1, 2$, $i \neq j$. That is, they are asymptotic orbits that depart from the periodic orbit γ around L_i and asymptotically approach the corresponding Lyapunov periodic orbit with the same energy around the Lagrangian point at the opposite end of the bar L_j, $i \neq j$ (Fig. 2b). We are interested in the homoclinic and heteroclinic orbits corresponding to the first intersection of the invariant manifolds with an appropriate surface of section. There exist also trajectories that spiral out from the region of the unstable periodic orbit and we refer to them as transit orbits (Fig. 2c). These three types of orbits are chaotic orbits since they fill part of the chaotic sea when we plot the Poincaré surface of section (e.g. the section $(x, \dot{x})$ near L_1).

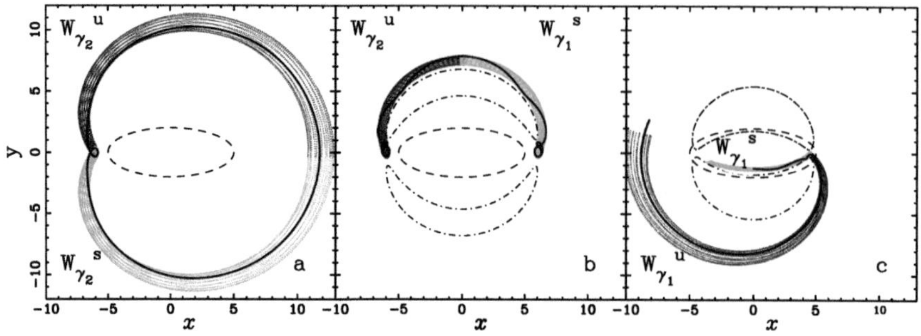

Fig. 2. Homoclinic (**a**), heteroclinic (**b**) and transit (**c**) orbits (*black thick lines*) in the configuration space. In *dark grey lines*, we plot the unstable invariant manifolds associated to the periodic orbits, while in *light grey* we plot the corresponding stable invariant manifolds. In *dashed lines*, we give the outline of the bar and, in (**b**) and (**c**), we plot the zero velocity curves in *dot-dashed lines*

4 Results

One of our goals is to check the influence of each main free parameter of the models introduced in Sect. 2. In order to do so, we make families of models in which only one of the free parameters is varied, while the others are kept fixed. Our results in [17] show that only the bar pattern speed and the bar strength have a considerable influence on the shape of the invariant manifolds and, thus, on the morphology of the galaxy. Having established this, we perform a two-dimensional parameter study for each bar potential and we obtain all types of rings and spiral arms.

In Fig. 3 we show the model rings and the spiral structure we obtain with our models. We plot the unstable (Fig. 3a,b,d) and the unstable and stable (Fig. 3c) invariant manifolds associated to one of the Lyapunov periodic orbits of the main family around L_1 and L_2. Note that we plot the projection of the invariant manifolds on the configuration space (x, y). Our results show that the morphologies obtained depend on dynamical factors, that is, on the presence of homoclinic or heteroclinic orbits of the first intersection of the corresponding invariant manifolds. If heteroclinic orbits exist, then the ring of the galaxy is classified as rR_1 (Fig. 3a). The inner branches of the invariant manifolds associated to γ_1 and γ_2 outline a nearly elliptical inner ring that encircles the bar. The outer branches of the same invariant manifolds form an outer ring whose principal axis is perpendicular to the bar major axis. If the model has neither heteroclinic, nor homoclinic orbits and only transit orbits are present, the barred galaxy will present two spiral arms emanating from the ends of the bar. The outer branches of the unstable invariant manifolds will spiral out from the ends of the bar and they extend azimuthally to more than $3\pi/2$ (Fig. 3d). If the outer branches of the unstable invariant manifolds intersect

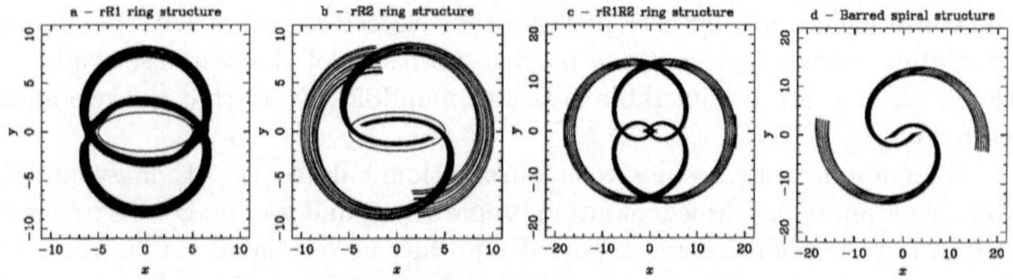

Fig. 3. Model rings and spiral arm structures. We plot characteristic examples of (**a**) rR_1 ring structure, (**b**) rR_2 ring structure, (**c**) R_1R_2 ring structure and (**d**) Spiral arms

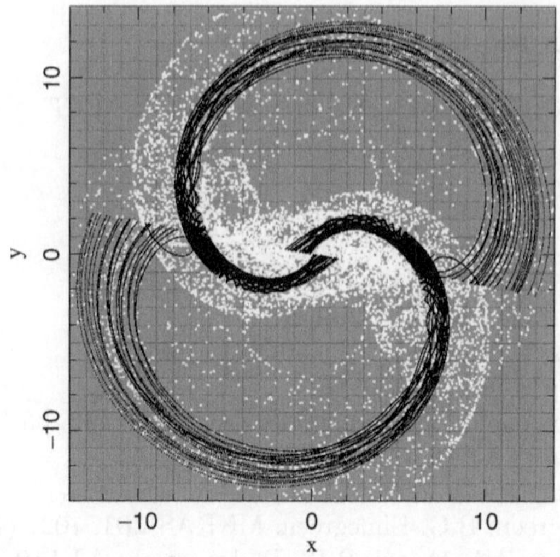

Fig. 4. Over-plot of the results obtained with the response simulation (*white dots*) and the invariant manifolds (*black lines*) in a model with spiral arms

in configuration space with each other,[2] then they form the characteristic shape of R_2 rings (Fig. 3b). That is, the trajectories outline an outer ring whose major axis is parallel to the bar major axis. The last possibility is if only homoclinic orbits exist. In this case, the inner branches of the invariant manifolds form an inner ring, while the outer branches outline both types of outer rings, thus the barred galaxy presents an R_1R_2 ring morphology (Fig. 3c).

We also study the response of an axisymmetric component to a bar perturbation. We use the same axisymmetric potential and the same bar potential as in our models and the bar is introduced gradually, to avoid transients. Once the bar has reached its maximum amplitude, we consider a snapshot of the response simulation and we compare its morphology to the corresponding structure we obtain with our models. In Fig. 4 we show the results for

[2] Note that they cannot intersect in phase space.

the spiral arms case, by over-plotting the selected snapshot with our model. The white points represent the particle positions of the response study and the black lines are the unstable invariant manifolds. Note that the two match perfectly.

We compare our results with observational data (E. Athanassoula, M. Romero-Gómez, J.J. Masdemont, in preparation) and we find good agreement. Regarding the photometry, the density profiles across radial cuts in rings and spiral arms agree with the ones obtained from observations. The velocities along the ring also show that these are only a small perturbation of the circular velocity.

Acknowledgement

MRG acknowledges her fellowship "Becario MAE-AECI".

References

1. E. Athanassoula: Phys. Rep. **114**, 319 (1984)
2. E. Athanassoula: MNRAS **259**, 328 (1992)
3. B. Barbanis, L. Woltjer: ApJ **150**, 461 (1967)
4. R. Buta: ApJS **96**, 39 (1995)
5. G. Contopoulos, Th. Papayannopoulos: A & A, **92**, 33 (1980)
6. J.M.A. Danby: AJ **70**, 501 (1965)
7. W. Dehnen: AJ **119**, 800 (2000)
8. D.M. Elmegreen, B.G. Elmegreen: MNRAS **201**, 1021 (1982)
9. P.B. Eskridge, J.A. Frogel, R.W. Podge, et al.: AJ **119**, 536 (2000)
10. N.M. Ferrers: Q. J. Pure Appl. Math. **14**, 1 (1877)
11. D.E. Kaufmann, G. Contopoulos: A & A, **309**, 381 (1996)
12. G. Kuzmin: Astron. Zh. **33**, 27 (1956)
13. B. Lindblad: Stockholms Observatorium Ann. **22**(5) (1963)
14. A. Lyapunov: Ann. Math. Studies **17** (1949)
15. P.A. Patsis: MNRAS **369**, L56 (2006)
16. M. Romero-Gómez, J.J. Masdemont, E. Athanassoula, C. García-Gómez: A & A **453**, 39 (2006)
17. M. Romero-Gómez, E. Athanassoula, J.J. Masdemont, C. García-Gómez: A & A **472**, 63 (2007)
18. M.P. Schwarz: ApJ **247**, 77 (1981)
19. M.P. Schwarz: MNRAS **209**, 93 (1984)
20. M.P. Schwarz: MNRAS **212**, 677 (1985)
21. A. Toomre: ApJS **138**, 385 (1963)
22. A. Toomre: ApJ **158**, 899 (1969)
23. N. Voglis, I. Stavropoulos, C. Kalapotharakos: MNRAS **372**, 901 (2006)
24. N. Voglis, P. Tsoutsis, C. Efthymiopoulos: MNRAS **373**, 280 (2006)

Secular Instabilities of Stellar Systems: Slow Mode Approach

E.V. Polyachenko[1], V.L. Polyachenko[1], and I.G. Shukhman[2]

[1] Institute of Astronomy, Russian Academy of Sciences, 48 Pyatnitskya St., Moscow 119017, Russia
epolyach@inasan.ru
[2] Institute of Solar-Terrestrial Physics, Russian Academy of Sciences, Siberian Branch, P.O. Box 291, Irkutsk 664033, Russia
shukhman@iszf.irk.ru

1 Introduction

The general account for the secular stability[1] is a very difficult problem with no universal treatment due to the diversity and complexity of the stellar systems [2, 5]. Here we follow [13] to derive a general integral equation for unstable eigenmodes. Let us consider a completely integrable stellar system in equilibrium with the Hamiltonian

$$H_0 = \frac{\mathbf{v}^2}{2} + \Phi_0(\mathbf{r}) = H_0(\mathbf{I}) \tag{1}$$

depending on action variables $\mathbf{I}$, but not on the angle variables $\mathbf{w}$. The motion of stars in this system is quasiperiodic with frequencies $\Omega_i = \partial H_0/\partial I_i$. Small perturbation of the system leads to a perturbed potential Φ, so that a full Hamiltonian is

$$H(\mathbf{I}, \mathbf{w}, t) = H_0(\mathbf{I}) + \Phi(\mathbf{I}, \mathbf{w}, t). \tag{2}$$

A stellar distribution function $f(\mathbf{I}, \mathbf{w}, t)$ obeys the collisionless Boltzmann equation (CBE)

$$\frac{\partial f}{\partial t} = [H, f] \equiv \sum_i \frac{\partial H}{\partial w_i}\frac{\partial f}{\partial I_i} - \sum_i \frac{\partial f}{\partial w_i}\frac{\partial H}{\partial I_i}. \tag{3}$$

The full DF can be represented as a sum of unperturbed and perturbed parts $f = \mathcal{F}_0 + \mathcal{F}$, so that

$$\Phi_0(\mathbf{r}) = -G\int d\mathbf{r}' d\mathbf{v}' \frac{\mathcal{F}_0(\mathbf{r}', \mathbf{v}')}{|\mathbf{r} - \mathbf{r}'|}, \qquad \Phi(\mathbf{r}) = -G\int d\mathbf{r}' d\mathbf{v}' \frac{\mathcal{F}(\mathbf{r}', \mathbf{v}')}{|\mathbf{r} - \mathbf{r}'|}. \tag{4}$$

[1] The term “secular instability” opposes here the term “violent instability”.

G. Contopoulos, P.A. Patsis, *Chaos in Astronomy.* Astrophysics and Space Science Proceedings, doi: 10.1007/978-3-540-75826-6,

Expansion of the perturbed functions into Fourier series in angle variables

$$\mathcal{F}(\mathbf{I}, \mathbf{w}, t) = \sum_{\mathbf{n}} \mathcal{F}_{\mathbf{n}}(\mathbf{I}) e^{i\mathbf{n}\cdot\mathbf{w} - i\omega t}, \qquad \Phi(\mathbf{I}, \mathbf{w}, t) = \sum_{\mathbf{n}} \Phi_{\mathbf{n}}(\mathbf{I})\, e^{i\mathbf{n}\cdot\mathbf{w} - i\omega t}, \tag{5}$$

and substitution into linearized CBE and (4) gives

$$[-\omega + \mathbf{n} \cdot \boldsymbol{\Omega}(\mathbf{I})]\mathcal{F}_{\mathbf{n}}(\mathbf{I}) = \left(\mathbf{n} \cdot \frac{\partial \mathcal{F}_0}{\partial \mathbf{I}}\right) \Phi_{\mathbf{n}}(\mathbf{I}) \tag{6}$$

and

$$\sum_{\mathbf{n}} \Phi_{\mathbf{n}}(\mathbf{I})\, e^{i\mathbf{n}\cdot\mathbf{w}} = -G \int d\mathbf{I}' d\mathbf{w}' \frac{1}{|\mathbf{r} - \mathbf{r}'|} \sum_{\mathbf{n}'} \mathcal{F}_{\mathbf{n}'}(\mathbf{I}')\, e^{i\mathbf{n}'\cdot\mathbf{w}'}, \tag{7}$$

where $\mathbf{n}$ is a vector with integer components n_i, ω is the frequency. Combining (6) and (7), one can obtain the general integral equation for eigenmodes

$$[\omega - \mathbf{n} \cdot \boldsymbol{\Omega}(\mathbf{I})]\mathcal{F}_{\mathbf{n}}(\mathbf{I}) = G\left(\mathbf{n} \cdot \frac{\partial \mathcal{F}_0}{\partial \mathbf{I}}\right) \int d\mathbf{I}' \sum_{\mathbf{n}'} \Pi_{\mathbf{n},\mathbf{n}'}(\mathbf{I}, \mathbf{I}')\mathcal{F}_{\mathbf{n}'}(\mathbf{I}'), \tag{8}$$

where the kernel functions Π are

$$\Pi_{\mathbf{n},\mathbf{n}'}(\mathbf{I}, \mathbf{I}') = \frac{1}{(2\pi)^N} \int d\mathbf{w} d\mathbf{w}' \frac{e^{i(\mathbf{n}'\cdot\mathbf{w}' - \mathbf{n}\cdot\mathbf{w})}}{|\mathbf{r} - \mathbf{r}'|}. \tag{9}$$

This equation, in principle, provides the eigenmodes of completely integrable stellar systems. We are interested in exponentially growing modes, for which $\operatorname{Im}\omega > 0$. In this case, in (8) the integration over real variables $\mathbf{I}$ is possible. For neutral and damping modes, one needs to change the integration contour to pass below the eigen-frequencies in the complex ω-plane.

However, the solution of this equation is rather costly. For example, consider the simplest case of disk-like systems. Introducing a grid in (I_1, I_2)-space consisting $N_1 \times N_2$ knots, and N_- and N_+ for the lowest and the highest Fourier harmonics of the expansion (5), one needs to compute eigenvalues and eigenvectors of a matrix with rank $N_1 \times N_2 \times (N_+ - N_- + 1)$.

If the frequencies of an unperturbed system are commensurable, i.e., if there is a resonance integer vector $\mathbf{n}$ such that $\sum_i n_i \Omega_i \approx 0$, additional integrals of motion can be found. One can eliminate fast angle variables, preserving only slowly varying combinations $\sum_i n_i w_i$. This leads to a simplification of the integral equation (8), in which the summation is over the resonance $\mathbf{n}$ only. Such a simplified equation is called a *slow integral equation*. In this paper, resonance combinations $\sum_i n_i \Omega_i$ correspond to precessing frequencies of orbits, whereas Ω_i correspond to usual stellar motion.

This technique is applied to two types of systems. The first one are stellar spherical clusters in the gravitational field of a heavy point mass (massive black hole or dense galactic center). The orbits in the clusters are asymmetric Keplerian ellipses, in which the azimuthal and radial frequencies are equal (1:1 orbits). The second type includes galactic stellar disks with smooth potentials, such as the Plummer potential or the isochrone potential. In the center, the potentials can be roughly approximated by the harmonic potential, so the orbits are symmetric ellipses in which the ratio of radial to azimuthal frequencies is equal 2 (2:1-orbits).

2 Spherical Stellar Clusters Around MBH

One of the key questions in physics of stellar clusters around a heavy point mass is connected with possible instability. It is easy to estimate the timescales of the processes specific to these systems (see, e.g., [18]). Let the stellar cluster consist of N stars, the total mass is M, the central heavy mass is M_c, so that $\epsilon \equiv M/M_c \ll 1$. The *collissional* two-body relaxation timescale is $t_{\rm col} \sim t_{\rm dyn} N/\epsilon^2$, where $t_{\rm dyn}$ is the typical dynamical timescale. *Resonant* relaxation timescale [17] is $t_{\rm res} \sim t_{\rm dyn} N/\epsilon$. The timescale of instability growing should be of the order of the precession time, $t_{\rm ins} \sim t_{\rm dyn}/\epsilon$, which is clearly the shortest. Therefore, if an instability exists, it can play the most important role in depositing matter to the black hole. Still, the instability development timescale is much longer than the dynamical time, so the instability is slow.

Tremaine [18] has examined spherical models of a low-mass stellar system with almost isotropic distribution and concluded that they are *stable* at least for modes with a number of spherical harmonics $l \leq 2$.

Study of $l \geq 3$ harmonics is difficult without simplifying assumptions. Here we assume a stellar cluster in which all stars possess equal energies, $E = E_0$, and show that instability is indeed possible for $l \geq 3$. This instability is an analog of the loss-cone instability in plasma traps (see, e.g., [11]), which is due to the peculiar anisotropy in the velocity distribution of plasma particles. The anisotropy is caused by departure of particles with sufficiently small velocity component transverse to the symmetry axis of the system. The presence of this *loss cone* produces deformation of the plasma distribution function in transverse velocities, giving it an unstable beam-like character. Similar deformation of the DF in angular momentum takes place in clusters in case of deficiency of stars with low angular momentum due to tidal disruption or direct absorption by the black hole. This deformation can trigger the instability which we shall call the *gravitational loss-cone instability.*

An additional condition should be met for the instability, which originates from the fundamental distinction between gravitating and plasma systems. In gravitating systems, particles have only one kind of "charge", and they attract each other. This fact ultimately leads to the Jeans instability substituting Langmuir oscillations in plasma [5]. In systems with nearly radial orbits we are going to study, there is a specific form of the Jeans instability called the radial orbit instability [5, 16]. It develops only in systems with prograde orbit precession, when the direction of orbit precession coincide with the direction of stellar revolution. Conversely, the gravitational loss-cone instability can occur only when orbit precession is *retrograde.* For spherical near-Keplerian systems, the retrograde precession is common.

2.1 The Equation for Slow Modes

Let the DF be

$$F(E, L) = A\,\delta(E - E_0)\,f(L). \tag{10}$$

All stars in the cluster have the same energy $E = E_0$. The normalization factor A in (10) is chosen so that the total mass of the cluster is equal to M_G,

$$A = \frac{\Omega_1 M_G}{16\pi^3 L_{\rm circ}^2}. \tag{11}$$

Here $\Omega_1(E_0) = (2|E_0|)^{3/2}/(GM_c)$ is the frequency of stellar radial oscillations, $L_{\rm circ}(E_0) = GM_c/(2|E_0|)^{1/2}$ is the angular momentum on circular orbits. It is convenient to introduce a dimensionless angular momentum $\alpha = L/L_{\rm circ}(E_0)$. For almost radial orbits, when $\alpha \ll 1$ or eccentricity $e \equiv \sqrt{1-\alpha^2} \approx 1$, the precession rate is

$$\Omega_{\rm pr}(\alpha) = -\frac{2\,\epsilon\,\Omega_1}{\pi^2}\,\alpha\,[1 + O(\alpha^2)]. \tag{12}$$

For the DF $f(\alpha)$ we shall assume that it satisfies the loss cone condition $f(0) = 0,\ f'(0) = 0$.

Typical frequencies of the slow perturbations under consideration are of the order of the precession frequency, $\epsilon\Omega_1$. Therefore, it is natural to turn to the dimensionless frequencies

$$\bar\omega = \frac{\omega}{\epsilon\,\Omega_1}, \qquad \nu(\alpha) = -\frac{\Omega_{\rm pr}(\alpha)}{\epsilon\,\Omega_1}. \tag{13}$$

For the spherical systems, the precession is retrograde, so $\nu(\alpha) > 0$. Then, the dimensionless integral equation for slow modes is [14, 15]:

$$\begin{aligned}\left[\bar\omega^2 - s^2\,\nu^2(\alpha)\right]\varphi_s(\alpha) = -2 \sum_{s'=s_{\min}}^{l} s'^2\, D_l^{s'} \\ \times \int_0^1 \nu(\alpha')\,\alpha'\,\frac{d\,f(\alpha')}{d\alpha'}\,K_{s,\,s'}^{(l)}(\alpha,\alpha')\,\varphi_{s'}(\alpha')\,d\alpha', \end{aligned} \tag{14}$$

where

$$D_l^s = \frac{1}{2^{2l}}\,\frac{(l+s)!(l-s)!}{\left[\left(\frac{1}{2}\,(l-s)\right)!\,\left(\frac{1}{2}\,(l+s)\right)!\,\right]^2} \tag{15}$$

for even $(l - s)$, and zero otherwise. Consequently, $s_{\min} = 1$ for odd l and $s_{\min} = 2$ for even l. To obtain the eigenfrequency spectrum for a model, it is necessary to compute preliminarily the kernels $\mathcal{K}_{s,s'}^{(l)}(\alpha, \alpha')$ (universal for all models), and the precession rate profile $\nu(\alpha)$ for the given model. The integration over the Keplerian orbits is most conveniently expressed using the variable τ (eccentric anomaly), which is connected with the current radius r and the true anomaly ζ of a star as follows:

$$r = \frac{1}{2}R\,(1 - e\cos\tau), \qquad \cos\zeta = \frac{\cos\tau - e}{1 - e\cos\tau}. \tag{16}$$

The kernels can be calculated as follows:

$$K^{(l)}_{s,\,s'}(\alpha,\alpha') = \frac{2}{(2l+1)\pi^2}\int\limits_0^{\pi} d\tau\, r\cos(s\zeta)\int\limits_0^{\pi} d\tau'\, r'\cos(s'\zeta')\,\mathcal{F}_l(r,r'), \tag{17}$$

where r' and ζ' specify the position of a star on the orbit with the eccentricity e' corresponding to the variable τ', and

$$\mathcal{F}_l(r,r') = \frac{\min(r,r')^l}{\max(r,r')^{l+1}}. \tag{18}$$

2.2 Monotonity of $f(\alpha)$

Due to the kernel functions symmetry

$$\mathcal{K}^{(l)}_{s,\,s'}(\alpha,\alpha') = \mathcal{K}^{(l)}_{s',\,s}(\alpha',\alpha), \tag{19}$$

from the integral equation (14) one can obtain

$$\mathrm{Im}(\bar{\omega}^2)\int\limits_0^1 g(\alpha)\,d\alpha \sum_{s=s_{\min}}^{l} s^2 D^s_l\,|\varphi_s(\alpha)|^2 \equiv 0. \tag{20}$$

Therefore, if the function $g(\alpha) \equiv df(\alpha)/d\alpha$ is the constant-sign, then the integral should be non-zero, and so $\mathrm{Im}(\bar{\omega}^2) = 0$. Otherwise, when $\mathrm{Im}(\bar{\omega}^2) \neq 0$, the integral must be equal to zero. Consequently, the function $g(\alpha)$ should change its sign, i.e., DF $f(\alpha)$ should be non-monotonic.

In the case of monotonic DF, the integral equation (14) can be interpreted mechanically in terms of positive and negative elastic forces. The positive force works for stabilization, whereas the negative force for destabilization of the system. If it exists the instability should be of *hydrodynamic* type, where a large portion of stars is involved in the perturbation. It can be shown that the negative force (the last term in (14)) decreases with the number of spherical harmonics l increasing. However, as it is shown by Tremaine [18], for $l = 1$, the system is neutrally stable (there is a lopsided mode with $\omega = 0$), and the systems with $l = 2$ are stable. Thus, the monotonic models are stable.

We checked numerically a possibility of instability development connected with the maximum on the edge of the distribution function's domain. For this purpose, a number of models with DFs vanishing quickly but smoothly near the circular orbits $\alpha = 1$, were computed. The computations show no sign of instability. The reason is that the kernels $\mathcal{K}$ of the integral operators in (14) vanish for the circular orbit $\alpha = 1$, so details of the distribution near circular orbits cannot considerably affect the solutions of the integral equation (14).

2.3 The Loss-Cone Instability

It is instructive to deduce an approximate integral equation for nearly radial orbits (spoke approximation limit). Let us consider non-monotonic DF; for definiteness, let the DF be

$$f(\alpha) = \frac{1}{\alpha_T^2}\tilde{f}(x), \quad \tilde{f}(x) = \frac{2}{n!}x^n \exp(-x), \quad x = \frac{\alpha^2}{\alpha_T^2}, \tag{21}$$

where α_T and n are the parameters, $\alpha_T \ll 1$. Setting

$$K_{s,s'}^{(l)}(\alpha,\alpha') \approx K_{s,s'}^{(l)}(0,0) = C_l \equiv \int\limits_0^\infty \frac{dz}{z}[J_{(l+1)/2}(z)J_{l/2}(z)]^2,$$

and $\phi_s(\alpha) = (\bar{\omega}^2 - s^2\nu(\alpha^2))\varphi_s(\alpha) \approx 1$, $\nu(\alpha) = 2\alpha/\pi^2$, $\bar{\omega} = 2\lambda\alpha_T/\pi^2$, one can obtain

$$\int\limits_0^\infty dx\,\frac{d\tilde{f}(x)}{dx}\sum_{s=s_{\min}}^{l}\frac{xD_l^s}{x-\lambda^2/s^2} = \frac{\alpha_T^2}{\pi^2 C_l}. \tag{22}$$

Since, the r.h.s. is $\mathcal{O}(\alpha_T^2)$, it can be set to zero. This is the equation for slow modes in *spoke approximation.*

To find unstable solutions for (22), one needs to find a critical value $\alpha_T = (\alpha_T)_c$, for which there is a neutral mode with the frequency $\lambda = \lambda_0$. If we find such a critical value $(\alpha_T)_c > 0$, then by variation of α_T we obtain instability.

Let us consider first the case when there is only one term in the sum in (22), i.e., $l = 1$ or $l = 2$. Obviously, since there is only one resonance, the frequency is connected with one of the extrema of the DF (see circles on Fig. 1, left panel), otherwise the integral would be a complex number. However, it is easy to show (see [14]) that these extrema do not give positive values of the integral. It means that instability is absent for $l \leq 2$.

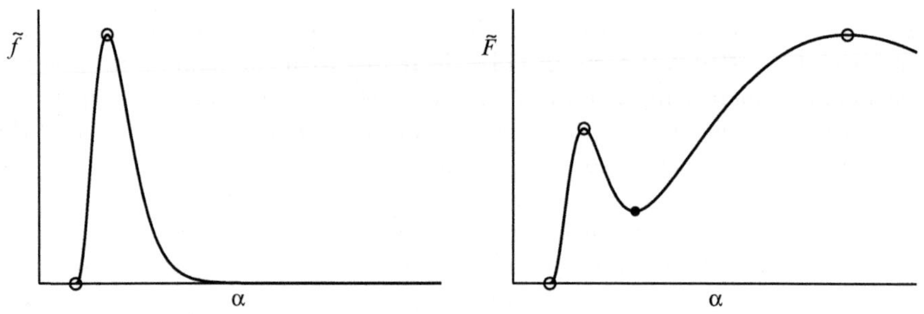

Fig. 1. *Left panel*: typical one-hump distribution $\tilde{f}(\alpha)$ of a stellar system; *right panel*: effective distribution $\tilde{F}(\alpha)$ for $l = 3$ according to (24)

For $l \geq 3$, there are several terms in the sum (22). Therefore, the positions of neutral modes must not coincide with extrema of $\tilde{f}$. For definiteness, one can consider $l = 3$ ($D_3^1 = \frac{3}{16}$, $D_3^3 = \frac{5}{16}$). For the neutral mode with the frequency $\lambda = \lambda_0$, there are two resonances, $x = x_1 = \lambda_0^2$, $x = x_2 = \lambda_0^2/9$. Setting the imaginary part of (22) to zero, one has an equation for λ_0 that potentially can be neutral frequencies corresponding to the critical values $(\alpha_T)_c$:

$$f_0'(\lambda_0^2) + \tfrac{5}{9}\, f_0'\left(\tfrac{1}{9}\,\lambda_0^2\right) = 0. \tag{23}$$

It should be also checked that the integral in l.h.s. of (22) is positive. Calculations show that for the models (21) the integral is positive if $n \geq 2$.

The slow mode equation in spoke approximation (22) for arbitrary l involving $[\frac{1}{2}\,(l+1)]$ terms can be reduced to a one-term equation similar to one obtained for $l = 1$ or $l = 2$ by using the substitution

$$F^{(l)}(x) = C_l \sum_{s\,=\,s_{\min}}^{l} D_l^s\, f(x/s^2). \tag{24}$$

transforms (22) into the equation

$$\int\limits_0^\infty dx\, x\, \frac{d\,F^{(l)}(x)}{dx}\, \frac{x}{x - \lambda^2} = \frac{\alpha_T^2}{\pi^2}. \tag{25}$$

However, now the integral involves the effective distribution $F^{(l)}(x)$, instead of the initial distribution $f(x)$ (with one maximum and tending to zero at $x = 0$ and infinity). Starting with modes $l = 3$, the new function, $F^{(l)}(x)$, can have a minimum (filled circle on the right panel of Fig. 1). Sufficiently deep minimum provides positive integral (22), and thus the valid critical value $(\alpha_T)_c$.

In Fig. 2, the results of our calculation of exact equation (14) and approximate equation (22) are presented. It shows the dependence of the growth rate $\mathrm{Im}\,\bar{\omega}$ vs. the dispersion α_T. For small α_T, the growth rates increase linearly with dispersion, in full agreement with spoke approximation approach. The instability becomes saturated at $\alpha_T \sim 0.3$. Note, that for both models ($n = 2$ and $n = 3$), the DFs are essentially non-monotonic for $\alpha_T \lesssim 0.3$.

To conclude this section, we would like to stress, that monotonicity of DF is the key in understanding the difference between our results and results obtained by Tremaine [18]. If the initial distribution of the spherical cluster is monotonic, the cluster is generally stable [18]. Details of the distribution behavior near circular orbits prove to be unimportant. Spherical clusters with the non-monotonic DFs can be affected by the gravitational loss-cone instability.

While monotonic DF are well-accepted (see, e.g., [3]), the non-monotonic distributions are also real. If the cluster is formed, for example, as a result of the collisionless collapse (several free fall times), then it remains collisionless for a long timescale of collisional relaxation (see, e.g., [10]). In principle, the system can have almost arbitrary DF both in the energy and in the angular momentum.

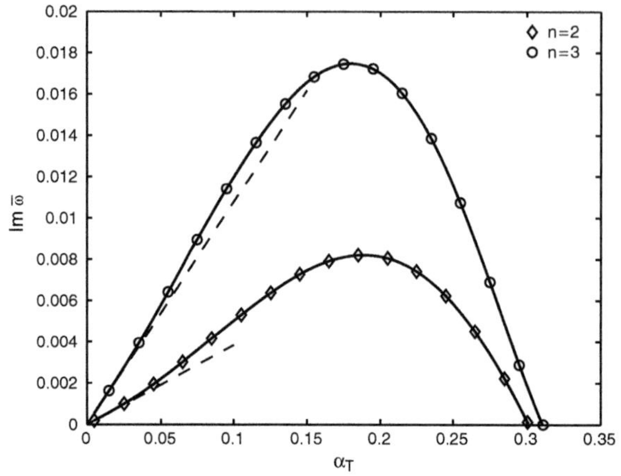

Fig. 2. The dependence of the growth rate Im $(\bar{\omega})$ vs. dimensionless angular momentum dispersion α_T of the mode $l = 3$ for the models $n = 2$ (*diamonds*) and $n = 3$ (*circles*). *Dashed lines* show the asymptotic behavior obtained using spoke approximation equation (22): $\mathrm{Im}(\bar{\omega}/\alpha_T) = (2/\pi^2)\,\mathrm{Im}\,\lambda = 0.189$ and 0.532 for $n = 2$ and 3 respectively (the exact solution for $\alpha_T = 0.003$ gives 0.185 and 0.529)

3 Galactic Disks

In this section we apply our theory to the Kuzmin–Toomre model of stellar galactic disk. The general integral equation for the azimuthal harmonics m is

$$\mathcal{F}_n(\mathbf{I})(\omega - n\Omega_1 - m\Omega_2) = G\mathcal{F}'_{0,n}(\mathbf{I}) \int d\mathbf{I}' \sum_{n'} \Pi_{n,n'}(\mathbf{I},\mathbf{I}')\mathcal{F}_{n'}(\mathbf{I}'). \tag{26}$$

where $\mathcal{F}'_{0,n}(\mathbf{I})$ denotes

$$\mathcal{F}'_{0,n}(\mathbf{I}) \equiv \left(\mathbf{n}\cdot\frac{\partial \mathcal{F}_0}{\partial \mathbf{I}}\right) = n\frac{\partial \mathcal{F}_0(\mathbf{I})}{\partial I_1} + m\frac{\partial \mathcal{F}_0(\mathbf{I})}{\partial L}, \tag{27}$$

I_1 and $I_2 \equiv L$ are the radial action and the angular momentum,

$$\Pi_{n,n'}(\mathbf{I},\mathbf{I}') = \frac{1}{2\pi}\int dw_1 dw'_1 \psi(r,r') e^{in'w'_1 - inw_1} e^{im\delta\varphi}, \tag{28}$$

where $\delta\varphi = \varphi(\mathbf{I}', w'_1) - \varphi(\mathbf{I}, w_1)$ and

$$\varphi(\mathbf{I}, w_1) = \Omega_2(\mathbf{I}) \int\limits_{r_{\min}(\mathbf{I})}^{r} \frac{dr'}{v_r} - L \int\limits_{r_{\min}(\mathbf{I})}^{r} \frac{dr'}{r'^2 v_r}. \tag{29}$$

The function

$$\psi(r, r') = \int_0^{2\pi} d\theta \frac{\cos m\theta}{[r^2 + r'^2 - 2rr' \cos\theta]^{1/2}} \tag{30}$$

has only a weak singularity $\propto \ln |r - r'|$, thus the expression for $\Pi_{n,n'}$ involves the improper integral (in the last equation r and r' are understood as functions of their respective action–angle variables: $r = r(\mathbf{I}, w_1)$, $r' = r'(\mathbf{I}', w_1'))$.

The set of equations (26) can be used for searching unstable eigen-modes in a 2D galactic stellar disc, so it is an alternative to the well-known matrix approach of Kalnajs (1977). The advantage of our approach is that we use linear eigen-value equations.

3.1 Slow Modes

Kuzmin–Toomre disks have the Plummer potential, $\Phi(r) = -(1 + r^2)^{-1/2}$, which is near harmonic in the center. Thus, the stellar orbits are symmetric ellipses slowly precessing with the rate $\Omega_{\rm pr} = \Omega_2 - \Omega_1/2$. For nearly circular orbits, the precession rate $\Omega_{\rm pr}(r) = \Omega(r) - \kappa(r)/2$ is shown in Fig. 3.

Fast stellar bars are composed of trapped, elongated, prograde orbits that precess at a rate close to $\Omega(r) - \kappa(r)/2$ [4], and their pattern speeds lie just above the maximum of $\Omega(r) - \kappa(r)/2$ curve, see the horizontal line in Fig. 3. For the stars involved in the perturbation, the following inequality holds

$$\epsilon \equiv |\Omega_p - \Omega_{\rm pr}|/\Omega \ll 1. \tag{31}$$

Therefore, the stellar orbit as a whole, but not the individual star, participates in the perturbations. The picture has points in common with Lynden-Bells [9] theory of bar formation in that it focuses on slowly precessing orbital rings, but

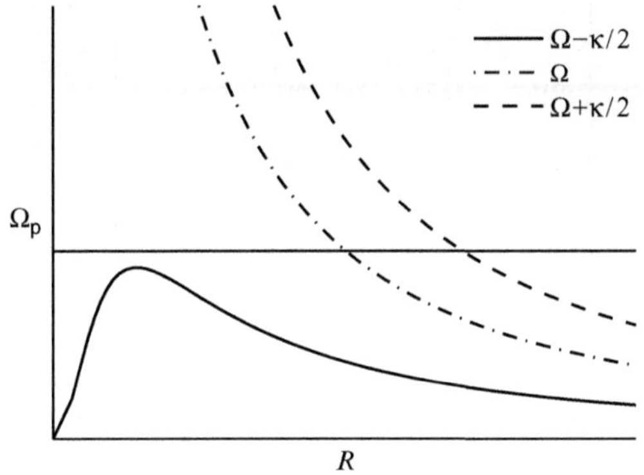

Fig. 3. Typical rotation curve $\Omega(R)$, and curves $\Omega(R) - \kappa(R)/2$, $\Omega(R) + \kappa(R)/2$ for smooth center (Kuzmin–Toomre disk)

it does not lead to the conclusion, as is the case for Lynden-Bells theory, that the bar rotates at a rate that lies between the fastest and slowest precession rates of the constituent rings. Instead we find that both bars and spirals, in common with many wave phenomena, move faster than the underlying medium. Several studies have shown that the rotation rates of bars in N-body models conform to our predictions rather than to those of Lynden-Bell [1, 12].

Assuming that the major part of the perturbation is located in the central parts of the disk, where (31) holds, one can omit all but one resonant equation from the system (26) corresponding to $n = -1$:

$$\mathcal{F}(\mathbf{I})(\omega - \Omega_{\rm pr}) = \frac{G}{m}\mathcal{F}'_{0,-1}(\mathbf{I}) \int d\mathbf{I}' \Pi_{-1,-1}(\mathbf{I},\mathbf{I}')\mathcal{F}(\mathbf{I}'). \tag{32}$$

Solution of this equation can be obtained numerically.

For example, we consider a Kalnajs (1976) model, which Athanassoula and Sellwood [1] have denoted by ($m = 6$, $\beta = 0$, $q = 1$, $J_c = 0.25$) (see their Table 1). The computed spectrum of the model is shown in Fig. 4. It consists of the continuous spectrum of van Kampen modes [19], located in the interval between the minimum and the maximum values of the precession speed $\Omega_{\rm pr}$, and possibly several discrete modes. Our primary interest is the neutral discrete modes to the right of the continuous spectrum. The discrete modes have different numbers of radial nodes (both the perturbed potential and surface density), the nodeless mode being the mode with the maximum pattern speed.

In the spectrum given in Fig. 4 one can see five discrete modes. The figure shows that the pattern speeds of the first and second modes (counting from

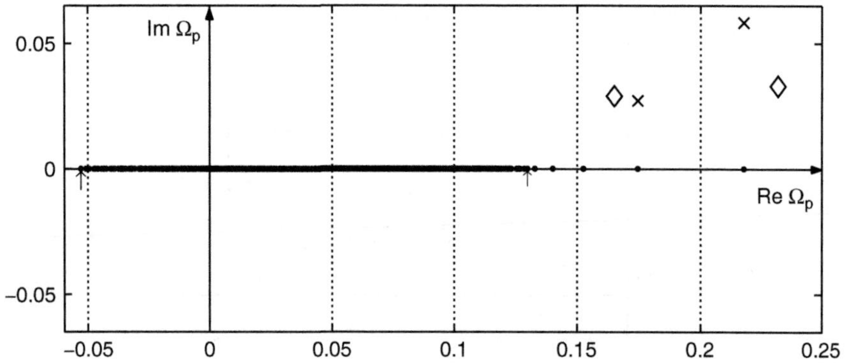

Fig. 4. The spectrum of pattern speeds (*filled circles*) computed as the eigenvalues of the problem (32), for one of the Kalnajs models. The *crosses* are the complex pattern speeds, with the growth rates estimated using formulas for weakly dissipating modes. The *diamonds* are the "experimental" complex pattern speeds according to Athanassoula and Sellwood [1]. The *arrows* point to the minimum and maximum values of the precession speed

the right) coincide with the pattern speeds obtained in the N-body simulations to within 6%. It is found (see below) that the growth rates of the other three modes are much lower, so they could not be observed in the numerical experiments.

Rough estimates for the most rapid modes [12], assuming that the major part – the body – of the mode is constituted by orbits in the center, and interaction with stars on corotation resonance and OLR gives rise to growth or damping of the modes, give the following values of the growth rates: $\gamma_1 = 0.117$, $\gamma_2 = 0.054$. Each of these values is composed of two parts corresponding to corotation (CR) and OLR. For these modes, the OLR contributions dominate. The mode with the maximum pattern speed has the highest growth rate. Obviously, the reason is that such a mode possesses the smallest corotation radius, where the disc surface density is high.

Let us sum up the results of this section. Some discrepancy between "experimental" data of Athanassoula and Sellwood [1] and our "theoretical" results was expected. There are several reasons for it:

1. The experimental values has been obtained for a finite value of the softening parameter, required for any N-body simulations. So, the experimental frequencies are systematically biased towards lower values.
2. The only ILR ($n = -1$) term in the Fourier expansion is insufficient for quantitative determination of eigenmodes.
3. Weak dissipative estimates obtained under the assumption that all stellar orbits are nearly circular.

Nevertheless, the slow mode approach is a good alternative to the well-known and widely accepted swing amplification (SA) mechanism for explanation of bar formation in disk galaxies. Taking into account additional terms in the Fourier expansion should not alter the mechanism of bar-mode excitation, which is analogous to inverse Landau damping. The new modal approach gives good estimates for mode pattern speeds and reasonable estimates for growth rates (recall that the growth rates values obtained by Athanassoula and Sellwood [1] using the swing amplification approach are smaller by a factor of 2 than their "experimental" values). Moreover, by contrast to the SA, it emphasizes the importance of the resonances other than corotation (see also [6]).

4 General Conclusions

The integral equation (8) for studying of secular stability of completely integrable Hamiltonian systems can be applied to astrophysical objects. We have reduced (8) to equations describing stellar clusters of spherical and disk-like geometry. Using some simplified assumptions, we were able to find a new gravitational instability similar to the loss-cone instability in plasma traps, and provide an alternative description for bar formation in galaxies.

Acknowledgements

The work was supported in part by Russian Science Support Foundation, RFBR grants No. 05-02-17874, 08-02-00928 and 07-02-00931, "Leading Scientific Schools" Grants No. 7629.2006.2, 900.2008.2 and "Young doctorate" Grant No. 2010.2007.2 provided by the Ministry of Industry, Science, and Technology of Russian Federation, and the "Extensive objects in the Universe" Grant provided by the Russian Academy of Sciences, and also by Programs of presidium of Russian Academy of Sciences No. 16 and OFN RAS No. 16.

References

1. L. Athanassoula, J. Sellwood: MNRAS **221**, 213 (1986)
2. J. Binney, S. Tremaine: *Galactic Dynamics* (Princeton University Press, Princeton, 1987)
3. H. Cohn, R.M. Kulsrud: ApJ **226**, 1087 (1978)
4. F. Combes, B.G. Elmegreen: A&A **271**, 391 (1993)
5. A.M. Fridman, V.L. Polyachenko: *Physics of Gravitating Systems* (Springer, New York, 1984)
6. M.A. Jalali, C.Hunter: ApJ **630**, 804 (2005)
7. A.J. Kalnajs: ApJ **205**, 751 (1976)
8. A.J. Kalnajs: ApJ **212**, 637 (1977)
9. D. Lynden-Bell: MNRAS, **187**, 101 (1979)
10. D. Merritt, J. Wang: ApJ, **621**, 101 (2005)
11. A.B. Mikhailovsky: *Theory of Plasma Instabilities, Vol. I* (Consultants Bureau, New York, 1974)
12. E.V. Polyachenko: MNRAS, **348**, 345 (2004)
13. E.V. Polyachenko: MNRAS, **357**, 559 (2005)
14. E.V. Polyachenko, V.L. Polyachenko, I.G. Shukhman: MNRAS, **379**, 573 (2007)
15. E.V. Polyachenko, V.L. Polyachenko, I.G. Shukhman: MNRAS, **386**, 1966 (2008)
16. V.L. Polyachenko, I.G. Shukhman: SvA **25**, 533 (1981)
17. K.P. Rauch, S. Tremaine: New Astron. **1**, 149 (1996)
18. S. Tremaine: ApJ **625**, 143 (2005)
19. P.O. Vandervoort: MNRAS, **339**, 537 (2003)

Stellar Velocity Distribution in Galactic Disks

Ch. Theis[1] and E. Vorobyov[2]

[1] Institute of Astronomy, University of Vienna, Türkenschanzstrasse 17, 1180 Vienna, Austria
theis@astro.univie.ac.at

[2] University of Western Ontario London, ON, Canada N6A 3K7
vorobyov@astro.uwo.ca

Summary. We present numerical studies of the properties of the stellar velocity distribution in galactic disks which have developed a saturated, two-armed spiral structure. In previous papers we used the Boltzmann moment equations (BME) up to second order for our studies of the velocity structure in self-gravitating stellar disks. A key assumption of our BME approach is the zero-heat flux approximation, i.e. the neglection of third order velocity terms. We tested this assumption by performing test particle simulations for stars in a disk galaxy subject to a rotating spiral perturbation. As a result we corroborated qualitatively the complex velocity structure found in the BME approach. An equilibrium configuration in velocity space is only slowly established on a typical timescale of 5 Gyrs or more. Since many dynamical processes in galaxies (like the growth of spirals or bars) act on shorter timescales, pure equilibrium models might not be fully appropriate for a detailed comparison with observations like the local Galactic velocity distribution. Third order velocity moments were typically small and uncorrelated over almost all of the disk with the exception of the 4:1 resonance region (UHR). Near the UHR (normalized) fourth and fifth order velocity moments are still of the same order as the second and third order terms. Thus, at the UHR higher order terms are not negligible.

1 Introduction

It is long known that the velocity distribution of stars in galactic disks is non-isotropic [13]. Kapteyn and Eddington invoked a superposition of (isotropic) stellar streams with different mean velocities, by this creating an anisotropic velocity distribution [9, 12]. However, an alternative interpretation by Schwarzschild became the general framework for describing the local velocity distribution of stars of equal age [22]. The Schwarzschild distribution is based on a single but anisotropic ellipsoidal distribution function. It is characterized by a Gaussian distribution in all three directions U (radial r), V (tangential ϕ) and W (vertical z) in velocity space, but with different velocity dispersions σ_{rr}, $\sigma_{\phi\phi}$, and σ_{zz}. In general, the velocity distribution is described by a velocity dispersion tensor

G. Contopoulos, P.A. Patsis, *Chaos in Astronomy.* Astrophysics and Space Science Proceedings, doi: 10.1007/978-3-540-75826-6,

$$\sigma_{ij}^2 \equiv \overline{(v_i - \bar{v}_i)(v_j - \bar{v}_j)}, \qquad (1)$$

where i and j denote the different coordinate directions and v_i gives the corresponding velocity. The bar denotes local averaging over velocity space. The principal axes of this tensor form an imaginary surface that is called the velocity ellipsoid. This ellipsoid is characterized by its anisotropy, measured by the ratio of the velocity dispersions along the principal axes and its orientation. Because the principal axes need not to be aligned with the coordinate axes, the vertex deviation l_{v} (which is defined as the angle between the (radial velocity) direction from the Sun to the Galactic centre and the direction of the major principal axis of the velocity ellipsoid) needs not to vanish.

In the case of stationary, axisymmetric systems and appropriate distribution functions (DF) the velocity ellipsoid is aligned with the coordinate axes, i.e. $l_{\mathrm{v}} = 0$. However, non-vanishing vertex deviations were found in the solar vicinity in many studies [16, 21, 25]. These vertex deviations could be explained by spiral structures [15, 26], by this supporting the importance of non-axisymmetric gravitational potentials.

A major step in analysing the local stellar velocity space was the astrometric satellite mission Hipparcos by ESA [10]. Earlier results concerning the general behaviour of the velocity ellipsoid, e.g. the dependence of the dispersions or the vertex deviations on $B - V$ were corroborated [2, 8, 11]. Moreover, the more numerous known proper motions allowed for a detailed mapping of the velocity space in the solar vicinity. Studies by Dehnen or by Alcobé and Cubarsi showed that the local velocity distribution of stars exhibits a rich substructure [1, 6]. For example, two major peaks in the velocity distribution were found where the smaller secondary peak is well detached by at least $30\,\mathrm{km\,s^{-1}}$ from the main peak (the "u anomaly"). Dehnen attributed this bimodality to the perturbation exerted by the Galactic bar assuming that its outer Lindblad resonance (OLR) is close to the Sun [7]. Mühlbauer and Dehnen showed that a central bar can also explain vertex deviations of about 10° [19]. Recent investigations about the influence of the Galactic bar on Oort's C constant corroborated these results and gave stronger limits on the bar's properties [18]. Minchev & Quillen showed that a *stationary* spiral cannot reproduce the "u anomaly" [17]. Though less likely, an alternative interpretation of the bimodality as a result of *non-stationary* spiral arms cannot be ruled out completely.

In general, a non-axisymmetric gravitational potential might lead to a misalignment of the velocity ellipsoid. In the case of spiral perturbations, this conclusion was made, e.g. by Mayor [15] (for a review see Kuijken and Tremaine [14]) and numerically confirmed recently by Vorobyov and Theis [23] (hereafter VT06). However, it is not clear if the non-axisymmetric gravitational field is entirely responsible for the observed vertex deviation. The existence of moving groups of stars was also shown to produce large vertex deviations [3]. The situation may become even more complicated because moving groups of stars may in turn be caused by the non-axisymmetric gravitational field of spiral arms. Therefore, a detailed numerical study of the vertex deviation in spiral galaxies is necessary.

In this paper, we present results of our analysis of the velocity structure in disk galaxies caused by spiral perturbations. In earlier papers (VT06, [24]) we presented numerical solutions of the Boltzmann moment equations (BME) up to second order for flat disks. In Sect. 2 we briefly describe this method and its results with respect to the vertex deviation. A characteristic of our BME approach is the assumption of a zero heat-flux, i.e. we neglect all velocity terms of higher order than 2. In order to test this assumption, we developed a test particle code which measures the velocity moments up to fifth order. This code and first results about the importance of higher order terms are presented in Sect. 3.

2 The Boltzmann Moment Equations: BEADS-2D

We developed a numerical code (BEADS-2D) for flat stellar disks based on the Boltzmann moment equations (BME) up to second order (VT06, [24]). This is basically a numerical solution of the Jeans equations. The advantage of the BME approach is twofold: first, it allows to follow perturbations growing from a very low perturbation amplitude up to the non-linear regime. Neither perturbation theory nor N-body simulations can do this, for different reasons. Secondly, observables like mean velocities or velocity dispersions can be calculated with a very high spatial resolution all over the disk.

We studied the evolution of an initially exponential disk which develops a two-armed spiral. Its vertex deviation shows a large spatial variation (Fig. 1). The values reach up to 90° in the central region. A strong variation of l_{v} has been found at the outer edges of the spiral structures: l_{v} can vary there from +40° to −40° within only a few kpc. The epicyclic approximation fails almost everywhere with respect to the vertex deviation (or the Oort ratio). For more details, see Vorobyov and Theis [24].

3 Test Particle Simulations

Our BEADS-2D code is based on the assumption of vanishing third order velocity moments. In the case of pressure-supported stellar systems like globular clusters third order terms are related to the heat flux which is controlled by two-body relaxation. Therefore, the zero-heat flux assumption (ZHFA) can be safely adopted on short timescales (e.g. a few dynamical timescales) due to long two-body relaxation timescales for stellar systems with more than 10^4 stars. However, for rotation-supported systems like disk galaxies the situation is less clear. Still, the two-body relaxation time is long, but now large-scale motions might result in non-negligible higher order moments.

In order to study the ZHFA, we performed test particle simulations for a galactic disk. We measured the velocity moments up to fifth order at different galactocentric radii. Assuming a constant pattern speed Ω_s we solved for the

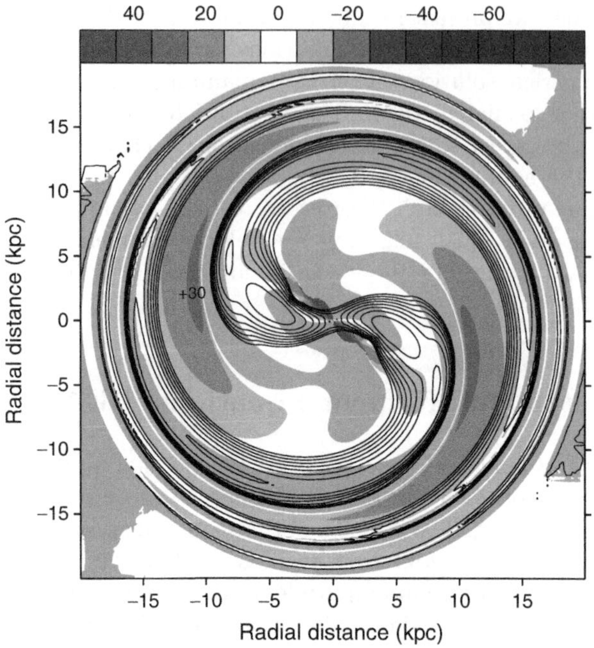

Fig. 1. Positive stellar density perturbation superimposed on the vertex deviation map at $t = 1.6\,\mathrm{Gyr}$. Positive and negative vertex deviations are shown in grey scale. The numbers indicate the maximum positive vertex deviation in the inter-arm region and maximum negative vertex deviation in the outer disk. The scale bar is in degrees. More details about this simulation as well as a color figure can be found in [24], Fig. 9

equations of motion of a set of test particles in a corotating frame in polar coordinates,

$$\ddot{R} = R\dot{\phi} - \frac{\partial \Phi}{\partial R} + 2R\dot{\phi}\Omega_s + \Omega_s^2 R, \qquad \ddot{\phi} = -2\frac{\dot{R}}{R}\dot{\phi} - \frac{1}{R^2}\frac{\partial \Phi}{\partial \phi} - 2\frac{\dot{R}}{R}\Omega_s\,.$$

R and ϕ describe the radial and azimuthal coordinate of the particle's trajectory, respectively. The gravitational potential $\Phi(\mathbf{r}, t)$ at position $\mathbf{r}$ and time t is split into two parts. The first part is a stationary, axisymmetric potential $\Phi_0(R)$ derived from a given rotation curve $v_c(R)$ with the galactocentric distance R. For a better comparison with other authors we chose the rotation curve suggested by Contopoulos and Grosbøl (cf. (2) in [5] and parameters therein). The second part of the gravitational potential was selected to be a time-dependent (rigidly rotating) two-armed spiral perturbation also suggested in [5]:

$$\Phi(R, \phi) = f(t) \cdot ARe^{-\epsilon_s R} \cdot \cos(m \ln R / \tan i_0 - 2\phi) \tag{2}$$

For our model we adopted a two-armed ($m = 2$) spiral with an amplitude $A = 200\,\mathrm{km}^2\,\mathrm{s}^{-2}\,\mathrm{kpc}^{-1}$, an inverse scale length $\epsilon_s = 0.1\,\mathrm{kpc}^{-1}$ and a pitch angle $i_0 = -30°$. We selected $\Omega_s = 12.5\,\mathrm{km\,s}^{-1}\,\mathrm{kpc}^{-1}$ which puts corotation

(CR) at a radius of about 23 kpc, the 1/4-resonance (UHR) at 12 kpc and the inner Lindblad resonance at 1.4 kpc. In order to avoid any initial kicks the perturbation was switched-on "adiabatically" on a timescale of 600 Myr. This procedure is attributed to the function $f(t)$ which we chose as in [4].

For our statistical analysis we adopted an approach similar to that of Blitz and Spergel [4]. Initially 2×10^6 stars were distributed in an exponential disk within a radial range from 4 to 40 kpc and a scale length of 10 kpc. The initial velocities were taken to be nearly circular (derived from the rotation curve) with a superimposed small perturbation: the radial velocity was derived from a Gaussian distribution with a dispersion of $20\,\mathrm{km\,s^{-1}}$. The azimuthal velocity perturbation was calculated from a Gaussian distribution, taking asymmetric drift and the ratio of the radial to the azimuthal velocity dispersions within the epicyclic approximation into account [4].

The velocity data are analysed on a polar grid ranging from 4 to 40 kpc with a radial grid spacing of 200 pc and an azimuthal spacing of 500 grid cells. Similar to [4] we measured the position of an orbit in intervals of 10^6 years and considered its contribution to the velocity moments of this cell. By this technique (which is based on the ergodicity assumption), a single orbit yields many contributions to the velocity moments, by this strongly increasing the sample size. E.g. a subset of about 2.3×10^5 particles hitting the radial range of 11.9–12.1 kpc yields about 3×10^7 contributions to the velocity moments in this annulus (measured within the last 2 Gyr of the simulation). In order to be able to detect evolutionary effects, we sampled usually the orbits in 10 different periods with a duration of 1 Gyr each.

4 Results

During the simulations we calculated the cumulative velocity moments $v_r^v v_\phi^w$ for each cell hit by a particle and for all sums $s = v + w$ of the exponents ranging from $s = 0$ up to $s = 5$. From the corresponding mean velocity moments we derived the corresponding velocity dispersion terms including higher order terms $\sigma_{ijk}, \sigma_{ijkl}$ and σ_{ijklm} with $i, j, k, l, m = r$ or ϕ like $\sigma_{rrr} \equiv \overline{(v_r - u_r)(v_r - u_r)(v_r - u_r)} = \overline{v_r^3} - 3\overline{v_r^2}u_r + 2u_r^3$ (with the mean radial velocity $u_r \equiv \overline{v_r}$). We controlled our dispersion term calculation by Mathematica.

Figure 2 shows the azimuthal distribution of the vertex deviation near the UHR (4:1 resonance) at 12 kpc. Similarly to our BEADS-2D result a four armed-structure is visible for the vertex deviation.[1] Already within the first Gyr four peaks are visible at about the final positions (a slight shift to larger angles can be seen for the second and the fourth peak). However, the absolute values of the vertex deviations vary considerably in time. It is interesting to note that the maximum values of l_v exceed the final values (at 10 Gyr) by about a factor of 2. After 6 Gyrs the final values are reached within 20%.

[1] Note: in the BEADS-2D model the UHR is at a radius of about 4.5 kpc.

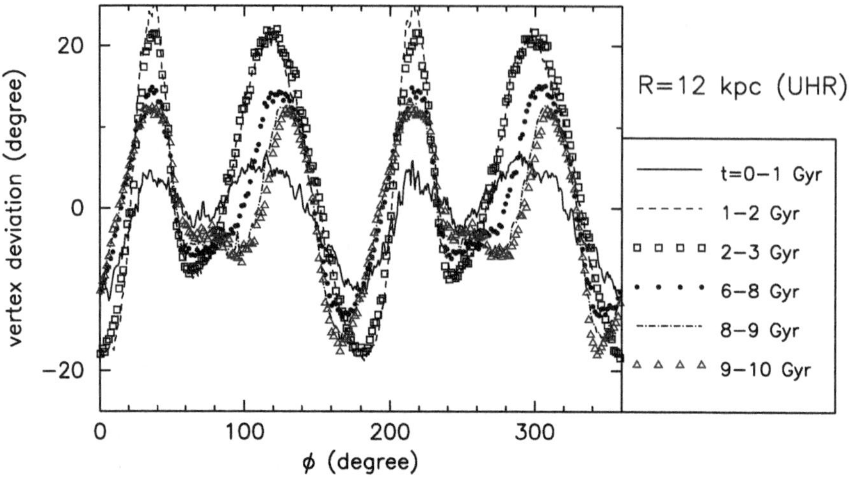

Fig. 2. Azimuthal distribution of the vertex deviation l_v sampled at different times at a galactocentric distance of 12 kpc (UHR) in a annulus of 200 pc thickness

This strong temporal variation is a caveat for test particle simulations: obviously the stellar system needs a long time (by far longer than 3 Gyr) to reach an equilibrium configuration. However, in reality, perturbations evolve on shorter timescales: some might be growing, others might be vanishing. Therefore matching absolute values derived from equilibrium configurations of test particle simulations to observational values can be misleading, because the stellar system might not have the time to establish an equilibrium. The latter strongly favors either self-consistent simulations (like the BEADS-2D approach) or non-equilibrium analyses of test particle models.

The applicability of the BEADS-2D models depends on the validity of the zero-heat flux approximation. Therefore, we calculated the third (and higher) order terms all over the disk. At the UHR most of the third order terms show a clear, well-defined fourfold structure characteristic for the 4:1 resonance region (e.g. [20]). Compared to the initial velocity dispersion (second order term) of $20\,\mathrm{km\,s^{-1}}$, most (normalized) values of the third order terms are smaller, but not necessarily negligible.

A view over the whole disk, however, shows that the third order terms are especially large at the UHR and become negligible elsewhere. E.g. Fig. 3 displays σ_{rrr} exhibiting clearly the four peaks associated with the 4:1 resonance. Beyond the UHR there are no significant third order terms discernible; it just looks like noise. Although one might speculate that some small structure can be identified by the broader "hills" at corotation, it is remarkable that corotation does not exhibit a more pronounced third order term. The strong peaks found at the lower and higher radial range are due to incomplete sampling: e.g. we did not consider stars inside 4 kpc which gives a bias in the velocity distribution near the inner edge.

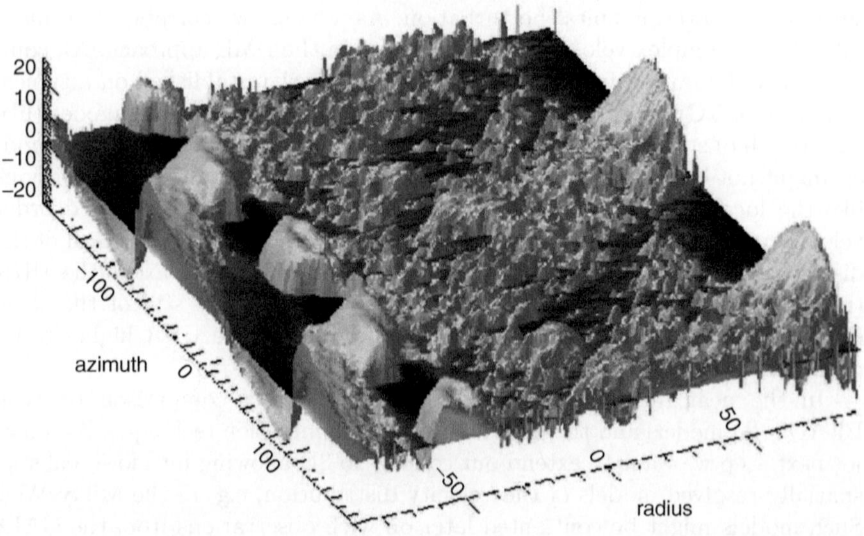

Fig. 3. Distribution of the normalized third order velocity dispersion term sign $(\sigma_{rrr}) \cdot (|\sigma_{rrr}|)^{1/3}$ in the galactic disks sampled in the period 9–10 Gyr. Both axes denote the indices of the corresponding polar coordinates. The radial index 0 corresponds to the corotation radius (23 kpc), whereas the index −90 denotes the inner edge of the disk (4 kpc). The UHR is at about −50. The z-axis is given in km s^{-1}; For better visualisation the data have been smoothed and filtered

It is also interesting to look for terms higher than third order. Near the UHR some of them become very large. E.g. $\sigma_{rr\phi\phi}$ reaches (normalized) peak values of 70 km s^{-1} and never drops below 20 km s^{-1}. Though other terms are smaller, they all are of the order of the initial velocity dispersion of the stars. Thus, there is no strong decay in magnitude for the higher order velocity dispersion terms near the UHR.

5 Summary and Future Plans

We analysed the structure of the velocity space in galactic disks which are subject to a spiral perturbation. In previous papers we used the Boltzmann moment equations up to second order in order to study the growth and saturation of spiral structure in self-gravitating stellar disks. From this analysis we derived the properties of the velocity ellipsoid all over the disk, namely the vertex deviations and the Oort ratios.

A key assumption of the BME approach is the zero-heat flux approximation, i.e. the neglection of third order velocity terms. We tested here this assumption by performing test particle simulations for stars in a disk galaxy subject to a rotating spiral perturbation. As a result we corroborated qualitatively the complex velocity structure found in the BME approach. An equilibrium configuration in velocity space is only slowly established on a typical timescale of 5 Gyrs or more. Since many dynamical processes in galaxies (like the growth of spirals or bars) act on shorter timescales, pure equilibrium models might not be fully appropriate for a detailed comparison with observations like the local Galactic velocity distribution. In our simulations third order velocity moments were typically small and uncorrelated over almost all of the disk with the exception of the 4:1 resonance region (UHR). Near the UHR (normalized) fourth and fifth order velocity moments are still of the same order as the second and third order terms. Thus, at the UHR higher order terms are not negligible.

In the near future we plan to do a more direct comparison between BEADS-2D models and the presented N-body simulation technique. As a major next step we want to extend our code(s) to 3D allowing for more realistic, spatially resolved models of the velocity distribution, e.g. in the Milky Way. Such models might be confronted later on with observations from the GAIA mission.

Acknowledgements

CT is very grateful to the organizers of the meeting for a very interesting and inspiring conference as well as for the financial support. Additionally, we would like to thank Christoph Lhotka for his advice in using Mathematica.

References

1. Alcobé S., Cubarsi R., 2005, A&A, 442, 929
2. Bienaymé O., 1999, A&A, 341, 86
3. Binney J., Merrifield M. S., 1998, *Galactic Astronomy*, Princeton University Press, Princeton
4. Blitz L., Spergel D.N., 1991, ApJ, 370, 205
5. Contopoulos G., Grosbøl P., 1986, A&A, 155, 11
6. Dehnen W., 1998, AJ, 115, 2384
7. Dehnen W., 2000, ApJ, 119, 800
8. Dehnen W., Binney J.J., 1998, MNRAS, 298, 387
9. Eddington A.S., 1906, MNRAS, 67, 34
10. ESA, 1997, The Hipparcos and Tycho Catalogues (ESA Sp-1200)
11. Hogg D.W., Blanton M.R., Roweis S.T., Johnston K.V., 2005, ApJ, 629, 268
12. Kapteyn J.C., 1905, Rep. Brit. Ass. Adv. Sci., p. 257
13. Kobold H., 1890, Astron. Nachr. 125, 65
14. Kuijken K., Tremaine S., 1991, in Proc. of *Dynamics of Disc Galaxies*, Göteborg, B. Sundelius (ed.), p. 71

15. Mayor M., 1970, A&A, 6, 60
16. Mayor M., 1972, A&A, 18, 97
17. Minchev I., Quillen A.C., 2007, MNRAS, 377, 1163
18. Minchev I., Nordhaus J., Quillen A.C., 2007, ApJ, 664, L31
19. Mühlbauer G., Dehnen W., 2003, A&A, 401, 975
20. Patsis P., 2006, MNRAS, 369, L56
21. Ratnatunga K.A., Upgren A.R., 1997, ApJ, 476, 811
22. Schwarzschild K., 1907, Nachr. Kgl. Ges. d. Wiss. zu Göttingen, Math. Phys. Klasse, 5, 614
23. Vorobyov E.I., Theis Ch., 2006, MNRAS, 373, 197 (VT06)
24. Vorobyov E.I., Theis Ch., 2008, MNRAS, 383, 817, cf. also astroph/0709.2768
25. Wielen R., 1974, Highlights of Astronomy, 3, 395
26. Yuan C., 1971, AJ, 76, 664

Dynamical Study of 2D and 3D Barred Galaxy Models

T. Manos[1,2] and E. Athanassoula[2]

[1] Department of Mathematics, Center for Research and Applications of Nonlinear Systems (CRANS), University of Patras, GR–26500, Greece
thanosm@master.math.upatras.gr

[2] LAM, OAMP, 38, rue Frederic Joliot-Curie, 13388 Marseille cedex 13, France
lia@oamp.fr

Summary. We study the dynamics of 2D and 3D barred galaxy analytical models, focusing on the distinction between regular and chaotic orbits with the help of the Smaller ALigment Index (SALI), a very powerful tool for this kind of problems. We present briefly the method and we calculate the fraction of chaotic and regular orbits in several cases. In the 2D model, taking initial conditions on a Poincaré (y, p_y) surface of section, we determine the fraction of regular and chaotic orbits. In the 3D model, choosing initial conditions on a cartesian grid in a region of the (x, z, p_y) space, which in coordinate space covers the inner disc, we find how the fraction of regular orbits changes as a function of the Jacobi constant. Finally, we outline that regions near the (x, y) plane are populated mainly by regular orbits. The same is true for regions that lie either near to the galactic center, or at larger relatively distances from it.

1 Introduction

The dynamical evolution of galactic systems depends crucially on their orbital structure and in particular on what fraction of their orbits is regular, or chaotic. Thus to permit further studies, it is essential to be able to distinguish between these two types of orbits in a manner that is both safe and efficient. This is not trivial and becomes yet more complicated in systems of many degrees of freedom. A summary of the methods that have been developed over the years can be found in [1].

In the present paper we use a method based on the properties of two deviation vectors of an orbit, the "Smaller ALingment Index" (SALI) [2]. It has been applied successfully in different dynamical systems [2–12], frequently also under the name Alignment Index (AI) [13–17] and has been shown to be a fast and easy to compute indicator of the chaotic or ordered nature of orbits. We first recall its definition and we then show its effectiveness in

G. Contopoulos, P.A. Patsis, *Chaos in Astronomy*. Astrophysics and Space Science Proceedings, doi: 10.1007/978-3-540-75826-6,

distinguishing between ordered and chaotic motion by applying it to a barred potential of 2 and 3 degrees of freedom. Recently, a generalization of the SALI, the "Generalized ALignment Index" (GALI) was introduced by Skokos et al. (2007) [18], which includes the full set of the k initially linearly independent deviation vectors of the system to determine if an orbit is chaotic or not.

2 Definition of the Smaller Aligment Index (SALI)

Let us consider the n-dimensional phase space of a conservative dynamical system, which could be a symplectic map or a Hamiltonian flow. We consider also an orbit in that space with initial condition $P(0) = (x_1(0), x_2(0), \dots, x_n(0))$ and two deviation vectors $\mathbf{w}_1(0)$, $\mathbf{w}_2(0)$ from the initial point $P(0)$. In order to compute the SALI for a given orbit one has to follow the time evolution of the orbit itself, as well as two deviation vectors $\mathbf{w}_1(t), \mathbf{w}_2(t)$ which initially point in two different directions. At every time step the two deviation vectors $\mathbf{w}_1(t)$ and $\mathbf{w}_2(t)$ are normalized by setting:

$$\hat{w}_i(t) = \frac{\mathbf{w}_i(t)}{\|\mathbf{w}_i(t)\|}, \quad i = 1, 2 \tag{1}$$

and the SALI is then computed as

$$SALI(t) = min\left\{\|\hat{w}_1(t) + \hat{w}_2(t)\|, \|\hat{w}_1(t) - \hat{w}_2(t)\|\right\}. \tag{2}$$

The properties of the time evolution of the SALI clearly distinguish between regular and chaotic motion as follows: In the case of Hamiltonian flows or dimensional symplectic maps with $n \geq 2$, the SALI fluctuates around a non-zero value for regular orbits [2, 3]. In general, two different initial deviation vectors become tangent to different directions on the torus, producing different sequences of vectors, so that SALI does not tend to zero but fluctuates around positive values. On the other hand, for chaotic orbits SALI tends exponentially to zero. Any two initially different deviation vectors tend to coincide in the direction defined by the nearby unstable manifold and hence either coincide with each other, or become opposite.

3 The Model

A 3D rotating model of a barred galaxy can be described by the Hamiltonian function:

$$H = \frac{1}{2}(p_x^2 + p_y^2 + p_z^2) + V(x, y, z) - \Omega_b(xp_y - yp_x). \tag{3}$$

The bar rotates around its z axis, while the x axis is along its major axis and the y axis is along its intermediate axis. The p_x, p_y and p_z are the canonically conjugate momenta. Finally, V is the potential, Ω_b represents the pattern speed of the bar and H is the total energy of the system in the rotating frame of reference (Jacobi constant). The corresponding equations of motion are

$$\dot{x} = p_x + \Omega_b y, \qquad \dot{y} = p_y - \Omega_b x, \qquad \dot{z} = p_z, \tag{4}$$
$$\dot{p_x} = -\frac{\partial V}{\partial x} + \Omega_b p_y, \qquad \dot{p_y} = -\frac{\partial V}{\partial y} - \Omega_b p_x, \qquad \dot{p_z} = -\frac{\partial V}{\partial z}.$$

The equations of the evolution of the deviation vectors and the calculation of the SALI are given by the corresponding variational equations.

The potential V of our model consists of three components:

1. A *disc*, represented by a Miyamoto potential [19]:

$$V_D = -\frac{GM_D}{\sqrt{x^2 + y^2 + (A + \sqrt{z^2 + B^2})^2}}, \tag{5}$$

where M_D is the total mass of the disc, A and B are the horizontal and vertical scale lengths and G is the gravitational constant.

2. A *bulge*, which is modeled by a Plummer sphere whose potential is

$$V_S = -\frac{GM_S}{\sqrt{x^2 + y^2 + z^2 + \epsilon_s^2}}, \tag{6}$$

where ϵ_s is the scale length of the bulge and M_S is its total mass.

3. A triaxial Ferrers *bar*, the density $\rho(x)$ of which is

$$\rho(x) = \begin{cases} \rho_c(1 - m^2)^2 & m < 1, \\ 0 & m \geq 1, \end{cases} \tag{7}$$

where $\rho_c = \frac{105}{32\pi}\frac{GM_B}{abc}$ is the central density, M_B is the total mass of the bar and

$$m^2 = \frac{x^2}{a^2} + \frac{y^2}{b^2} + \frac{z^2}{c^2}, \qquad a > b > c > 0, \tag{8}$$

with a, b and c being the semi-axes. The corresponding potential is

$$V_B = -\pi Gabc \frac{\rho_c}{n+1} \int_\lambda^\infty \frac{du}{\Delta(u)} (1 - m^2(u))^{n+1}, \tag{9}$$

where

$$m^2(u) = \frac{x^2}{a^2 + u} + \frac{y^2}{b^2 + u} + \frac{z^2}{c^2 + u}, \tag{10}$$

$$\Delta^2(u) = (a^2 + u)(b^2 + u)(c^2 + u), \tag{11}$$

n is a positive integer (with $n = 2$ for our model) and λ is the unique positive solution of

$$m^2(\lambda) = 1, \tag{12}$$

outside of bar ($m \geq 1$) and $\lambda = 0$ inside the bar.

This model has been used extensively for orbital studies [20–25] and we will refer to it hereafter as the Fererrs model. We adopt the following values of parameters: $G = 1$, $\Omega_b = 0.054$ ($54 \, \frac{Km}{sec \cdot Kpc}$), $a = 6$, $b = 1.5$, $c = 0.6$, $A = 3$, $B = 1$, $\epsilon_s = 0.4$, $M_B = 0.1$, $M_S = 0.08$, $M_D = 0.82$. The units we use, are: 1 kpc (length), 1 Myr (time) and $2 \times 10^{11} M_{\odot}$ solar masses (mass). The total mass $G(M_S + M_D + M_B)$ is set to be equal to 1.

4 Results in the 2D and 3D Ferrers Model

The 2D Ferrers model is a subcase of the general 3D one and it can be described by the Hamiltonian equation (3) by setting $(z, p_z) = (0, 0)$:

$$H = \frac{1}{2}(p_x^2 + p_y^2) + V(x, y) - \Omega_b(xp_y - yp_x). \quad (13)$$

Fixing $H = -0.335$, $x = 0$, we chose 50,000 initial conditions on the (y, p_y)-plane of the Poincaré surface of section (PSS), while $p_x = H(x, y, p_y)$ and we calculated the final values of the SALI for $t = 5{,}000$. We were able to detect very small regions of instability that can not be visualized easily by the PSS method. In Fig. 1, on the left panel, we have plotted the PSS for this Hamiltonian value. On the right panel we attributed to each grid point a color according to the value of the SALI at the end of the evolution. The

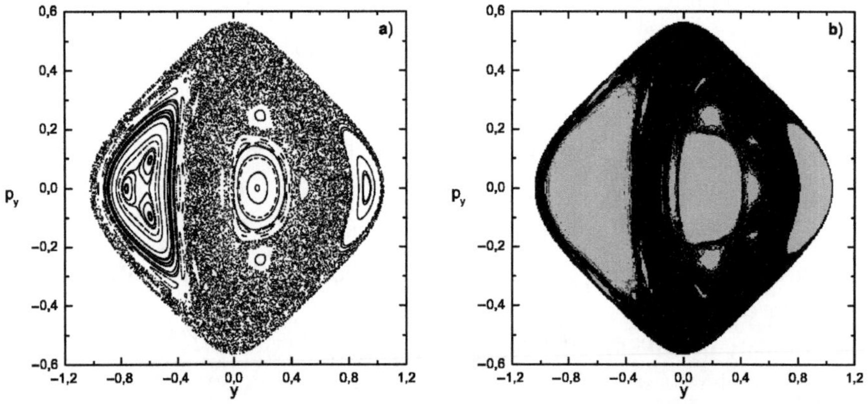

Fig. 1. Agreement between the results of the Poincaré surface of section (PSS) and SALI. (**a**) The Poincaré surface of section for the 2D Ferrers model and $H = -0.335$. (**b**) Regions of different values of the SALI for 50,000 initial conditions on the (y, p_y)-plane integrated up to $t = 5{,}000$ for the same value of the Hamiltonian. The *light grey colored areas* correspond to regular orbits, while the *dark black ones* to chaotic. Few *gray and dark gray colored points* correspond to "sticky" orbits. Note the excellent agreement between the two methods as far as the gross features are concerned, as well as the fact the SALI can easily pick out small regions of stability which the PSS has difficulties detecting

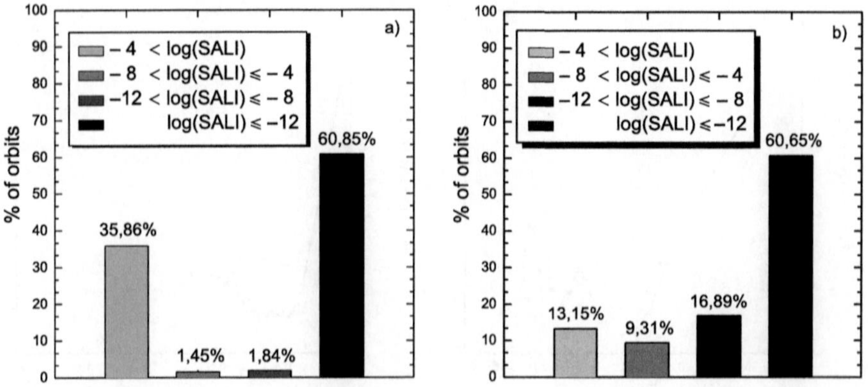

Fig. 2. (**a**) Percentages of regular (*light gray bar*), chaotic (*black bar*) and sticky orbits (two intermediate bars – *gray and dark gray bars*) for the 2D Ferrers model from a mesh of 50,000 orbits on the (y, p_y) plane of Fig. 1 and final time of integration $t = 5{,}000$. (**b**) The percentages for 3D, using 50,000 orbits and the same classification. The initial conditions are taken as described in the text

light grey color corresponds to regular orbits and to the areas that host them while the black color represents the chaotic ones. The intermediate shades of gray color represent the so-called "sticky" orbits, i.e. orbits whose nature is chaotic but need more time to show their behavior (weak-chaotic orbits). The distinction between them is done by measuring SALI at $t = 5,000$: orbits with SALI$\leq 10^{-12}$ correspond to strongly chaotic, orbits with SALI$> 10^{-4}$ to regular, while orbits with $10^{-8} <$SALI$\leq 10^{-4}$ and $10^{-12} <$SALI$\leq 10^{-8}$ are "sticky". In Fig. 2a we present the percentages of orbits from this PSS, according to this classification. We find that $\simeq 37,3\%$ of the orbits on this PSS are regular.

For the 3D case of the model we used a sample of 50,000 orbits, equally spaced on a cartesian grid in the (x, z, p_y) space, with $x \in [0.0, 7.0]$, $z \in [0.0, 1.5]$ and $p_y \in [0.0, 0.45]$, while $(y, p_x, p_z) = (0, 0, 0)$. In this way, we attempted to create initial conditions that could support the bar and be mainly trapped by the x_1 tree, i.e. the x_1 family and the x_1v_i, $i = 1, \ldots$ families that bifurcate from it and extend vertically well above the disc region [21]. Although these initial conditions cover all the available energy interval, they are not spread uniformly over it. Note also that a few of these have an energy value beyond the escape energy, and we dismiss them. These, however, constitute less than 0.5% of the total, so that they influence very little our study and the statistics. In Fig. 2b we present the corresponding percentages of regular, chaotic and sticky orbits. We find that, using these initial conditions and parameters, the phase space of the model is dominated by chaotic orbits, since 77.54% of them are chaotic.

In order to check the variation of the percentages of regular and chaotic orbits as the energy of the model varies, we did the following: We first sorted

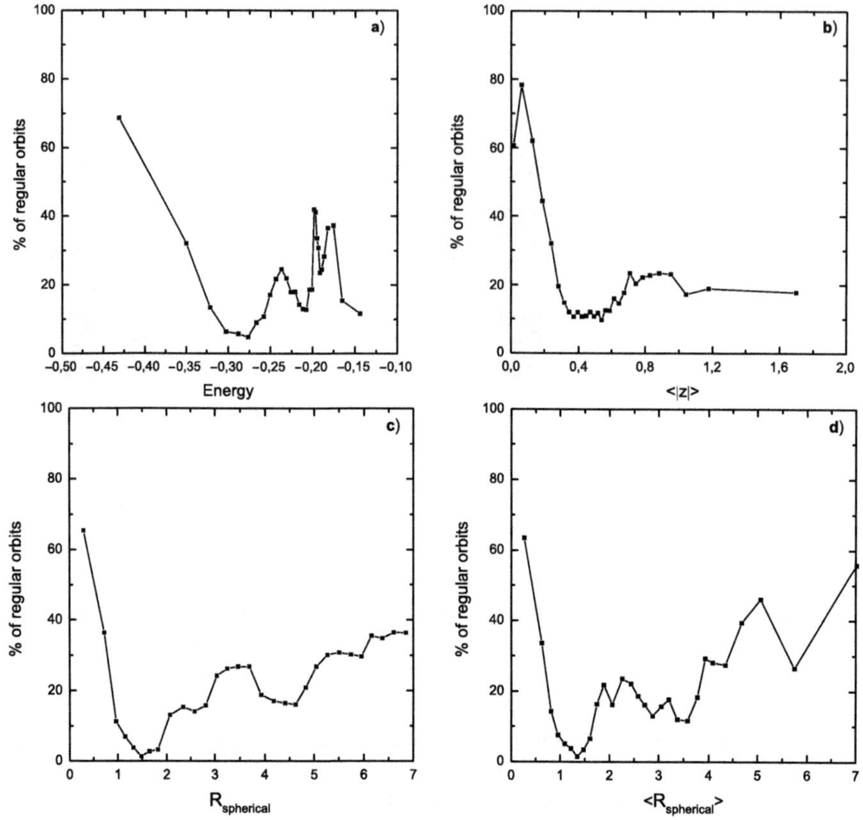

Fig. 3. Percentages of regular orbits as a function of: the Jacobi constant (**a**), the absolute value of their z coordinate, $<| z |>$, mean z value over their orbital evolution (**b**), their initial spherical radius, $R_{spherical}$ (**c**) and $< R_{spherical} >$, mean spherical radius over their motion (**d**)

the energy values for all the initial conditions. Then, we created 30 energy intervals containing equal number of orbits, since our way of giving initial conditions does not imply their uniform distribution in the total energy interval. In every energy interval we calculated the percentages of regular and chaotic orbits, considering as chaotic all the orbits with SALI$< 10^{-8}$. In Fig. 3a we show the percentage of regular orbits in each energy interval, as a function of the mean energy in that interval. Generally, the percentage of regular orbits decreases as the energy increases, but before and after the escape energy (where the Jacobi constant value is $H \simeq -0.20$) there are two peaks. This non-monotonic behavior is related to the appearance or disappearance of stable periodic orbits in the phase space and the size variation of the stability regions around them.

We also attempted to explore the way that regular and chaotic orbits are distributed along the z-direction of the configuration space. Following the evolution of each orbit, we calculated the mean of the absolute value of their z coordinate ($<|\ z\ |>$). Then, we divided the available $<|\ z\ |>$-interval in 30 slices with equal number of orbits in each one of them. This restriction gives us better samples for the estimation of the percentages, implying at the same time that these slices are not equally sized necessarily. For every slice separately we calculated the fraction of regular orbits and in Fig. 3b we plot these percentages as a function of the $<|\ z\ |>$ in that slice. It reveals that the slices "near" the (x, y)-plane ($<| z |> \ < 0.35$) contain mainly "regular" orbits. Contrarily, slices for larger values of z host mainly chaotic motion.

Furthermore, we looked at these percentages as a function of the initial spherical radius ($R_{spherical}$) and the mean spherical radius over the evolution ($< R_{spherical} >$). Again, dividing in 30 slices the total range of the $R_{spherical}$, in a similar manner with the $<|\ z\ |>$, we calculated the percentages of regular orbits. We plot this for every slice, as a function of the mean $R_{spherical}$ of that slice. We see that the fraction of regular orbits decreases strongly with increasing $R_{spherical}$ up to $R_{spherical} < 1.5$ where it reaches a minimum, while for $1.5 < R_{spherical} < 7$ this percentage starts increasing gradually. This result is in good agreement with the results in Fig. 3d, where the horizontal axis corresponds to the value of the mean spherical radius over time during the evolution.

5 Conclusions

We used SALI to study the dynamical behavior of Hamiltonian models of 2D and 3D barred galaxies. We found that in both cases there is a significant amount of chaotic orbits. In the 2D model, we were able to chart a subspace of the phase space, to identify rapidly even tiny regions of regular motion and measure their percentages. In the 3D model, apart from computing the global percentages of regular and chaotic orbits, we calculated these percentages as a function of the energy and found that low values of the energy are mainly dominated by "regular" orbital motion. We also followed the distribution of the chaotic and regular orbits in the configuration space and we found that orbits which lie near the (x, y)-plane with relatively small mean deviations in z-direction are generally regular. Finally, we monitored the variation of their percentages as a function of their initial spherical radius and their mean spherical radius. We find that the fraction of regular orbits is dominant in regions near the center, as well as at relatively larger distances from it.

Acknowledgments

T. Manos was partially supported by the "Karatheodory" graduate student fellowship No B395 of the University of Patras, the program "Pythagoras II" and the Marie Curie fellowship No HPMT-CT-2001-00338. We acknowledge financial support from grant ANR-06-BLAN-0172.

References

1. Contopoulos, G., 2002, Order and chaos in dynamical astronomy, Springer, Berlin
2. Skokos Ch., 2001, J. Phys. A: Math. Gen., 34, 10029
3. Skokos Ch., Antonopoulos Ch., Bountis T. and Vrahatis M., 2003, Prog. Theor. Phys. Suppl., 150, 439
4. Skokos Ch., Antonopoulos Ch., Bountis T. and Vrahatis M., 2004, J. Phys. A, 37, 6269
5. Panagopoulos P., Bountis T. and Skokos Ch., 2004, J. Vib. & Acoust., 126, 520
6. Antonopoulos Ch. and Bountis T., 2006, Phys. Rev. E, 73, 056206
7. Antonopoulos Ch., Bountis T. and Skokos Ch., 2006, Int. J. Bif. Chaos, 16(6), 1777
8. Bountis T. and Skokos Ch., 2006, Phys. Lett. A, 358, 126
9. Manos T. and Athanassoula E., 2005, In: SF2A-2005: Semaine de l'Astrophysique Francaise, edited by Casoli F., Contini T., Hameury J.M. and Pagani L., Edp-Sciences, Conference Series, 631
10. Manos T. and Athanassoula E., 2005, In: 5th International Cosmology Conference: The Fabulous destiny of galaxies: Bridging past and present, edited by Le Brun V., Mazure A., Arnouts S., Burgarella D., Frontier Group
11. Manos T. and Athanassoula E., 2006, In: AIP Conference Proceedings, Recent Advances in Astronomy and Astrophysics: 7th International Conference of the Hellenic Astronomical Society, Vol. 848, 662
12. Manos T. Skokos Ch., Athanassoula E. and Bountis T., 2007, In: 19th Panhellenic Conference/Summer School "Nonlinear Science and Comlexity", Thessaloniki, Greece, (nlin/0703037)
13. Voglis N., Kalapotharakos C. and Stavropoulos I., 2002, MNRAS, 337, 619
14. Voglis N., Kalapotharakos C. and Stavropoulos I., 2006, MNRAS, 372, 901
15. Voglis N., Harsoula M. and Contopoulos G, 2007, MNRAS, 381, 757
16. Kalapotharakos C., Voglis N. and Contopoulos G., 2004, MNRAS, 428, 905
17. Kalapotharakos C., Efthymiopoulos C. and Voglis N., 2008, MNRAS, 383, 971
18. Skokos Ch., Bountis T. and Antonopoulos Ch., 2007, Physica D, 231, 30
19. Miyamoto M. and Nagai R., 1975, PASJ, 27, 533
20. Pfenniger D., 1984, A&A, 134, 373
21. Patsis P. A., Skokos Ch. and Athanassoula E., 2002, MNRAS, 337, 578
22. Patsis P. A., Skokos Ch. and Athanassoula E., 2003, MNRAS, 342, 69
23. Patsis P. A., Skokos Ch. and Athanassoula E., 2003, MNRAS, 346, 1031
24. Skokos Ch., Patsis P. A. and Athanassoula E., 2002, MNRAS, 333, 847
25. Skokos Ch., Patsis P. A. and Athanassoula E., 2002, MNRAS, 333, 861

Chaos in the Mergers of Galaxies

P.O. Vandervoort

Department of Astronomy and Astrophysics, The University of Chicago,
5640 S. Ellis Avenue, Chicago, IL 60637-1433, USA
voort@oddjob.uchicago.edu

Summary. This paper begins with an account of a current investigation of chaos in the mergers of galaxies. It concludes with the formulation of a conjecture regarding chaotic decay of stationary oscillations in galaxies formed by mergers.

1 Introduction

In the workshop "Galaxies and Chaos. Theory and Observations," which was held in Athens in 2002, Henry Kandrup argued that chaotic mixing is the fundamental mechanism that underlies the violent relaxation of a galaxy to equilibrium [1]. In particular, he described a scenario of *transient chaos*, an episode in the evolution of a galaxy in which the prevailing gravitational field would be time-dependent, a substantial population of stellar orbits would be chaotic, and chaotic mixing would drive the approach of the system to equilibrium. At equilibrium, the field would be time-independent, most orbits would be regular, and chaotic mixing would be suppressed. Kandrup and his collaborators, most recently Ileana Vass and Ioannis Sideris [2] and Balša Terzić [3], found support for this scenario in their investigations of the chaotic behavior of orbits in potentials that undergo damped oscillations.

More recently, Monica Valluri, Ileana Vass, Stelios Kazantzidis, Andrey Kravtsov, and Courtlandt Bohn analyzed N-body simulations of mergers of galaxies in order to investigate the mechanisms responsible for driving the mixing of stellar orbits in phase space and the evolution of merged systems to equilibrium [4]. Those authors recognized that chaotic mixing might be one possible mechanism of relaxation. In their simulations, however, they found that the relaxation of a pair of merging galaxies occurs mostly during their pericenter passages and that relaxation is mainly a consequence of tidal shocking and redistribution of energy by dynamical friction. Valluri et al. concluded that chaotic mixing does not contribute significantly to relaxation.

G. Contopoulos, P.A. Patsis, *Chaos in Astronomy.* Astrophysics and Space Science Proceedings, doi: 10.1007/978-3-540-75826-6,

2 A Model of Chaos in the Mergers of Galaxies

The use of N-body calculations has become customary in the study of chaotic behavior and other aspects of the dynamics of stellar systems, and that approach is, of course, quite reasonable. However, a more elementary approach to the study of chaotic behavior might also yield significant insights. For it often happens in dynamics that chaotic behavior in a system with many degrees of freedom is well represented by a model with only a few degrees of freedom. In galactic dynamics, the virial equations provide a model of a many-body system with only a few degrees of freedom. The work described in what follows is a study of chaotic behavior in a model of mergers of galaxies in which the internal dynamics of the galaxies is described by the tensor virial equations.

2.1 Construction of the Model

We consider mergers of collisionless stellar systems. The governing equations of the present model can be derived from the collisionless Boltzmann equation, or they can be constructed with the aid of the equations of the gravitational many-body problem.

We let $\hat{M}$ and $\tilde{M}$ denote the masses of two merging galaxies, and we describe the system in terms of the separation $\mathbf{x}$ of their centers of mass and coordinates of stars $\hat{\mathbf{r}}$ and $\tilde{\mathbf{r}}$, respectively, relative to those centers of mass. Here and in what follows, we make use of a circumflex ^ and a tilde ~ in order to label quantities that refer to the more massive galaxy and less massive galaxy, respectively.

The relative motion of the centers of mass of the galaxies is governed by the equation

$$\mu \frac{\mathrm{d}^2\mathbf{x}}{\mathrm{d}t^2} = -\frac{\partial W}{\partial \mathbf{x}} , \tag{1}$$

where μ is their reduced mass, and

$$W(\mathbf{x}, t) = -G \int_{\hat{V}} \int_{\tilde{V}} \frac{\hat{\rho}(\hat{\mathbf{r}}, t)\tilde{\rho}(\tilde{\mathbf{r}}, t)}{|\mathbf{x} + \hat{\mathbf{r}} - \tilde{\mathbf{r}}|} \mathrm{d}\hat{\mathbf{r}}\mathrm{d}\tilde{\mathbf{r}} \tag{2}$$

is the gravitational potential energy of their mutual interaction. We let $\hat{\rho}(\hat{\mathbf{r}}, t)$ and $\tilde{\rho}(\tilde{\mathbf{r}}, t)$ denote the densities in the two galaxies.

The system admits of the classical integrals of the motion of the gravitational many-body problem. The energy integral

$$E = \frac{1}{2}\mu \left|\frac{\mathrm{d}\mathbf{x}}{\mathrm{d}t}\right|^2 + W + \hat{T} + \hat{W} + \tilde{T} + \tilde{W} \tag{3}$$

is composed of the kinetic energy of the relative motion of the galaxies, the gravitational potential energy of their mutual interaction, the internal kinetic energies $\hat{T}$ and $\tilde{T}$ of the two galaxies, and their gravitational potential energies $\hat{W}$ and $\tilde{W}$. Likewise, the total angular momentum integral

$$\mathbf{L} = \mu \mathbf{x} \times \frac{\mathrm{d}\mathbf{x}}{\mathrm{d}t} + \hat{\mathbf{L}} + \tilde{\mathbf{L}} \tag{4}$$

is composed of the orbital angular momentum of the two galaxies and their spin angular momenta $\hat{\mathbf{L}}$ and $\tilde{\mathbf{L}}$. The remaining classical integrals describe the motion of the center of mass of the entire system. We are writing the governing equations for the present model in the frame of reference in which that center of mass is at rest, so the remaining integrals are already implicit in the construction of the model.

The tensor virial equations that describe the internal dynamics of the more massive galaxy are of the form

$$\frac{1}{2} \frac{\mathrm{d}^2 \hat{I}_{ij}}{\mathrm{d}t^2} = 2\hat{T}_{ij} + \hat{W}_{ij} + \hat{V}_{ij} \ , \tag{5}$$

where $\hat{I}_{ij}$, $\hat{T}_{ij}$ and $\hat{W}_{ij}$ are, respectively, the moment of inertia tensor, the kinetic energy tensor, and the gravitational potential energy tensor of the galaxy [5–7]. Equation (5) also includes the tensor virial

$$\begin{aligned} \hat{V}_{ij} = \ & \tfrac{1}{2} G \int_{\hat{V}} \int_{\tilde{V}} \frac{\hat{\rho}(\hat{\mathbf{r}},t)\tilde{\rho}(\tilde{\mathbf{r}},t)\hat{r}_i(\tilde{r}_j - \hat{r}_j - x_j)}{|\mathbf{x} + \hat{\mathbf{r}} - \tilde{\mathbf{r}}|^3} \mathrm{d}\hat{\mathbf{r}}\mathrm{d}\tilde{\mathbf{r}} \\ & + \tfrac{1}{2} G \int_{\hat{V}} \int_{\tilde{V}} \frac{\hat{\rho}(\hat{\mathbf{r}},t)\tilde{\rho}(\tilde{\mathbf{r}},t)\hat{r}_j(\tilde{r}_i - \hat{r}_i - x_i)}{|\mathbf{x} + \hat{\mathbf{r}} - \tilde{\mathbf{r}}|^3} \mathrm{d}\hat{\mathbf{r}}\mathrm{d}\tilde{\mathbf{r}} \end{aligned} \tag{6}$$

of the gravitational attraction of the less massive galaxy. The tensor virial equations that govern the internal dynamics of the less massive galaxy are obtained by interchanging the circumflex and the tilde in (5) and (6) and the accompanying text. The virials $\hat{V}_{ij}$ and $\tilde{V}_{ij}$ are the sources of tidal distortion of the galaxies in the virial equations, and those terms couple the internal motions of the two galaxies.

For the sake of brevity, we have omitted expressions for the energies and their tensor generalizations, the moment of inertia tensors, and the spin angular momenta in the presentation of (3)–(5) above. The definitions of those quantities are well known [5–7], and explicit expressions in terms of $\mathbf{x}$, $\hat{\mathbf{r}}$, and $\tilde{\mathbf{r}}$ are readily constructed with the aid of those definitions.

For applications of the virial equations, we must know the density distributions in the two galaxies in order to evaluate the moment of inertia tensors, the gravitational potential energy tensors, and the tensor virials. We must also know enough about the stellar kinematics to be able to evaluate the kinetic energy tensors. In other words, we must adopt models for the density distributions and kinetic energy tensors of the two galaxies in order to complete the system of governing equations.

We model each galaxy as a heterogeneous ellipsoid with a Gaussian stratification of the density

$$\rho(\mathbf{r}) = \frac{M}{\pi^{3/2} a_1 a_2 a_3} \exp(-m^2) \qquad \left(m^2 = \sum_{k=1}^{3} \frac{{r_k}^2}{{a_k}^2} \right) \ , \tag{7}$$

where a_1, a_2, and a_3 are the semi-axes of an ellipsoid that provides a measure of the characteristic size and shape of the galaxy represented, and we have resolved the position $\mathbf{r}$ into Cartesian components r_1, r_2, and r_3 aligned with those principal axes. (In applications more general than what is described below, it would be necessary to allow for an arbitrary orientation of the principal axes with respect to the coordinate axes.) We label quantities in (7) with a circumflex or a tilde, according as that equation represents the density of the more massive or less massive galaxy. We require that the evolution of the system preserve the functional form of the density distributions. That requirement reduces the virial equations to equations of motion for the semi-axes of the characteristic ellipsoids.

In the model of the kinetic energy tensors, we require that the distribution of the peculiar velocities of the stars be isotropic in each galaxy. That requirement reduces the kinetic energy tensors to $\hat{T}_{ij} = \delta_{ij}\hat{T}/3$ and $\tilde{T}_{ij} = \delta_{ij}\tilde{T}/3$. Now (3) determines $\hat{T} + \tilde{T}$ in terms of E, $\mathbf{x}$, $\mathrm{d}\mathbf{x}/\mathrm{d}t$, and the semi-axes of the ellipsoids that characterize the galaxies. In order to determine $\hat{T}$ and $\tilde{T}$ separately, we have adopted an *ad hoc* prescription that incorporates two conditions. (1) At large separations of their centers of mass, the galaxies are mutually isolated and their energies $\hat{E} = \hat{T} + \hat{W}$ and $\tilde{E} = \tilde{T} + \tilde{W}$ are conserved separately. (2) At small separations of the centers of mass, changes in the kinetic energy of each galaxy are consistent with an adiabatic condition that can be derived from the equations of stellar hydrodynamics (i.e., the Jeans equations).

2.2 Capture of a Dwarf Galaxy by a Giant Galaxy

The simplest version of the present model of a merger describes a head-on encounter of galaxies with very different masses. The present calculations are "free-fall experiments" in which initially non-rotating, spherical galaxies drop into each other from rest at assigned, initial separations of their centers of mass. In other words, at time $t = 0$ in a typical experiment, the galaxies are spherical systems with characteristic radii $\hat{a}_1 = \hat{a}_2 = \hat{a}_3 = \hat{a}_0$ and $\tilde{a}_1 = \tilde{a}_2 = \tilde{a}_3 = \tilde{a}_0$, and they are at rest with their centers of mass at a separation r_0. During such a merger, the relative motion of the centers of mass is rectilinear, and the two galaxies remain axisymmetric with respect to the line connecting their centers of mass. In particular, the orbital angular momentum and the spin angular momenta on the right-hand side of (4) all vanish. The governing equations of this model of a merger describe an oscillator in five degrees of freedom in a coordinate x_3 describing the separation of the centers of mass and the characteristic semi-axes $\hat{a}_1 = \hat{a}_2 \neq \hat{a}_3$ and $\tilde{a}_1 = \tilde{a}_2 \neq \tilde{a}_3$ of the two galaxies.

By assumption, the dwarf galaxy is much smaller and much less massive than the giant, and it is a consequence of that assumption that the tidal distortion of the giant by the dwarf is slight. Accordingly, we solve the governing equations in the lowest orders of approximation in the quantities

$$\frac{\tilde{a}_0}{\hat{a}_0} \ll 1 \quad \text{and} \quad \left|1 - \frac{\hat{a}_3}{\hat{a}_1}\right| \ll 1 . \tag{8}$$

We include effects of dynamical friction in the model in the prescription introduced by Tremaine, Ostriker, and Spitzer [8]. Accordingly, we replace (1) with

$$\mu \frac{\mathrm{d}^2\mathbf{x}}{\mathrm{d}t^2} = -\frac{\partial W}{\partial \mathbf{x}} + \mathbf{F}^{(\mathrm{Drag})} , \tag{9}$$

where

$$\mathbf{F}^{(\mathrm{Drag})} = -16\pi^2 G^2 \hat{m}_* (\tilde{M} + \hat{m}_*) \ln \Lambda \frac{\mathbf{V}}{|\mathbf{V}|^3} \tilde{M} \int_0^{|\mathbf{V}|} \hat{f}(-\mathbf{x}, |\mathbf{u}|)|\mathbf{u}|^2 \mathrm{d}|\mathbf{u}| , \tag{10}$$

$\hat{m}_*$ is the mass of a typical star in the giant galaxy, Λ is the ratio of the maximum and minimum impact parameters for encounters of such stars with the dwarf galaxy, $\mathbf{V} = \mathrm{d}\mathbf{x}/\mathrm{d}t$, and the distribution function $\hat{f}$ is the density of stars in the phase space of a single star in the giant galaxy. The drag force $\mathbf{F}^{(\mathrm{Drag})}$ is derived by treating the dwarf galaxy as a Brownian particle interacting with a background of field stars, which are the stellar constituents of the giant galaxy. In the evaluation of Λ we let the maximum and minimum impact parameters be equal to the initial radii of the giant and dwarf, respectively. In the integral on the right-hand side of (10), we model the distribution function $\hat{f}$ as the product of the number density $\hat{\rho}/\hat{m}_*$ and a Maxwellian distribution of the velocities with a dispersion assigned consistently with the virial theorem for the equilibrium of the giant galaxy.

In the present calculations, the adopted units of mass, length, and time are

$$M = \hat{M} + \tilde{M} , \quad \hat{a}_0 , \quad \text{and} \quad \left[\frac{G(\hat{M} + \tilde{M})}{\hat{a}_0^3}\right]^{-1/2} , \tag{11}$$

respectively. The unit of time is essentially a dynamical time defined in terms of the units of mass and length.

Our criterion for chaotic behavior is sensitivity to initial conditions. For a given pair of systems with slightly different initial conditions, our measure of sensitivity is the growth of the separation of the representative points of the two systems in an appropriate phase space. We study the growth of the separation

$$\xi = \left[(\tilde{a}_1^{(1)} - \tilde{a}_1^{(2)})^2 + (\tilde{a}_3^{(1)} - \tilde{a}_3^{(2)})^2\right]^{1/2} \tag{12}$$

of pairs of systems in the phase plane defined by the characteristic semi-axes of the dwarf galaxy, where the superscripts (1) and (2) distinguish evolutions with slightly different initial conditions. If the growth is exponential in time, then the motion is judged to be chaotic. If the growth is no more than algebraic in time, then the motion is judged to be regular. We have concentrated on a search for chaotic behavior in the oscillations of the dwarf galaxy, because it is the dwarf that suffers the greater tidal distortion during a merger.

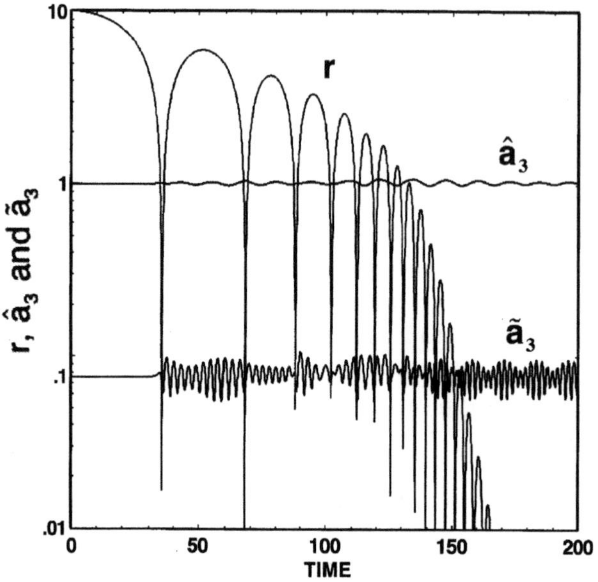

Fig. 1. Dependence of r, $\hat{a}_3$, and $\tilde{a}_3$ on time in a free-fall experiment in which $\hat{M} = 0.99$, $\tilde{M} = 0.01$, $r_0 = 10.0$, $\hat{a}_0 = 1.0$, and $\tilde{a}_0 = 0.075$

Figures 1–3 illustrate free-fall experiments in which the masses of the galaxies are $\hat{M} = 0.99$ and $\tilde{M} = 0.01$ and the initial radius of the giant galaxy is $\hat{a}_0 = 1.0$. Figure 1 shows the runs of the separation $r = |x_3|$ of the centers of mass of the galaxies and their semi-axes $\hat{a}_3$ and $\tilde{a}_3$ as functions of time in the case that the initial separation of the centers of mass is $r_0 = 10.0$, and the initial radius of the dwarf is $\tilde{a}_0 = 0.075$. We omit here a figure showing the runs of the semi-axes $\hat{a}_1$ and $\tilde{a}_1$, because that figure is similar to Fig. 1. The relative motion of the centers of mass is an oscillation, which decays under the influence of dynamical friction. Tidal interactions excite oscillations in the two galaxies during the first pericenter passage and modulate those oscillations during subsequent pericenter passages. The merger is complete between time $t \approx 130$, when the amplitude of the relative motion of the centers of mass is comparable with the size of the giant galaxy, and $t \approx 150$, when that amplitude is comparable with the size of the dwarf. The oscillations of the dwarf are apparently irregular during the merger but quite regular after the merger is complete.

Figure 2 illustrates the sensitivity of the oscillations of the dwarf galaxy to initial conditions in terms of the growth of the separation of two systems defined in (12). Here ξ is the separation in the $(\tilde{a}_1, \tilde{a}_3)$-plane of the system described in the preceding paragraph and a second system, which is identical except that the initial radius of the dwarf galaxy is now $\tilde{a}_0 = 0.07499$. Between the first and second pericenter passages, at times $t = 35$ and $t = 68$, respectively, the growth of ξ is approximately exponential in time. During the

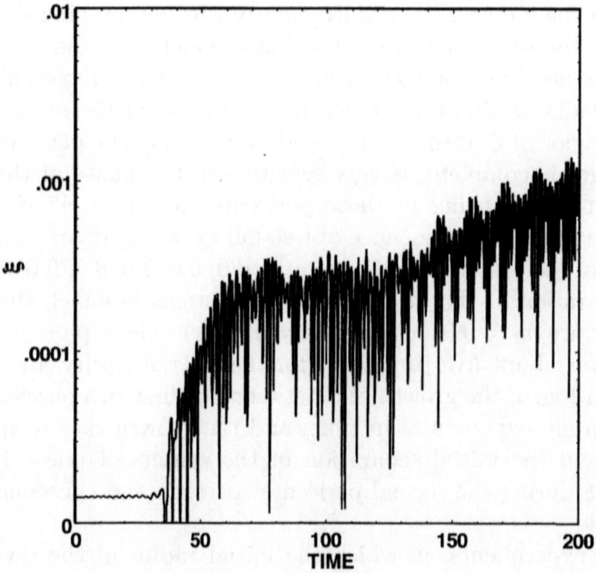

Fig. 2. Growth of the separation of two systems in the $(\tilde{a}_1, \tilde{a}_3)$-plane in the case that $\hat{M} = 0.99$, $\tilde{M} = 0.01$, $r_0 = 10.0$, and $\hat{a}_0 = 1.0$. The initial radii of the dwarf galaxy in the two systems are $\tilde{a}_0 = 0.07500$ and $\tilde{a}_0 = 0.07499$

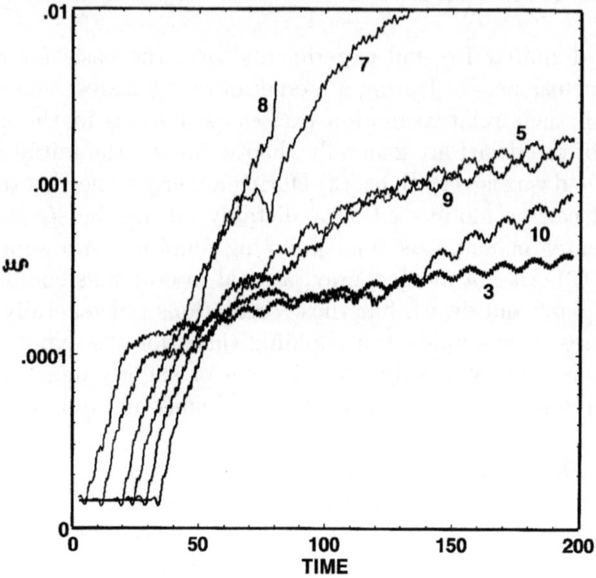

Fig. 3. Growth of the separation of two systems in the $(\tilde{a}_1, \tilde{a}_3)$-plane for different initial separations r_0 in the case that $\hat{M} = 0.99$, $\tilde{M} = 0.01$, and $\hat{a}_0 = 1.0$. The initial radii of the dwarf galaxy in the two systems are $\tilde{a}_0 = 0.07500$ and $\tilde{a}_0 = 0.07499$. The values of r_0 label the different curves

remainder of the merger, ξ varies irregularly without growing significantly. In a plot of the run of ξ out to time $t = 2{,}000$, which we omit here, the growth of ξ after the merger is complete is linear in time. Thus, the oscillations of the dwarf are chaotic during the merger and regular after the merger.

The behavior of ξ during the second and subsequent pericenter passages, until a merger is complete, is very sensitive to the phase of the oscillations of the dwarf at the times of those pericenter passages. That sensitivity is illustrated in Fig. 3 for a sequence of free-fall experiments in which the initial separations of the centers of mass are $r_0 = 3.0$, 5.0, 7.0, 8.0, 9.0, and 10.0, and ξ is the separation in the $(\tilde{a}_1, \tilde{a}_3)$-plane of systems in which the initial radii of the dwarf are $\tilde{a}_0 = 0.07500$ and $\tilde{a}_0 = 0.07499$. These plots of ξ have been smoothed over about five time units for the sake of clarity. In each example illustrated in Fig. 3, the growth of ξ between the first two pericenter passages is approximately exponential in time, and the growth rate is approximately independent of the initial separation of the centers of mass. However, the behavior of ξ during the second pericenter passage and subsequently is quite unpredictable.

In other experiments, in which the initial radius of the dwarf galaxy is very small, the separation ξ of two systems grows linearly in time during a merger as well as afterward. This result suggests that, if the gravitational binding of the dwarf galaxy is sufficiently tight, then the tidal interaction of the two galaxies is too weak a perturbation to make the oscillations of the dwarf chaotic. Additional experiments will be required in order to delineate this trend.

These and similar free-fall experiments form the basis for the following tentative conclusions. (1) During a merger of two galaxies, tidal pumping associated with their relative motion excites oscillations in the galaxies. The oscillations of the dwarf are generally chaotic, unless the initial gravitational binding of the dwarf is very tight. (2) During a merger, the chaotic oscillations of the dwarf can be modulated unpredictably during the second and subsequent pericenter passages as tidal pumping amplifies and suppresses those oscillations. (3) After a merger, gravitational interactions couple the oscillations of the giant and dwarf, but those oscillations are generally not chaotic. (4) In the present examples, the e-folding times for the exponential growth of ξ are of the order of 7–8 dynamical times or slightly more than 10^8 years in systems in which our unit of mass is 10^{12} suns and our unit of length is 10 Kpc.

In this model, integral properties of the system vary chaotically during the capture of a dwarf galaxy by a giant galaxy. Presumably, the chaotic behavior of those integral properties would be communicated to individual stellar orbits in the system. However, the study of chaotic behavior of the individual orbits is beyond the scope of the model.

3 Chaotic Decay of Stationary Oscillations in Galaxies Formed by Mergers: A Conjecture

We now consider the following conjecture: A galaxy formed by mergers might pass through states of stationary oscillation as it approaches equilibrium. A stationary oscillation of a galaxy is a periodic oscillation with a constant amplitude. Stationary oscillations of galaxies are a generalization of certain nonlinear plane waves, so-called BGK waves, in an infinite and homogeneous electrostatic plasma [9]. States of stationary oscillation appear to be well established as possible states of galaxies [10, 11]. BGK waves and stationary oscillations are nonlinear generalizations of van Kampen modes of oscillation [7, 9, 11, 12] in plasmas and galaxies. A state of stationary oscillation might be expected to decay, and the galaxy to approach equilibrium, as a consequence of chaotic behavior of resonant orbits.

This conjecture is inspired by the following considerations. In models and simulations of the formation of structure in the universe, galaxies appear to form in sequences and hierarchies of mergers. Tidal streams of stars and other substructure in the halo of the Milky Way have been interpreted as fossil remains of material accreted in mergers [13]. The conjecture rests on a speculation that material accreted in a merger could be trapped in resonances and observed as star streams in a galaxy and that such captures could leave the galaxy in a state of stationary oscillation.

3.1 Construction of a Stationary Oscillation of a Galaxy

As formulated in [11], the theory of stationary oscillations of a galaxy describes slightly nonlinear oscillations in which the density and gravitational potential are of the forms

$$\rho(\mathbf{x},t) = \rho_0(\mathbf{x}) + \epsilon \mathrm{Re}\left[\rho_1(\mathbf{x}) \exp(-i\sigma t)\right] \tag{13}$$

and

$$V(\mathbf{x},t) = V_0(\mathbf{x}) + \epsilon \mathrm{Re}\left[V_1(\mathbf{x}) \exp(-i\sigma t)\right] , \tag{14}$$

respectively, where $\mathbf{x}$ and t denote the position and time, respectively, $\rho_0(\mathbf{x})$ and $V_0(\mathbf{x})$ are given functions of position, the frequency σ is a given constant, and the constant ϵ is a small parameter. The functions $\rho_1(\mathbf{x})$ and $V_1(\mathbf{x})$ are to be determined together with the distribution function $f(\mathbf{v},\mathbf{x},t)$, where $\mathbf{v}$ is the velocity of a star, as a self-consistent solution of the collisionless Boltzmann equation and Poisson's equation. The construction of that solution is described in detail in [11], so it suffices here to describe only those properties of the solution that bear on the conjecture stated above.

The solutions for $\rho_1(\mathbf{x})$, $V_1(\mathbf{x})$, and $f(\mathbf{v},\mathbf{x},t)$ are constructed in [11] as series in powers of $\epsilon^{1/2}$ through terms of order ϵ. It is assumed that the motion of a star in the potential $V_0(\mathbf{x})$ alone would be integrable and therefore describable in terms of a set of actions $\mathbf{J} = (J_1, J_2, J_3)$ and their conjugate angles $\mathbf{w} = (w_1, w_2, w_3)$. A resonance occurs at each point in the phase space at which $\mathbf{n} \cdot \boldsymbol{\omega} - \sigma = 0$, where the constants $\mathbf{n} = (n_1, n_2, n_3)$ are sets of three integers corresponding to the three degrees of freedom of motion in the potential $V_0(\mathbf{x})$ and the quantities $\boldsymbol{\omega}(\mathbf{J}) = (\omega_1, \omega_2, \omega_3)$ are the frequencies of the motion in those degrees of freedom.

The orbits in the region of a given resonance are described in terms of a certain set of transformed actions $\mathbf{G} = (G_\mathrm{R}, G_2, G_3)$, say, and their conjugate angles $\boldsymbol{\theta} = (\theta_\mathrm{R}, \theta_2, \theta_3)$. Through terms of order ϵ, the transformed Hamiltonian in that representation is a function only of the actions $\mathbf{G}$ and the single angle θ_R. It follows immediately that the actions G_2 and G_3 are integrals of the motion. Moreover, the transformed Hamiltonian is a third integral, because the time dependence appearing originally in (13) and (14) is absorbed in the definition of θ_R.

Apart from terms of order $\epsilon^{3/2}$ and higher, the transformed Hamiltonian governing motion in the $(G_\mathrm{R}, \theta_\mathrm{R})$-plane, for assigned values of the integrals G_2 and G_3, is of the form of the Hamiltonian for a physical pendulum. It follows that each resonance region can be described in terms of a stable periodic orbit and an unstable periodic orbit, which are defined by fixed points of the Hamiltonian in the $(G_\mathrm{R}, \theta_\mathrm{R})$-plane. Moreover, a separatrix contains the fixed point of the unstable periodic orbit and forms the boundary of the resonance region in the $(G_\mathrm{R}, \theta_\mathrm{R})$-plane. The solution for the distribution function in each resonance region has the property that the resonant stars concentrate near that separatrix, i.e., near the unstable periodic orbit.

These and other results described in [11] reproduce results obtained earlier by Louis and Gerhard in an investigation of a nonlinear, radial oscillation of a spherically symmetric galaxy [10]. Louis and Gerhard constructed a numerical model of a galaxy in such a state of stationary oscillation with the aid of a generalization of Schwarzschild's method [14] of the superposition of orbits. Their paper contains a detailed analysis of the orbit structure in their model in terms of surfaces of section, spectral stellar dynamics, and a pendulum model of the resonant orbits. Louis and Gerhard thus provide a complete interpretation and practical validation of the theory of stationary oscillations described in [11]. In particular, they also find that the populations of resonant stars tend to concentrate near the unstable periodic orbits. Their investigation includes an extensive delineation of regular and chaotic orbits, which is not possible within the analytic framework of [11]. (The canonical perturbation theory employed in [11] effectively assumes that all orbits are regular.) The successful construction of a numerical model in [10] indicates that states of stationary oscillation are possible even in a galaxy in which there is a population of chaotic orbits.

3.2 Chaos, the Decay of Stationary Oscillations, and an Approach to Equilibrium

The present conjecture postulates that galaxies formed in mergers pass through states of stationary oscillation as a result of the capture of accreted material into resonant orbits. However, the resonant orbits that are populated in a state of stationary oscillation concentrate near unstable periodic orbits (i.e., near the separatrices that form the boundaries of resonance regions in the phase space). Such orbits are particularly vulnerable to transitions from order to chaos.

When an integrable system (e.g., the approximate model of resonant orbits described in [11]) is perturbed, the transition from order to chaos occurs in regions of the phase space in the neighborhoods of the separatrices [15–17]. In a galaxy in a state of stationary oscillation, chaotic orbits can occur in two ways. (1) The size of a typical resonance region in the phase space is of order $\epsilon^{1/2}$, where ϵ is the small parameter in (13) and (14) [10, 11]. If the amplitude of the oscillation of the galaxy were sufficiently large and the sizes of the resonance regions accordingly large, then resonance regions in the phase space, as constructed in [11], would overlap. (2) Under the influence of subsequent mergers or other external perturbations, the asymptotic curves of the main unstable periodic orbit would form homoclinic intersections, and they would also form heteroclinic intersections with the asymptotic curves of higher order periodic orbits. In either case, chaotic orbits would appear in the neighborhoods of the separatrices in place of regular, resonant orbits.

Resonant stars contribute significantly to the gravitational field that supports a stationary oscillation of a galaxy. Stars on chaotic orbits might temporarily mimic stars on resonant orbits in their contributions to the prevailing gravitational field; however, it is unlikely that the gravitational field that is required for the self-consistency of a stationary oscillation can persist if a substantial population of orbits is chaotic. In other words, the oscillation of a galaxy containing a substantial population of stars on chaotic orbits might be only "quasi-stationary," and that oscillation could decay. Thus, the second part of the present conjecture is that such a "chaotic decay" of a stationary oscillation in a galaxy formed by mergers is a possible route in the approach of the galaxy to equilibrium.

Well established aspects of this conjecture are that stationary oscillations are possible states of galaxies and that the resonant stars in such states are vulnerable to chaotic behavior. However, verification of the conjecture would require a demonstration that capture of material into resonances during a merger would leave the galaxy thus formed in a state of stationary oscillation, at least approximately. Verification would also require a demonstration that the appearance of a substantial population of chaotic orbits in the system would cause the oscillation to decay.

4 Epilogue: Transient Chaos in the Mergers of Galaxies

In the model of mergers investigated in Sect. 2 and according to the conjecture presented in Sect. 3, transient episodes of chaotic behavior occur during the approach of merged systems to their final states. A connection between such behavior and transient chaos in the sense of Kandrup might be a subject meriting further investigation.

Acknowledgments

I thank Monica Valluri for stimulating my interest in the mergers of galaxies in our discussions of the work described in [4]. I also thank Richard Miller for asking challenging questions that encouraged improvements in the present effort. I am grateful to George Contopoulos, Panos Patsis, and their colleagues for this opportunity to present and publish early accounts of the work described in Sect. 2.

References

1. H. E. Kandrup: Chaos and chaotic phase mixing in galaxy evolution and charged particle beams. In: *Galaxies and Chaos*, ed by G. Contopoulos, N. Voglis (Springer, Berlin Heidelberg New York, 2003) pp 154–168
2. H. E. Kandrup, I. M. Vass, I. V. Sideris: Mon. Not. R. Astr. Soc. **341**, 927 (2003)
3. B. Terzić, H. E. Kandrup : Mon. Not. R. Astr. Soc. **347**, 957 (2004)
4. M. Valluri, I. M. Vass, S. Kazantzidis et al: Astrophys. J. **658**, 731 (2007)
5. S. Chandrasekhar, E. P. Lee: Mon. Not. R. Astr. Soc. **139**, 135 (1968)
6. S. Chandrasekhar, D. D. Elbert: Mon. Not. R. Astr. Soc. **155**, 435 (1972)
7. J. Binney, S. Tremaine: *Galactic Dynamics* (Princeton University Press, Princeton, 1987)
8. S. D. Tremaine, J. P. Ostriker, L. Spitzer Jr: Astrophys. J. **196**, 407 (1975)
9. I. B. Bernstein, J. M. Greene, M. D. Kruskal: Phys. Rev. **108**, 546 (1957)
10. P. D. Louis, O. E. Gerhard: Mon. Not. R. Astr. Soc. **233**, 337 (1988)
11. P. O. Vandervoort: Mon. Not. R. Astr. Soc. **339**, 537 (2003)
12. N. G. van Kampen: Physica **21**, 949 (1955)
13. S. R. Majewski: Substructure in the galactic halo. In: *Satellites and Tidal Streams*, ed by F. Prada, D. Martinez Delgado, T. J. Mahoney (Astronomical Society of the Pacific, San Francisco, 2004) pp 63–79
14. M. Schwarzschild: Astrophys. J. **232**, 236 (1979)
15. G. Contopoulos: *Order and Chaos in Dynamical Astronomy* (Springer, Berlin Heidelberg New York, 2002)

16. A. J. Lichtenberg, M. A. Lieberman: *Regular and Stochastic Motion* (Springer, Berlin Heidelberg New York, 1983)
17. F. Diacu, P. Holmes: *Celestial Encounters: The Origins of Chaos and Stability* (Princeton University Press, Princeton, 1996) pp 37–42

Hamiltonian Normal Forms and Galactic Potentials

G. Pucacco

Dipartimento di Fisica – Università di Roma "Tor Vergata",
Via de lla Ricerca Scientifica, 1 – 00133 Rome, Italy
pucacco@roma2.infn.it

1 Introduction

The study of self-gravitating stellar systems has provided important hints to develop tools of analytical mechanics. We may cite the ideas of Jeans [1] about the relevance of conserved quantities in describing the phase-space structure of large N-body systems and his introduction of the concept of *isolating integral.* Later important contributions are those of Contopoulos [2], who applied a direct approach to compute approximate forms of the isolating integrals of motion, of Hénon and Heiles [8] with a paradigmatic example of non-integrable system derived from a simple galactic model and of Hori [7], who introduced the theory of Lie transforms in the field of canonical perturbation theory. These and other cues contributed to the body of methods and techniques that we use today to study regular and chaotic dynamics of non-integrable systems.

The direct approach applied by Contopoulos aims at solving the equation for the conserved quantity along the classical procedure developed early in the last century [3]. The method of the Lie transform [7], subsequently improved by several authors [9–13], has some technical advantages and has gradually become a standard method in the perturbation theory of Hamiltonian dynamical systems [4]. However, its application in galactic dynamics, with a few remarkable exceptions [14, 15], has not been as systematic and productive as it could be.

Hamiltonian normal forms constructed in this way [16–18] are a powerful tool to investigate the orbit structure of galactic potentials and to gather several qualitative informations concerning the near integrable dynamics below the stochasticity threshold (if any) of the system. Results obtained in the same class of systems by the averaging method [19] are easily overtaken. As a matter of fact, with a normal form truncated to an order sufficient to incorporate the main resonance, one can also make reliable *quantitative* predictions. In the present contribution we review how to exploit detuned resonant normal forms to extract information on several aspects of the dynamics in systems

G. Contopoulos, P.A. Patsis, *Chaos in Astronomy*. Astrophysics and Space Science Proceedings, doi: 10.1007/978-3-540-75826-6,

with self-similar elliptical equipotentials. In particular, using energy and ellipticity as parameters, we compute the instability thresholds of axial orbits, bifurcation values of low-order boxlets and phase-space fractions pertaining to the families around them. We also show how to infer something about the singular limit of the potential.

A remarkable side-effect of expressing the stability–instability threshold as a series expansion, is that its predictive ability goes well beyond the radius of convergence of the perturbing expansion. Exploiting asymptotic properties of the series constructed via the normal form [20, 21], we may try to estimate an optimal truncation order.

2 The Hamiltonian Normal Form

The subject of our investigation is the class of 2-dof natural systems

$$H(\mathbf{p}, \mathbf{r}) = \frac{1}{2}(p_x^2 + p_y^2/q) + V(s(x, y)). \tag{1}$$

V is a uniformly increasing function of the variable

$$s = x^2 + y^2/q, \tag{2}$$

with an absolute regular minimum ($V(0,0) = V'(0,0) = 0$), so that the energy E may take any non-negative value. Two simple examples are

$$V_L = \frac{1}{2}\log(1 + s), \tag{3}$$

$$V_C = \sqrt{1 + s} - 1. \tag{4}$$

The parameter q gives the "ellipticity" of the equipotentials and ranges in the interval

$$0.6 < q < 1. \tag{5}$$

Lower values of q can in principle be considered but correspond to an unphysical density distribution if V is a gravitational potential. Values greater than unity are included in the treatment by reversing the role of the coordinate axes. With respect to the standard "physical" notation, the scaling transformation

$$p_y \longrightarrow \sqrt{q}\, p_y, \quad y \longrightarrow y/\sqrt{q} \tag{6}$$

is implicit in the Hamiltonian written in the form (1).

2.1 Series Expansions

To investigate the dynamics of system (1), we look for a new Hamiltonian given by the series expansion in the new canonical variables $\mathbf{P}, \mathbf{R}$,

$$K(\mathbf{P},\mathbf{R}) = \sum_{n=0}^{\infty} K_n(\mathbf{P},\mathbf{R}), \tag{7}$$

with the prescription that

$$\{H_0, K\} = 0. \tag{8}$$

In these and subsequent formulas we adopt the convention of labeling the first term in the expansion with the index zero: in general, the "zero order" terms are quadratic homogeneous polynomials and terms of *order* n are polynomials of degree $n+2$. The zero order (unperturbed) Hamiltonian,

$$H_0(\mathbf{P},\mathbf{R}) \equiv K_0 = \frac{1}{2}(P_X^2 + X^2) + \frac{1}{2q}(P_Y^2 + Y^2), \tag{9}$$

with unperturbed frequencies $\omega_1 = 1$ and $\omega_2 = 1/q$, is expressed in terms of the new variables found at each step of the normalizing transformation. It is customary to refer to the series constructed in this way as a "Birkhoff" normal form [5]. The presence of terms with small denominators in the expansion, forbids in general its convergence. It is therefore more effective to work since the start with a *resonant normal form* [6], which is still non-convergent, but has the advantage of avoiding the small divisors associated to a particular resonance. To catch the main features of the orbital structure, we therefore approximate the frequencies with a rational number plus a small "detuning"

$$\frac{\omega_1}{\omega_2} = q = \frac{m_1}{m_2} + \delta. \tag{10}$$

We speak of a *detuned* (m_1/m_2) *resonance*, with $m_1 + m_2$ the *order* of the resonance. In order to implement the normalization algorithm, also the original Hamiltonian (1) has to be expressed as a series expansion around the equilibrium: performing the rescaling

$$\mathcal{H} := \frac{m_2 H}{\omega_2} = m_2 q H, \tag{11}$$

we redefine the Hamiltonian as the series

$$\mathcal{H} = \sum_{k=0}^{\infty} \mathcal{H}_k = \frac{1}{2}\left[m_1(p_x^2+x^2)+m_2(p_y^2+y^2)\right] + \tfrac{1}{2}m_2\delta(p_x^2+x^2)+\sum_{k=1}^{\infty} b_k(q)s^{k+1}, \tag{12}$$

with expansion coefficients b_k depending only on the ellipticity in view of the restriction imposed by the choice of the potentials. The procedure is now that of an ordinary resonant "Birkhoff–Gustavson" normalization [22, 23] with two variants: the coordinate transformations are performed through the Lie transform and the detuning quadratic term is treated as a term of higher order and put into the perturbation. This is analogous to the strategy of the "nearly resonant construction" of Contopoulos and Moutsoulas [24] in the context of the direct approach and is implemented in the program by Giorgilli [25].

2.2 Lie Transform Normalization

Considering a generating function g, the new coordinates $\mathbf{P}, \mathbf{R}$ result from the canonical transformation

$$(\mathbf{P}, \mathbf{R}) = M_g(\mathbf{p}, \mathbf{r}). \tag{13}$$

The *Lie transform operator* M_g is defined by [4]

$$M_g \equiv \sum_{k=0}^{\infty} M_k \tag{14}$$

where

$$M_0 = 1, \quad M_k = \sum_{j=1}^{k} \frac{j}{k} L_{g_j} M_{k-j}. \tag{15}$$

The functions g_j are the terms in the expansion of the generating function ($g_0 = 1$) and the linear differential operator L_g is defined through the Poisson bracket, $L_g(\cdot) = \{g, \cdot\}$.

The terms in the hew Hamiltonian are determined through the recursive set of linear partial differential equations [4]

$$K_n = \mathcal{H}_n + \sum_{j=0}^{n-1} M_{n-j} \mathcal{H}_j, \quad n = 1, 2, \ldots \tag{16}$$

"Solving" the equation at the n-th step consists of a twofold task: to find K_n *and* g_n. We observe that, in view of the reflection symmetries of the Hamiltonian (1), the chain (16) is composed only of members with even index and so the normal form itself is composed of even-index terms only. The unperturbed part of the Hamiltonian, $\mathcal{H}_0$, determines the specific form of the transformation. In fact, the new Hamiltonian K is said to be *in normal form* if, analogously to (8),

$$\{\mathcal{H}_0, K\} = 0, \tag{17}$$

is satisfied. The function

$$\mathcal{I} = K - \mathcal{H}_0 \tag{18}$$

can be used as a second integral of motion. For practical applications (for example to compare results with numerical computations) it is useful to express approximating functions in the original physical coordinates. Inverting the coordinate transformation, the new integral of motion can be expressed in terms of the original variables. Denoting it as the power series

$$I = \sum_{n=0}^{\infty} I_n, \tag{19}$$

its terms can be recovered by means of the set of equations

$$\sum_{j=1}^{n} M_{n-j}\left[\mathcal{H}_j - I_j\right] - K_n = 0, \qquad n \geq 1, \tag{20}$$

that is obtained from (16) and (18) by exploiting the nice properties of the Lie transform with respect to inversions [4].

2.3 Effective Order of the Detuning

We have to discuss how to treat the detuning term: it is considered as a higher order term and the most natural choice is to put it into $\mathcal{H}_2$. However, there is no strict rule for this and one may ask which is the most "useful" choice, always considering that applications are based on series expansions with coefficients depending on q. We remark that, different choices of the *effective order*, say d, of the detuning, lead to different terms of higher order in the normal form. We also observe that, whatever the choice made, the algorithm devised to treat, step by step, the system (16) must be suitably adapted to manage with polynomials of several different orders. In practice, since at each step the actual order of terms associated to detuning is lower than the corresponding effective order, the algorithm is adapted by incorporating routines already used at previous steps.

In practice, at step say j, we have an equation of the form

$$K_j = \mathcal{H}_j + A_j + \delta B_{j-d} + \delta^2 B_{j-2d} + \cdots + L_{g_j}(\mathcal{H}_0), \tag{21}$$

where A_i, B_i are homogeneous polynomials of degree $i+2$ coming from previous steps. As usual, the algorithm is designed to identify in all terms with the exclusion of

$$L_{g_j}(\mathcal{H}_0) \equiv -L_{\mathcal{H}_0}(g_j), \tag{22}$$

monomials in the kernel of the linear operator $L_{\mathcal{H}_0}$. These monomials are used to construct K_j: the remaining terms are used to find g_j in the standard way. It is clear that both the normal form and the generating function are affected by the effective order of the detuning term.

In both cases (3,4) investigated [26], with the detuning treated as a term of order 2, the next appearance of a related term is in K_6. Rather, if it is treated as a term of order 4, the next appearance of a related term is in K_8. Truncating at order 6 (polynomials of degree 8) is therefore sufficient to make a comparison with other predictions not sensitive to the detuning.

2.4 Structure of the Normal Form

In principle, the recursion process to solve the system (16) can be carried out to arbitrary order. In practice we have to truncate it at some finite order N. The ideal choice would be an *optimal* truncation order N_{opt}, to evaluate which one can work with the formal integral (19): minimizing its failure to commute with the Hamiltonian, one truncates the series at the order giving the best conservation [13, 20, 21]. However, in general this is a costly procedure.

On the other hand, a very conservative strategy can be that of truncating at the lowest order adequate to convey some non-trivial information on the system. In the resonant case, it can be shown [27] that the lowest order to be included in the normal form in order to *capture* the main effects of the m_1/m_2 resonance with double reflection symmetries is $N_{\min} = 2 \times (m_1 + m_2 - 1)$. Truncating at this level is enough to study the resonance and the main periodic orbits associated to it [16, 17]. Using "action-angle"–*like* variables $\mathbf{J}, \boldsymbol{\theta}$ defined through the transformation

$$X = \sqrt{2J_1} \cos\theta_1, \quad P_X = \sqrt{2J_1} \sin\theta_1, \tag{23}$$

$$Y = \sqrt{2J_2} \cos\theta_2, \quad P_Y = \sqrt{2J_2} \sin\theta_2, \tag{24}$$

the typical structure of the doubly-symmetric resonant normal form truncated at $N_{\min}$ is [2, 6]

$$K = m_1 J_1 + m_2 J_2 + \sum_{k=2}^{m_1+m_2} \mathcal{P}^{(k)}(J_1, J_2) + a J_1^{m_2} J_2^{m_1} \cos[2(m_2\theta_1 - m_1\theta_2)], \tag{25}$$

where $\mathcal{P}^{(k)}$ are homogeneous polynomials of degree k whose coefficients may depend on δ and a is a constant. In these variables the second integral is

$$\mathcal{E} = m_1 J_1 + m_2 J_2 \tag{26}$$

and the angles appear only through the resonant combination

$$\psi = m_2\theta_1 - m_1\theta_2. \tag{27}$$

Introducing its conjugate variable

$$\mathcal{R} = m_2 J_1 - m_1 J_2, \tag{28}$$

the new Hamiltonian can be expressed in the *reduced* form $K(\mathcal{R}, \psi; \mathcal{E}, \delta)$ that is a family of 1-dof systems parametrized by $\mathcal{E}$ and δ.

3 Applications

We now analyze a sample of problems that can be addressed and solved with the tools developed so far. The main periodic orbits and the families of quasi-periodic orbits parented by them give the structure of the phase space, at least in the regular regime. The study of existence and stability of normal modes and periodic orbits in general position admitted by (25) or its higher-order generalization is therefore of outmost importance.

3.1 Stability of Normal Modes

In systems of the form (1) the orbits along the symmetry axes are simple periodic orbits. It can be readily verified that these orbits correspond to the two solutions in which either J_1 or J_2 vanish. If the axial orbit is stable it parents a family of "box" orbits. A case that is both representative of the state of affairs and useful in galactic applications is that of the stability of the x-axis periodic orbit (the "major-axis orbit", if q is in the range (5)). Among possible bifurcations from it, the most prominent is usually that due to the 1:2 resonance between the frequency of oscillation along the orbit and that of a normal perturbation, producing the "banana" and "anti-banana" orbits [28]. The inclusion of detuning allows one to catch the passage of the system through the resonance due to the nonlinear coupling between the two degrees of freedom: the strength of the coupling depends on energy and we expect that the onset of the resonance is described by one (or more) curves on the (δ, E)-plane. To investigate this problem in the potentials (3,4), we construct the normal form with $m_1 = 1, m_2 = 2$ and study the nature of the critical points of the function $K^{(\mu)} = K + \mu \mathcal{H}_0$, where μ has to be considered as a *Lagrange multiplier* to take into account that there is the constraint $\mathcal{H}_0 = \mathcal{E}$. The condition for a change in the nature of the critical point corresponding to the normal mode is given by the solutions of the algebraic equation

$$\det[\mathrm{d}^2 K^{(\mu)}(\mathcal{E}, \delta)]|_{J_2=0} = 0 \tag{29}$$

of degree N in $\mathcal{E}$: each transition of the kind extremum $\rightarrow$ saddle is equivalent to the onset of an instability and to the bifurcation of the banana (or of the anti-banana).

However, in order to get a form usable in comparison with other results (for example coming from a numerical treatment) it is necessary to use a "physical" energy variable rather than the parameter $\mathcal{E}$. The conversion is possible if the physical energy E appears explicitly [17]. According to the rescaling (11), we assume that $m_2 q E$ is the constant "energy" value assumed by the truncated Hamiltonian K. In the present instance $m_2 = 2$ so that, on the x-axis orbit, the new Hamiltonian is a series of the form

$$K = 2q\mathcal{E} + cq\mathcal{E}^2 + \cdots = 2qE. \tag{30}$$

This series can be inverted to give

$$\mathcal{E} = E - \frac{c}{2}E^2 + \cdots \tag{31}$$

and this can be used in the treatment of stability to replace $\mathcal{E}$ with E. Recalling that, in this case, (10) gives $q = 1/2 + \delta$, every solution can be expanded as

$$E_{\mathrm{crit}}(\delta) = \sum_{k=1}^{N/2} c_k \delta^k \tag{32}$$

Table 1. Coefficients in the expansion (32) with $N = 14$ for the logarithmic potential (banana, 2nd column and anti-banana, 3rd column) and the conical potential (banana, 4th column and anti-banana, 5th column)

k	Potential V_L		Potential V_C	
	Banana	Anti-banana	Banana	Anti-banana
1	8	8	16	16
2	$-\frac{20}{3}$	$\frac{28}{3}$	$\frac{248}{3}$	$\frac{536}{3}$
3	$\frac{268}{9}$	$\frac{460}{9}$	$\frac{3,608}{9}$	$\frac{18,584}{9}$
4	$-\frac{1,724}{27}$	$\frac{3,928}{27}$	$\frac{43,328}{27}$	$\frac{657,848}{27}$
5	$\frac{79,184}{405}$	$\frac{267,404}{405}$	$\frac{525,704}{81}$	$\frac{23,668,304}{81}$
6	$-\frac{567,178}{1,215}$	$-\frac{510,200,857}{405}$	$\frac{28,118,794}{1,215}$	$\frac{4,304,374,384}{1,215}$
7	$-\frac{30,991,946}{25,515}$	$\frac{615,376,795,556}{8,505}$	$\frac{309,430,864}{3,645}$	$\frac{31,575,390,356}{729}$

and in this form they can be used for quantitative predictions.

In Table 1. we list the coefficients of the series (32) giving these bifurcations for the logarithmic potential (3) and the "conical" potential (4). They have been obtained [26] with a normal form truncated at order $N = 14$ and with the detuning treated as a term of order 2. There is a complete agreement with the analytical approach based on the Poincarè–Lindstedt method [29] and, as discussed below, there is a striking agreement with the numerical approach based on the Floquet method. The agreement of all fractional coefficients is complete up to $N/2 = 7$. On the other hand, if the detuning is treated as a term of order $d = 4$ or greater, we get a disagreement in the coefficients starting from c_3. This result confirms the analysis made above on the "propagation" of the detuned terms in the normal form and show that the choice $d = 2$ is the optimal one.

What is remarkable in the quality of the prediction with regard to "experimental" numerical data is that numerical computations are performed with the *exact* logarithmic (or conical) potentials (3,4), whereas the analytical predictions are, in any case, based on the series expansions of these potentials that appear in (12) with limited convergence radii. The reliability of these predictions in a range wider than foreseen can be explained if we interpret the series of the form (32) as *asymptotic series* and evaluate their truncations by computing the successive partial sums

$$E_n(q) = \sum_{k=1}^{n} c_k \delta^k, \quad n = 1, \ldots, N/2. \tag{33}$$

Table 2. Subsequent truncations of expansion (32) with $N = 14$ for the logarithmic potential (banana). E_B is the value obtained by means of the Floquet method

n	δ			
	0.1	0.2	0.3	0.4
1	0.800000	1.60000	2.40000	3.20000
2	0.733333	1.33333	1.80000	2.13333
3	0.763111	1.57156	2.60400	4.03911
4	0.756726	1.46939	2.08680	–
5	0.758681	1.53196	2.56190	–
6	0.758214	1.50208	2.22160	–
7	0.758336	1.51763	2.48724	–
E_B	0.758	1.513	2.401	3.646

Minimizing the difference between the "exact" value and its approximations provides an estimate of the optimal truncation. As an example, in Table 2. we report these partial sums for the banana bifurcation in the logarithmic potential (3), with $0.1 < \delta < 0.4$ $(0.6 < q < 0.9)$ and compare them with the values obtained by means of the Floquet method [17, 28] given in the last row. The numerical values of the partial sums are given with six digits just to show more clearly the asymptotic behaviour: we can see that, up to $\delta = 0.3$, the predictions are apparently still (slowly) converging at $n = 7$. Only at the rather extreme value $\delta = 0.4$ we get an "optimal" truncation order $n_{\text{opt}} = 3$, with a 10% error on the exact value of the critical energy. We may wonder if there is a way to speed up the convergence rate: this can be done with a resummation method like the continued fraction [26]. It can be shown that, for all values of δ up to 0.3, $n = 6$ *is enough to reach a precision comparable to the numerical error.* For $\delta = 0.4$ we get an optimal truncation order $n_{\text{opt}} = 5$, with a 3% error on the exact value of the critical energy.

3.2 Periodic Orbits in General Position and Boxlets

In addition to the normal modes, each resonant normal form of the type (25) admits a double family of resonant periodic orbits in general position usually called *boxlets* in galactic dynamics [28]. They can be easily identified using the fact that the two "angles" have a fixed phase relation given either by $\psi = 0$ or $\psi = \pm\pi$. In addition to the already mentioned 1:2 resonant banana, we have the 1:1 "loop", the 2:3 "fish" and so forth [28]. Each of them, if stable, is surrounded by a family of quasi-periodic orbits usually inheriting the same nickname. In a given potential in the class of (1), several boxlets can be present at the same time, whereas each resonant normal form is able to correctly render only one type: even with this limitation, a knowledge of the corresponding family is very useful. In particular, we can analytically compute the phase-space fraction occupied by the given family, an important information in the process of constructing self-consistent models. We illustrate the idea in the case of the loop family: the principle is the same for higher resonating boxlets but the computations much more involved.

Usually the loop bifurcates from the minor-axis periodic orbit at energy lower than that of the banana bifurcation from the major axis [16]: there is a regime in which the loop family and the boxes around the stable major-axis orbit coexist. To identify loops, we impose the condition that the Hamiltonian flow generated by the 1:1 version of the reduced normal form $K(\mathcal{R},\psi;\mathcal{E},\delta)$ has a fixed point in $\mathcal{R}=\mathcal{R}_L, \psi=\pi$. Using relations (26) and (28) and the value of the detuning $\delta = q-1$, this solution fixes the actions on the closed loop: $J_{1L}(\mathcal{E},q)$ and $J_{2L}=\mathcal{E}-J_{1L}(\mathcal{E},q)$. On the periodic orbit, it is possible to find a relation between $\mathcal{E}$ and the true energy E in a form analogous to the expansion (30). We can then express the actions as a series in E and, exploiting their geometric meaning, produce an estimate of the fraction of phase space occupied by the loops *and* the boxes. Truncating the series at first order, in the logarithmic case the results are [17]

$$f_{Loop} = \frac{J_{1L}(E,q)}{\mathcal{E}(E)} = \frac{2(-3+3q-5q^2+5q^3)+E(9-9q+11q^2-3q^3)}{(3-2q+3q^2)(-2(1-q)^2+E(3-2q+3q^2))} \tag{34}$$

and $f_{Box} = 1 - f_{Loop}$. These predictions and the corresponding one for the banana family [17] agree very well with numerical estimates [28] up to energy values much greater than that corresponding to the harmonic core of the potential.

3.3 Singular Limits

The potentials considered up to now are assumed to be analytic in the origin. However, we know that realistic models should include a singularity related to a density "cusp" and/or a central point mass [30]. In the examples of the form (3,4) is implicitly assumed the use of adimensional coordinates by introducing a "core radius" R_c which can be put equal to 1 without less of generality. In the limit $R_c \to 0$, those examples reduce to members of the family of *singular scale-free* potentials [31]

$$V_\alpha(s) = A\, s^\alpha, \quad -1/2 \le \alpha \le 1, \quad A\alpha > 0. \tag{35}$$

The singular conical potential is given by $\alpha = 1/2$ and $A = 1$ while the singular logarithmic potential corresponds to the limit $\alpha \to 0$ with $A = 1/2$.

It is tempting to try to extract information concerning the scale-free singular limit from our analytical setting based on series expansions. Formally, this operation should be hindered by the lack of a series representation of the singular potential. However, we may nonetheless "force" our approximate integrals of motion to play their role in the singular limit too and try our chance by constructing a Poincaré surface of section by using the approximate integral $I(x, y, p_x, p_y; q)$ given by (20). In view of the scale invariance, we fix the energy level E_0 and construct, e.g., a y-p_y surface of section by means of the intersection of the function $I(x, y, p_y; E_0, q)$ with the $x = 0$ hyperplane. The level curves of the function

$$F(y, p_y) = I(0, y, p_y; E_0, q) \tag{36}$$

give the invariant curves on the section. In the singular logarithmic case with $E_0 = 0$ and $q = 0.7$ we get, quite surprisingly, acceptable results [17]. In the section constructed by using the approximate integral $I^{(1:1)}$ related to the 1:1 resonance and obtained by truncating the series (20) at order 6, it gives the family of loops around the stable periodic orbit at $y \simeq 0.56$ in good agreement with numerical data [28]. A similar result is given with the section obtained by using the approximate integral $I^{(1:2)}$ again truncated at order 6: it gives the family around the stable banana at $y \simeq 0.16$ and boxes around it.

4 Comments and Outlook

As any analytical approach, this method has the virtue of embodying in (more or less) compact formulas simple rules to compute specific quantities, giving a general overview of the behavior of the system. In the case in which a non-integrable system has a regular behavior in large part of its phase space, a very conservative strategy, like that of truncating at a low order including the resonance, provides sufficient qualitative and quantitative agreement with other more accurate but less general approaches. In our view, the most relevant limitation, common to all perturbation methods, is due to the intrinsic structure of a *single-resonance* normal form. However, we remark that the regular dynamics of a non-integrable system can be imagined as a superposition of very weakly interacting resonances. If we are not interested in thin stochastic layers, each portion of phase space associated with a given resonance has a fairly good alias in the corresponding normal form.

There are several lines of development of this line of research; we mention a few of them: to extend the asymptotic analysis of perturbation series representing the building blocks of phase space (actions, frequencies, etc.); to devise suitable coordinate transformations to enable the investigation of cuspy potentials and/or central "black holes"; to apply the normalization algorithm to three degrees of freedom systems with and without rotation.

Acknowledgments

I wish to thank George Contopoulos, Christos Efthymiopoulos, Giuseppe Gaeta, Luigi Galgani, Antonio Giorgilli and Ferdinand Verhulst for very useful discussions.

References

1. J. H. Jeans: *Astronomy & Cosmogony*, 2nd edn (Cambridge University Press, Cambridge, 1929)
2. G. Contopoulos: *Order and Chaos in Dynamical Astronomy*, (Springer, Berlin, 2004)
3. E. T. Whittaker: *A Treatise on the Analytical Dynamics of Particles and Rigid Bodies*, 4th edn (Cambridge University Press, Cambridge 1937)
4. D. Boccaletti & G. Pucacco: *Theory of Orbits*, Vol. 2, (Springer, Berlin, 1999)
5. G. D. Birkhoff: *Dynamical Systems*, Amer. Math. Soc. Coll. Publ., **9** (New York, 1927)
6. J. A. Sanders & F. Verhulst: *Averaging Methods in Nonlinear Dynamical Systems* (Springer, Berlin, 1985).
7. G. I. Hori: Pub. Astr. Soc. Japan, **18**, 287 (1966)
8. M. Hénon & C. Heiles: Astron. J. **69**, 73 (1964)
9. A. Deprit: Cel. Mech. **1**, 12 (1969)
10. A. Kamel: Cel. Mech. **1**, 190 (1969)
11. M. Kummer: Lecture Notes in Physics, **93**, 57 (1977)
12. A. Giorgilli & L. Galgani: Cel. Mech. **17**, 267 (1978)
13. A. Giorgilli: *Notes on Exponential Stability of Hamiltonian Systems*, (Centro di Ricerca Matematica E. De Giorgi, Pisa, 2002)
14. O. Gerhard & P. Saha: MNRAS **251**, 449 (1991)
15. P. Yanguas: Nonlinearity, **14**, 1 (2001)
16. C. Belmonte, D. Boccaletti & G. Pucacco: Cel. Mech. Din. Astr. **95**, 101 (2006)
17. C. Belmonte, D. Boccaletti & G. Pucacco: Astrophys. J. **669**, 202 (2007)
18. C. Belmonte, D. Boccaletti & G. Pucacco: QTDS, in press (2008)
19. T. de Zeeuw & D. Merritt: Astrophys. J. **267**, 571 (1983)
20. G. Contopoulos, C. Efthymiopoulos, C. & A. Giorgilli: J. Phys. A: Math. Gen. **36**, 8639 (2003)
21. C. Efthymiopoulos, A. Giorgilli & G. Contopoulos: J. Phys. A: Math. Gen. **37**, 10831 (2004)
22. F. Gustavson: Astron. J. **71**, 670 (1966)
23. J. Moser: Mem. Am. Math. Soc. **81**, 1 (1968)
24. G. Contopoulos & M. Moutsoulas: Astron. J. **71**, 687 (1966)
25. A. Giorgilli: Comp. Phys. Comm. **16**, 331 (1979)

26. G. Pucacco, D. Boccaletti & C. Belmonte: Cel. Mech. Din. Astr., DOI 10.1007/s10569-008-9141-x (2008)
27. J. M. Tuwankotta & F. Verhulst: SIAM J. Appl. Math. **61**, 1369 (2000)
28. J. Miralda-Escudé & M. Schwarzschild: Astrophys. J. **339**, 752 (1989)
29. R. Scuflaire: Cel. Mech. Din. Astr. **61**, 261 (1995)
30. K. Gebhardt et al.: Astron. J. **112**, 105 (1996)
31. J. Touma & S. Tremaine: MNRAS **292**, 905 (1997)

Local Phase Space: Shaped by Chaos?

D. Chakrabarty

School of Physics and Astronomy, University of Nottingham,
Nottingham NG7 2RD, UK
dalia.chakrabarty@nottingham.ac.uk

1 Introduction

The exploration of the nature of the phase space that we live in, is a naturally attractive endeavour. On astronomical scales, this translates to an exercise in understanding the state of the phase space in the neighbourhood of the Sun. Of course, for a long time, this was impeded by the dearth of sufficiently large samples of proper motion data of the nearby stars. However, with the information available in Woolley's catalogue [1], Agris Kalnajs concluded the local velocity space to be basically bimodal [2] and attributed this behaviour to the rough proximity of the Sun to the Outer Lindblad Resonance of the central bar in the Galaxy (OLR_b). The observational domain expanded with the availability of the transverse velocity information of stars in the vicinity of the Sun, as measured by *Hipparcos*. Since then many workers have attempted to chart the local phase space distribution DF [3–5] and investigate the origin of the structure in the same [4, 6–9].

Here we pursue the hypothesis that the observed state of the local phase space owes to the dynamical effect of Galactic features such as the central bar or the spiral pattern. However, dynamical modelling of the Solar neighbourhood is incomplete without the inclusion of the effects of *both* the bar and the spiral pattern; after all, a major resonance of the bar occurs in the vicinity of the Sun, (as suggested by the Kalnajs Mechanism) and the inter-arm separation in the Galactic spiral pattern falls short of the average epicyclic excursion of a star at the Solar radius [10].

However, this joint perturbation has been considered only occasionally [8, 10]. That too, it is contentious if the Galactic spiral arms should attach themselves to the ends of the bar or rotate with a pattern speed (Ω_s) that is distinct from that of the bar (Ω_b). The former scenario was invoked to produce self-consistent models of the galaxies NGC 3992, NGC 1073, and NGC 1398 [11] while the best-fit model for NGC 3359 was reported to manifest distinct pattern speeds for the bar and the spiral [12]. Such dissimilar pattern speeds were considered for the first time by [10] in her modelling of the Solar

G. Contopoulos, P.A. Patsis, *Chaos in Astronomy.* Astrophysics and Space Science Proceedings, doi: 10.1007/978-3-540-75826-6,

neighbourhood. The effect of the relative separation between the locations of the resonances due to the Galactic bar and the spiral pattern is expected to bear interesting dynamical consequences.

In this article, we report the results of modelling of the local phase space, undertaken with the aim of disentangling the mystery about the origin of the observed phase space structure. In particular, we address the question of the imperativeness of the presence of chaos, in order to explain this structure.

2 Local Phase Space

The exploration of the physics responsible for the observed DF, follows the exercise of density estimation in phase space, given the observed velocity data. Assuming the Galactic disk to be ideally flat, we approximate the phase space as given by four coordinates: the two plane of the disk spatial coordinates (radius R and azimuth ϕ) and the (heliocentric) radial and transverse velocities (U and V, respectively). U is measured positive in the direction of the Galactic centre while V is positive along the sense of Galactic rotation. Thus, our simulated velocity space is 2D in nature.

The stellar U–V distribution is noted to be highly non-linear and multi-modal. [10] uses the same distribution that was used in [4]; this construction uses V of single stars with distance $d < 100\,\mathrm{pc}$ and dispersion of parallax $\sigma(\pi) < 0.1\pi$ from the Hipparcos Catalogue and U of the 3,481 stars in the *Hipparcos* Input Catalogue. This measured kinematic data, when smoothed by the bisymmetric adaptive kernel estimator, (as used by [5]) suggests five major clumps in the local velocity U–V plane (see Fig. 2 in [10]); these have been identified with the Hercules, Hyades, Pleiades, Coma Berenecius and Sirius stellar streams or moving groups. It is the Hercules stream that corresponds to the smaller mode while the bigger mode constitutes the other four groups.

It is accepted that other choices of the kernel might have given rise to a different picture of the U–V plane. This is expected to be more the case at the outer parts of the distribution, where there are relatively fewer stars than near the central parts (between $-50\,\mathrm{km\,s^{-1}} < V < 10\,\mathrm{km\,s^{-1}}$ and $-50\,\mathrm{km\,s^{-1}} < U < 50\,\mathrm{km\,s^{-1}}$). To check if the suggestion of the five streams is an idiosyncracy of the density estimation procedure, other kernel estimators were tried; it was concluded that this did not affect the above mentioned central part of the distribution in question, though the outer parts were indeed affected.

It merits mention that in the course of the density estimation procedure, as suggested by [5], the calculation of the local bandwidth λ_i, at the ith velocity data point, involves achieving convergence, by iteratively toggling between the distribution estimate at this point and λ_i. The effect of varying the initial seed for the density estimate, on the result of this iteration, has however not been thoroughly investigated or reported; subsequent workers [4, 10] have used values provided by [5].

Thus, accepting that the observed structure in the local velocity space has been robustly estimated, test particle simulations (2-D by nature) were carried out to constrain the model that best fit the observations.

3 Simulations

The simulations are test-particle in nature. This form of constricted N-body sampling of the stars is sufficient in the outer parts of disk galaxies and is useful when there is a multi-dimensional parameter space that need to be scanned. In the simulations, a bunch of initial phase space coordinates (~3,500, i.e. the number of stars used to construct the local U–V diagram) are evolved in the potential of the Galactic disk, as perturbed by the potentials of the bar and/or the spiral pattern. The disk is assigned an initial phase space distribution function given by the doubly cut-out forms used in [13]; this ensures an exponential surface mass density profile. We choose to ascribe a Mestel potential to the disk, in order to recover a flat rotation curve; the warmth parameter in this potential is maintained sufficiently high to ensure the values of radial and transverse velocity dispersions and vertex deviations of stars in the vicinity of the Sun. The choice of the perturbation potential is motivated by their geometries: the bar is modelled as a rigidly rotating quadrupole while the spiral potential is approximated by a logarithmic spiral.

3.1 Spiral Characteristics

The spiral pattern that we use in our simulations is considered detached from the ends of the bar. In line with the gas dynamical model of the galaxy, [14], we choose the bar pattern speed to be more than double that of the spiral. This is corroborated by [15] who also suggests an upper-limit of 25 km s^{-1} kpc^{-1} on Ω_s. We experiment with three distinct ratios of $\Omega_s : \Omega_b$ – 18/55, 21/55, 25/55.

We choose this spiral to be a 4-armed, tightly wound structure (pitch angle = 15°), along the lines of [16]. This implies that the ILR of this 4-armed spiral (ILR_s) occurs outside, on top of and inside OLR_b, for the choices of $\Omega_s : \Omega_b$ – 18/55, 21/55 and 25/55, respectively. Naturally, the 21/55 model raises interest, given that in this case, resonance overlap occurs, ensuring global chaos. Quillen [8] had invoked this scenario to explain the observed stellar streams in the vicinity of the Sun.

3.2 Some Technicalities

In our scale-free disk, all lengths are expressed in units of the co-rotation radius of the bar (R_{CR}). All pattern speeds are also expressed in terms of the bar pattern speed, which is set to unity.

We choose to record orbits in an annular region between $R/R_{CR} = 1.7$ to $R/R_{CR} = 2.3$. The significance of the value of 1.7 is that the Outer Lindblad Resonance of an $m = 2$ perturbation occurs at $R/R_{CR} = 1.7$, in a background Mestel potential.

All orbits are recorded in the frame that is rotating with the bar. Azimuth $\phi = 0$ coincides with the bar major axis. In the bar+spiral simulations, where the figure of the spiral is not static in the frame of the bar, we record the orbits only when, at $R = R_{CR}$, the potential of the spiral is maximum. At the end of the simulations, each orbit is sampled (1,000 times) in time, which is equivalent to sampling in phase, assuming ergodicity to be valid, to provide us the output.

4 Results

In our simulations, we work with five models. These are:

- Bar only model with no perturbation from the spiral: model *bar-only*.
- Bar and spiral, with $\Omega_s : \Omega_b = 18/55$: model $bar + sp_{18}$.
- Bar and spiral, with $\Omega_s : \Omega_b = 21/55$: model $bar + sp_{21}$.
- Bar and spiral, with $\Omega_s : \Omega_b = 25/55$: model $bars + p_{25}$.
- Spiral only model with no perturbation from the bar: model *spiral-only*.

The output orbits are first put on a regular polar grid; each of our $R - \phi$ bins are chosen to be $0.025R_{CR}$ wide in radius and 10° wide in azimuth. At each relevant $R - \phi$ cell, the recorded velocities are put on a regular Cartesian U–V grid. With this stored kinematic data, velocity distributions are prepared at each $R - \phi$ bin. When the simulated velocity distribution at a given disk location ($f_s(R, \phi)$)is found to match the observed distribution (f_o) well, we refer to such a velocity distribution as a "good" distribution and the corresponding $R - \phi$ bin is called a "good" location.

The quality of overlap between f_o and $f_s(R, \phi)$ is quantified by a goodness-of-fit technique (discussed below). All this is undertaken individually for each of the five models.

4.1 Goodness of Fit Technique

We test the null hypothesis: the observed U–V data is drawn from $f_s(R, \phi)$. This testing is done by estimating the p-value of a test statistic S. p-value measures the probability of how unlikely it is that a null hypothesis is true by fluke; thus, low p-values indicate bad fits to the data. Though extensive literature exists to suggest various options for S, we find that the definition of S as the reciprocal of the likelihood L, suffices in our case. Here, the likelihood of data D, given the simulated distribution $f_s(R, \phi)$ is

$$L(D|f_s) = \prod \nu_{R,\phi}(U_i, V_j), \tag{1}$$

where $\nu_{R,\phi}(U_i, V_j)$ is the value of $f_s(R, \phi)$, in the (i, j)th U–V bin and the product is performed over all the U and V velocity bins. We decide to choose the pre-set significance level of 5%, for the acceptance of the null hypothesis.

4.2 Identification of "Good" Locations

For a given model, the quality of overlap between f_o and $f_s(R, \phi)$ is given by the p-value of the test statistic in that $R - \phi$ bin. This is referred to as $p-value(R, \phi)$. Distributions of $p-value(R, \phi)$ are shown in Fig. 1, for models $bar-only$ and $spiral-only$. It is to be noted that the function $p-value(R, \phi)$ for the resonance overlap ($bar + sp_{21}$) and $spiral - only$ models is distinct from that for the other three models. This is attributed to the chaos inducing efficacy of the spiral perturbation.

From the distribution p-value(R, ϕ), "good" locations are identified as those $R - \phi$ addresses, at which the p-value attains the maximum, i.e. 100%. This distribution of the "good" locations can be marginalised over ϕ to allow us to quantify the $\pm 1-\sigma$ range that defines the best radial locations. Similarly, marginalising over R gives the constraints on the best azimuthal location for the observer, in a frame where $\phi = 0$ implies bar major axis. Of course, these are then constraints on our location, i.e. the Solar position, in the frame of the bar. Such marginalised distributions (with R and ϕ) are shown in Fig. 2, for model $bar + sp_{25}$. This tells us that:

- The best radial location for the observer is $R/R_{\rm CR}$= $2.0875^{0.15}_{0.25}$. Equating the Solar radius (of 8 kpc) with the medial value, we scale the bar pattern speed to $57.4^{4.1}_{6.9}$ km s^{-1} kpc^{-1}(using circular velocity at Sun=220 km s^{-1}).
- The best fit values for the bar angle is [0°, 43°], with a medial value of 6°.

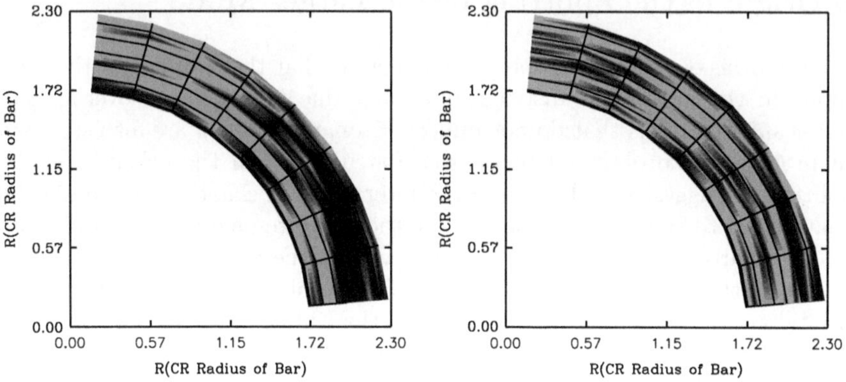

Fig. 1. The distribution p-value(R, ϕ) over the part of the annulus that we record our orbits in, for models *bar-only* (*left*) and *spiral-only* (*right*). The $R - \phi$ cells in the *darker colours* indicate higher p-values than those in *lighter shading*

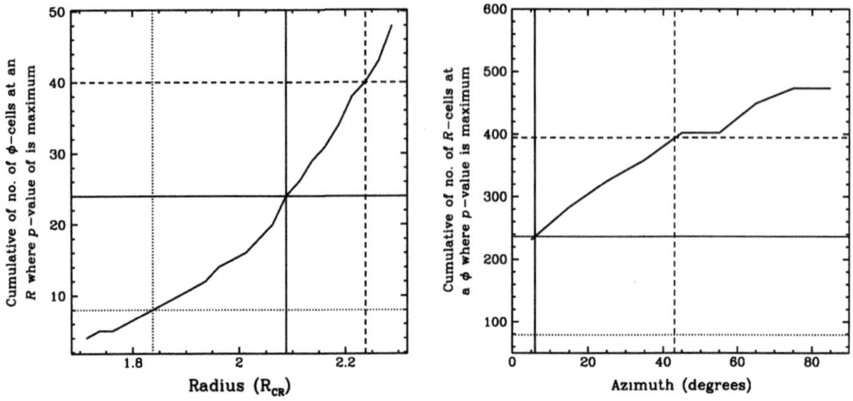

Fig. 2. The distribution of the maximal p-value locations is marginalised, over azimuth and radius, to yield best-fit values of observer radial and azimuthal location (respectively), in order for the observer's local velocity distribution to best match the observed distribution

We find that all the models that include the influence of the bar and the spiral pattern are able to produce the observed stellar streams in the vicinity of the Sun, at different disk locations. Though the *bar-only* model is also capable of the same, it has to be discarded in light of the dynamical reasoning that no model of the Solar neighbourhood is complete without taking the effect of the spiral into account [10]. The *spiral-only* model is rejected since it fails to warm up the local patch in the disk enough, to guarantee velocity dispersions of the order of those observed in the Solar neighbourhood.

5 Origin of the Splitting of the Larger Mode

On the basis of the results above, we conclude that the splitting of the larger mode in the local U–V distribution is not due to resonance overlap since *bar* + *spiral* models that do not impose resonance overlap are also successful in producing the observed phase space structure. On the other hand, this splitting is suggested to be due to the interaction of chaotic orbits and orbits that belong to families corresponding to the minor resonances due to the spiral and the bar (such as the outer 1:1 resonance of the bar).

As for the overall bimodal character of the local phase space, a rudimentary orbital analysis confirms that in line with the Kalnajs Mechanism, the overall bimodality is indeed due to scattering off OLR_b; the Hercules stream appears to be built of quasi-periodic orbits belonging to the anti-aligned family while the anti-aligned family was not spotted in the other larger mode. (The other mode manifests orbits from the aligned family.)

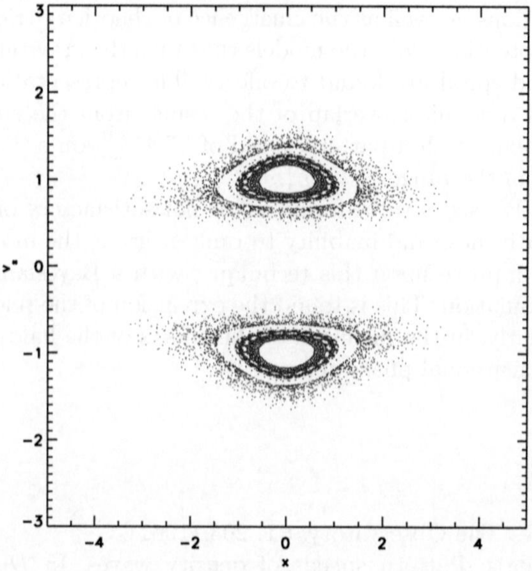

Fig. 3. Surface of section of orbits at effective energy value of -1, from the *bar*$+$*sp*$_{18}$ model. The strongly chaotic orbits are in *medium grey shading*, while the regular orbits are in the *darkest shade*. The *intermediate shading* corresponds to the weakly chaotic orbits

6 The Role of Chaos

[4] has suggested bar-induced chaos as the cause for the local streams. However, in a chaos quantification exercise that is currently underway (Chakrabarty and Sideris, *submitted to A&A*), the orbits in the *bar-only* model are found to be regular; it is noted that it is the presence of the spiral that triggers the onset of chaos. Figure 3 indicates the relative fraction of chaotic orbits, over regular and weakly chaotic orbits, in the *bar* + *sp*$_{18}$ models. Similar surfaces of section for the *bar-only* models, at even higher energies indicate no irregularity.

It is of course possible that a stronger bar would induce greater irregularity than in our work, but the point is that the local phase space structure can be emulated with the bar strength used herein.

7 Summary and Future Work

In this contribution, we have attempted to understand the state of the local phase space via dynamical modelling that includes the influence of the bar and the spiral pattern in the Galaxy. The clumpy structure of the 2-D local velocity distribution was attributed to the major and minor resonances of these perturbations, as well as the emergence of chaoticity triggered primarily by the spiral potential. Only the models that take the gravitational potentials of both bar and spiral are found to offer viable representations of the Solar neighbourhood dynamics; overlap of the results from the successful models indicate a fast bar, with a pattern speed of $57.4^{+3.0}_{-2.5}$ km s^{-1} kpc^{-1}and a bar angle that lies in the range [0°, 30°].

The currently used p-value formalism is not satisfactory on account of the lack of automativeness and inability to rank or grade the models. In fact, we are hoping to improve upon this technique, with a Bayesian measure of the quality of a simulation. This is to aid the expansion of the relevant parameter space, particularly, in the inclusion of the effects of the halo and progression to a full six dimensional phase space.

References

1. Woolley, R.: The Observatory, **81**, 203 (1961).
2. A. J. Kalnajs: Pattern speeds of density waves. In *Dynamics of Disk Galaxies*, ed by B. Sundelius (Göetberg, Sweden, 1991) pp. 323.
3. Dehnen, W. and Binney, J.: Monthly Notices of the Royal Astronomical Society, **298**, 387 (1998).
4. Fux, R.: Astrophysical Journal, **373**, 511 (2001).
5. Skuljan, J., Hearnshaw, J. B. and Cottrell, P. L.: Monthly Notices of the Royal Astronomical Society, **308**, 731 (1999).
6. Dehnen, W.: Astronomical Journal, **119**, 800 (2000).
7. De Simone, R., Wu, X. and Tremaine, S.: Monthly Notices of the Royal Astronomical Society, **350**, 627 (2004).
8. Quillen, A. C.: Astrophysical Journal, **125**, 785 (2003).
9. Famaey, B., Jorissen, A., Luri, X., Mayor, M., Udry, S., Dejonghe, H. and Turon, C.: Astronomy & Astrophysics, **430**, 165 (2005).
10. Chakrabarty, D.: Astronomy & Astrophysics, **467**, 145, (2007).
11. Kaufmann, D. E. and Contopoulos, G.: Astronomy & Astrophysics, **309**, 381 (1996).
12. Boonyasait, V., Patsis, P. A. and Gottesman, S. T.: New York Academy Sciences Annals, 1045, 2 (2005).
13. Evans, N. W. and Read, J. C. A.: Monthly Notices of the Royal Astronomical Society, **300**, 83 (1998).
14. Bissantz, N., Englmaier, P. and Gerhard, O.: Monthly Notices of the Royal Astronomical Society, **340**, 949 (2003).
15. Melnik, A.: Astronomy Letters, **32**, 7 (2006).
16. Johnston, S., Koribalski, B., Weisberg, J. M. and Wilson, W.: Monthly Notices of the Royal Astronomical Society, **322** , 715 (2001).

Barred Galaxies: An Observer's Perspective

Dimitri A. Gadotti

Max Planck Institute for Astrophysics, Karl-Schwarzschild-Strasse 1,
D-85741, Garching bei München, Germany
dimitri@mpa-garching.mpg.de

Summary. I review both well established and more recent findings on the properties of bars, and their host galaxies, stemming from photometric and spectroscopic observations, and discuss how these findings can be understood in terms of a global picture of the formation and evolution of bars, keeping a connection with theoretical developments. In particular, I show the results of a detailed structural analysis of ≈300 barred galaxies in the Sloan Digital Sky Survey, providing physical quantities, such as bar length, ellipticity and boxyness, and bar-to-total luminosity ratio, that can either be used as a solid basis on which realistic models can be built, or be compared against more fundamental theoretical results. I also show correlations that indicate that bars grow longer, thinner and stronger with dynamical age, and that the growth of bars and bulges is connected. Finally, I briefly discuss open questions and possible directions for future research.

1 Some Basic Facts

It is well established now that bars are very often found in disc galaxies. One usually finds a bar fraction of ∼2/3 [32], considering both prominent and weak bars, unless one is looking for bars in images observed at too short wavelengths [86], as the stellar content of bars is usually dominated by old, red stars. In fact, most bars fade away in ultra-violet images [44], although some bars can still be recognised (cf. NGC 1097). Even though the remaining ∼1/3 of disc galaxies do not seem to harbour identifiable bars, they might still have less prominent non-axisymmetric distortions.

In addition, bars are seen in galaxies with a wide range of bulge-to-total ratio and mass, i.e. from lenticulars to irregulars. Thus, secular evolution processes induced by bars occur not only in disc-dominated galaxies with inconspicuous bulges. They also happen in bulge-dominated galaxies, which suggests the coexistence of classical, merger-built bulges, with bulges built from disc dynamical instabilities [51]. But secular evolution can also happen without bars. Oval distortions in discs can also induce a substantial exchange of angular momentum from the inner to the outer parts of galaxies [51].

G. Contopoulos, P.A. Patsis, *Chaos in Astronomy.* Astrophysics and Space Science Proceedings, doi: 10.1007/978-3-540-75826-6,

Finally, a recent development suggests that the total amount of mass within stars that reside in bars at $z \approx 0$ is similar to that kept in stars belonging to classical bulges. A similar amount is confined to elliptical galaxies. Approximately 15% of the total mass in stars at $z \approx 0$ is located in bars. Classical bulges and elliptical galaxies contain each a comparable fraction. This means that, as far as the stellar mass budget in the local universe is concerned, bars are as relevant as classical bulges and elliptical galaxies. The other $\sim 1/2$ of the stellar mass content at $z \approx 0$ is in galaxy discs [24, 36].

2 Observed Properties of Barred Galaxies

A number of theoretical studies indicate that bars redistribute the disc gas content, driving it through torques and dynamical resonances into spiral arms, rings and dust lanes, where it usually changes phase from neutral to molecular as it is submitted to higher pressures. Gas lying beyond the bar ends is driven outwards, whereas gas lying within the bar ends is driven to the central regions (e.g. [2, 18, 33, 75, 81]; see also [51, 82]). This rearrangement of the gas can erase chemical abundance radial gradients [35]. Accordingly, barred galaxies have flatter O/H gradients than unbarred galaxies [97], and the stronger the bar the flatter the gradient [61]. Furthermore, it has also been observed that the central concentration of molecular gas in barred galaxies is higher than in unbarred galaxies [80].

What happens to the gas funnelled by the bar to the centre? Several observational results point out that star formation activity is enhanced in the central regions of barred galaxies as compared to unbarred galaxies, for spirals of early and intermediate types [1, 15, 46, 47, 49, 83, 84], arguing that the gas driven to the centre by the bar is relatively efficiently transformed in young stars. Consequently, barred galaxies also show flatter *colour* gradients than unbarred galaxies, due to bluer central regions [41]. The average *global* star formation, however, seems to be comparable in barred and unbarred galaxies of intermediate types, since their integrated colours are very similar. In addition, bars in late type spirals show star forming regions all along the bar, whereas bars in early type spirals have star formation concentrated at the centre and/or at the bar ends [74]. There are suggestions that the former bars are also dynamically younger (i.e. formed later) than the latter ([34, 60]; see also [73, 94, 95]).

Theoretical work indicates that not only the disc gas content is rearranged by the effects induced by a bar, but that the distribution of *stars* in the disc also changes (see, e.g. [70, 71, 79]). Furthermore, due to an exchange of angular momentum between the disc and the dark matter halo, mediated by the bar, the disc becomes more centrally concentrated [3, 7]. The inflow of disc material (gas and stars) to the centre, and the star formation episodes associated with it, seem to often produce sub-structures such as nuclear rings, nuclear discs and spiral arms, and secondary bars, as shown by observations (see, e.g. [31]).

The increase in central mass concentration can also make stars migrate to orbits out of the disc plane [10]. These structures built at the centre of the disc by the bar might naturally generate an excess of luminosity in the central part of the disc luminosity profile, as compared to an extrapolation to the centre of its outer part. Likewise, classical bulges can also be defined as the extra light on top of the inner disc profile, and also extend out of the disc plane. Thus, to distinguish these different components, formed through distinct processes and having dissimilar physical characteristics, structures built through the inflow of disc material are called pseudo-bulges [51], or disc-like bulges [4]. There are also observational evidences that bars might affect the distribution of stars in the outer disc [30].

Our understanding of yet another type of bulge also benefits largely from a successful connection between theory and observation. Theoretical studies show that the vertical structure of a bar grows in time through dynamical processes, generating a central structure out of the plane of the disc that can have a boxy or peanut-shaped morphology (e.g. [19, 62, 76]). Several observational evidences, mostly based on theoretical expectations, argue that such box/peanut bulges are indeed just the inner part of bars seen edge-on (see [13, 14, 16, 22, 52]).

The gas brought to the centre by the bar can also end up fuelling a super massive black hole and AGN activity. The basic idea is that the primary bar brings gas from scales of ~5 kpc to ~1 kpc and a secondary bar instability brings gas further inwards to ~100 pc. The gas still needs to loose further angular momentum, and at that scales other physical processes, such as viscosity, come to help [88, 89]. However, studies comparing the fraction of barred galaxies in quiescent and active galaxies show contradictory results [46, 50, 53, 57, 68]. Such comparisons are plagued by issues such as AGN classification, sample selection and bar definition, which in turn depends on wavelength and spatial resolution. Should one expect to see a clear distinction in bar fraction in quiescent and active galaxies? Even though there are more clear evidences of bars at least building up a fuel reservoir at the galaxy centre for star formation and AGN activity ([20, 58, 80]; see also [96]), the answer to this question is, for several reasons, more likely, no. To begin with, we saw that bars are very often seen in disc galaxies, and thus any difference is likely to have low statistical significance, unless samples are large enough. More important are factors such as the availability of gas (quiescent barred galaxies might just lack gas to fuel the black hole) and the strength of the bar (or its ability to bring gas inwards). The issue gets more complicated when one considers in detail the role of inner spiral arms and rings, secondary bars and dynamical resonances near the centre. Inner rings might prevent gas to reach the centre, and inner spiral arms could either remove or give further angular momentum from (to) the gas if they are trailing (leading) [17]. Hydrodynamical simulations show that, at least in some cases, secondary bars might not help (and might even prevent) the gas inflow to the nucleus ([56]; but see [45]). Finally, one must keep in mind that the typical time-scale of

AGN activity episodes is likely to be much shorter than the typical time-scale for funnelling the gas and the life time of bars. This means that even if a bar is bringing gas to the nucleus one has to be looking at the right time to see the black hole accreting the gas. Thus, one should not be too surprised (or rather confused) to see results like the similar fractions of secondary bars found in quiescent and active galaxies [63], or the fact that galaxies with the strongest bars are mostly quiescent [55, 90]. Things are just not that simple! The nuclear region of barred galaxies shows complex structures that follow several distinct patterns, which might be related to the strength of the bar and the final destiny of the gas collected by it [72].

From theoretical work, it is still unclear whether primary and secondary bars are long-lived or have short lives. Earlier studies suggested that primary bars could be destroyed by a central mass concentration, such as a super massive black hole. However, it turned out that the central densities needed are unrealistically high, and that at most a weakening of the bar can result (see [5, 85] and references therein). This suggests that primary bars are long-lived. Models that include accretion of gas from the halo to the disc, however, suggest that primary bars can be destroyed due to a transfer of angular momentum from the gas to the bar [11, 12], but this is contested (see [10]). In addition, a new bar could be formed after the previous one vanished, since the accreted gas could turn the disc unstable again, by replenishing the disc with new stars. A key point in models that predict the demise and rebirth of bars is the availability of external gas. They are thus just relevant to gas-rich galaxies. Still on theoretical ground, also the life of secondary bars remains uncertain, with results suggesting both long and short life times (see [21, 28, 64]).

A natural and direct way to assess the life time of bars is to study bars at different redshifts, but this is as yet also not free from opposing results. Early studies had pointed out a sharp drop in the fraction of barred galaxies at $z \sim 1$ (see [92]). Not long afterwards, some studies showed that the apparent lack of bars was only caused by band shifting and poor spatial resolution [26, 48, 86, 93]. The most recent studies, however, do not seem to agree on whether there is a rapid decline in the bar fraction with redshift to $z \sim 1$ [87], or if this fraction is fairly constant [9]. A constant fraction of barred galaxies from $z \sim 1$ to $z \sim 0$ might put models that predict bar dissolution and reformation in trouble, as this would require a fine tuning of the corresponding time-scales. There are recent indications that bars in early and intermediate type spirals are long-lived [27]. If this is correct, it indicates that, in fact, models of bar reformation might concern only late type spirals (later than Sc). This agrees with the findings in [39, 40], where a methodology to estimate the dynamical ages of bars is introduced. This will be discussed in more detail in the next section, including implications for secular evolution scenarios and AGN fuelling by bars.

3 Estimating the Dynamical Ages of Bars

From the previous discussion it is possible to see that there is a number of questions whose answers are still escaping from us. For instance:

- How fast bars drive gas to the galaxy centre?
- For how long are bars modifying the overall evolution of their host galaxies?
- What is the fraction of stars in bulges that comes from secular processes?
- When did the first bars appear?
- Are bars recurrent? If yes, in which conditions?

It is evident that a method through which one could estimate the dynamical ages of bars would greatly improve tentative answers to these questions. One possibility, explored in [39], is to use the vertical thickening of bars as a clock. As pointed out above, this is predicted by theoretical work and it is in very good agreement with observations. The vertical extent of bars translates directly into the vertical velocity dispersion of its stars, σ_z. Thus, one can take spectra of face-on barred galaxies from the bar and from the disc and determine and compare the corresponding values of σ_z. Recently formed bars should have values of σ_z similar to that of the disc, from which it just formed. Evolved bars should have σ_z substantially higher than the disc. Two parameters have thus been defined in [39]:

- $\sigma_{z,\mathrm{bar}}$ – the vertical velocity dispersion of stars in the bar at a galactocentric distance of about half the bar length
- $\Delta\sigma_z$ – the difference between $\sigma_{z,\mathrm{bar}}$ and the vertical velocity dispersion of *disc* stars at the same galactocentric distance

Using spectra obtained for a sample of 14 galaxies, and considering only clear cases, the authors find that young bars have typically $\sigma_{z,\mathrm{bar}} \sim 30\,\mathrm{km\,s^{-1}}$ and $\Delta\sigma_z \sim 5\,\mathrm{km\,s^{-1}}$, whereas evolved bars have typically $\sigma_{z,\mathrm{bar}} \sim 100\,\mathrm{km\,s^{-1}}$ and $\Delta\sigma_z \sim 40\,\mathrm{km\,s^{-1}}$. Statistical tests indicate that these young and evolved bars are indeed different populations at 98% confidence level.

Furthermore, with measurements of the length L_{Bar} and the colour $(B-I)_{\mathrm{Bar}}$ of these bars, presented in [40], it is found that young bars have, on average, $L_{\mathrm{Bar}} \approx 5.4\,\mathrm{kpc}$ and $(B-I)_{\mathrm{Bar}} \approx 1.5$, whereas evolved bars have, on average, $L_{\mathrm{Bar}} \approx 7.5\,\mathrm{kpc}$ and $(B-I)_{\mathrm{Bar}} \approx 2.2$. Evolved bars are both longer and redder than young bars. The difference in length also holds when it is normalised by the galaxy diameter. The bar colour was measured close to the bar ends, but *outside* star forming regions.

The fact that evolved bars are longer than young, recently formed bars is in agreement with theoretical results [3, 7, 66]. These works indicate that, while bars evolve, they capture stars from the inner disc, redistribute angular momentum along the disc and dark matter halo, and get longer and thinner in the process. NGC 4608 and NGC 5701 might represent cases where the capture of disc stars by the bar is substantial. Recent results suggest that the bar in NGC 4608 had an increase in mass of a factor of ≈1.7, through the capture of ≈13% of disc stars ([36]; see also [38, 54]).

The difference in colour between young and evolved bars represents a difference in the age of their stars of ≈10 Gyr. As seen in the previous section, bars seem to follow two different patterns of star formation, which might be related to the dynamical age. A recently formed bar seems to form stars along its whole extent, whereas an evolved bar seems to form stars mostly at its centre and/or its ends. This indicates that, when one carefully measures the colour of stars in the middle of the bar, one is probing mainly the first generation of stars formed in the bar. That seems to be the reason why those bars which are *dynamically* old, as estimated from their stellar kinematics, are also redder than the dynamically young bars. Altogether, these results also indicate that at least some bars are very old, and thus most likely not recurrent (unless the first generation of bars has very short life times).

Interestingly, dynamically young bars are found preferentially in gas-rich, late type spirals. This suggests that bar recurrence is restricted to this class of galaxies, as expected from such models (see discussion in previous section). In addition, it was also found that galaxies hosting AGN typically have young bars, which possibly means that the funnelling of gas to feed the black hole at the nucleus occurs on short time-scales. A similar conclusion is reached in [69] after the finding of a significantly higher bar fraction in narrow-line Seyfert 1 galaxies, which are supposedly in an early stage of black hole evolution.

These results come, however, from the analysis of a small sample of galaxies. It is highly desirable to have the dynamical ages of bars measured for a much larger sample, and assess the validity of these results. Furthermore, at the current stage, one can only discriminate between recently formed and evolved bars. It is now difficult to measure with more precision the dynamical age of the bar. Finally, one would like to be able to give more stringent numbers to this parameter without having to rely on the age of the bar stellar population. The accuracy in estimates of the dynamical age of a bar has to be improved. This might be accomplished by an approach involving both observations (e.g. with a large scale 2D mapping of σ_z in barred galaxies) and theory (e.g. with a more detailed analysis of the vertical evolution of bars with time).

4 The Structural Properties of Bars

Although bars are ubiquitous and might account for a significant fraction of a galaxy total luminosity, only recently studies dedicated to a more detailed modelling of the structural properties of bars started to come out more often (see, e.g. [36, 54, 78]). This is partially because of the significant increase in complexity when one includes another component in the structural modelling of disc galaxies. These studies usually make use of the full data contained in

2D galaxy images, rather than only 1D surface brightness profiles, in order to obtain more accurate results. The rise in complexity thus usually means that automated procedures become much less reliable. In [36], BUDDA v2.1 [23] is used to individually fit galaxy images with model images that include up to three components: a Sérsic bulge, an exponential disc and a bar. Bars are modelled as a set of concentric generalised ellipses [8], with same position angle and ellipticity:

$$\left(\frac{|x|}{a}\right)^c + \left(\frac{|y|}{b}\right)^c = 1, \tag{1}$$

where x and y are the pixel coordinates of the ellipse points, a and b are the extent of its semi-major and semi-minor axes, respectively, and c is a shape parameter. Bars are better described by boxy ellipses (i.e. with $c > 2$). The surface brightness profile of the model bar is described as a Sérsic profile, as bulges. The Sérsic index of bars n_{Bar} is often in the range $\approx 0.5 - 1$, with lower values representing flatter profiles. Another bar parameter fitted by the code is the length of the bar semi-major axis L_{Bar}, after which the bar light profile is simply truncated and drops to zero.

Similar fits were individually done to ≈1,000 galaxies in a sample carefully drawn from the Sloan Digital Sky Survey (SDSS – see [42]). The sample spans from elliptical to bulgeless galaxies, with stellar masses above 10^{10} $\mathrm{M}_{\odot}$ (k-corrected z-band absolute magnitudes ≤ -20 AB mag), at a typical redshift of 0.05, and includes ≈300 barred galaxies. All galaxies are very close to face-on (axial ratio ≥ 0.9) and do not show morphological perturbations, thus assuring that dust extinction is minimised and that the sample is suitable for image decomposition. This also avoids the uncertainties in obtaining deprojected quantities. Fits were done in g, r and i-band images. The distributions of several bar structural parameters obtained in this work from the i-band images are shown in Figs. 1 and 2.

Models of bar formation and evolution should be in agreement with these results. Conversely, models where bar properties are imposed can use these results as a guide in the adjustment of the bar properties. One must note, however, that, due to the relatively poor spatial resolution of the SDSS, these results are biased against bars shorter than $L_{\mathrm{Bar}} \approx 2$–3 kpc, typically seen in very late type spirals (later than Sc [25]). These results are thus representative of the prototypical, bonafide bars seen mostly in early type spirals (earlier than Sc) and lenticulars. Interestingly, the median bar ellipticity is ≈20% higher than the value found via ellipse fitting to galaxy images in [59]. This is exactly what was predicted in [36] as ellipse fits systematically underestimate the true bar ellipticity due to the dilution of the bar isophotes by the rounder, axisymmetric light distribution of bulge and disc. When fitting ellipses to galaxy images one does not separate the contributions to the total galaxy light distribution from the different components, but this is done in image fitting with different models for each component. This shows that results based on the

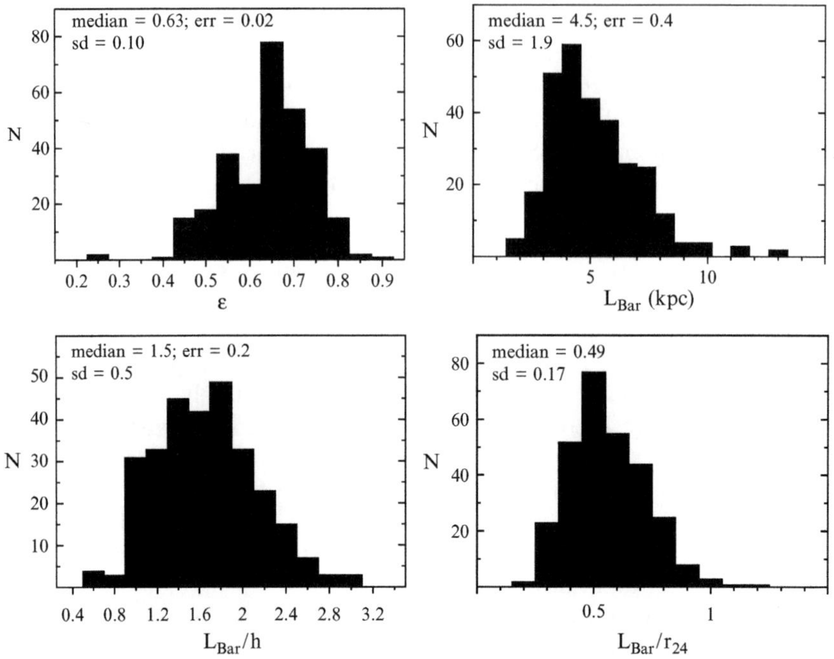

Fig. 1. Distributions of bar structural parameters obtained from 2D bar/bulge/disk i-band image decomposition of ≈300 barred galaxies in the Sloan Digital Sky Survey (SDSS). The sample spans from lenticular to bulgeless galaxies at a typical redshift of 0.05, and excludes galaxies with stellar masses below 10^{10} $M_{\odot}$. From *top* to *bottom* and *left* to *right*: bar ellipticity, bar length (semi-major axis in kpc – $H_0 = 75\,\mathrm{km\,s^{-1}\,Mpc^{-1}}$), bar length normalised by disc scalelength h, and bar length normalised by r_{24} (the radius at which the galaxy surface brightness reaches $24\,\mathrm{mag\,arcsec^{-2}}$ in the r-band). Noted at each panel are the median and standard deviation values of the corresponding distribution, as well as the mean 1σ error of a single measurement, when available. Bin sizes are $\approx 1 - 2\sigma$

ellipticity of bars measured via ellipse fitting should be considered with this caveat in mind. For instance, there is an indication from ellipse fits that, for faint galaxies, disk-dominated galaxies have more eccentric bars than bulge-dominated galaxies [9]. It it is not clear, however, if this result holds if the axisymmetric light contribution from bulge and disk is taken into account. Figure 1 also shows that most of these bars have a semi-major axis between 3 and 6 kpc (but with a long tail to longer bars), and that bars do not extend further than ≈3 times the disc scalelength h, or $\approx 1 r_{24}$ (the radius at which the galaxy surface brightness reaches 24 mag arcsec^{-2} in the r-band – see also [29]). Figure 2 shows that a typical bar is responsible for ≈10% of the total galaxy light, and has a quite flat luminosity profile. Interestingly, one also sees that, indeed, bars have very boxy shapes.

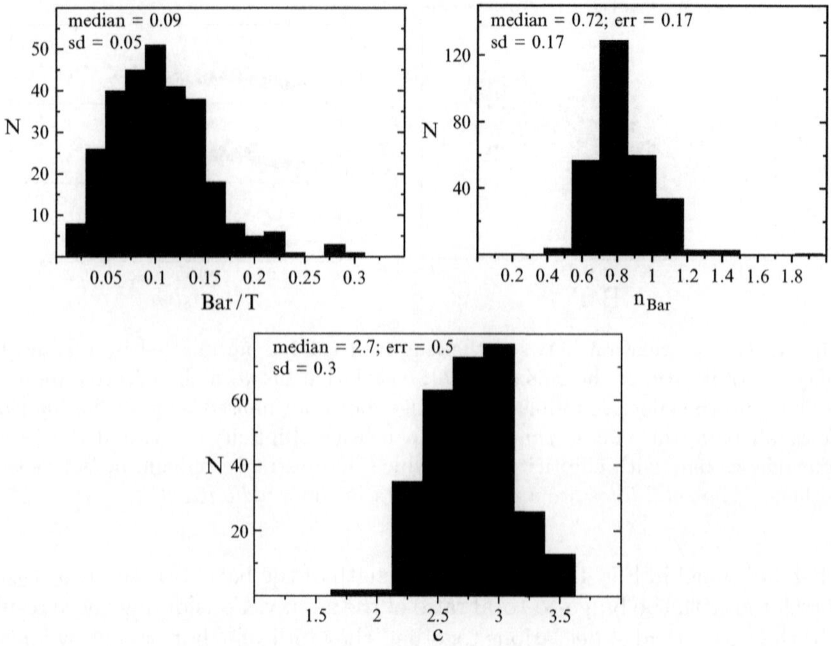

Fig. 2. Same as Fig. 1, but for bar-to-total luminosity fraction, bar Sérsic index, and bar boxyness [parameterised as c – see (1)]

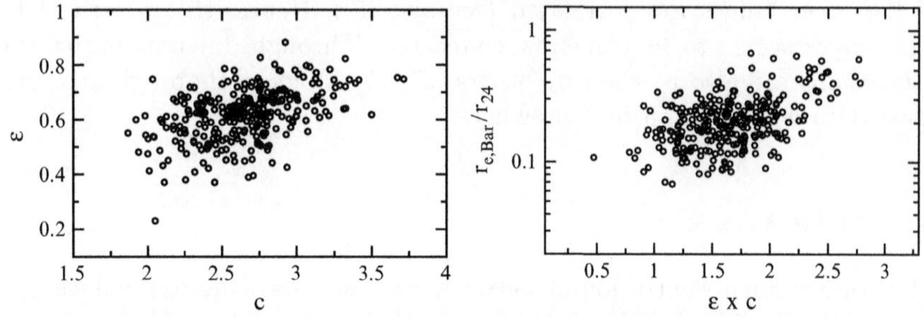

Fig. 3. *Left*: correlation between the bar ellipticity and boxyness. *Right*: correlation between the effective radius of the bar normalised by r_{24} and the product of the bar ellipticity and boxyness

These data also reveal interesting correlations. The left-hand panel in Fig. 3 shows a correlation between bar ellipticity and boxyness, i.e. more eccentric bars are also more boxy. Both quantities contribute to the strength of the bar. Thus, the product of both, $\epsilon \times c$, can be used as a measure of bar strength. The right-hand panel shows that the effective radius of the bar, normalised by r_{24}, is correlated with $\epsilon \times c$ (see also [65]). Thus, longer bars are thinner and stronger, as expected from the theoretical models mentioned above. The

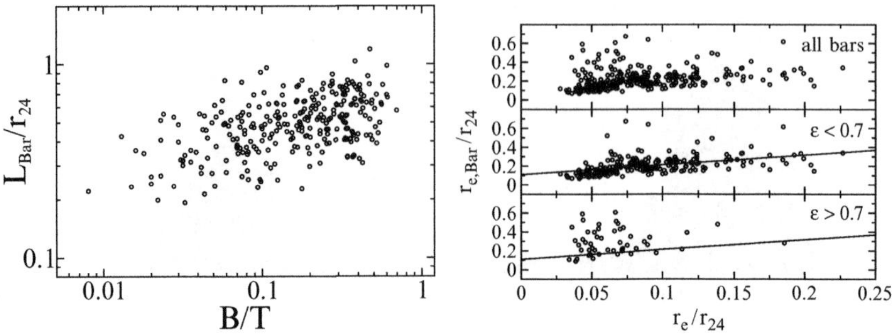

Fig. 4. *Left*: correlation between the length of the bar normalised by r_{24} and the bulge-to-total ratio of the galaxy. *Right*: correlation between the effective radius of the bar and the effective radius of the bulge, both normalised by r_{24}. The *top panel* shows all bars, the *middle panel* shows bars with ellipticity <0.7, and the *bottom panel* shows bars with ellipticity ≥ 0.7 (which, interestingly enough, include several outliers). The *solid lines* are a fit to the data in the *middle panel*

left-hand panel in Fig. 4 shows that the length of the bar, normalised by r_{24}, is correlated with the bulge-to-total ratio of the galaxy. Considering these results and the theoretical expectations together, they indicate that bars grow longer, thinner and stronger with age, as a result of angular momentum exchange, and that bars have had more time to evolve in galaxies with more massive bulges. In fact, the right-hand panel in Fig. 4 shows that the normalised effective radii of bars and bulges are correlated (see also [6]). Hence, the growth of bars and bulges seems to be somehow connected. Through different paths, these conclusions are also reached by others [27, 87]. A more thorough analysis of these data will be published elsewhere.

5 Future Work

The connection recently found between bars and dark matter haloes opens the possibility of indirectly assessing the physical properties of haloes through the observed properties of bars within them. With this aim, one first needs to make a detailed comparison between real and simulated bars. If simulations can successfully reproduce the structural properties of barred galaxies, then they might indeed give us useful estimates of physical properties of real haloes, via comparisons of observations of barred galaxies to models with known halo properties. Such a project has been started (see [37]), and preliminary results are encouraging: n-body snapshots are being used as real galaxy images as input to BUDDA, and a careful comparison of the structural parameters so obtained with those of real barred galaxies shows that simulations are able to generally reproduce the observed quantities.

It is evident the need of further work on the methodology to estimate the dynamical ages of bars. I already mentioned above some possible ways in this direction, and the need to enlarge the sample for which this parameter is measured. Another important parameter that has to be measured for larger samples is the bar pattern speed (see, e.g. [43, 77, 91]). Models of bar formation and evolution make clear predictions on the behaviour of this parameter, which seems to be related to a number of other physical properties: the central concentration of the dark matter halo, the exchange of angular momentum between disc and halo, bar age, and even bar generation. It will be very useful to have estimates, for a large sample, of *both* bar age and pattern speed.

On the theory side, it is important now to obtain an updated criterion for the onset of the bar instability in discs, accounting for the role of the halo. This will be very useful for semi-analytic models (see [67]) which consider bar instability based on earlier studies. Such models can also benefit largely from detailed prescriptions for the transport of disc material to the centre, and the building of disc-like and box/peanut bulges. It is only recently that n-body simulations dedicated to study the formation and evolution of bars started to use responsive, cosmologically motivated haloes, and it is naturally expected that significant progress will come from such studies. Finally, the observed dichotomy between bars in early and late type disc galaxies has to be better understood. It is very likely that the availability of a large gas content in the disc plays a key role. Theoretical work focused on the effects of gas in bar formation and evolution can substantially improve our understanding on this subject.

Acknowledgement. I am very grateful to the organisers, in particular Nikos Voglis and Panos Patsis, for this wonderful opportunity. I benefitted from discussions with several authors, especially Lia Athanassoula and Peter Erwin. I thank Guinevere Kauffmann for letting me present here previously unpublished results from our structural analysis of SDSS images. Comments from the reviewer, Preben Grosbøl, were greatly appreciated and helpful to improve this paper. The author is supported by the Deutsche Forschungsgemeinschaft priority program 1177 ("Witnesses of Cosmic History: Formation and evolution of galaxies, black holes and their environment"), and the Max Planck Society.

References

1. Alonso-Herrero, A., Knapen, J.H.: AJ **122**, 1350 (2001)
2. Athanassoula, E.: MNRAS **259**, 345 (1992)
3. Athanassoula, E.: MNRAS **341**, 1179 (2003)
4. Athanassoula, E.: MNRAS **358**, 1477 (2005)
5. Athanassoula, E., Lambert, J.C., Dehnen, W.: MNRAS **363**, 496 (2005)
6. Athanassoula, E., Martinet, L.: A&A 87, L10 (1980)
7. Athanassoula, E., Misiriotis, A.: MNRAS **330**, 35 (2002)
8. Athanassoula, E., et al.: MNRAS **245**, 130 (1990)
9. Barazza, F.D., Jogee, S., Marinova, I.: ApJ **675**, 1194 (2008)

10. Berentzen, I., Shlosman, I., Martinez-Valpuesta, I., Heller, C.H.: ApJ **666**, 189 (2007)
11. Bournaud, F., Combes, F.: A&A **392**, 83 (2002)
12. Bournaud, F., Combes, F., Semelin, B.: MNRAS **364**, L18 (2005)
13. Bureau, M., Athanassoula, E.: ApJ **626**, 159 (2005)
14. Bureau, M., Freeman, K.C.: AJ **118**, 126 (1999)
15. Carollo, C.M., Stiavelli, M., de Zeeuw, P.T., Mack, J.: AJ **114**, 2366 (1997)
16. Chung, A., Bureau, M.: AJ **127**, 3192 (2004)
17. F. Combes: In: *Advanced Lectures on the Starburst-AGN Connection*, ed. by I. Aretxaga D. Kunth, R. Mújica, p223
18. Combes, F., Gerin, M.: A&A **150**, 327 (1985)
19. Combes, F., Sanders, R.H.: A&A **96**, 164 (1981)
20. Davoust, E., Contini, T.: A&A, **416**, 515 (2004)
21. Debattista, V.P., Shen, J.: ApJL, **654**, L127 (2007)
22. de Souza, R.E., dos Anjos, S.: A&AS **70**, 465 (1987)
23. de Souza, R.E., Gadotti, D.A., dos Anjos, S.: ApJS **153**, 411 (2004)
24. Driver, S.P., Allen, P.D., Liske, J., Graham, A.W.: ApJL **675**, L85 (2007)
25. Elmegreen, B.G., Elmegreen, D.M.: ApJ **288**, 438 (1985)
26. Elmegreen, B.G., Elmegreen, D.M., Hirst, A.C.: ApJ **612**, 191 (2004)
27. Elmegreen, B.G., Elmegreen, D.M., Knapen, J.H., Buta, R.J., Block, D.L., Puerari, I.: ApJL **670**, L97 (2007)
28. El-Zant, A.A., Shlosman, I.: ApJL **595**, L41 (2003)
29. Erwin, P.: MNRAS **364**, 283 (2005)
30. Erwin, P., Pohlen, M., Beckman, J.E.: AJ **135**, 20 (2008)
31. Erwin, P., Sparke, L.S.: AJ **124**, 65 (2002)
32. Eskridge, P.B., et al: AJ **119**, 356 (2000)
33. Friedli, D., Benz, W.: A&A **268**, 65 (1993)
34. Friedli, D., Benz, W.: A&A **301**, 649 (1995)
35. Friedli, D., Benz, W., Kennicutt, R.: ApJL **430**, 105 (1994)
36. Gadotti, D.A.: MNRAS **384**, 420 (2008)
37. Gadotti, D.A., Athanassoula, E., Carrasco, L., Bosma, A., de Souza, R., Recillas, E.: MNRAS **381**, 943 (2007)
38. Gadotti, D.A., de Souza, R.E.: ApJL **583**, L75 (2003)
39. Gadotti, D.A., de Souza, R.E.: ApJ **629**, 797 (2005)
40. Gadotti, D.A., de Souza, R.E.: ApJS **163**, 270 (2006)
41. Gadotti, D.A., dos Anjos, S.: AJ **122**, 1298 (2001)
42. D.A. Gadotti, G. Kauffmann: Multi-Band Bar/Bulge/Disk Image Decomposition of a Thousand Galaxies. In: *Stellar Populations as Building Blocks of Galaxies*, IAU Symp. 241, ed. by A. Vazdekis, R. Peletier, p507
43. Gerssen, J., Kuijken, K., Merrifield, M.R.: MNRAS **345**, 261 (2003)
44. Gil de Paz, A., et al: ApJS **173**, 185 (2007)
45. Heller, C.H., Shlosman, I., Athanassoula, E.: ApJL **657**, L65 (2007)
46. Ho, L.C., Filippenko, A.V., Sargent, W.L.W.: ApJ **487**, 579 (1997)
47. Huang, J.H., Gu, Q.S., Su, H.J., Hawarden, T.G., Liao, X.H., Wu, G.X.: A&A **313**, 13 (1996)

48. Jogee, S., et al.: ApJ **615**, L105 (2004)
49. Jogee, S., Scoville, N., Kenney, J.D.P.: ApJ **630**, 837 (2005)
50. Knapen, J.H., Shlosman, I., Peletier, R.F.: ApJ **529**, 93 (2000)
51. Kormendy, J., Kennicutt, R.C., Jr.: ARA&A **42**, 603 (2004)
52. Kuijken, K., Merrifield, M.R.: ApJL **443**, L13 (1995)
53. Laine, S., Shlosman, I., Knapen, J.H., Peletier, R.F.: ApJ **567**, 97 (2002)
54. Laurikainen, E., Salo, H., Buta, R.: MNRAS **362**, 1319 (2005)
55. Laurikainen, E., Salo, H., Rautiainen, P.: MNRAS **331**, 880 (2002)
56. Maciejewski, W., Teuben, P.J., Sparke, L.S., Stone, J.M.: MNRAS **329**, 502 (2002)
57. Maia, M.A.G., Machado, R.S., Willmer, C.N.A.: AJ **126**, 1750 (2003)
58. Maiolino, R., Risaliti, G., Salvati, M.: A&A **341**, L35 (1999)
59. Marinova, I., Jogee, S.: ApJ **659**, 1176 (2007)
60. Martin, P., Friedli, D.: A&A **326**, 449 (1997)
61. Martin, P., Roy, J.R.: ApJ, **424**, 599 (1994)
62. Martinez-Valpuesta, I., Shlosman, I., Heller, C.: ApJ **637**, 214 (2006)
63. Martini, P., Pogge, R.W., Ravindranath, S., An, J.H.: ApJ **562**, 139 (2001)
64. Masset, F., Tragger, M.: A&A **322**, 442 (1997)
65. Menéndez-Delmestre, K., Sheth, K., Schinnerer, E., Jarrett, T.H., Scoville, N.Z.: ApJ **657**, 790 (2007)
66. Michel-Dansac, L., Wozniak, H.: A&A **452**, 97 (2006)
67. Mo, H.J., Mao, S., White, S.D.M.: MNRAS **295**, 319 (1998)
68. Mulchaey, J., Regan, M.: ApJL, **482**, L135 (1997)
69. Ohta, K., Aoki, K., Kawaguchi, T., Kiuchi, G.: ApJS **169**, 1 (2007)
70. Patsis, P.A.: MNRAS **369L**, L56 (2006)
71. Patsis, P.A., Skokos, Ch., Athanassoula, E.: MNRAS **346**, 1031 (2003)
72. Peeples, M.S., Martini, P.: ApJ **652**, 1097 (2006)
73. Pérez, I., Sánchez-Blázquez, P., Zurita, A.: A&A *465*, L9 (2007)
74. A.C. Phillips: Star Formation in Barred Galaxies. In: *ASP Conf. Ser.*, vol 91, ed. by R. Buta, D.A. Crocker, B.G. Elmegreen (ASP, San Francisco, 1996) p44
75. Piner, B.G., Stone, J.M., Teuben, P.J.: ApJ **449**, 508 (1995)
76. Raha, N., Sellwood, J.A., James, R.A., Kahn, F.D.: Nature **352**, 411 (1991)
77. Rautiainen, P., Salo, H., Laurikainen, E.: ApJL **631**, L129 (2005)
78. Reese, A.S., Williams, T.B., Sellwood, J.A., Barnes, E.I., Powell, B.A.: AJ **133**, 2846 (2007)
79. Romero-Gómez, M., Athanassoula, E., Masdemont, J.J., García-Gómez, C.: A&A **472**, 63 (2007)
80. Sakamoto, K., Okumura, S.K., Ishizuki, S., Scoville, N.Z.: ApJ **525**, 691 (1999)
81. Schwarz, M.P.: ApJ **247**, 77 (1981)
82. Sellwood, J.A., Wilkinson, A.: Rep. Prog. Phys. **56**, 173 (1993)
83. Sérsic, J.L., Pastoriza, M.: PASP **77**, 287 (1965)

84. Sérsic, J.L., Pastoriza, M.: PASP **79**, 152 (1967)
85. Shen, J., Sellwood, J.A.: ApJ **604**, 614 (2004)
86. Sheth, K., Regan, M.W., Scoville, N.Z., Strubbe, L.E.: ApJL **592**, L13 (2003)
87. Sheth, K., et al.: ApJ **675**, 1141 (2008)
88. Shlosman, I., Begelman, M.C., Frank, J.: Nature **345**, 679 (1990)
89. Shlosman, I., Frank, J., Begelman, M.C.: Nature **338**, 45 (1989)
90. Shlosman, I., Peletier, R.F., Knapen, J.H.: ApJL **535**, L83 (2000)
91. Treuthardt, P., Buta, R., Salo, H., Laurikainen, E.: AJ **134**, 1195 (2007)
92. van den Bergh, S., Abraham, R.G., Ellis, R.S., Tanvir, N.R., Santiago, B., Glazebrook, K.G.: AJ **112**, 359 (1996)
93. van den Bergh, S., Abraham, R.G., Whyte, L.F., Merrifield, M.R., Eskridge, P.B., Frogel, J.A., Pogge, R.: AJ **123**, 2913 (2002)
94. Verley, S., Combes, F., Verdes-Montenegro, L., Bergond, G., Leon, S.: A&A **474**, 43 (2007)
95. Wozniak, H.: A&A **465**, L1 (2007)
96. Wyse, R.F.G.: ApJL **612**, L17 (2004)
97. Zaritsky, D., Kennicutt, R.C., Jr., Huchra, J.P.: ApJ, **420**, 87 (1994)

Invariant Manifolds and the Spiral Arms of Barred Galaxies

C. Efthymiopoulos, P. Tsoutsis, C. Kalapotharakos, and G. Contopoulos

Research Center for Astronomy, Academy of Athens,
Soranou Efessiou 4, GR-11527, Athens, Greece
cefthim@academyofathens.gr, ptsoutsi@phys.uoa.gr, ckalapot@phys.uoa.gr, gcontop@academyofathens.gr

Summary. The unstable invariant manifolds of the short period family of unstable periodic orbits around the Lagrangian points L_1 and L_2 of a rotating barred galaxy generate a pattern in configuration space which essentially reproduces the observed spiral pattern of the galaxy. Examples of this phenomenon are given in the self-consistent models of barred galaxies treated by Kaufmann and Contopoulos [1]. Our main finding is that, despite the fact that the bar potential is dominant over the spiral potential, the invariant manifolds produced by the bar potential alone cannot fit the spiral pattern. The self-consistent response potential of the spirals must be taken into account in order to produce a good fit. An iterative model is proposed that reconstructs the spiral arms starting from a pure bar potential.

1 Introduction

The present work is a continuation of the work started by N. Voglis in 2006 [2, 3]. Voglis et al. [2] found that the spiral arms in an N-body model of a rotating barred galaxy were supported almost entirely by *chaotic orbits*. An explanation of this phenomenon was provided in Voglis et al. [3].

The main constituents of Voglis' theory are explained in the article by Contopoulos in the present volume of proceedings. A brief summary is the following: many chaotic orbits wander stochastically inside and outside the corotation region. However, the successive apocentric positions of these chaotic orbits remain correlated for times much longer than the period of the pattern's rotation. This correlation is caused by the delineation of the apocenters of the chaotic orbits along the *unstable invariant manifolds of the short period family* of periodic orbits around the Lagrangian points L_1 or L_2. This family, called hereafter the "PL1 family", exists for Jacobi constant values larger than the value at $L_{1,2}$. The invariant manifolds of the PL1 family yield a pattern that fits well the observed spiral pattern of the galaxy. This property of the manifolds, that produce a spiral pattern, can be understood on the basis of

G. Contopoulos, P.A. Patsis, *Chaos in Astronomy.* Astrophysics and Space Science Proceedings, doi: 10.1007/978-3-540-75826-6,

a normal form representation of the Hamiltonian in action-angle variables locally, near corotation [3, 4]. The robustness of the phase correlations of the chaotic orbits can be explained qualitatively by considering the time flow of the angles of the successive apocenters for initial conditions along an unstable manifold, which is given by a soliton-type partial differential equation [3].

Other authors have also pointed out the role of the chaotic orbits in barred galaxies in general and in their spiral structure in particular. Some references are [1, 5–9]. In particular, Romero-Gomez et al. [8] explored the morphology of the spiral arms induced by the invariant manifolds of the PL1 family in different models of bars. By varying the parameters of a bar model, it was found that the manifolds are able to reproduce a range of observed morphologies, from rings to open spirals. However, we show below that a main assumption of that study, namely that the spiral arms are determined only by the bar potential, is questionable. Finally, we recently explored the contribution to the spiral structure of the invariant manifolds of families of unstable periodic orbits other than the PL1 family (see the article by Tsoutsis and Efthymiopoulos in the present volume). In that study we show that the manifolds of practically *all* the unstable families contribute constructively to the spiral pattern, by creating a collective pattern termed the "coalescence of invariant manifolds".

In our present work we applied the theory of the invariant manifolds to a number of realistic models of barred-spiral galaxies, in order to probe the applicability of our results obtained so far. To this end, we worked with three self-consistent models of Kaufmann and Contopoulos [1], which represent fittings to the real galaxies NGC 3992, NGC1073 and NGC1398. This investigation yielded an important new result. The non-axisymmetric part of the imposed potential contains both a bar and a spiral term. In all three galaxies the bar term is dominant over the spiral term. Nevertheless, despite this dominance, we found that the invariant manifolds of the PL1 family could fit correctly the form of the galaxies' observed spiral arms only when the calculation of the manifolds was made by *keeping the spiral potential term turned on.* In fact, a calculation with a pure bar, i.e., with the spiral term turned off, yielded manifolds resembling to rings rather than to spirals. A spiral pattern could be found in pure bar cases only by pushing the bar's amplitude to unphysically high values.

In the sequel we present a summary of our results together with a qualitative explanation of them. Details are deferred elsewhere [10].

2 Model

The model of Kaufmann and Contopoulos [1] consists of a number of potential/density terms representing various components of a barred galaxy. In particular we have:

– A halo density term given by a Plummer sphere

$$\rho_h(r) = \frac{3M_h}{4\pi b_h^3}\left(1 + \frac{r^2}{b_h^2}\right)^{-5/2} \quad . \tag{1}$$

– A disk surface density given by an exponential law

$$\Sigma_d(r) = \Sigma_0 \exp(-\epsilon_d r) \quad . \tag{2}$$

– A Ferrers bar

$$\rho_b(x, y, z) = \rho_c\left(1 - \frac{x^2}{b^2} - \frac{y^2}{a^2} - \frac{z^2}{c^2}\right)^2 , \qquad a > b > c \tag{3}$$

– A spiral perturbation in the potential

$$V_s(r, \theta) = A(r) r \exp(-\epsilon_s R)\left(\cos 2\Phi + a_4 \cos 4\Phi\right), \tag{4}$$

where

$$\Phi = \frac{\ln(r/a_s)}{\tan i}(\theta - \theta_0) \tag{5}$$

and $A(r)$ is a hyperbolic tangent function yielding a nearly constant plateau for $A(r)$ truncated at two radii r_1, r_2 (see [1]). In the above formulae r, θ are polar coordinates on the disk plane, in which we set $z = 0$ for the calculation of the bar's potential or force. The bar's major axis is always aligned on the y-axis. A model is specified by a set of values for the parameters M_h, b_h, Σ_0, ϵ_d, ρ_c, a, b, c, ϵ_s, a_s, i, a_4, the function $A(r)$ as well as the pattern angular speed Ω_p. In [1], "best fit" parameters were specified for three models corresponding to three real barred galaxies. The criterion for a best fit was that the "response density", i.e., the density obtained by the superposition of many orbits in the fixed potential should match as closely as possible the imposed density represented by the above equations. The matching refers to (a) the amplitudes of the surface density map on the disk plane, and (b) the phases of the maxima of the bar and of the spiral arms, in the imposed and in the response models. When such a matching is found we have a "self-consistent" model of the galaxy under study.

Our choice to study such models is motivated by the fact that the fulfilment of the self-consistency condition implies that the values of the parameters in each model are reliable in the sense that they provide a realistic representation of the true potential/density of a galaxy. Thus, any structure formed by the orbits under the best fit imposed potential should compare well with an observed morphological structure of the galaxy. In particular, our main question in the present context was to see whether the invariant manifolds emanating from the PL1 family of unstable periodic orbits produce spiral patterns fitting well the observed spiral patterns in each galaxy.

3 Results

The orbits are numerically integrated in the two degrees of freedom Hamiltonian:

$$H(r,\theta,p_r,p_\theta) \equiv \frac{1}{2}\left(p_r^2+\frac{p_\theta^2}{r^2}\right) - \Omega_p p_\theta + V(r,\theta), \tag{6}$$

where $V(r,\theta)$ is the total potential resulting from the sum of the potential terms of the various above mentioned components of the galaxy. Our choice of surface of section is (θ, p_θ), $\dot{r}=0$, $\ddot{r}<0$, i.e., we plot the points (θ, p_θ) whenever an orbit reaches an apocentric position, where the density is maximum. The variable p_θ yields the angular momentum of an orbit in a frame of reference *at rest*, i.e., non-rotating, centered at the origin. One readily finds that a higher value of p_θ implies also a higher value of the radial distance r at an apocentron. Thus, all surface of section plots (θ, p_θ) have nearly isomorphic counterparts in the configuration space (r,θ). In the subsequent plots we show both types of plots for the chosen surface of section.

Figure 1 shows the invariant manifolds emanating from the PL1 and PL2 unstable orbits in the case of the self-consistent model of the galaxy NGC3992 (see [1] for the model parameters). In Fig. 1a,b the invariant manifolds are shown in the plane (θ, p_θ) of the surface of section when the spiral potential is turned off (Fig. 1a) or on (Fig. 1b). The phase space structure is also shown in the background of the same figures. The value of the Jacobi constant is $E_J=-1.9\times10^5$ in Fig. 1a, and $E_J=-1.911\times10^5$ in Fig. 1b (small variations of the Jacobi constant do not change these figures appreciably).

A main phenomenon is immediately seen in these figures:

In both cases, the nearly round white domains correspond to energetically prohibited regions of the stars' motion. On the other hand, for values $p_\theta<1$ both figures show large chaotic domains which are limited by librational KAM curves marking the edge of the domain of regular motions inside the bar, at values $p_\theta \approx 0.4$. The chaos in the region $0.4 \leq p_\theta \leq 1$ is due mainly to the nonlinear phenomena introduced by the strong bar forcing near corotation.

However, at values $p_\theta>1$ the two phase portraits are no longer similar. Namely, in the case of Fig. 1a (spiral potential turned off) the region beyond $p_\theta=1$ is almost entirely filled with KAM curves up to an outer limit $p_\theta \simeq 1.5$ beyond which higher order resonances are present. On the contrary, in Fig. 1b (spiral potential turned on), a layer of outer KAM curves has been destroyed, and the chaotic domain extends up to values of $p_\theta \simeq 1.5$.

This difference in the phase portraits induces a dramatic difference in the form and extension of the unstable invariant manifolds emanating from the PL1 orbits in the two cases. Thus, in the case of Fig. 1a the manifolds occupy a very thin chaotic layer and they practically have the form of a separatrix. On the contrary, in the case of Fig. 1b the manifolds are well developed both inside and outside corotation (below and above PL1). The invariant manifolds form conspicuous lobes and foldings which are typical of systems with a large degree of homoclinic chaos.

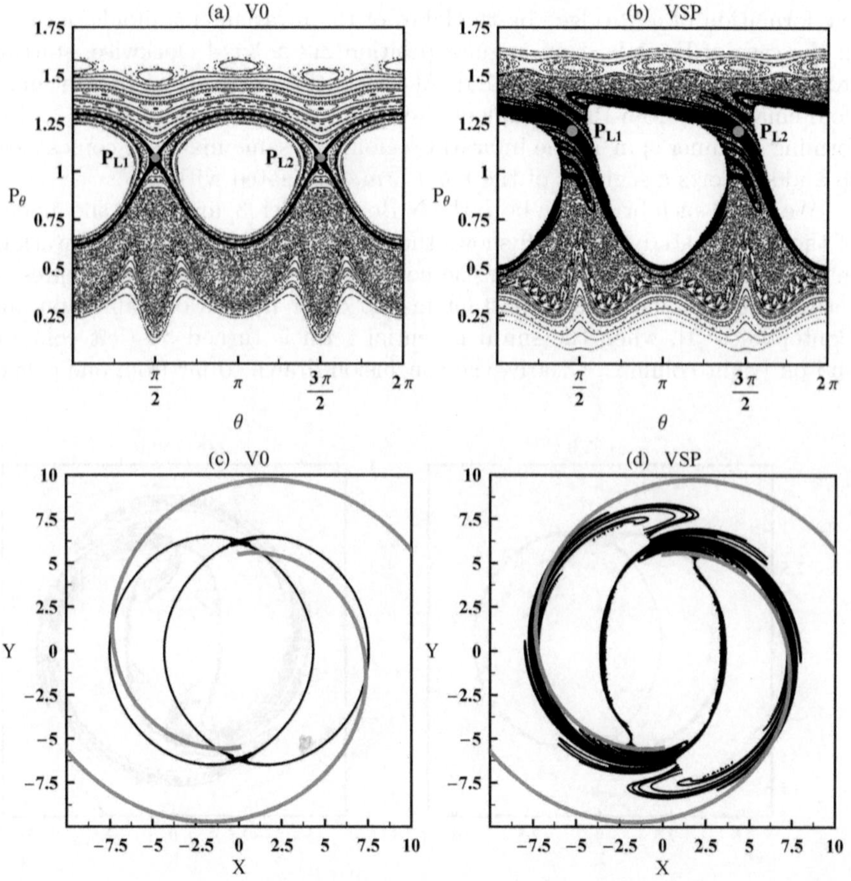

Fig. 1. (**a**) The phase portrait, the manifolds (*black line*) and the $PL_{1,2}$ fixed points (*gray*) in the surface of section for $E_J = -1.9 \times 10^5$ when the spiral potential is turned off. (**b**) Same as in (a) but for $E_J = 1.911 \times 10^5$ and the spiral potential turned on. (**c**), (**d**) same manifolds as in (a), (b) respectively, in the configuration space. The theoretical spiral of the best-fit self-consistent model (*gray thick line*) is also included

The consequences of such a difference appear even more important when the manifolds are plotted in the configuration space (Fig. 1c, spiral potential turned off, and Fig. 1d, spiral potential turned on). We see that the manifolds produced by the bar alone yield only an outer ring structure that deviates considerably from the spiral structure of the galaxy (gray solid lines). On the contrary, when the contribution of the spiral potential is taken into account, the lobes of the invariant manifolds become considerably wider, and they are also angularly deformed so as to closely follow the spiral arms of the galaxy.

Some details of Fig. 1d are worth noticing. In particular, we stress a feature of the manifolds which is commonly observed in all studied cases. This is the formation of a "bridge" by the lobes of the invariant manifolds, located, in the case of Fig. 1d, at an angular position $\Delta\theta \simeq 3\pi/4$ clockwise, starting from L_1 (and symetrically from L_2). At nearly that angle, the invariant manifold emanating from PL1 abandons the spiral arm associated with L_1. After forming an inner spur in the interarm region, the same manifold comes close to and supports a segment of the *other* arm, associated with L_2.

We found such bridges in both the N-Body model [3] and in all the models of the present study. Figure 2 shows the projection of the unstable invariant manifolds of the PL1 family on the configuration space for some representative values of the Jacobi constant in two other models of Kaufmann and Contopoulos [1], when the spiral potential term is turned off (left column) and on (right column). Clearly, the conclusion drawn so far from one galaxy

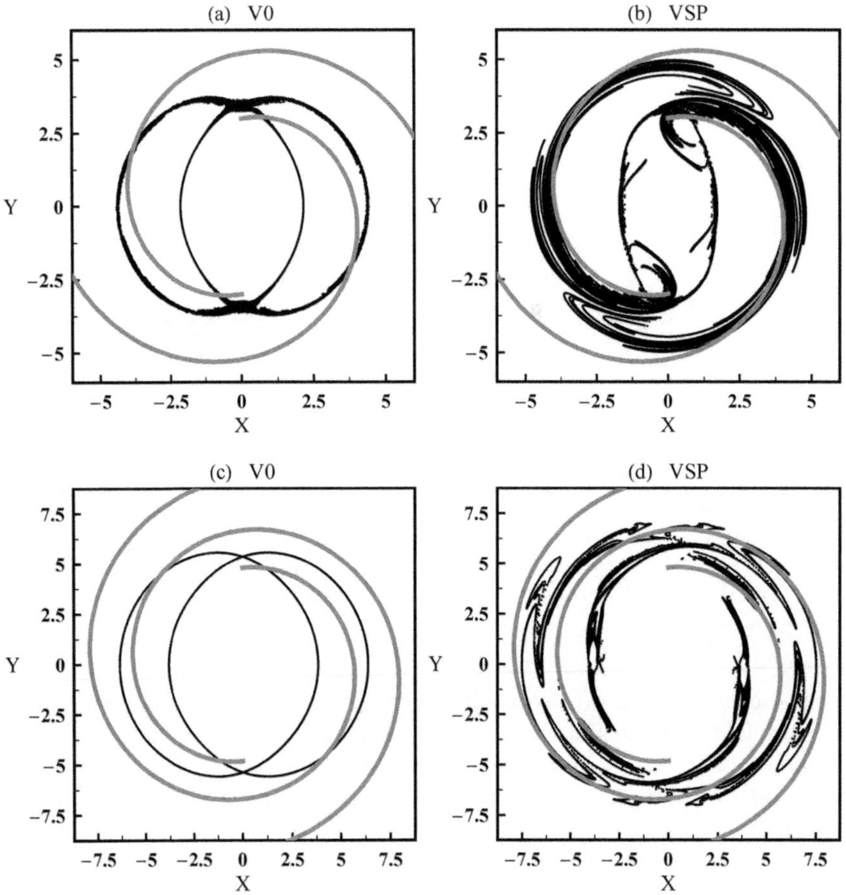

Fig. 2. (**a**), (**b**) Manifolds for the best-fit self-consistent model of Kaufmann and Contopoulos for the galaxy NGC1073. (**c**), (**d**) Same for the galaxy NGC1398. The Jacobi constants used are (a) $E_J = -2.94 \times 10^4$, (b) $E_J = -2.97 \times 10^4$, (c) $E_J = -2.895 \times 10^5$, (d) $E_J = -2.88 \times 10^5$

appears to be generic, i.e., in neither case can the bar potential alone produce manifolds that follow the real spiral patterns of the galaxies. Conversely, when the spiral potential terms are turned on, the fittings of the real spiral patterns with the invariant manifolds is quite satisfactory in all three cases.

Bridges are seen in all the panels of the right column of Fig. 2. A case of particular interest is shown in Fig. 2d. In this case, referring to the galaxy NGC1398, the spirals are tightly wound and a significant extension of the spiral arms is found up to angles $\theta = \pi$ (spiral from L_1) or 2π (spiral from L_2). The invariant manifolds of the PL1 family in that case show a remarkable behavior (Fig. 2d). We see that initially the manifold from PL1 is transverse to the inner part of the spiral arm associated with L_1, and it quickly joins a part of the *outer* spiral arm associated with L_2, forming a bridge outwards with $\Delta\theta \simeq \pi/4$. Later (for larger length), the manifold forms two consecutive bridges, in which it joins consecutively the inner part of the spiral arm associated with L_1 (at an angle $\Delta\theta \simeq \pi/2$) and, then, the inner part of the spiral arm associated with L_2 (at an angle $\Delta\theta \simeq \pi$). We conclude that, although the manifolds are always found to support the spiral structure of the galaxies, this can be achieved by different sequences of bridges which depend on the model considered.

An immediate consequence of the previous analysis is that one cannot find the morphology of the spiral arms that correspond to a particular type of bar by calculating the invariant manifolds of the PL1 family in only the bar potential. In fact, although our theory predicts that the spirals are dynamically coupled to the bar, the self-consistent response of the spiral must be taken into account in all calculations related to the morphology of the spiral arms. The question is thus whether a simple model can be constructed to account for this type of the bar–spiral interaction.

Figure 3 shows the result of an attempt to construct such a simple model for the galaxy NGC3992. Our basic assumption is that the unstable mani-

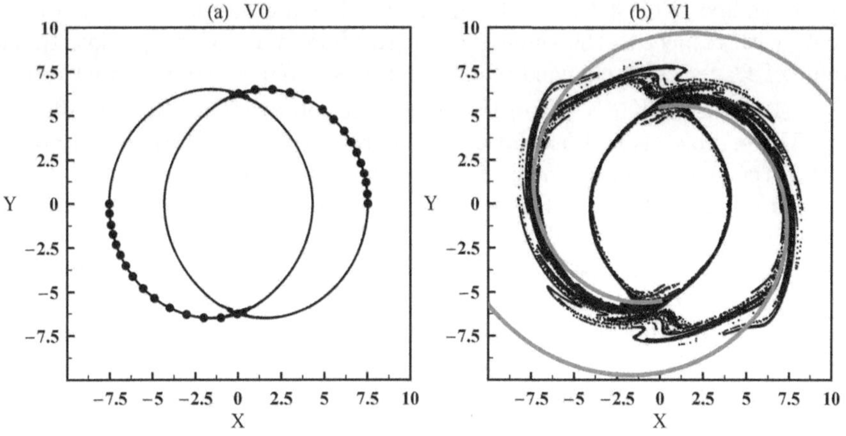

Fig. 3. (**a**) Same manifold as in Fig. 1c. The black points represent the spiral mass distribution used. (**b**) the response model of the spiral mass distribution of (a)

folds produced by the bar alone "trigger" the formation of a ring-like pattern, which however is non-uniformly populated by stars, the parts closer to the Lagrangian points $L_{1,2}$ being more densely populated. Such a process is roughly modelled by distributing a total mass equal to

$$M_s = \frac{1}{2}\int_{r_1}^{r_2}\int_0^{2\pi}\int_{-\infty}^{\infty} |\rho_s(r,\theta,z)| r dr d\theta dz \tag{7}$$

along the invariant manifolds produced by the bar alone, in the angular intervals $[0,\pi/2)$ (manifold from PL1) and $[\pi,3\pi/2)$ (manifold from PL2, Fig. 3a). In (7), the quantity $\rho_s(r,\theta,z)$ is an approximate expression for the density perturbation corresponding to the potential (4) given by the WKB ansatz:

$$\rho_s(r,\theta,z) = -\frac{|k\Delta|(|k\Delta|+1)}{4\pi G\Delta^2} V_s(r,\theta) sech^{2+|k\Delta|}(z/\Delta), \tag{8}$$

where $k = 2/(r\tan i)$, according to the formula given by Vandervoort (see [11]). Thus, the quantity M_s is an estimate of the total mass contained in the maxima of the spiral arms of the real galaxy. However, contrary to the model of Sect. 2, in the present case the mass M_s is nonzero, and it is initially distributed along the invariant manifolds produced by the bar alone.

Figure 3b shows a calculation of the invariant manifolds obtained after the addition of the contribution of the mass M_s in the total potential. Clearly, after only one iteration the obtained manifolds are much wider and much more extended in the radial direction than in Fig. 3a. In fact, these manifolds are already very close to the manifolds of Fig. 1d, and they also fit well the real spiral pattern of the galaxy.

Finally, we examined whether it is possible to produce manifolds fitting the observed spiral pattern with a pure bar, by altering the bar parameters with respect to the values given in one model of Kaufmann and Contopoulos [1]. Figure 4 shows such an example referring to the model for the galaxy NGC3992. In this figure the bar amplitude was increased by 300% with respect to the value given in Kaufmann and Contopoulos [1]. The value of Ω_p was also changed to $\Omega_p' = 60\,\mathrm{km\,s^{-1}\,Kpc^{-1}}$ (the original value is $\Omega_p = 43.6\,\mathrm{km\,s^{-1}\,Kpc^{-1}}$) in order to keep the corotation at the same distance $\simeq 5.5\,\mathrm{Kpc}$, which is an observational constraint corresponding roughly to the end of the bar. Now, while the fitting of the spirals by the manifolds of the new pure bar model (Fig. 4) is still not as good as in the original model with the spiral potential turned on (compare Figs. 1d and 4), the changes in the model parameters necessary to produce even this result are very unrealistic. We should notice that the parameters of the original model were obtained by the requirement of self-consistency. In particular, the bar perturbation in the original model is of order 100% with respect to the axisymmetric background. Thus, with the new parameters the bar perturbation should rise to 400% of the axisymmetric background, and this bar should rotate about 1.5 times faster than the bar in the best-fit model of Kaufmann and Contopoulos [1].

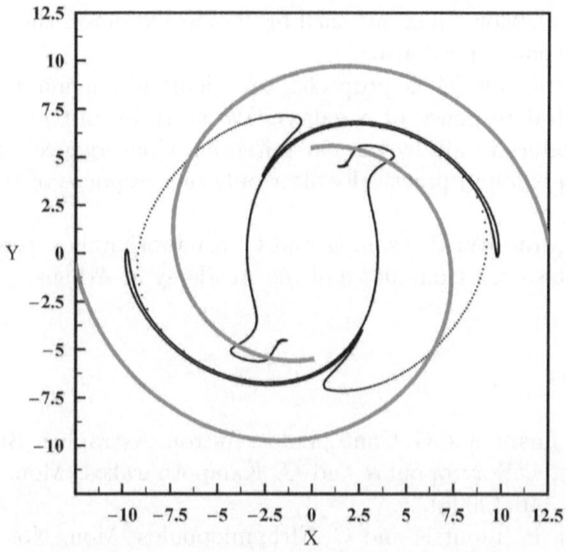

Fig. 4. The unstable invariant manifolds of the PL1 orbits in a modification of the model of Kaufmann and Contopoulos [1] for the galaxy NGC3992, in which a pure bar is considered with mass $M_b' = 3M_b$. The pattern speed was also changed, $\Omega_p' \simeq 1.5\Omega_p$, so that the corotation radius remains unaltered

4 Conclusions

We calculated the unstable invariant manifolds of the family of the short period unstable periodic orbits around the Langrangian points L_1 or L_2 of the three self-consistent models of real barred galaxies treated by Kaufmann and Contopoulos [1]. Our aim was to examine the applicability of the theory of Voglis et al. [3] regarding the reproduction of the observed spiral pattern by the invariant manifolds. Our conclusions are the following:

(1) When both the bar and spiral components of the self-consistent models are taken into account, the projection of the invariant manifolds on the configuration space produces a pattern that follows closely the observed spiral pattern of the galaxies.

(2) When only the bar component is taken into account, the manifolds no longer support the spiral arms but they only form outer ring-like structures. This lack of spiral arms is due to the existence of a critical set of invariant tori in the corotation region which pose obstructions to the development of the invariant manifolds beyond corotation.

(3) Only if we increase the bar perturbation to 400% of the axisymmetric background (and increase also the pattern velocity by 50%) we find that these invariant curves are destroyed and outer spiral arms can be formed.

(4) The invariant manifolds form bridges that transfer the flow of the chaotic orbits from one spiral arm to the other via spurs developed in the

interarm region. More than one such bridges are found in the case of systems with tightly wound spiral arms.

(5) A simple model is proposed to calculate the manifolds in a self-consistent spiral response of a galaxy. We start by putting matter on the manifolds produced only by the bar potential. Convergence towards the real spiral pattern is found practically after only one response of the model.

Acknowledgements: P. Tsoutsis and C. Kalapotharakos were supported in part by the Research Committee of the Academy of Athens.

References

1. D.E. Kaufmann and G. Contopoulos. Astron. Astrophys. **309**, 381 (1996).
2. N. Voglis, I. Stavropoulos and C. Kalapotharakos. Mon. Not. R. Astr. Soc. **372**, 901 (2006).
3. N. Voglis, P. Tsoutsis and C. Efthymiopoulos. Mon. Not. R. Astr. Soc. **373**, 280 (2006).
4. G. Contopoulos. Astron. Astrophys. **64**, 323 (1978).
5. P.A. Patsis. Mon. Not. R. Astr. Soc. **369**, 56 (2006).
6. D. Pfenniger and D. Friedli. Astron. Astrophys. **252**, 75 (1991).
7. M. Romero-Gomez, J.J. Masdemont, E.M. Athanassoula and C. Garcia-Gomez. Astron. Astrophys. **453**, 39 (2006).
8. M. Romero-Gomez, E.M. Athanassoula, J.J. Masdemont and C. Garcia-Gomez. Astron. Astrophys. **472**, 63 (2007).
9. L.S. Sparke and J.A. Sellwood. Mon. Not. R. Astr. Soc. **225**, 653 (1987).
10. P. Tsoutsis, C. Kalapotharakos, C. Efthymiopoulos and G. Contopoulos, submitted (2008).
11. G. Contopoulos and P. Grosbol. Astron. Astrophys. **197**, 83 (1988).

Collisional N-Body Simulations and Time-Dependent Orbital Complexity

N.T. Faber

Observatoire Astronomique, Université de Strasbourg and CNRS UMR, 7550, 11 rue de l'Université 67000, Strasbourg, France
nfaber@astromaster.u-strasbg.fr

1 Introduction

In this contribution, we present an accurate, easy-to-implement and time-resolved complexity-detection tool for individual orbits in collisional N-body simulations. A state of dynamical equilibrium is assumed throughout. We implement a specially adapted method of time series analysis that is based on a discrete wavelet transform. The method post-processes the orbital data of a simulation and provides a time-resolved measure of complexity. The present contribution mainly focuses on a basic description of the method and is structured as follows. In Sect. 2 we introduce the identification technique for time-dependent complexity. Section 3 presents an application to the $N = 3$ Pythagorean problem. A brief conclusion is given in Sect. 4.

2 Method

2.1 Wavelet Transforms

Definitions

The complexity of motion is analyzed by means of a wavelet transform and is described in detail in [5]. The use of wavelet transforms has found applications in a wide range of domains such as seismology, financial time series processing or medical electrocardiogram studies; see [6] for further references. A wavelet transform provides a time–frequency representation of a time series $f(t_q)$ by fitting a wavelet Ψ to a set of points $\{t_q\}$. Whereas Fourier-based methods decompose a signal into infinite sine and cosine functions, effectively losing information at individual times t_q, the wavelet transform offers precise localization in both the frequency- and time-domain. Wavelet transforms remain band-limited however; they are made up of not one but a limited range of several frequencies.

G. Contopoulos, P.A. Patsis, *Chaos in Astronomy.* Astrophysics and Space Science Proceedings, doi: 10.1007/978-3-540-75826-6,

A wavelet family $\Psi_{a,\tau}$ is defined by the set of elemental functions generated by scaling and translating a mother wavelet $\Psi(t)$:

$$\Psi_{a,\tau}(t) = |a|^{-\frac{1}{2}}\Psi\left(\frac{t-\tau}{a}\right), \tag{1}$$

where a represents the scale variable and τ the translation variable ($a, \tau \in \mathcal{R}, a \neq 0$).

The continuous wavelet transform (CWT) is defined as the correlation between a signal $S(t) \in L^2(\mathcal{R})$ (the space of square summable functions) and the wavelet family $\Psi_{a,\tau}(t)$ for each a and τ (see e.g. [4]):

$$\langle S(t), \Psi_{a,\tau}\rangle \equiv |a|^{-\frac{1}{2}} \int_{-\infty}^{+\infty} S(t)\bar{\Psi}\left(\frac{t-\tau}{a}\right) \mathrm{d}t. \tag{2}$$

Here $\langle S, \Psi_{a,\tau}\rangle$ denotes the wavelet coefficients and $\bar{\Psi}$ is the complex conjugate of Ψ. Equation (2) can be inverted to reconstruct the original time series. The CWT is known to produce a large amount of wavelet coefficients which implies considerable CPU execution times (see e.g. [13]). In addition, the information the CWT displays at closely spaced scales or at closely spaced time points is highly correlated and thus unnecessarily redundant.

For these reasons we instead compute a *discrete* wavelet transform (DWaT). The DWaT offers a highly efficient wavelet representation that can be implemented with a simple recursive filter scheme [4, 7]. Unlike the numerical CWT implementation which easily produces more than 10^5 coefficients for a single orbital time series of $Q = 8,192$ data points, the DWaT only produces as many coefficients as there are samples in the time series, i.e. Q. This property of the DWaT of avoiding redundant wavelet coefficients serves in defining a proper measure of complexity, as we will show in Sect. 2.2. For a given choice of the mother wavelet function $\Psi(t)$ and for the discrete set of parameters $a_j = 2^j$ and $\tau_{j,k} = 2^j k$ ($j,\ k \in \mathcal{Z}$), the wavelet family

$$\Psi_{j,k}(t) = |a_j|^{-\frac{1}{2}}\Psi\left(\frac{t-\tau_{j,k}}{a_j}\right) = 2^{-\frac{j}{2}}\Psi(2^{-j}t - k) \tag{3}$$

defines an orthonormal basis of $L^2(\mathcal{R})$. The time series $f(t_q)$ is sampled at $Q = 2^J$ ($J \in \mathcal{N}$) constant time intervals of size $\Delta = t_{q+1} - t_q$. The discrete wavelet expansion then reads (see e.g. [12], (36))

$$f(t) = \sum_{j=1}^{J} \sum_{k=1}^{2^{j-1}} \langle f(t_q), \Psi_{j,k}\rangle_{\text{discrete}}\, \Psi_{j,k}(t)\Delta^{-1}. \tag{4}$$

For simplicity we set $\Delta = 1$ throughout Sect. 2. Here $f(t)$ is the reconstructed signal and $J = \log_2 Q$ is the number of scales over which the time series $f(t_q)$ is analyzed. The DWaT coefficients $\langle f(t_q), \Psi_{j,k}\rangle_{\text{discrete}}$ can be understood as a representation of the wavelet power spectrum (or, energy) at scale j and time t_q, associated to the time series $f(t_q)$. They represent the local residual errors between successive signal approximations at scales j and $j-1$. In what follows, we set

$$P_j(k) \equiv \langle f, \Psi_{j,k} \rangle_{\text{discrete}} \tag{5}$$

for a more convenient notation of the DWaT coefficients. As mentioned earlier, there are Q coefficients $P_j(k)$ and the number of coefficients computed for resolution level j is 2^{j-1}. The frequency band over which the $P_j(k)$ are computed is limited by the frequency $1/(Q\Delta)$ (scale $j=1$) in the low-frequency domain and by the Shannon–Nyquist critical frequency $f_{cr} = 1/(2\Delta)$ in the high-frequency domain (scale $j = J$; see also [10], Sects. 12.1 and 13.10). We refer to [13] for further details about the wavelet representation.

We use bi-orthogonal cubic spline functions as mother wavelets ([3], see case $\{N, \tilde{N}\} = \{3, 9\}$ in their Table 6.1),

$$\Psi(t) = \sum_l g_l \Theta_3(2t - l), \tag{6}$$

where the g_ls are known as basic spline coefficients and

$$\Theta_3(t) = \begin{cases} (t+1)^2/2, & -1 \le t \le 0 \\ -(t-1/2)^2 + 3/4, & 0 \le t \le 1 \\ (t-2)^2/2, & 1 \le t \le 2 \\ 0, & \text{otherwise.} \end{cases} \tag{7}$$

This choice is motivated by three arguments. First and most importantly, spline functions provide an excellent time–frequency localization when compared to other mother wavelet candidates [1, 17]. Instantaneous changes in the dynamics are accurately singled out by the DWaT (see [5] for further information). Second, the use of splines is computationally inexpensive [16] and provides further desirable properties such as, e.g. compact support and smoothness. (For an exhaustive discussion on spline interpolation, see [16, 17].) Finally, the use of a bi-orthogonal spline mother wavelet also implies reduced border effects, an undesired artifact of the wavelet transform algorithm (see [5]).

We analyze the velocity time series

$$f(t_q) = v_{\alpha_i}(t_q) \qquad (\alpha = x, y, z), \tag{8}$$

i.e. the v_{x_i}, v_{y_i} and v_{z_i} velocity components of each particle i ($i = 1, ..., N$). We do not use the information available on the positions of the bodies since there may be important differences in magnitude between the beginning and the end of these time series. Positional information is then likely to produce pronounced DWaT border effects [5].

2.2 Measures of Complexity

Our goal is to obtain from the DWaT a quantitative estimate of the time-dependent complexity of the frequency spectrum of a trajectory. Exploiting some notions of information theory, we here present the discrete wavelet transform information measure (DWaTIM) as an efficient indicator for complexity. In what follows, we provide a succinct overview of the concept. We follow closely the approach of [12] and [8]. We refer the reader to these works for a more extended discussion.

Discrete Wavelet Transform Information Measure (DWaTIM)

For a chosen time window of size $\kappa\Delta$ we compute the wavelet energy at each resolution level j,

$$E_j = \sum_{k=1}^{2^{j-1}} |P_j(k)|^2, \tag{9}$$

and the total wavelet energy,

$$E = \sum_j E_j, \tag{10}$$

to obtain the so-called relative wavelet energy $p_j = E_j/E$ at scale j. The DWaT then provides a probability distribution

$$\mathcal{P} = \{p_j\}, \quad (j = 1, ..., J), \tag{11}$$

which weighs the base frequency j in the reconstruction of the original signal $v_{\alpha_i}(t_q)$. By definition we have $\sum_{j=1}^{J} p_j = 1$.

The amount of disorder present at time t in an orbital time series can be quantified by determining the information needed to describe the orbit at that time [15]. An information measure (hereafter IM) can be seen as a quantity that describes the characteristics of the time-scale probability distribution $\mathcal{P}$ of (11) [12]. The IM gives the amount of information required per time unit to specify the state of the system up to a given accuracy. The IM we use in this work is the Shannon entropy [11, 14, 15].

$$W_S[\mathcal{P}] = -K \sum_j p_j \log_2(p_j), \tag{12}$$

where $K = 1$ is an arbitrary numerical constant. A minimum of information entropy $\min(W_S[\mathcal{P}]) = 0$ is obtained if $p_j = 1$ for some scale j and 0 for all the remaining scales. This situation only occurs for the ordered dynamics of a periodic orbit with a single base frequency. Likewise, the state of highest complexity is obtained for the case of a white noise signal. For such an orbit, the entire band of base frequencies is sampled by the DWaT in equal proportions, and $\mathcal{P}$ is characterized by the uniform probability distribution

$$\mathcal{P}_u = \{\frac{1}{J}, ..., \frac{1}{J}\}. \tag{13}$$

In this case the IM is $\max(W_S[\mathcal{P}]) = W_S[\mathcal{P}_u] = \log_2 J$. For a given velocity component v_{α_i} ($\alpha = x, y, z$; $i = 1, ..., N$) we define the DWaTIM to be the measure

$$H_{\alpha_i} \equiv \left(\frac{W_S[\mathcal{P}]}{W_S[\mathcal{P}_u]} \right)_{\alpha_i}, \tag{14}$$

i.e. the normalized Shannon entropy for velocity component α of particle i. The generalized DWaTIM of particle i is then obtained by taking the arithmetic mean over the three components,

$$\Upsilon_i \equiv \frac{H_{x_i} + H_{y_i} + H_{z_i}}{3}. \tag{15}$$

Finally, the overall DWaTIM $\Upsilon_{\rm tot}$ of the system is computed by averaging over all particles i,

$$\Upsilon_{\rm tot} \equiv \frac{\sum_{i=1}^{N} \Upsilon_i}{N}. \tag{16}$$

In what follows we will argue that an increase in DWaTIM correlates with an increase in the complexity of the underlying orbital dynamics at that instant.

3 Application to the $N = 3$ Pythagorean Problem

We show a small application of the method to the $N = 3$ Pythagorean problem. We refer to [5] for a more exhaustive explanation of this analysis and for further illustrations of the efficiency of the method. The well-studied Pythagorean configuration [2] is a classic example of long-term complex behavior. The initial conditions consist of three particles at rest, placed at the vertices of a Pythagorean triangle. The initial conditions are given in Table 1 and the configuration is depicted in Fig. 1.

Orbital integrations were performed with the *kira* integrator [9] over 1,500 time units by repeatedly re-running the initial configuration with a reduced per-step integration error until convergence of the result was reached. In this way, the final trajectories showed a total absolute energy error $|E_{\rm final} - E_{\rm initial}|$ of less than 10^{-10}. The analysis is restrained to the first 1,024 time units of integration, during which we found by visual inspection that the system spent

Table 1. Pythagorean problem: initial conditions

Body	mass	x	y	z	v_x	v_y	v_z
No. 1	3	1	3	0	0	0	0
No. 2	4	−2	−1	0	0	0	0
No. 3	5	1	−1	0	0	0	0

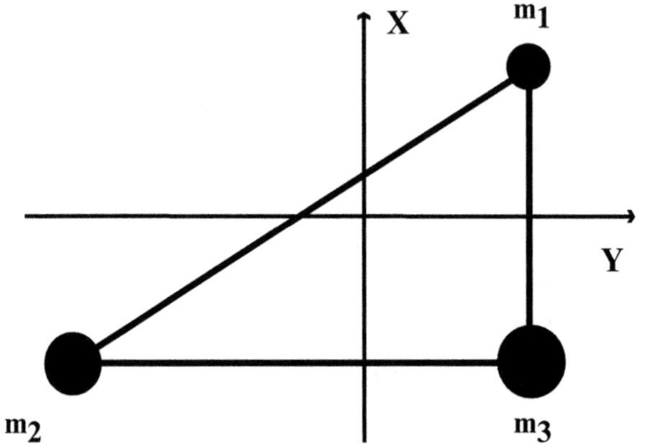

Fig. 1. Pythagorean problem: initial configuration

a comparable amount of time in trivial and in more complicated states. The time series comprise $Q = 8,192$ data points. Results of the respective time series analysis for particle no. 1 and for the entire Pythagorean problem are shown in the left-hand and right-hand panels of Fig. 2, respectively. The v_y-coordinate of body no. 1 is shown in the left-hand Fig. 2a. In the right-hand panel of Fig. 2a we illustrate the dynamics of the whole Pythagorean problem by showing the position x of all the three bodies. The DWaT scalogram of body no. 1 and the average DWaT of the Pythagorean problem (the average magnitude of base frequency $1/\Pi_j$ at integration time t), computed by averaging over the three particles and over the x- and y-components, are shown in the left-hand and right-hand panels of Fig. 2b, respectively. The DWaTIM for body no. 1, H_{y_1}, and the overall DWaTIM of the Pythagorean problem, Υ_{tot} (see Sect. 2.2), are shown in Fig. 2c. The dash-dotted line depicts the DWaTIM as computed by taking into account all the scales $j = 1, \ldots, 13$. The solid line highlights the dramatic improvement of the diagnostic, obtained when ignoring the first four scales ($j = 1, \ldots, 4$) in the computation. Broadly speaking, the DWaTIM measure captures the time-dependent complexity of the dynamics of the Pythagorean problem. The complexity is gauged on a well-defined scale between 0 and 1 defined by the two limiting cases of a unique base frequency sinusoid (DWaTIM = 0) and of a white noise signal (DWaTIM = 1). We refer to [5] for a detailed discussion of Fig. 2.

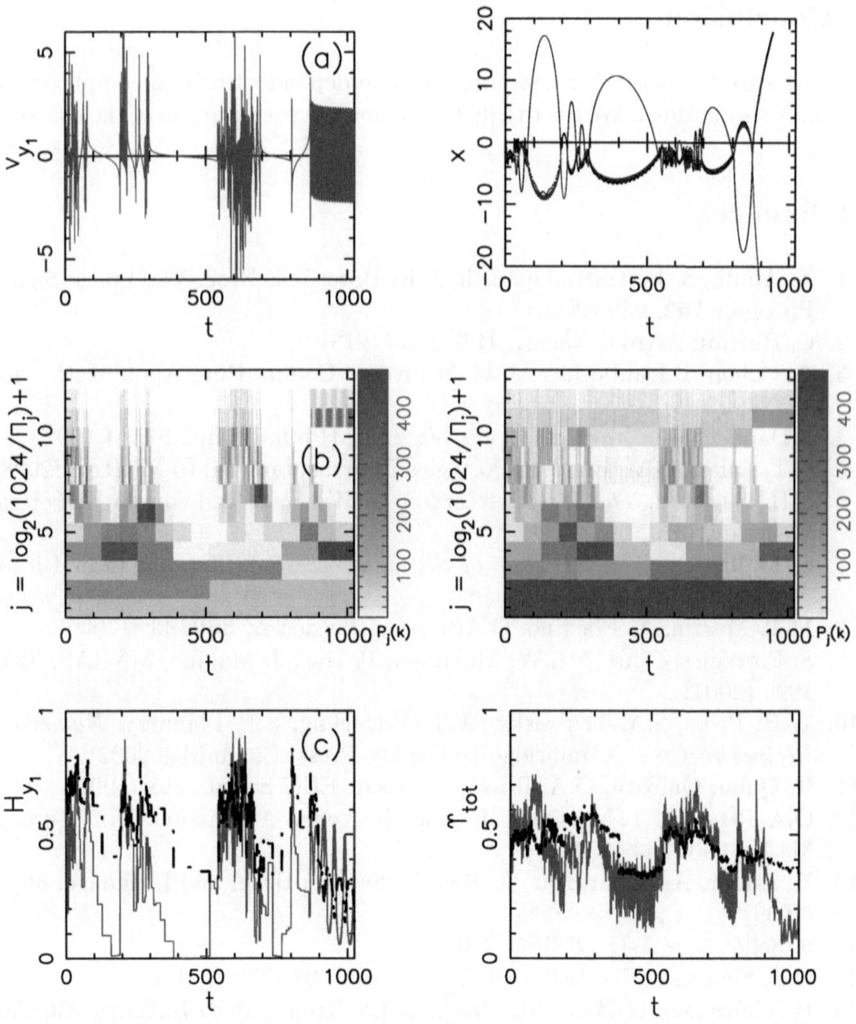

Fig. 2. Complexity measures of the x component of particle no. 1 (*left-hand panels*) and averaged overall complexity (*right-hand panels*) of the Pythagorean problem (see Table 1, Fig. 1 and text for further detail). From top to bottom: (**a**) velocity time series (*left-hand panel*) and x position time series of the three particles (*right-hand panel*), (**b**) discrete wavelet transform (DWaT) (*left-hand panel*) and average DWaT (*right-hand panel*), (**c**) discrete wavelet transform information measure (DWaTIM) H_{y_1} and average DWaTIM Υ_{tot}. The *dash-dotted lines* on (**c**) are obtained by computing the measures using the complete bandwidth of the DWaT. The *solid lines* show the same result computed by excluding the contribution of scales $1, \ldots, 4$

4 Conclusion

We presented a method to compute the time-dependent orbital complexity in N-body simulations. We refer to [5] for a complete presentation of the concept.

References

1. N. Ahuja, S. Lertrattanapanich, N.K. Bose: IEE Proc.-Vis. Image Signal Process., **152**, 659 (2005)
2. C. Burrau: Astron. Nachr., **195**, 113 (1913)
3. A. Cohen, I. Daubechies, J.-C. Feauveau: Comm. Pure Appl. Math., **45**, 485 (1992)
4. I. Daubechies: *Ten Lectures on Wavelets*, (Philadelphia: SIAM 1992)
5. N.T. Faber, C.M. Boily, S. Portegies Zwart: submitted to MNRAS (2008)
6. B.B. Hubbard: *The World According to Wavelets*, 2nd edn (A.K. Peters, Wellesley, MA 1998)
7. S. Mallat: *A Wavelet Tour of Signal Processing*, 2nd edn (San Diego, Academic Press 1999)
8. M.T. Martin, A. Plastino, O.A. Rosso: Physica A, **369**, 439 (2006)
9. S. Portegies Zwart, S.L.W. McMillan, P. Hut, J. Makino: MNRAS, **321**, 1999 (2001)
10. W.H. Press, S.A. Teukolsky, W.T. Vetterling, B.P. Flannery: *Numerical Recipes in C++* (Cambridge University Press, Cambridge 2002)
11. R. Quian Quiroga, O.A. Rosso, E. Basar: EEG Suppl., 298 (1999)
12. O.A. Rosso, M.T. Martin, A. Figliola, K. Keller, A. Plastino: J. of Neurosc. Meth., **153**, 163 (2006)
13. V. Samar, A. Bopardikar, R. Rao, K. Swartz: Brain and Language, **66**, 7 (1999)
14. S. Sello: New Astr., **8**, 105 (2003)
15. C.E. Shannon: The Bell Technical Journal, **27**, 379 (1948)
16. P. Thévenaz, T. Blue, M. Unser: IEEE Trans. Med. Imaging, **19**, 739 (2000)
17. M. Unser: IEEE Signal Process. Mag., **16**, 22 (1999)

Resonances in Galactic and Circumstellar Disks

A.C. Quillen

Department of Physics and Astronomy, University of Rochester,
Rochester, NY, USA
aquillen@pas.rochester.edu

Summary. I contrast the theory of Lindblad resonances for the stars in a galaxy and mean motion resonances for bodies in a circumstellar disk. Though these resonances are present in different astrophysical settings, they are identical mathematically. I introduce some recent and rich applications of first order resonance theory including interpretations of circumstellar disks in terms of predicted but unseen planets and tight constraints on Galactic structure.

1 Introduction

In this paper I first review the theory of first order mean motion and Lindblad resonances. Though these resonances are present in different astrophysical settings, they are identical mathematically. Thus some applications that have been commonly used in celestial mechanics settings can be applied toward Galactic systems and vice versa.

In galaxies perturbations from non-axisymmetric structures such as spiral arms or bars have strength (in units of force divided by that from the axisymmetric component) of order a few percent. In circumstellar disks planet masses are at most 1/1,000 of that of the central star. Because of their small sizes these perturbations are important only at resonances. Thus predictions for planets residing in circumstellar disks can depend on a deep understanding of the effects of these resonances. A major difference between the two settings is the timescale over which the resonances can act. For example only 40 rotation periods comprise a Hubble time at the Sun's radius in the galaxy, however 4 billion years (earth orbits about the Sun) comprise the age of the solar system.

Resonance overlap leading to chaotic behavior can occur in both settings, for example near corotation of a planet and when both spiral and bar perturbations are present in a galaxy [1]. Resonances can be particularly important if they capture particles. In this case they can sculpt the disk leaving evidence of their passing even if they are weak and narrow. While most applications of

G. Contopoulos, P.A. Patsis, *Chaos in Astronomy.* Astrophysics and Space Science Proceedings, doi: 10.1007/978-3-540-75826-6,

resonance capture theory have been applied to drifting planetary or satellite systems, bar growth and evolution provides a setting where Lindblad resonances may capture stars [2]. Better understanding of resonance capture in the galactic setting may yield constraints on the previous structure of our Galaxy.

Resonances are of current interest in interpreting circumstellar disks as they may allow one to place constraints on unseen planets based on observations of disk morphology. For example the Fomalhaut debris disk contains a central clearing with a sharp inner edge. The inner clearing and truncation of the disk could be a sign post of a planetary system. Its inner edge has eccentricity of 0.1 and semi-major axis 133 AU [3]. The collision timescale for the dust particles is estimated from the observed opacity and is $\sim 10^3$ orbital periods. At this collision timescale spiral density waves cannot be driven into the disk, thus Quillen [4] adopted a diffusion and clearing mechanism for disk truncation. This led to the prediction that a Neptune mass planet with an eccentricity equal to that of the dust ring and aligned with it lies just interior to the dust ring. The disk edge is located at the edge of the chaotic zone associated with the proposed planet's corotation region. Hence a resonance overlap criterion was used to estimate the distance between the planet and the disk edge and so constrain the semi-major axis and mass of the proposed planet [5].

Understanding of the dimensional scaling of the first order resonances has been very useful, leading to generalization of capture theory to include the non-adiabatic limit [6] and an estimate of when spiral density waves can be driven into a collisional disk by a planet [4].

When resonances provide a good explanation for observed phenomena then tight constraints can be placed on the pattern speed of the perturbations. For example, this has resulted in a very tight constraint (at accuracy of a few percent) on the Galactic bar's pattern speed [7]. Hence dynamical models for forthcoming radial velocity and proper motions surveys may provide tight constraints on spiral and bar models for our Galaxy.

2 Lindblad and Mean Motion Resonances

A star orbiting in a disk galaxy or a small body orbiting in a circumstellar disk can experience resonant gravitational perturbations with an external perturber. In a circumstellar disk the perturber is often a planet in orbit with a mean motion n_p (akin to an angular rotation rate but taking into account motion if the planet is on an eccentric orbit). In a galaxy stars are perturbed by non-axisymmetric features such as spiral arms or a bar moving at an angular rotation rate Ω_p. Here the pattern speed of spiral or bar perturbation is analogous to the mean motion of the planet.

2.1 The Resonant Angle

Lindblad resonances occur where the angular rotation rate Ω, and epicylic frequency, κ, of the particle and pattern speed of the spiral or bar pattern, Ω_p can be related by a relation

$$i\Omega + j\kappa \approx k\Omega_p \tag{1}$$

that depends on integers i, j, k. Mean motion resonances occur when the particle mean motion, n, can be related to the mean motion of the planet, n_p.

$$in \approx jn_p \tag{2}$$

with i, j integers. This relation is simpler than (1) because the epicylic frequency is very close to the angular rotation rate, $\kappa \sim \Omega$, in the Keplerian system.

If we integrate (1) we find that

$$\phi = i\theta_2 + j\theta_1 - k\Omega_p t \approx \text{constant}, \tag{3}$$

where $\Omega_p t$ is an angle associated with the moving bar or spiral arm that is fixed in the rotating frame. We can integrate (2) finding

$$\phi = i\lambda - j\lambda_p \approx \text{constant}, \tag{4}$$

where λ_p is the planet's mean longitude. We refer to the slowly varying angle ϕ as a resonant angle.

A particle in resonance can be considered one in which a resonant angle ϕ librates about a fixed value (typically 0 or π) rather than cycles, constantly increasing or decreasing, going from 0 to 2π over and over again. This is analogous to a pendulum with Hamiltonian

$$H(p, \phi) = \frac{1}{2}p^2 + A\cos\phi, \tag{5}$$

which exhibits two types of motion, each separated by a limiting curve called the separatrix. Motion about the fixed point, $p = 0$, $\phi = \pi$ is similar to that of a harmonic oscillator with period $2\pi/\sqrt{A}$. Here the angle ϕ librates about its equilibrium value. But for large p all motion involves continuously increasing values of ϕ. Physically this corresponds to the pendulum swinging in a large circle rather than gently oscillating about its equilibrium state. See Fig. 1 for trajectories in phase space for the pendulum model.

Near Lindblad and mean motions resonances a resonant angle given as ϕ in (3), (4) varies slowly. Hence these are regions where small perturbations, such as from spiral arms or from a planet, can become important and cause the resonant angle to librate about a fixed value rather than oscillate. We consider a particle trapped in resonance when the resonant angle librates about a particular value. A particle is not in resonance when the resonant angle continuously increases or decreases.

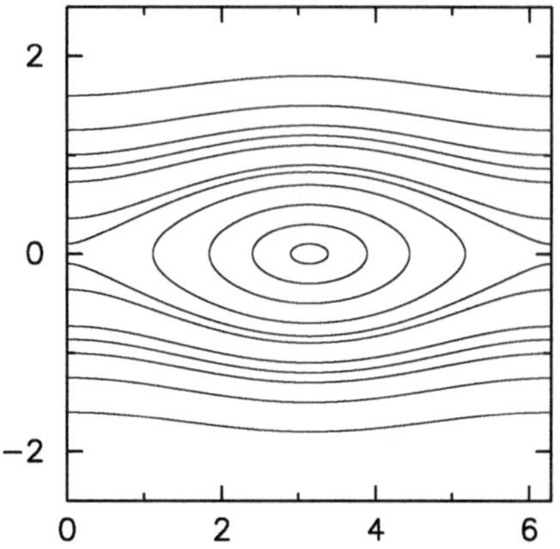

Fig. 1. Orbits move along the level contours shown here for a pendulum with Hamiltonian shown in (5). Here the x-axis is ϕ and the y-axis is p. At high and low p orbits oscillate with continuously increasing angle ϕ. Near the fixed point at $p = 0, \phi = \pi$ orbits librate about the fixed or equilibrium point. A separatrix divides the two types of solutions. This separatrix is particularly prone to instability in the presence of additional perturbations.

2.2 Motion Without perturbations

As shown by Contopoulos [8], the dynamics of stars confined to the galactic plane moving in a smooth Galactic potential lacking non-axisymmetric perturbations is described by a Hamiltonian which can be written in a third order post epicyclic approximation as

$$H_0(I_1, \theta_1; I_2, \theta_2) = h + \omega_1 I_1 + \omega_2 I_2 + aI_1^2 + 2bI_1I_2 + cI_2^2... \tag{6}$$

The action variables $I_1 = \frac{1}{2\pi}\int \dot{r}dr$ and $I_2 = J_0 - J_c$, where I_2 is the difference between the angular momentum of the particle and that in a circular orbit. The action variables are integrals of motion when the Hamiltonian is unperturbed. The radial action variable, I_1, is conjugate to the angle θ_1 describing the phase of the epicylic motion. Consequently $\omega_1 \approx \kappa$ the epicyclic frequency. The action variable, I_2, dependent primarily on the particle angular momentum, is conjugate to θ_2 which is approximately the azimuthal angle in the plane of the galaxy. Consequently $\omega_2 \approx \Omega$ with Ω the angular rotation rate for a particle in a circular orbit. The coefficients of the above Hamiltonian and expressions for the action angle variables in terms of coordinates and their momenta were derived by Contopoulos [8].

For the Keplerian system restricted to the plane we employ the Poincaré coordinates $\lambda = M + \varpi$, $\gamma = -\varpi$ and their associated momenta

$$L = \sqrt{GM_*a}, \qquad \Gamma = \sqrt{GM_*a}(1 - \sqrt{1 - e^2}),$$

where M_* is the mass of the star, λ is the mean longitude, M is the mean anomaly, ϖ is the longitude of pericenter, a is the semimajor axis, and e is the eccentricity. These variables are those describing the orbit of a particle or planetesimal in a plane. In analogy to the Galactic system, the momentum Γ describes epicyclic perturbations and is akin to I_1 whereas L is similar to the angular momentum variable I_2. For the Keplerian system the unperturbed Hamiltonian or that lacking the disturbing function describing perturbations from a planet is $H_0(L, \lambda; \Gamma, \gamma) = -1/2L^2$. When expanded near a mean motion resonance, the Hamiltonian is

$$K_0(\Lambda, \psi; \Gamma, \gamma) = a'\Lambda^2 + b'\Lambda + \text{constant} \tag{7}$$

with coefficients given for example in [5]. Here the angles for a $j : j-k$ exterior resonance (to the planet) are $\psi = j\lambda - (j-k)\lambda_p$ and $\gamma = -\varpi$. Subtracting the factor depending on the mean longitude of the planet, λ_p (as seen in the ψ angle), is akin to working in the frame rotating with the planet.

We note that (6) and (7) have been expanded to second order in the momenta. Consider a dynamical system like the pendulum expanded containing only a first order term in p,

$$H(p, \phi) = \kappa p + A \cos \phi. \tag{8}$$

This only has solution with $\dot{\phi} = \kappa$. Hence ϕ increases forever, solutions only oscillate and never librate about a fixed value. To exhibit both types of motion, libration and oscillation, an expansion to second order in the momenta variables is required. This was pointed out by Contopoulos [8] for the galactic setting, and the reason the high order epicyclic expansion was used. Another problem with lower order expansions is that solutions can incorrectly blow up near resonances. To exhibit the richer dynamics present in resonances at least a second order (in momenta) expansion is required.

2.3 Form of Perturbations

For a bar like perturbation we assume that the non-axisymmetric component of the gravitational potential depends on radius and can be expanded in Fourier components

$$V_1(r, \theta) = \sum_m B_m(r) \cos[m(\theta - \Omega_b t)], \tag{9}$$

where Ω_b is the angular rotation rate of the bar and r, θ are coordinates in the plane of the galaxy. For spiral structure we assume that the radial variations depend on angle and the amplitude is nearly constant with radius.

Spiral and bar perturbations look similar once transferred to action angle coordinates. Contopoulos [8] showed to first order in $I_1^{1/2}$ the potential perturbation

$$V_{1,m}(I_1, \theta_1; I_2, \theta_2) = \left(\tfrac{2I_1}{\kappa}\right)^{\frac{1}{2}} \epsilon_m \times \tag{10}$$
$$[\cos(\theta_1 - m(\theta_2 - \Omega_p t + \gamma_{m+})) + \cos(\theta_1 + m(\theta_2 - \Omega_p t + \gamma_{m-})),]$$

where the strength ϵ_m depend on the potential perturbation strength and form, the angles γ are angular offsets and the pattern speed, Ω_p is that of the bar or spiral pattern.

For the Keplerian system perturbations are described by the disturbing function. Taking into account only resonant first order terms (and neglecting secular terms) this can be written for a first order resonance

$$R = g_0 \Gamma^{1/2} \cos(\psi - \varpi) + g_1 \Gamma_p^{1/2} \cos(\psi - \varpi_p), \tag{11}$$

where we have used the approximation $e^2 \sim 2\Gamma/L \sim 2\Gamma\alpha^{1/2}$ where α is the ratio of the particle and planet semi-major axis. Here the perturbation strengths, g_0, g_1 are functions of semi-major axis, j and Laplace coefficients. The second term is sometimes called the corotation resonance and is often ignored. The resonant angle for the dominant resonance term is $\phi = \psi - \varpi$, where ψ depends on the mean longitude of the planet and particle.

2.4 The Full Hamiltonian for First Order Resonances

The perturbations for first order mean-motion and first order Lindblad resonances are both proportional to square root of the action variable times the cosine of a resonant angle. The pendulum Hamiltonian contains a term proportional to the cosine of the resonant angle. Hence the Hamiltonian of these first order resonances is more complex than that of the pendulum. Following canonical transformation to new coordinates, specifying one to be the resonant angle, ϕ, for both first order Lindblad and mean motion resonances, the Hamiltonian can be written

$$H(J, \phi) = a' J^2 + J\delta - \beta J^{1/2} \cos\phi, \tag{12}$$

(e.g., see [1, 5, 6, 8]). Following unit changes it can be rewritten in dimensionless form as

$$H(j, \phi) = j^2 + j\bar{\delta} - \bar{\beta} j^{1/2} \cos\phi. \tag{13}$$

Here $\bar{\delta}$ describes the distance from resonance. This is dependent on mean radius in the galactic setting and on the semi-major axis in the Keplerian setting. The parameter $\bar{\beta}$ is proportional to the bar or spiral pattern strength in the galactic setting and to the planet mass in the Keplerian setting. Figure 2 shows the orbits in phase space for this system in three different regimes. Either one or three fixed points are present. When three are present only two are stable. Fixed points correspond to closed or periodic orbits in the system. When there are two stable fixed points one corresponds to orbits aligned with the bar or spiral arm and the other corresponds to orbits perpendicular to the other (as discussed in [8]). Only when $\frac{2\bar{\delta}}{3|\bar{\beta}|^{\frac{2}{3}}} < -1$ is there a separatrix and two stable fixed points. Additional perturbations are most likely to cause instability or chaotic motion near a separatrix. We note that the above Hamiltonian neglects the second momentum variable that is a conserved quantity but could

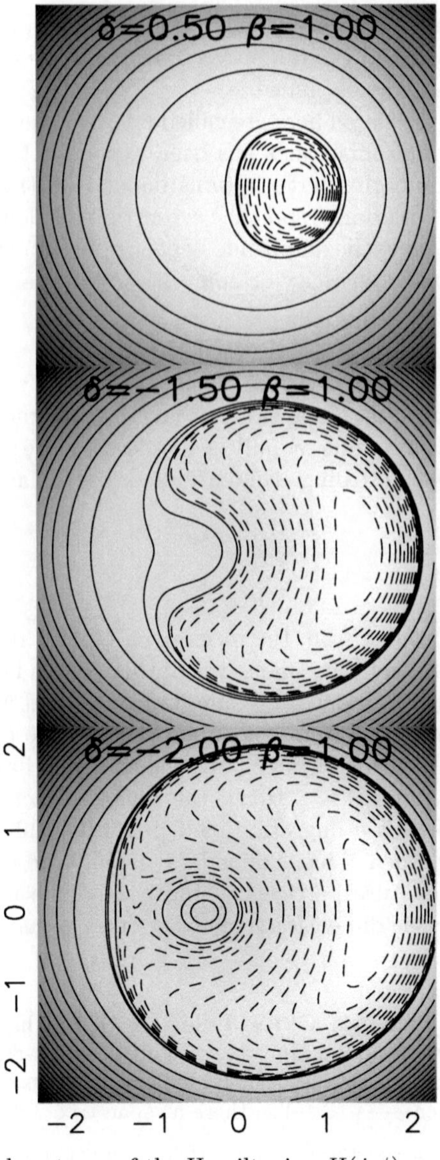

Fig. 2. We plot level contours of the Hamiltonian $H(j,\phi) = j^2 + j\bar{\delta} - \bar{\beta} j^{1/2} \cos\phi$. Dashed *contours* are negative. The axes are x, y are defined by $x = \sqrt{2j}\cos\phi, y = \sqrt{2j}\sin\phi$. Contopoulos [8] was first to show that first order resonances with a bar or spiral pattern perturbation have this structure. Orbits follow the contours and fixed points correspond to periodic orbits. The critical value for the resonance to bifurcate happens in the middle panel where $\frac{2\bar{\delta}}{3|\bar{\beta}|^{\frac{2}{3}}} = -1$. The top of the panel contains a fixed point which would correspond to a closed or periodic orbit aligned with the bar so that $\phi = 0$. This is the situation outside the bar's outer 2:1 Lindblad

influence the values of the coefficients. We have also dropped other perturbation terms that depend upon non-resonant angles as they will only weakly perturb the system near resonance.

The resonances discussed here are called first order even though we require an unperturbed Hamiltonian that has been expanded to second order in the momenta variable and the perturbations depend on the square root of the momenta. For the Keplerian system the eccentricity depends on $\sqrt{\Gamma}$ and in the galactic system the epicyclic amplitude depends on $\sqrt{I_1}$. Hence the resonances are first order in the eccentricity or epicyclic amplitude.

2.5 Dimensional Scaling and Applications

It is useful to estimate dimensions from the parameters describing the Hamiltonian. The parameter δ only sets the distance to resonance, so we can neglect it during rescaling. Dimensionless coordinates can be chosen as j, τ with

$$j = |\beta|^{2/3} |a'|^{-2/3} J, \tag{14}$$

$$\tau = |\beta|^{2/3} |a'|^{1/3} t, \tag{15}$$

reducing the Hamiltonian from the form given by (12) to that in (13).

The understanding of the dimensions of this system is useful as every first order resonance can be understand simultaneously and related to the others via rescaling. The distance from resonance is set by the parameter δ and this is in units of t^{-1}. A drifting or migrating system would have a time dependent δ with drift rate in units of t^{-2}. Hence the square of factors given in (15) can be used to predict when a migration or drift rate is sufficiently fast that the adiabatic limit is reached. This is done by Quillen [6] to predict when drifting resonances fail to capture particles in the non-adiabatic limit. Because the prediction was done via dimensional analysis, criteria were give for all first and

Fig. 2. *(continued)* resonance, OLR$_B$. Inside the OLR$_B$ the resonance bifurcates, and both periodic orbit families are present. This situation is shown in the bottom panel. The fixed points at $\phi = \pi$ correspond to periodic orbits oriented perpendicular to the bar. Only when $\frac{2\bar{\delta}}{3|\bar{\beta}|^{\frac{2}{3}}} < -1$ is there a separatrix. Additional perturbations are most likely to cause instability when there is a separatrix. Changing the sign of β is equivalent to adding π to ϕ. The Lindblad resonances associated with spiral arms have this same structure, though outside the spiral patterns inner Lindblad resonance, ILR$_S$, the resonance looks like the bottom panel and two classes of closed or periodic orbits exist. Inside the ILR$_S$ the resonance looks like the topmost panel and only one class of periodic orbits exists. When the system evolves from top to bottom, resonance capture is possible. When a particle is captured in resonance it jumps from the island on the left to that on the right (in bottom panel) and particles are pumped to higher eccentricity or epicyclic amplitude as the system continues to drift.

second order resonances simultaneously. For example,the square of the unit of time in (15) depends on the perturbation strength (β) to the 4/3 power. Thus the theory quickly predicts that critical migration rates allowing resonance capture scale with the planet mass to the 4/3 power, confirming previous but more complex predictions [9] and scaling seen in numerical studies [10].

The timescale dominant in the resonance is also given by the scaling factor shown in (15). For spiral density waves to be driven in a particulate disk, the resonant libration time must exceed the collision timescale. Thus this scaling parameter can be used to estimate when resonances are capable of driving spiral density waves in a collisional disk. Since spiral density waves cannot be driven into Fomalhaut's dusty debris disk by a planet another process is required to maintain the disk's sharp edge [4].

Resonant widths depend on the units of the momentum shown in (14). Hence critical spacing of resonances for resonance overlap can be estimated using the scaling parameter given in this equation. The 2/7 law [11] describing the distance to the chaotic zone boundary at corotation of a planet was re-estimated in [5] using dimensions given in (14). Quillen [5] also estimated the velocity dispersion just exterior to the chaotic zone boundary using the assumption that the dispersion would depend on the size of librations in the nearest resonance that does not overlap the chaotic zone. The dispersion estimate was then used to place an upper limit on the mass of the predicted planet.

Mean motion resonances have two terms, the common one with resonant angle $\psi - \varpi$ and the corotation term with resonant angle $\psi - \varpi_p$. The strength of the corotation term depends on the planet's eccentricity. As these two resonances can overlap the system can be chaotic. Little work on resonance capture has been done for chaotic resonances. Quillen [6] showed that the corotation resonance term could prevent a drifting system from capturing and using dimensional scaling proposed that Neptune is nearly eccentric enough to prevent capture into the 2:1 mean motion resonance. This provides a possible explanation for the low fraction of twotinos compared to plutinos in the Kuiper belt.

When secular terms are kept in the Hamiltonian, slow variations in eccentricity and angle of perihelion are predicted. At a given semi-major axis only one orbit remains fixed in eccentricity and angle of perihelion. This orbit has an eccentricity known as the forced eccentricity. Other orbits oscillate about this fixed orbit with non-zero free eccentricity. Thus secular perturbations are one way to account for eccentric disks such as Fomalhaut through the "pericenter glow" model [12]. When the Hamiltonian for an eccentric planet is expanded about the zero free eccentricity fixed point, the first order resonances look identical to those near a zero eccentricity planet. This establishes a close connection between the dynamics of particles at low free eccentricity near an eccentric planet and those at low eccentricity near a zero eccentric planet. This close connection was used to account for the insensitivity of the corotation chaotic zone width to planet eccentricity seen in numerical simulations [5].

3 Chaos Due to Resonance Overlap in Galaxies

Interpretations of structure in external galaxies has involved Lindblad resonances, for example the 4:1 Lindblad resonance may cause changes in the morphology of spiral arms [13]. Finding evidence for resonances in the phase space distribution of stars in the solar neighborhood has been more challenging. Resonant conditions are set by orbital period, hence the angular rotation rate in a galaxy and the semi-major axis in a circumstellar disk. The asteroid belt shows little structure when plotted in real space (radius vs. angle) but a lot of structure when plotted on an eccentricity vs. semi-major axis plot. The same is likely true for stars in the solar neighborhood. Stars can be plotted according to their space motions u, v where u is the radial component toward the Galactic center and v is the tangential component with respect to a star in a circular orbit. On such a plot structure is observed and much of it remains unexplained [14]. However, within an epicyclic approximation, v sets the mean radius of the orbit [15]. Thus breaks in the stellar velocity distribution at particular v values can be interpreted in terms of resonant perturbations. This has led to proposed interpretations for the Hercules stream involving the 2:1 outer Lindblad Resonance with the Galactic bar [16] and the division between the Pleiades/Hyades and Coma Berenices moving groups involving a 4:1 inner Lindblad resonance with local spiral structure [15]. As these models are very sensitive to the pattern speed of the perturbation, they are associated with very tight constraints on these speeds. For example Minchev et al. [7] has constrained the bar pattern speed to an impressive accuracy of few percent.

As the Galactic bar's 2:1 outer Lindblad resonance affects the solar neighborhood and local spiral structure is simultaneously present it is interesting to consider what happens when stars are affected by two perturbations each with a different pattern speed. Quillen [1] used a one-dimensional Hamiltonian in the following form to model this situation

$$h(j_1, \phi) \approx j_1^2 + \bar{\delta} j_1 + \bar{\beta}_2 j_1^{\frac{1}{2}} \cos(\phi) + \bar{\epsilon}_m j_1^{\frac{1}{2}} \cos(\phi + \bar{\nu} t + \gamma). \qquad (16)$$

Here $\bar{\epsilon}_m$ is set by the spiral perturbation with m arms and $\bar{\beta}_2$ is the perturbation from the $m = 2$ component of the bar potential. The frequency $\bar{\nu}$ depends on the bar and spiral pattern speeds. This Hamiltonian is only approximate nevertheless illustrates interesting phenomena. It is similar to the forced pendulum studied in [17] and some trajectories exhibit chaos. The forced pendulum does not have perturbation terms proportional to the square root of the action variable and so is less complex than this model.

The above Hamiltonian is not time independent so the Hamiltonian itself is not conserved. However area preserving or Poincaré maps can be constructed by plotting points very timestep $P_\nu = \frac{2\pi}{\bar{\nu}}$. This procedure generates maps that are like surfaces of section and can be used to study the stability of the orbits. In such a map, orbits are either area filling or curves. We denote area filling orbits as chaotic (in the sense that orbits diverge exponentially) and the curved orbits as quasiperiodic. See Fig. 3 for examples of maps generated

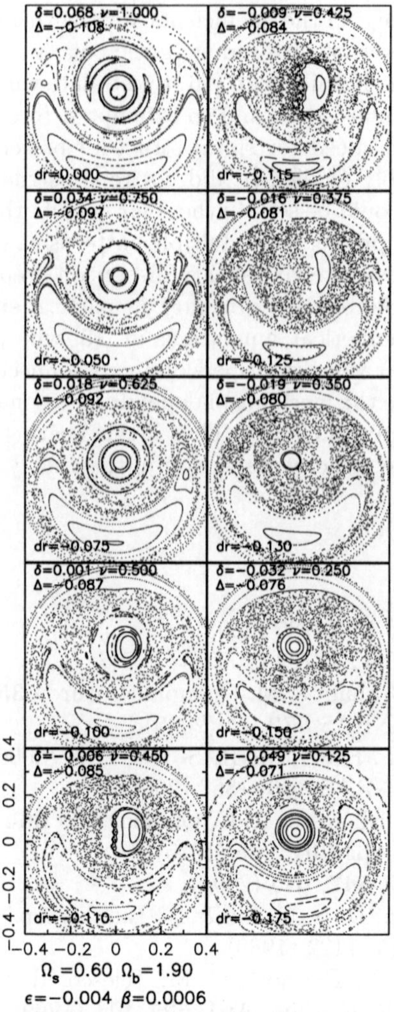

Fig. 3. Poincare maps made by integrating the Hamiltonian of (16) with $m = 2$ and with a time step of $\frac{2\pi}{\bar{\nu}}$. The panels are at different values of dr, the distance to the solar galactocentric radius in units of the Sun's galactocentric radius. The top left panel has the smallest value of dr corresponding to a location near the solar neighborhood. The lower and righter most panel has the most negative value of dr corresponding to a radius closer to the Galactic center. All panels have the same values of $\bar{\epsilon}$, describing the strength of spiral structure, Ω_s the pattern speed of the spiral structure in units of the solar neighborhood angular rotation rate (Ω_0), $\bar{\beta}$, describing the strength of the bar, and Ω_b, the pattern speed of the bar in units of Ω_0. Phase space, as illustrated by the structure in these maps, has two types of orbits; curved linear structures, corresponding quasiperiodic orbits, and area filling orbits corresponding to chaotic regions. The radial distance on these plots is approximately the same as the radial epicyclic amplitude in units of r_0, or the distance the orbit reaches from r_c. The spiral pattern speed considered here would result from a two-armed spiral pattern with the solar neighborhood just within the 4:1 ILR$_S$.

from this Hamiltonian. As the coefficients vary with radius the morphology of the Poincar'e maps vary with radius in the galaxy.

Quillen [1] used the Hamiltonian (16) to estimate where the chaotic regions prevented nearly circular orbits and so attempted to exclude some values of for the spiral pattern speed for spiral structure present in the solar neighborhood. Unfortunately Quillen [1] did not relate structure predicted by the model to that which could be seen in the solar neighborhood velocity distribution. The model does predict that multiple patterns can cause chaotic motion and so an increase in the stellar velocity dispersion even when spiral patterns are fixed (do not grow or decay or vary in pattern speed). This was later confirmed with the numerical simulations in [18] who proposed a new heating mechanism (stellar velocity dispersion increase mechanism) based on this phenomenon. Because the mechanism depends on resonances it is strongly dependent on position in the Galaxy. Thus if this heating mechanism operates, large variations in the velocity dispersion in the galaxy are predicted.

References

1. A. C. Quillen, AJ **125**, 785 (2003)
2. A. C. Quillen, AJ **124**, 722 (2002)
3. P. Kalas, J. R. Graham & M. Clampin, Nature **435**, 1067 (2005)
4. A. C. Quillen, MNRAS **372**, 14 (2006)
5. A. C. Quillen & P. Faber, MNRAS **373**, 1245 (2006)
6. A. C. Quillen, MNRAS **365**, 1367 (2006)
7. I. Minchev, J. Nordhaus & A. C. Quillen. ApJ **664**, L31 (2007)
8. G. Contopoulos, ApJ **201**, 566 (1975)
9. L. Friedland, ApJ **547**, L75 (2001)
10. S. Ida, G. Bryden, D. N. C. Lin & H. Tanaka, ApJ **534**, 428 (2000)
11. J. Wisdom, AJ **85**, 1122 (1980)
12. M. C. Wyatt, S. F. Dermott, C. M. Telesco, R. S. Fisher, K. Grogan, E. K. Holmes & R. K. Pina, ApJ, **527**, 918 (1999)
13. P. A. Patsis, N. Hiotelis, G. Contopoulos & P. Grosbol, A&A **286**, 46 (1994)
14. W. Dehnen, W. AJ **115**, 2384 (1998)
15. A. C. Quillen & I. Minchev, AJ, **130**, 576 (2005)
16. W. Dehnen, W. AJ **119**, 800 (2000)
17. M. J. Holman & N. W. Murray, AJ **112**, 127 (1996)
18. I. Minchev & A. C. Quillen, MNRAS **368**, 623 (2006)

Regular and Chaotic Motion in Elliptical Galaxies

J.C. Muzzio

Facultad de Ciencias Astronómicas y, Geofísicas de la Universidad Nacional de La Plata and Instituto de Astrofísica de La Plata (CCT-CONICET La Plata and UNLP), La Plata, Argentina
jcmuzzio@fcaglp.fcaglp.unlp.edu.ar

Summary. Here I review recent work, by other authors and by myself, on some particular topics related to the regular and chaotic motion in elliptical galaxies. I show that it is quite possible to build highly stable triaxial stellar systems that include large fractions of chaotic orbits and that partially and fully chaotic orbits fill different regions of space, so that it is important not to group them together under the single denomination of chaotic orbits. Partially chaotic orbits should not be confused with weakly fully chaotic orbits either, and their spatial distributions are also different. Slow figure rotation (i.e., rotation in systems with zero angular momentum) seems to be always present in highly flattened models that result from cold collapses, with the rotational velocity diminishing or becoming negligibly small for less flattened models. Finally, I comment on the usefulness and limitations of the classification of regular orbits via frequency analysis.

1 Introduction

It is fitting, in this conference in memory of Voglis, to recall that I became interested in the investigation of regular and chaotic motions in elliptical galaxies thanks to a paper of his [1]. By that time, I had been working on N-body problems for two decades, and on regular and chaotic motion for 7 or 8 years, but I had never been involved in research on elliptical galaxies. The paper by Voglis and his coworkers showed me that, with the computers and the numerical tools I had at my disposal, I might be able to contribute significantly to a very interesting subject and, in fact, I have been devoted to that subject ever since.

Having worked in this field for a few years only, it would be presumptuous from my part to attempt to present here a complete review of the subject. Alternatively, to be a relative newcomer to the field has the advantage of bringing to it views and opinions different from the prevailing ones: they may be wrong, but they stimulate progress.

G. Contopoulos, P.A. Patsis, *Chaos in Astronomy.* Astrophysics and Space Science Proceedings, doi: 10.1007/978-3-540-75826-6,

Therefore, I will limit the scope of this review to a few items that have been of particular interest to me and which I have strived to clarify with my research: (1) Can we build stable triaxial models of stellar systems that contain high fractions of chaotic orbits? (2) Is the distinction between partially and fully chaotic orbits of any use? (3) Is figure rotation significant in triaxial stellar systems? (4) Which are the usefulness and limitations of frequency analysis for the classification of large numbers of regular orbits in model stellar systems?

Since galactic dynamics is not the only subject of this conference, which includes other fields like celestial mechanics, it may be useful to recall that the time scales pertinent to galaxies are completely different from those that rule the Solar System. While the age of the latter is of the order of 10^8 orbital periods, galactic ages are of the order of 10^2 orbital periods only. Thus, the chaotic orbits we will refer here are much more strongly chaotic (i.e., their Lyapunov times measured in orbital periods are much shorter) than those of the Solar System. Technical tools, such as frequency analysis, should also be considered with this fact in mind.

2 Highly Chaotic Triaxial Stellar Systems

2.1 Building Self-Consistent Triaxial Stellar Systems

A popular method to build a self-consistent triaxial stellar system is the one due to Schwarzschild [2]. One chooses a density distribution and obtains the potential that it creates; a library of thousands of orbits is then computed in that potential and weights are assigned according to the time that a body on that orbit spends in different regions of space; finally, those weights are used to compute the relative numbers of those orbits that are needed to obtain the original density distribution.

Another way to proceed is to use an N-body code to build a triaxial stellar system (e.g., through the collapse of an N-body system initially out of equilibrium), then to smooth and to freeze the potential fitting it with adequate formulae, to use these formulae, together with the positions and velocities of the bodies as initial conditions, to compute a representative sample of orbits in that potential and, finally, to classify those orbits to get the orbital structure of the system [1].

Those two methods should be regarded as complementary. Schwarzschild's one allows a very precise definition of the density distribution of the system one wants to study; alternatively, some properties of the models dictated by mathematical simplicity (e.g., constant axial ratios over the whole system) might bias its results, while the N-body method is free of that problem.

2.2 The Problem of Chaotic Orbits in Schwarzschild's Method

Schwarzschild [3] found it necessary to include chaotic orbits in his models but, then, these were not fully stable. He built several models using orbits computed over a Hubble time and, subsequently, followed those orbits for two additional Hubble times. When he computed the axial ratios obtained using the data for the third Hubble time, he found significant differences with respect to the ratios computed over the first Hubble time, from a low of about 4% for his second and fourth models, to a high of about 17% for his fifth model.

The cause of that evolution is that chaotic orbits change their behaviour with time, resembling that of regular orbits at certain intervals, behaving more chaotically at other intervals and exploring different regions of space in the meantime. Moreover, that weaker or stronger chaotic behaviour can be traced with Lyapunov exponents computed over finite intervals which decrease and increase their values accordingly [4, 5]. Merritt and his coworkers tried to solve this problem using what they called "fully mixed solutions" [6] and, more recently, integrating orbits over five Hubble times [7]. In the former work, they found solutions for the weak cusp model, but not for their strong cusp model; the subsequent evolution of these models to test their stability was not investigated, however. In the latter work they indicate that there is "a slight evolution toward a more prolate shape", but they provide no quantitative estimates other than indicating that differences in velocity dispersions are "almost always below 10%". Clearly, it is very difficult to incorporate chaotic orbits in Schwarzschild's method: as some chaotic orbits begin to behave more chaotically, one needs to have other chaotic orbits that behave more regularly as compensation; such a delicate equilibrium cannot be attained simply obtaining the weights of chaotic orbits over longer integration times and, moreover, the relatively low number of orbits used (typically a few thousands) makes even more difficult that task. Finally, the usual imposition of constant axial ratios over the whole system in Schwarzschild's method prevents the existence of a rounder halo of chaotic orbits that seems to be a necessary condition to have highly chaotic triaxial stellar systems [1, 8, 9].

2.3 The Stability of Highly Chaotic Triaxial Stellar Systems

The models of the N-body method are built self-consistently from the start and typically contain hundreds of thousands, or even millions, of bodies so that they should be free of the difficulties that plage the construction of highly chaotic triaxial stellar systems with Schwarzschild's method. In fact, stable models with high fractions of chaotic orbits were obtained with the N-body method, using about 10^5 particles [1, 8]; moreover, the fractions of the different types of orbits were not significantly altered when the potential was fitted to the N-body distribution at different times. A stable cuspy model that was mildly triaxial and made up of 512,000 particles, plus several others with 128,000 particles, were also built [10]; later on, it was shown that the introduction of a black hole, although affecting the inner regions of the model, did not alter the triaxiality at larger radii and the authors concluded that the triaxiality of elliptical galaxies is not inconsistent with the presence of supermassive black holes at their centers [11].

Highly stable models of 10^6 particles were built by us with the N-body method [9, 12]: all of them have decreasing flattening from center to border, which arised naturally from the N-body evolution during the generation of the systems; they have different degrees of flattening and triaxiality, two of them are moderately cuspy ($\gamma \approx 1.0$), and all have high fractions (between 36 and 71%) of chaotic orbits. When integrated with the N-body code, our models suffer changes in their central density and minor semiaxis values which do not exceed, respectively, about 4 and 2% over a Hubble time. Nevertheless, these changes are most likely due to collisional effects of the N-body code [13] because, when the number of bodies is reduced by a factor of 10 (and their masses are increased by the same factor), those changes increase by factors between 3 and 10. Alternatively, integrating the motion of the bodies in the fixed smooth potential, which suppresses the collisional effects (and which, by the way, is what Schwarzschild did) reduces those changes to 0.1% only (i.e., between one and two orders of magnitude smaller than those found by Schwarzchild [3]).

Thus, we may conclude that highly stable triaxial models with large fractions of chaotic orbits can be built with the N-body method. The difficulties to build such models with Schwarzschild's method should thus be attributed to the method itself and not to physical reasons.

3 Partially and Fully Chaotic Orbits

Since we are dealing with stationary systems, the orbits of the particles that make them up obey the energy integral, but they need two additional isolating integrals to be regular orbits. Thus, we distinguish between partially chaotic orbits (one additional integral besides energy) and fully chaotic orbits (energy is the only integral they obey). A practical way to make the distinction is to compute the six Lyapunov exponents: they come in three pairs of equal value and opposite signs, due to the conservation of phase space volume, and each isolating integral makes zero one pair. Thus, in our case, two Lyapunov exponents are always zero (due to energy conservation); of the remaining four, if two are positive the orbit is fully chaotic, if only one is positive the orbit is partially chaotic and, finally, if all are zero the orbit is regular.

It was noted in [14] that orbits obeying two isolating integrals have smaller fractal dimension than orbits obeying only one, but earlier hints of the differences between them can also be found in [15] (whose semi-stochastic orbits are probably what we now call partially chaotic orbits) and in [16] (whose orbits in their big and small seas can be identified, respectively, with the fully and partially chaotic orbits).

The reason why distinguishing partially from fully chaotic orbits in galactic dynamics is important is that, since they obey different numbers of isolating integrals, they have different spatial distributions as shown in [8, 9, 12, 17, 18]. In triaxial systems, partially chaotic orbits usually exhibit a distribution intermediate between those of regular and of fully chaotic orbits, and a possible explanation is that some of the partially chaotic orbits lie in the stochastic layer surrounding the resonances and thus behave similarly to regular orbits [19]. Nevertheless, that is not the whole story as some partially chaotic orbits seem to obey a global integral, rather than local ones [9].

Partially chaotic orbits should not be confused with fully chaotic orbits with low Lyapunov exponents, which also tend to have distributions more similar to those of regular orbits than those of fully chaotic orbits with high Lyapunov exponents [8, 18]. It is worth recalling that, no matter how small their Lyapunov exponents are, fully chaotic orbits obey only one isolating integral while partially chaotic orbits obey two so that, from a theoretical point of view, they are indeed different kinds of orbits. From a practical point of view, it is also easy to see that they have different distributions: Table 1 gives the axial ratios of the distributions of different kinds of orbits for models E4, E5 and E6 from [9] and E4c and E6c from [12]; the x, y and z axes are parallel, respectively, to the major, intermediate and minor axes of the models. The third column gives the axial ratios for the distributions of partially chaotic orbits, and the fourth and fifth columns give the same ratios for weakly fully chaotic orbits for two choices of the limiting value of the Lyapunov exponents used to define "weakly", 0.050 and 0.100. Although for some models (e.g., E4 and E4c) the possible differences are masked by the rather large statistical errors, it is clear from the table that the distributions of partially chaotic orbits are significantly different from those of weakly fully chaotic orbits (at the 3σ level) for the other models.

At any rate, it is clear that the distributions of partially and fully chaotic orbits differ significantly and that they should not be bunched together as a

Table 1. Axial ratios of the different classes of orbits in our models

Ratio	System	Partially Ch.	W.F.Ch. (0.050)	W.F.Ch. (0.100)
y/x	E4	0.896 ± 0.064	0.692 ± 0.027	0.745 ± 0.019
	E5	0.808 ± 0.036	0.764 ± 0.024	0.797 ± 0.017
	E6	0.658 ± 0.035	0.789 ± 0.027	0.845 ± 0.019
	E4c	0.748 ± 0.027	0.733 ± 0.016	0.730 ± 0.013
	E6c	0.528 ± 0.020	0.693 ± 0.013	0.700 ± 0.010
z/x	E4	0.790 ± 0.054	0.802 ± 0.035	0.826 ± 0.024
	E5	0.477 ± 0.018	0.684 ± 0.021	0.708 ± 0.014
	E6	0.286 ± 0.013	0.644 ± 0.022	0.673 ± 0.015
	E4c	0.692 ± 0.024	0.762 ± 0.017	0.757 ± 0.013
	E6c	0.334 ± 0.010	0.466 ± 0.007	0.490 ± 0.006

single group of chaotic orbits. The problem is that the computation of the Lyapunov exponents demands long computation times and there are not yet faster methods that allow to distinguish partially from fully chaotic orbits. The fact that many chaotic orbits can be frequency analyzed and are found to lie in regions of the frequency map corresponding to regular orbits [20] might, perhaps, lead to a faster method of separation in the future. Nevertheless, many fully chaotic orbits can be frequency analyzed, while many partially chaotic orbits cannot [9], so that much remains to be done before a workable method based on frequency analysis can be designed.

4 Figure Rotation in Triaxial Systems

Although the system investigated in [8] had been regarded as stationary, integrations much longer than those used in that work revealed that, in fact, it was very slowly rotating around its minor axis [21]. The total angular momentum of the system was zero, so that this was an unequivocal case of figure rotation.

Figure rotation was also found in most of the models studied in [9, 12], and it is clear that the rotational velocity increases with the flattening of the system; only model E4 from [9], which is almost axially symmetric, prolate and with axial ratio close to 0.6 has no significant rotation. It should be stressed, however, that even the highest rotational velocities found thus far are extremely low: the systems can complete only a fraction of a revolution in a Hubble time or, put in a different way, the radii of the Lindblad and corotation resonances are at least an order of magnitude larger than the systems themselves.

It had been suggested that figure rotation might produce important changes in the degree of chaoticity [6] and it turned out that, in spite of the extremely low rotational velocity, a significant difference in the fraction of chaotic orbits was found between the models of [8, 21], which only differ in that the former is stationary and the latter is rotating. Alternatively, no significant difference was found for the different kinds of regular orbits in those two models. The most likely explanation is that, although the rotational velocity is too low to produce a measurable effect on the regular orbits, the break of symmetry caused by the presence of rotation suffices to increase chaos significantly.

5 Musings on Orbital Classification Through Frequency Analysis

5.1 Classification Methods

The spectral properties of galactic orbits were investigated by Binney and Spergel [22] and, more recently, Papaphilippou and Laskar [23, 24] applied to stellar systems the frequency analysis techniques developed by the latter for celestial mechanics. Following the ideas of Binney and Spergel, Carpintero and Aguilar [25] developed an automatic orbit classification code. Kalapotharakos and Voglis [20] developed a classification system based on the frequency map of Laskar and, later on, I [21] improved it somewhat.

Having used extensively both the Carpintero and Aguilar [8, 26–30], and the Kalapotharakos and Voglis methods [9, 12, 21], I strongly prefer the latter. The main advantage of the Kalapotharakos and Voglis method is that one can see what is happening throughout the process. It is very easy to detect problems from the anomalous positions that the corresponding frequency ratios yield on the frequency map and, thus, to improve the method. This is an aspect that deserves to be emphasized: the need to use frequencies different from those corresponding to the maximum amplitudes had not been noted in [20], but it was in [21], probably because a somewhat cuspier potential was investigated in the latter work; similarly, that distinction was unnecessary for the long axis tubes (LATs hereafter) of [21], but had to be made for those of the almost axially symmetric E4 system of [9]. In other words, as one explores different stellar system models (cuspier, closer to axisymmetry, and so on) the orbital classification system may need to be improved and that need is quite evident with the Kalapotharakos and Voglis method. Thanks to these improvements, virtually all the regular orbits can be classified with the frequency map, while usually between 10 and 15% of them remain unclassified with the other method [8, 31].

Besides, separation of chaotic from regular orbits with the method of Carpintero and Aguilar is erratic, at least in rotating systems [30]. Since the problem seems to arise from the presence of nearby lines in the spectra, which is worse in rotating systems but not exclusive of them, I strongly suspect that orbit classification in non-rotating systems may also be affected. That is why in our last work with that method [8] we used it only to classify regular orbits, previously selected using Lyapunov exponents.

5.2 Which Frequency to Choose?

Frequency analysis is usually performed on complex variables formed taking one coordinate as the real part and the corresponding velocity as the imaginary part. One thus gets the frequencies Fx, Fy and Fz corresponding, respectively, to motion along the (x, y, z) axes which, in turn, are parallel to the main axes of the stellar system. The frequencies usually selected for the frequency map are those corresponding to the maximum amplitudes in each coordinate [20, 32], but it has been known since 1982 [22] that, due to a libration effect, one should not always take those. Besides, another effect linked to very highly elongated orbits also demands to adopt frequencies which are not the ones corresponding to the maximum amplitudes [9, 21]. Nevertheless, it is just fair to note that these exceptions are not too common: out of 17,103 orbits investigated in [9, 12] only 265 (1.5%) needed the former correction and 153 (0.9%) the latter one. These fractions vary considerably from one model to another, however, and as the affected orbits tend to concentrate at low absolute values of energy and/or are extremely elongated, not taking these effects into account might bias the sample of classified orbits.

5.3 The Usefulness of The Energy Vs. Frequency Plane

Regular orbits obey two additional isolating integrals, besides energy, and the values of the orbital frequencies are related to these integrals. For a given energy, different frequencies imply different values of the other integrals and, thus, different types of orbits. Inner and outer LATs, short axis tubes, boxes and even different resonant orbits can be separated on the energy vs. frequency (or frequency ratio) plane, but that does not mean that it is practical to use it, because those separations are more easily done on the frequency map.

Nevertheless, some insight can be gained from the use of the energy vs. frequency plane. Figure 1 of [21] offers a good example, because that plane was used there to show that one should not always use the frequency corresponding to the maximum amplitude as the principal frequency. Besides, while in [20] it was correctly stated that outer LATs had larger Fx/Fz values than inner LATs, no indication of which was the separating value was provided there. Actually, as shown in Fig. 2 of [9], one has to use the energy vs. frequency ratio plane to separate inner from outer LATs, because the separating value varies with the energy of the orbit.

Figure 1 presents the (x, y) projections of several LATs from model E4 of [9]. The orbits on the left column have similar energy values, close to the minimum energy of -5.96 and, although their Fx/Fz values range from 0.6516 to 0.8118, they are all inner LATs, as evidenced by their concave upper and lower limits. We also notice that their extension along the x axis is reduced as their Fx/Fz values increase and, in fact, the regions of space occupied by orbits 0872 and 0009 resemble more those occupied by outer LATs than those occupied by inner LATs. We found a similar effect on the x extension of the orbits at other energy values although, when the separation shown in Fig. 2 of [9] is crossed, there is also of course a change from inner to outer LATs. The upper and middle parts of the right column of Fig. 1 correspond to orbits 3141 and 1513 that are virtually face to face at each side of the separation on the energy vs. frequency ratio plot: they have similar Fx/Fz values but, due to their energy difference, the former is an outer, and the latter an inner, LAT. Notice also that the Fx/Fz value of (outer LAT) orbit 3141 is lower than that of (inner LAT) orbit 0872. Finally, the lower right section of Fig. 1 corresponds to (outer LAT) orbit 3307, whose Fx/Fz value is lower than those of (inner LAT) orbits 0009 and 0872.

Interestingly, the shortening of the x axis as the Fx/Fz ratio increases, shown above for the LATs, affects the boxes as well. Figure 2 presents the (x, y) and (x, z) projections of orbits 0104 and 0097 from model E4 of [9], which have both essentially the same energy. Nevertheless, while the former, with $Fx/Fz = 0.6332$, lies straight on the line occupied by the boxes on the energy

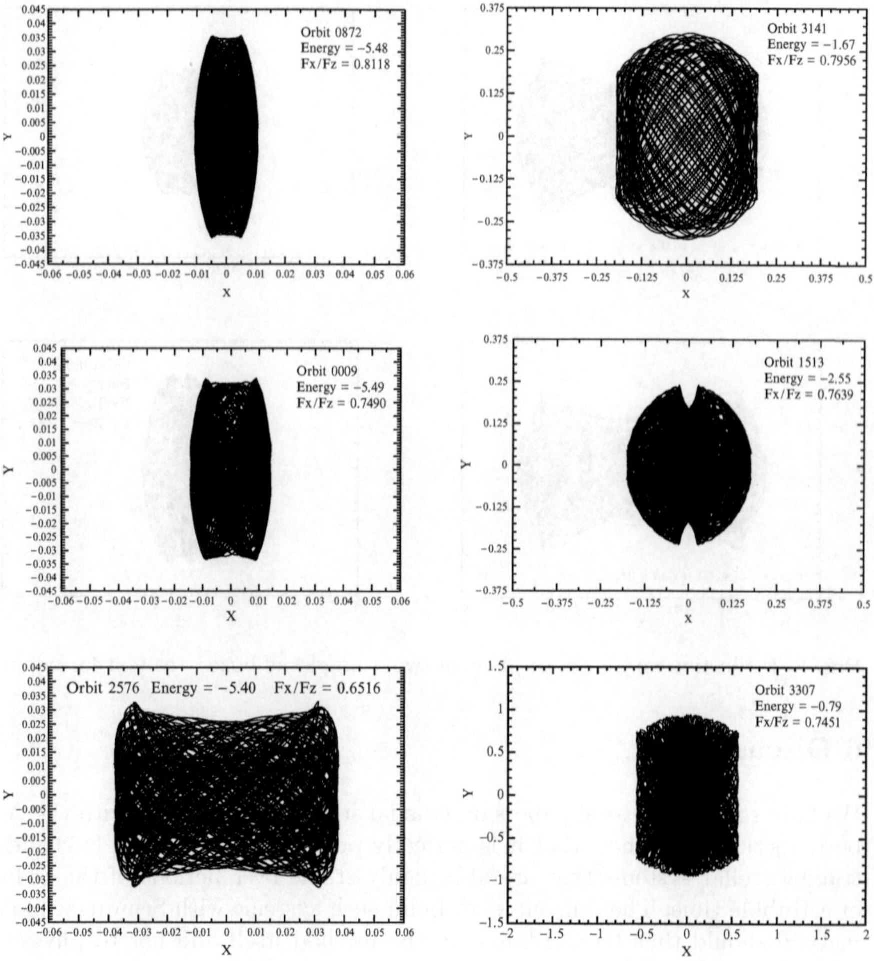

Fig. 1. Projection on the $(x,\ y)$ plane of several examples of LATs; see text for details

vs. frequency ratio plane, the latter, with $Fx/Fz = 0.8076$, lies well above that line. We see on the left part of Fig. 2 that 0104 is indeed a typical box, but the right part shows that 0097, although still a box, is strongly compressed along the x axis.

Due to their elongation along the major axis, inner LATs and boxes are usually considered as the main building blocks of highly elongated triaxial systems, but we now see that there are inner LATs and boxes that are, in fact, strongly compressed along that axis. To put things in the proper perspective we should emphasize, however, that these orbits were found in the almost rotationally symmetric model E4 of [9] and that they are not very abundant.

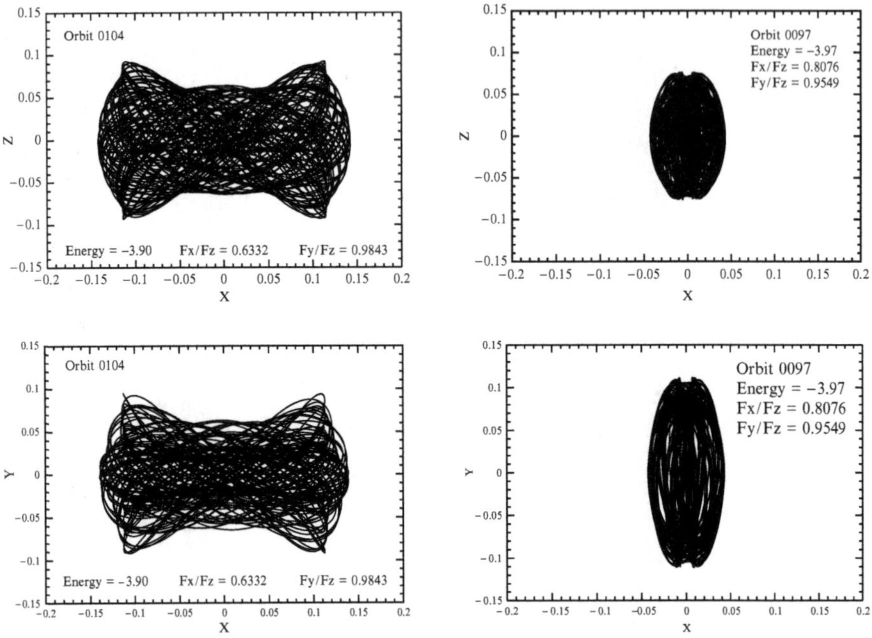

Fig. 2. Projection on the (x, y) plane of two examples of boxes; see text for details

6 Discussion

We have reviewed several papers on triaxial stellar systems built with the N-body method that show that it is perfectly possible to have strongly chaotic triaxial stellar systems that are also highly stable over periods of the order of a Hubble time. The difficulties to build such systems with Schwarzschild's method should thus be attributed to the method itself and not to physical reasons.

It is clear, both from a theoretical and from a practical point of view, that partially and fully chaotic orbits populate different regions of space and should not be bunched together under the single banner of chaotic orbits. The main problem here is that the single method thus far available to separate them, that of Lyapunov exponents, is very slow and faster methods are wanted. We also showed that the distribution of partially chaotic orbits is different from that of weakly fully chaotic orbits, in accordance with the fact that the former obey two isolating integrals of motion and the latter only one.

Very slow figure rotation seems to be an ordinary trait of strongly elongated triaxial stellar models formed through the collapse of cold N-body systems. The rotational velocity diminishes, and even disappears entirely, as one goes to less elongated and less triaxial models.

Frequency analysis offers a very useful tool for the classification of large numbers of regular orbits. I strongly favor the use of the method of Kalapotharakos and Voglis [20], with the improvements we introduced in [9, 21]. Since the need for those improvements became apparent when models with different characteristics (cuspiness, approximate rotational symmetry) were considered, it would not be surprising that further refinements will be necessary as the method is applied to other systems. Nevertheless, a nice feature of this method is that, when there is such need, it becomes plainly evident. Besides, plots of known integrals, such as energy, and the orbital frequencies (or frequency ratios), that are related to the values of the integrals, are very useful to reveal peculiarities of the orbits as one explores different models; a good example of this is provided by the compression along the major axis of some LATs and boxes from an almost axisymmetric system, shown in our Figs. 1 and 2.

7 Acknowledgements

I am very grateful to Héctor R. Viturro and to Ruben E. Martínez for their technical assistance, and to Lilia P. Bassino and to an anonymous referee for carefully reading the first version of this paper and suggesting some language improvements. This work was supported with grants from the Consejo Nacional de Investigaciones Científicas y Técnicas de la República Argentina, the Agencia Nacional de Promoción Científica y Tecnológica and the Universidad Nacional de La Plata.

References

1. N. Voglis, C. Kalapotharakos and I. Stavropoulos, Mon. Not. R. Astron. Soc. **337**, 619 (2002)
2. M. Schwarzschild, Astrophys. J. **232**, 236 (1979)
3. M. Schwarzschild, Astrophys. J. **409**, 563 (1993)
4. H.E. Kandrup and M.E. Mahon, Astron. Astrophys. **290**, 762 (1994)
5. J.C. Muzzio, Estructura orbital regular y caótica en sistemas estelares triaxiales (Informe Invitado). In *Memorias del Workshop sobre Astronomía Teórica en Argentina: Problemas y Perspectivas*, ed by S.A. Cellone and P. Benaglia. Asociación Argentina de Astronomía, La Plata, pp. 17–28 (2007)
6. D. Merritt and T. Fridman, Astrophys. J. **460**, 136 (1996)
7. R. Capuzzo-Dolcetta, L. Leccese, D. Merritt and A. Vicari, Astrophys. J. **666**, 165 (2007)
8. J.C. Muzzio, D.D. Carpintero and F.C. Wachlin, Celest. Mech. Dynam. Astron. **91**, 173 (2005)
9. R.O. Aquilano, J.C. Muzzio, H.D. Navone and A.F. Zorzi, Celest. Mech. Dynam. Astron. **99**, 307 (2007)

10. K. Holley-Bockelmann, J.C. Mihos, S. Sigurdsson and L. Hernquist, Astrophys. J. **549**, 862 (2001)
11. K. Holley-Bockelmann, J.C. Mihos, S. Sigurdsson, L. Hernquist and C. Norman, Astrophys. J. **567**, 817 (2002)
12. J.C. Muzzio, H.D. Navone and A.F. Zorzi, in preparation (2008)
13. L. Hernquist and J. Barnes, Astrophys. J. **349**, 562 (1990)
14. M. Pettini and A. Vulpiani, Phys. Lett. **106A**, 207 (1984)
15. J. Goodman and M. Schwarzschild, Astrophys. J. **245**, 1087 (1981)
16. G. Contopoulos, L. Galgani and A. Giorgilli, Phys. Rev. A **18**, 1183 (1978)
17. J.C. Muzzio, Bol. Asoc. Argentina Astron. **45**, 69 (2003)
18. J.C. Muzzio and M.E. Mosquera, Celest. Mech. Dynam. Astron. **88**, 379 (2004)
19. N.P. Maffione, Comparación de indicadores de la dinámica, Tesis de Licenciatura, Universidad Nacional de La Plata, La Plata (2007)
20. C. Kalapotharakos and N. Voglis, Celest. Mech. Dynam. Astron. **92**, 157 (2005)
21. Muzzio, J.C., Celest. Mech. Dynam. Astron. **96**, 85 (2006)
22. J. Binney and D. Spergel, Astrophys. J. **252**, 308 (1982)
23. Y. Papaphilippou and J. Laskar, Astron. Astrophys. **307**, 427 (1996)
24. Y. Papaphilippou and J. Laskar, Astron. Astrophys. **329**, 451 (1998)
25. D.D. Carpintero and L.A. Aguilar, MNRAS **298**, 1 (1998)
26. D.D. Carpintero, J.C. Muzzio, and F.C. Wachlin, Celest. Mech. Dynam. Astron. **73**, 159 (1999)
27. J.C. Muzzio, D.D. Carpintero and F.C. Wachlin, Regular and chaotic motion in galactic satellites. In: *The Chaotic Universe, Proceedings of the Second ICRA Network Workshop, Advanced Series in Astrophysics and Cosmology*, vol. 10, ed by V.G. Gurzadyan and R. Ruffini. World Scientific, Singapur, pp. 107–114 (2000)
28. J.C. Muzzio, F.C. Wachlin and D.D. Carpintero, Regular and chaotic motion in a restricted three-body problem of astrophysical interest. In: *Small Galaxy Groups, ASP Conference Series*, vol. 209, ed by M. Valtonen and C. Flynn. Astronomical Society of the Pacific, Provo, pp. 281–285 (2000)
29. S.A. Cora, M.M. Vergne and J.C. Muzzio, Astrophys. J. **546**, 165 (2001)
30. D.D. Carpintero, J.C. Muzzio, M.M. Vergne and F.C. Wachlin, Celest. Mech. Dynam. Astron. **85**, 247 (2003)
31. R. Jesseit, T. Naab and A. Burkert, Mon. Not. R. Astron. Soc. **360**, 1185 (2005)
32. F.C. Wachlin and S. Ferraz-Mello, Mon. Not. R. Astron. Soc. **298**, 22 (1998)

Orbital Distributions and Self-Consistency in Elliptical Galaxies

C. Kalapotharakos

Research Center for Astronomy, Academy of Athens,
Soranou Efessiou 4, GR-11527, Athens, Greece
ckalapot@phys.uoa.gr

1 Introduction

It is well known that a galaxy in equilibrium with a specific configuration can be built by many different orbital distributions. The difference between these distributions is in their associated velocity distributions. This property is used properly by Schwarzschild's method [22]. The Schwarzschild method assumes a density distribution and a corresponding potential (via Poisson equation) yielding self-consistent solutions, i.e. proper combinations of orbits that reproduce the original spatial density. There are some problems, however, with such solutions: (a) there is no guarantee about the solutions' stability and (b) there is no evidence that one solution should be preferable over the others.

On the other hand, *N-body methods*, by their nature, provide stable self-consistent solutions for galaxies. Some of these methods (e.g. direct summation, tree) are not suitable for orbital analysis because the corresponding potential is not a smooth function. On the other hand, the self-consistent field (SCF) method [1, 7] is an appropriate N-body method for orbital studies of isolated systems. In fact, the goal of an SCF code is precisely to reproduce a smooth potential corresponding to the N-body system under study.

Studies based on Schwarzschild's method [18–21, 23] showed the existence of self-consistent solutions of smooth centre elliptical galaxies containing only regular orbits. At the same time, the study of the so-called "perfect ellipsoid" [3, 27] showed the existence of triaxial smooth centre models of elliptical galaxies which are fully integrable. These works led to a dominant viewpoint in the 1980s that regular orbits prevail in elliptical galaxies. This viewpoint changed, partly, in the 1990s when *chaos* was found to play a significant role in elliptical galaxies with central masses [6, 15, 16, 24] (for a review see [4, 14]).

In [9, 17, 30], the authors studied smooth centre models of elliptical galaxies by N-body methods, showing that such systems have significant fractions of chaotic orbits reaching up to 30–50%. They also showed that the existence of a significant chaotic component has observable consequences since such a

G. Contopoulos, P.A. Patsis, *Chaos in Astronomy*. Astrophysics and Space Science Proceedings, doi: 10.1007/978-3-540-75826-6,

component forms a different configuration than the regular component, being more isotropic and more uniformly distributed.

In the present study we use an improved SCF code in order to make a detailed orbital analysis for a wide range of systems representing smooth centre elliptical galaxies in equilibrium. We reveal the role of chaos and the significance of the various types of regular orbits in these systems. A detailed presentation can be found in [11].

In the following sections we discuss briefly the SCF code used and its sensitivity on the classification of the orbits. We also describe the studied models and the methods used for the classification of orbits. Finally, we present our results and summarize our conclusions.

2 SCF Code

A detailed analysis of the SCF code used in our simulations can be found in [10]. Briefly, following [31] we consider potential-density pairs (Φ, ρ) written in the form

$$\rho(r,\theta,\phi) = \sum_{l=0}^{l_{max}} \sum_{m=-l}^{l} \sum_{n=0}^{n_{max}} b_{nlm}\rho_{lm}(r)u_{nlm}(r)Y_l^m(\theta,\phi), \tag{1}$$

$$\Phi(r,\theta,\phi) = \sum_{l=0}^{l_{max}} \sum_{m=-l}^{l} \sum_{n=0}^{n_{max}} c_{nlm}\Phi_{lm}(r)u_{nlm}(r)Y_l^m(\theta,\phi), \tag{2}$$

where $Y_l^m(\theta,\phi)$ are spherical harmonics and $\rho_{lm}(r)$, $\Phi_{lm}(r)$ are arbitrarily chosen functions of the radius r. The basis functions $u_{nlm}(r)$ are solutions of a Sturm–Liouville problem

$$-\frac{\mathrm{d}}{\mathrm{d}r}\left[p(r)\frac{\mathrm{d}u}{\mathrm{d}r}\right] + q(r)u = \lambda w(r)u \tag{3}$$

determined by the functions $p(r), q(r), w(r)$ which depend on the functions $\Phi_{lm}(r), \rho_{lm}(r)$. The coefficients b_{nml}, c_{nml} and the eigenvalue λ satisfy the relation $\lambda = b_{nlm}/c_{nlm}$. Using the orthogonality relation for u_{nlm} we obtain an approximation for the coefficients b_{nlm} as a sum over the system's particles

$$b_{nlm} \simeq \sum_{i=1}^{N} -4\pi\Phi_{lm}(r_i)u_{nlm}(r_i)Y_l^{m*}(\theta_i,\phi_i). \tag{4}$$

This calculation is a Monte-Carlo estimation of b_{nlm} since the particles' positions are a discrete realization of the smooth underlying density function $\rho(r)$.

In the previous analysis the choice of $\Phi_{lm}(r)$ and $\rho_{lm}(r)$ is important. If $\Phi_{lm}(r)$ and $\rho_{lm}(r)$ are close to the true smooth potential and density of the studied system, the corresponding basis sets provide correct representations with few terms.

For some special choice of $\Phi_{lm}(r)$ and $\rho_{lm}(r)$ the corresponding solutions u_{nlm} are given analytically by series of polynomials. Such examples are the basis sets of Clutton–Brock [2], Allen, Palmer and Papaloizou [1] and Hernquist and Ostriker [7].

We developed a modified version of the Hernquist–Ostriker basis set which is more flexible for the representation of systems with various density slopes at the centres [10]. Our choice of $\Phi_{lm}(r)$ is

$$\Phi_{lm}(r) = -\sqrt{4\pi}\frac{r^l}{(a+r)^{2l+1}}, \tag{5}$$

which is exactly the same as in the HO basis set, but the chosen function $\rho_{lm}(r)$ is

$$\rho_{lm}(r) = \sqrt{4\pi}\frac{1}{2\pi}\frac{(2l+1)(l+1)}{(r^2+\varepsilon^2)^{1/2}}\frac{ar^l}{(a+r)^{2l+3}}, \tag{6}$$

which coincides with the respective HO function only for $\varepsilon = 0$. In [10] we describe an analytical procedure for the derivation of the basis set u_{nlm} in the case of $\varepsilon \neq 0$.

The following results have been produced using the basis set corresponding to $\varepsilon = 0.06$ (the half mass radius is equal to 1). Furthermore we have used expansions up to $n_{max} = 10$ and $l_{max} = 2$.

3 Models

Starting from two different models produced by cosmological initial conditions [30] we created a series of models that cover the entire domain of possible configurations of observed elliptical galaxies. The two original models are triaxial (nearly prolate) and have maximum ellipticity $E_{max} \simeq 7$ and $E_{max} \simeq 4$, respectively. Starting from a specific model, a new model is obtained by transferring small portions of kinetic energy from one axis to another. For example, if we want to produce a system that has its principal axis a smaller than that of the original system, we take from all the particles (or from a fraction of the particles) small portions of kinetic energy from the axes b and/or c and give this energy to the motion along the axis a. After such a transfer we have new initial conditions in which each particle's energy as well as the system's total energy distribution remain unchanged. Evolving these initial conditions by the SCF code, after some time (at most half a Hubble time) the system reaches a new equilibrium state. We then check the system's new morphological properties. We keep those systems belonging morphologically to our domain of interest, or run a experiment with a different redistribution of the energy along the principal axes of the previous system.

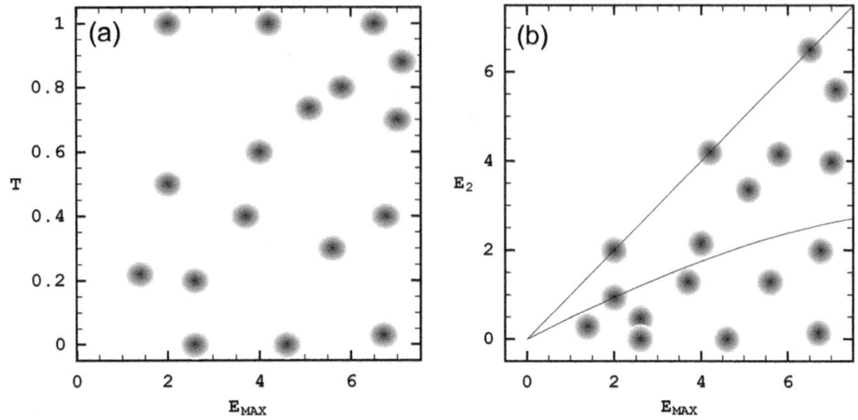

Fig. 1. (**a**) The *big dots* show the positions of the studied models on the plane $E_{max} - T$. (**b**) The same as (a) but on the plane $E_{max} - E_2$. In (b) the diagonal $E_{max} = E_2$ denotes the area of prolate configurations ($T = 1$) while the *curved line* denotes the area of maximally triaxial configurations ($T = 0.5$). Note, that the degeneracy we have between oblate and prolate configurations for low values of E_{max} is raised in diagram (a)

Working in this way, we finally constructed 17 models with morphologies covering almost the entire domain of observable morphologies. Figure 1 shows with big dots all the systems in the diagrams (a) maximum ellipticity (E_{max}) vs triaxiality (T) and (b) maximum ellipticity (E_{max}) vs ellipticity on the plane of intermediate-long axis (E_2). Note, that the triaxiality parameter T of an ellipsoid with axes ($a \leq b \leq c$) is defined by the relation $T = (c^2 - b^2)/(c^2 - a^)$. In these diagrams each point corresponds to one ellipsoid with specific axial ratios. The difference between these two diagrams is that the first raises the existing degeneracy between the oblate ($T = 0$) and the prolate ($T = 1$) morphology in the area of low values of maximum ellipticity. In Fig. 1b the diagonal $E_2 = E_{max}$ is the area of prolate configurations, the horizontal line $E_2 = 0$ is the area of oblate configurations and the curved line denotes the area of maximally triaxial configurations ($T = 0.5$).

4 Results

When the systems reach their equilibrium state the coefficients b_{nlm} of the expansion (2) remain stabilized with small variations due to Poisson noise effects. Hence, at each snapshot of the equilibrium stage we can write a time-independent Hamiltonian

$$H = \frac{1}{2}\mathbf{v}^2 + \Phi(r, \theta, \phi). \qquad (7)$$

and perform its orbital analysis, i.e. identify the regular and the chaotic orbits. Regular orbits are further classified into various basic types (box – short axis tube (SAT) – inner long axis tube (ILAT) – outer long axis tube (OLAT)). Chaotic orbits are also further classified according to an appropriate numerical measure of their chaoticity.

For the identification of regular and chaotic orbits we used the combination of various methods: (a) Lyapunov exponents [30], (b) Alignment indices [26, 28–30], (c) Fast Lyapunov indicator [5] and (d) Average power law exponent [13]. For the detailed classification of regular orbits we used the code of [25] which is an improved version of the Numerical Analysis of Fundamental Frequencies method proposed by Laskar [12].

Figure 2a shows the regular orbits in the plane of rotation numbers $\frac{\nu_z}{\nu_x}, \frac{\nu_y}{\nu_x}$ for the system (E_{max}, T)=(6, 0.8). Figure 2b is the same as Fig. 2a but for the chaotic orbits. Along the axes x, y, z are the small, intermediate and long axis of each system respectively. Figure 2a shows the loci corresponding to the various types of regular orbits. The SAT orbits lie along the diagonal $\nu_z = \nu_y$. The ILAT orbits lie along the vertical line $\nu_y = \nu_x$ for low values of $\frac{\nu_z}{\nu_x}$, and the OLAT orbits lie along the same line but for higher values of $\frac{\nu_z}{\nu_x}$. The box orbits and high order resonant tube – HORT (or boxlets) orbits lie on a wider area centred at (0.85,0.65). In Fig. 2b we observe that the chaotic orbits create a scattered distribution, although many of them are trapped along the various resonance lines following thereupon a behavior similar to the behavior of the corresponding regular orbits.

Performing the same analysis in all the systems we collect the results into successive figures. Figure 3a,b shows the percentages of chaotic orbits

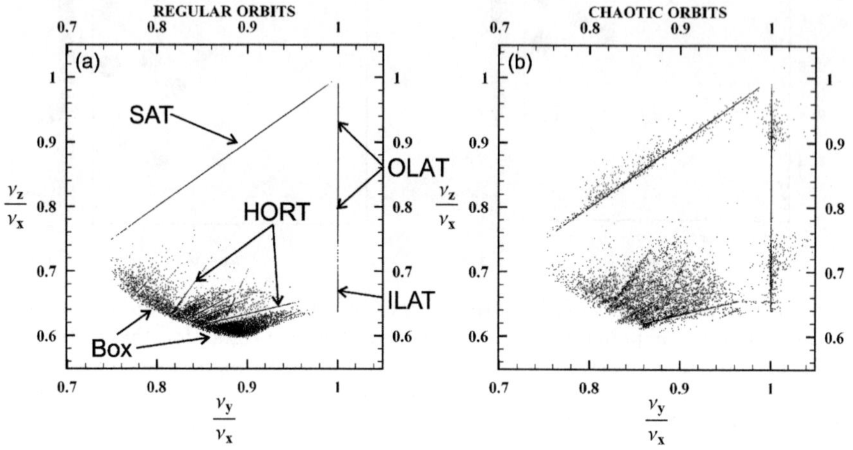

Fig. 2. (**a**) The regular orbits of the system $(E_{max}, T) = (6, 0.8)$ on the plane of rotational numbers $\frac{\nu_z}{\nu_x}, \frac{\nu_y}{\nu_x}$. The loci of the various types of orbits are indicated (see the text for more details). (**b**) The same as (a) but for the chaotic orbits. We see that many chaotic orbits are trapped along the various resonance lines

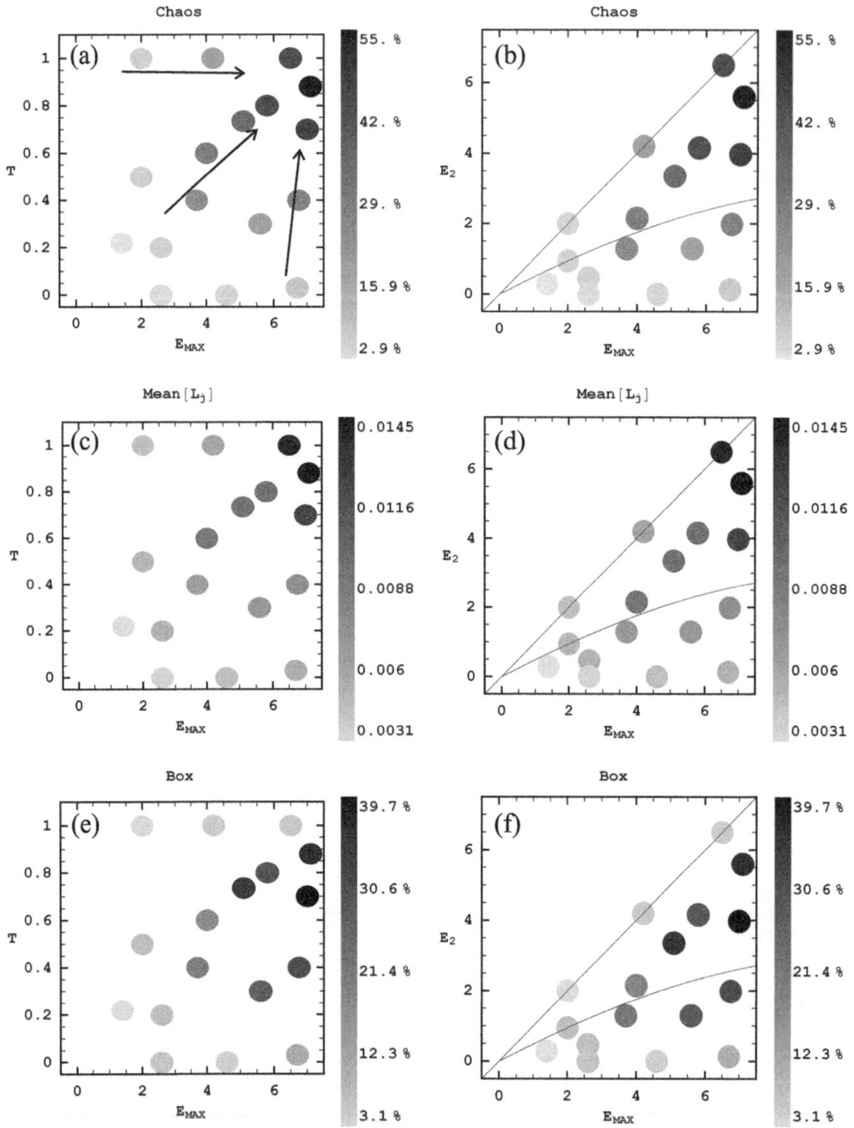

Fig. 3. The percentages of the various types of orbits as well as the mean values of the Lyapunov exponents (of the identified chaotic orbits) for the systems we studied. Note, that nearly prolate systems ($T \simeq 1$) with high values of E_{max} are 'more' chaotic in a double sense, i.e. not only they have high percentages of chaotic orbits but these chaotic orbits have also high values of the Lyapunov exponents. The three arrows in (a) indicate the directions leading to higher percentages of chaotic orbits

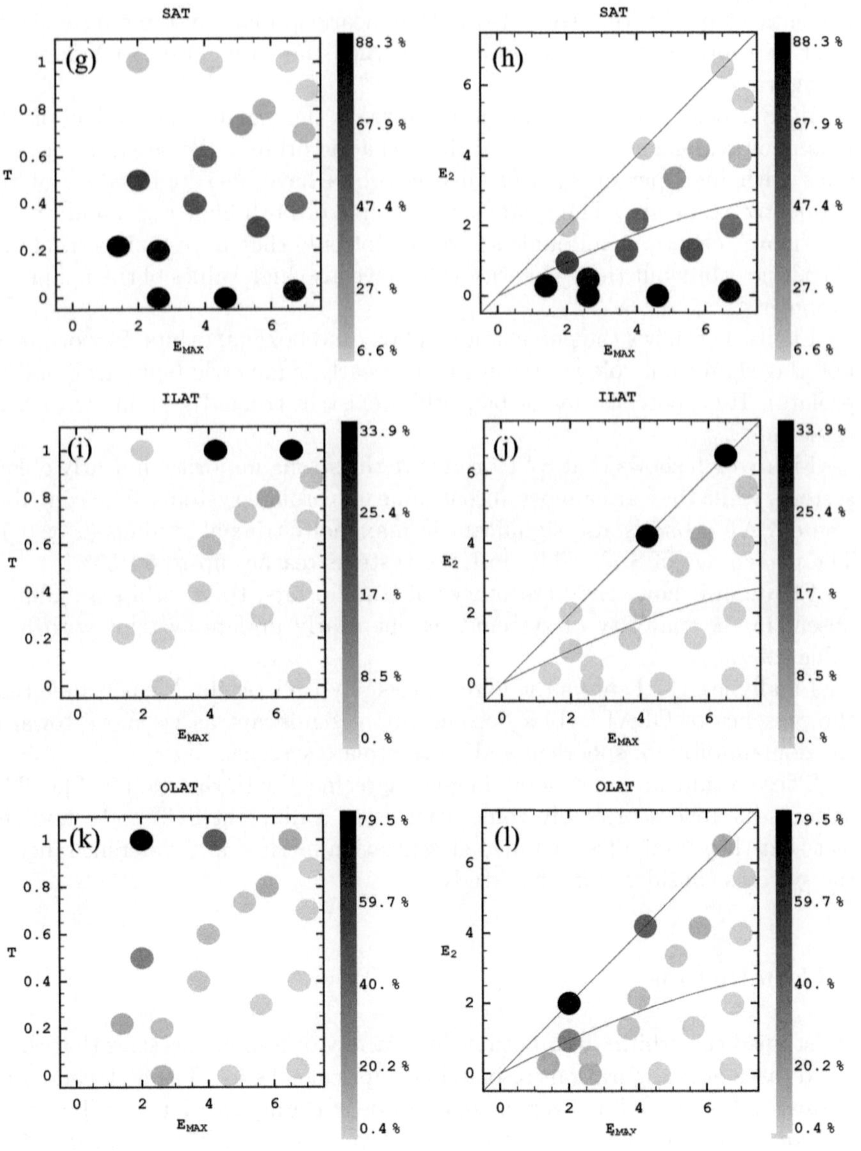

Fig. 3. Continued.

in grey-color scale for all the systems on both the diagrams ($E_{max} - T$) and ($E_{max} - E_2$). We see that the chaotic orbits can reach percentages up to $\simeq 55\%$. The nearly oblate ($T \simeq 0$) systems and the nearly spherical systems ($E_{max} \simeq 0$) have the lowest percentages of chaotic orbits. The highest percentages of chaotic orbits were found in nearly prolate systems with high values of E_{max}. The arrows in Fig. 3a clearly show the directions leading to higher percentages of chaotic orbits. Note, that nearly prolate systems have significant percentages of chaotic orbits even when they are close to a symmetric configuration.

In Fig. 3c,d the grey-color scale reveals the mean level value of the Lyapunov exponents (for the identified chaotic orbits). We see that the systems with high percentages of chaotic orbits have also high values of the Lyapunov exponents. The systems in the area of both high E_{max} and high T are "more" chaotic in a double sense, i.e. not only they have high percentages of chaotic orbits but these chaotic orbits have also high values of the Lyapunov exponents.

Figure 3e,f shows the percentages of box (and boxlets) orbits. Box orbits do not play significant role in systems being nearly symmetric (spherical, oblate, prolate). High percentages of box orbits exist in triaxial systems with high values of E_{max}.

Figure 3g,h shows that SAT orbits are the strong majority in nearly oblate systems while they are almost absent in nearly prolate systems. The contribution of SAT orbits is also significant in maximally triaxial systems ($T \simeq 0.5$). The percentage of SAT orbits in these systems reaches up to 50–60%.

Figure 3i,j shows the percentages of ILAT orbits. ILAT orbits are almost absent in the majority of systems, except nearly prolate systems with high values of E_{max}.

Finally, Fig. 3k,l shows the percentages of OLAT orbits. We conclude that the presence of OLAT orbits becomes more significant as we move towards the domain of more spherical and more prolate systems.

These results are, in general, in good agreement with the results of [8]. The various differences probably come from using a different SCF code, and the fact that they used a less detailed classification method and different range of the systems considered in that study.

5 Conclusions

We studied the orbital distributions in a variety of N-body systems that cover a wide area of configurations resembling elliptical galaxies. The SCF code used in this study is a modified version of the code of Hernquist–Ostriker. This code is flexible and suitable for the study of systems with central density slope $\lesssim 1$.

Our main conclusions are the following:

(1) The chaotic component of the mass distribution plays a significant role in elliptical galaxies, especially in nearly prolate systems with high values of E_{max}, in which the percentage of chaotic orbits reaches up to 55%. The chaotic orbits in these systems have in addition large values of the Lyapunov exponents.

(2) The box (and boxlet) orbits are a significant building block of triaxial and triaxial-prolate systems with high values of E_{max}. Their percentage reaches up to 40%.

(3) The SAT orbits prevail in nearly oblate systems and have a strong presence ($\approx 50\%$) in maximally triaxial systems (in some cases surpass the population of boxes). In nearly prolate systems, they are, on the other hand, a rather small minority ($< 10\%$).

(4) The importance of ILAT and OLAT orbits seems to be rather small in the majority of systems. Such orbits appear mostly in nearly prolate systems. ILAT orbits are more important for systems with high values of E_{max} while OLAT orbits are important only in more spherical systems.

References

1. Allen A., Palmer P., Papaloizou J., Mon. Not. R. Astr. Soc., **242**, 576 (1990)
2. Clutton-Brock M., Astroph. & Sp. Sc., **23**, 55 (1973)
3. de Zeeuw T., Lynden-Bell D., Mon. Not. R. Astr. Soc., **215**, 713 (1985)
4. Efthymiopoulos C., Voglis N., Kalapotharakos C.: Special Features of Galactic Dynamics. In: *Topics in Gravitational Dynamics*, vol 729 of Lecture Notes In Physics ed by Benest D., Froeschlé C., Lega E., (Springer, Berlin 2008), p. 295
5. Froeschlé, Cl., Gonczi, R., Lega, E., Planetary and Space Science, **45**, 881 (1997)
6. Gerhard O. E., Binney J., Mon. Not. R. Astr. Soc., **216**, 467 (1985)
7. Hernquist L., Ostriker J., Astrophys. J., **386**, 375 (1992)
8. Jesseit R., Naab T., Burkert A., Mon. Not. R. Astr. Soc., **360**, 1185 (2005)
9. Kalapotharakos C., Voglis N., Cel. Mech. Dyn. Astron., **92**, 157 (2005)
10. Kalapotharakos C., Efthymiopoulos C., Voglis N., Mon. Not. R. Astr. Soc., **383**, 971 (2008)
11. Kalapotharakos C., in preparation (2008)
12. Laskar J., Cel. Mech. Dyn. Astron., **56**, 191 (1993)
13. Lukes-Gerakopoulos, G., Voglis, N., Efthymiopoulos, C., Physica A, **-**, - (2008)
14. Merritt D., Proc. Astr. Soc. Pacific, **111**, issue 756, 129 (1999)
15. Merritt D., Fridman T., Astrophys. J., **460**, 136 (1996)
16. Merritt D., Valluri M., Astrophys. J., **471**, 82 (1996)
17. Muzzio J. C., Carpintero D. D., Wachlin F. C., Cel. Mech. Dyn. Astron., **91**, 173 (2005)
18. Richstone D., Tremaine S., Astrophys. J., **286**, 27 (1984)
19. Richstone D. O., Astrophys. J., **238**, 103 (1980)
20. Richstone D. O., Astrophys. J., **252**, 496 (1982)
21. Richstone D. O., Astrophys. J., **281**, 100 (1984)
22. Schwarzschild M., Astrophys. J., **232**, 236 (1979)
23. Schwarzschild M., Astrophys. J., **263**, 599 (1982)
24. Schwarzschild M., Astrophys. J., **409**, 563 (1993)
25. Sidlichovsky, M., Nesvorny, D., Cel. Mech. Dyn. Astron., **65**, 137 (1997)

26. Skokos Ch., J. Phys. A, **34**, 10029, (2001)
27. Statler T. S., Astrophys. J., **321**, 113 (1987)
28. Voglis N., Contopoulos G., Efthymiopoulos C., Physical Review E, **57**, 372 (1998)
29. Voglis N., Contopoulos G., Efthymiopoulos C., Cel. Mech. Dyn. Astron., **73**, 211 (1999)
30. Voglis N., Kalapotharakos C., Stavropoulos I., Mon. Not. R. Astr. Soc., **337**, 619 (2002)
31. Weinberg M., Astron. J., **117**, 629 (1999)

The Connection Between Orbits and Isophotal Shape in Elliptical Galaxies

R. Jesseit, T. Naab, and A. Burkert

Universitätssternwarte München, Scheinerstrasse 1, 81679 München, Germany,
jesseit@usm.lmu.de

We examine the origin of photometrical properties of N-body merger remnants from the perspective of their orbital content. We show that disc mergers alone are unlikely to form a boxy-discy dichotomy in isophotal shape, because of the survival of disc-like orbits even in violent mergers. The shape of the different orbit families can vary strongly, depending on how violent the merger was. Minor axis tubes can become boxy and box orbits can be round. However, for edge-on projections isophotal shape, orbital content and line-of-sight velocity are well connected.

1 Introduction

The original classification of elliptical galaxies, which goes back to Edwin Hubble, is based on their apparent ellipticity. The projected ellipticity, however, is also a function of inclination, and not only of the intrinsic ellipticity. Bender [4] showed that the isophotes of elliptical galaxies deviate significantly from perfect ellipses. The deviations can be quantified by the a_4-parameter, where negative a_4 values signify boxy deviations and positive a_4 values discy deviations from a perfect ellipse. But it was also found that ellipticals which have boxy isophotes, are also X-ray bright, more luminous, rotate slowly and have cored inner surface density slopes, while discy ellipticals are less luminous, rotate fast and have power-law surface density profiles [5]. This led Kormendy and Bender [11] to revise the Hubble classification of ellipticals using isophotal shape instead of ellipticity as a galaxy family membership criterium. One of the most successful models of elliptical galaxy formation is the merging of disc galaxies. Barnes [3] proposed that the isophotal shape depends on the violence of the merging and subsequently, Naab et al. [12] showed that mergers of galaxies with comparable mass form more likely boxy ellipticals, while mergers of galaxies with differing mass form discy ellipticals. However, disc–disc mergers do not form a perfect dichtomy [13], while elliptical–elliptical mergers seem to disconnect merging ratio and isophotal

G. Contopoulos, P.A. Patsis, *Chaos in Astronomy*. Astrophysics and Space Science Proceedings, doi: 10.1007/978-3-540-75826-6,

shape, i.e. they are always boxy [16]. In contrast to this gas seems to play an important role in fast-rotating, discy ellipticals [9, 14]. However, recently the SAURON sample, a survey of 48 elliptical and lenticular galaxies, Emsellem et al. [6] showed that isophotal shapes are not well connected to rotation, i.e. fast rotating galaxies can have boxy isophotes [7]. In the following we want to shed some light on the connection of photometric and kinematic parameters in N-body merger remnants and how they relate to the internal orbital structure.

2 Superposition of Orbit Classes and Isophotal Shape

The orbital structure is the backbone of any galaxy. The properties of different orbit classes determines the final shape and kinematic structure of the galaxy. It is much less clear how the population of orbits is connected to the formation history of a galaxy and if we can detect the traces today. We examined a sample of 96 collisioneless disc–disc mergers, with merging mass ratios from 1:1 to 4:1. Most of the orbits on which the simulation particles move, belong to one of the following classes: box orbits, minor axis tubes or major axis tubes. In general the abundance of box orbits, which are non-rotating is decreasing with increasing merging mass ratio [8] and vice versa for minor axis tubes. This means that the less violent the merger is the more particles which have been in the disc of the progenitor survive. If we correlate the effective isophotal shape with the balance between minor axis tube and box orbits, we see that some mergers can appear discy, but have a sizable box orbit component (see Fig. 1).

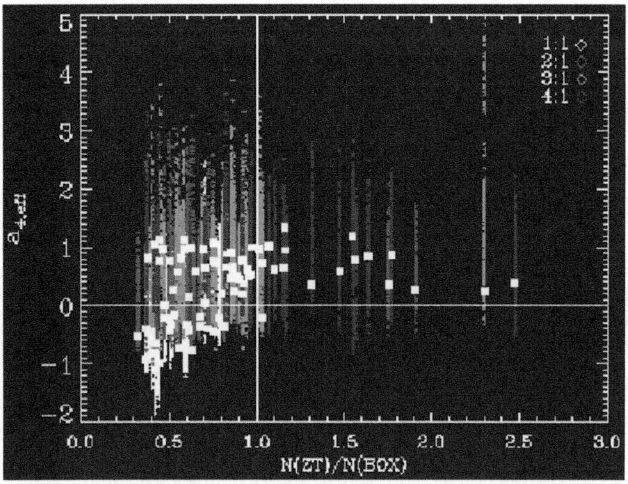

Fig. 1. Relation between effective isophotal shape and internal orbital structure. *White square*: most probable projection of a given remnant. Two hundred random projections for each remnant

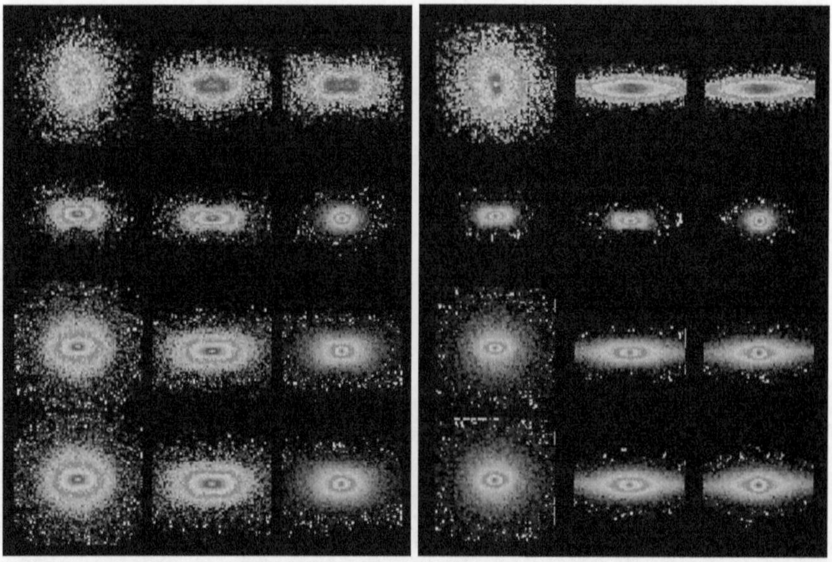

Fig. 2. Two-dimensional shape distribution of particles belonging to a certain orbital class. *From top to bottom*: Minor axis tubes, box orbits, box orbits + minor axis tubes, all particles. *Left panel*: Equal-mass merger remnant with boxy isophotes. *Right panel*: 4:1 discy merger remnant. Both remnants have approximately 40% box orbits

How can that be? We chose two remnants, one is very boxy and the other is discy, but both have a similar fraction of box orbits, i.e. about 40%. We use the classification to dissect the remnants into orbit classes. Regular orbits or at least orbits who behave regular over very long time scales, respect certain symmetries even if they move in a complicated triaxial potential, therefore we show the projections along the principal axes of the inertia tensor (see Fig. 2). It becomes immediately clear that the spatial distribution of particles belonging to the same orbit class are very different in each remnant. The difference is particularly striking for the minor axis tubes, which can appear peanut-shaped in the 1:1 remnant and disc-like in the 4:1 remnant.

Although a_4 can be an ambiguous indicator for the presence of certain orbit classes, the kinematic properties, are more clear-cut, e.g. while a box orbit can appear more round or more boxy depending on the overall shape of the potential the star moves in, the population of stars moving on box orbits will never show a netto rotation. However, even large rotation can be hidden if we observe a galaxy perpendicular to the line-of-sight, i.e. face-on. If we observe our sample of merger remnants edge-on we can probably learn the most of their internal structure. In Fig. 3 (left) we see the abundance of box orbits with respect to the location of the remnant in the a_4–ϵ plane. The isoabundance lines cross the $a_4 = 0$ line, indicating that we can hide a lot of box orbits in a discy remnant. In the v/σ–ϵ plane the orbital content can more

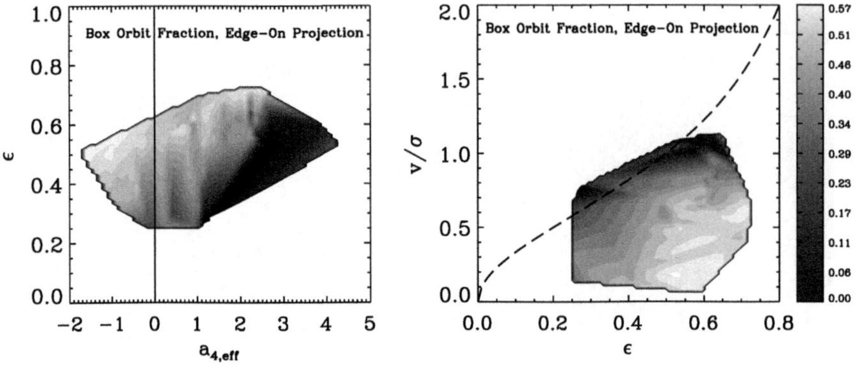

Fig. 3. Box orbit content of N-body merger remnants in the a_4–ϵ plane and in the v/σ–ϵ plane. The observational parameters are determined when the merger remnants are seen edge on

reliably be deduced. Remnant with little amount of box orbits lie closer to the isotropic rotator line. The gradient of box orbit content is approximately parallel to this line and peaks at $\epsilon = 0.5$ and $v/\sigma = 0$, where one would expect that triaxial galaxies are located.

3 Discussion and Conclusion

The trends seen here are typical for disc–disc merger remnants, while they are probably representative for low to intermediate luminosity elliptical galaxies, they cannot explain the whole population of elliptical galaxies. The analysis of the orbital fine structure should be extended to a broader range of formation mechanisms, e.g. E/S0 galaxies with boxy isophotes, will certainly have bars [2] which have not been covered by our present analysis. Dry merging will have certainly played a role in the formation of the most massive stellar systems in the universe and will lead to round or boxy systems [10, 16], while gas physics is important to form remnants with realistic LOSVDs [9, 14]. Monolithic collapse can also form triaxial self-gravitating systems which can have higher fractions of semi-stochastic orbits than our remnants [1]. Finally binary merger remnants can be compared to elliptical galaxies which formed in cosmological simulations and will probably have very different orbital structures [15].

References

1. Aquilano, R. O., Muzzio, J. C., Navone, H. D., & Zorzi, A. F. 2007, Celestial Mechanics and Dynamical Astronomy, 99, 307
2. Athanassoula, E. 2005, MNRAS, 358, 1477

3. Barnes, J. E. 1998, Saas-Fee Advanced Course 26: Galaxies: Interactions and Induced Star Formation, 275
4. Bender, R. 1988, A&A, 193, L7
5. Bender, R., Doebereiner, S., & Moellenhoff, C. 1988, A&AS, 74, 385
6. Emsellem, E., et al. 2004, MNRAS, 352, 721
7. Emsellem, E., et al. 2007, MNRAS, 379, 401
8. Jesseit, R., Naab, T., & Burkert, A. 2005, MNRAS, 360, 1185
9. Jesseit, R., Naab, T., Peletier, R. F., & Burkert, A. 2007, MNRAS, 376, 997
10. Khochfar, S., & Burkert, A. 2005, MNRAS, 359, 1379
11. Kormendy, J., & Bender, R. 1996, ApJ, 464, L119
12. Naab, T., Burkert, A., & Hernquist, L. 1999, ApJ, 523, L133
13. Naab, T., & Burkert, A. 2003, ApJ, 597, 893
14. Naab, T., Jesseit, R., & Burkert, A. 2006, MNRAS, 372, 839
15. Naab, T., Johansson, P. H., Ostriker, J. P., & Efstathiou, G. 2007, ApJ, 658, 710
16. Naab, T., Khochfar, S., & Burkert, A. 2006, ApJ, 636, L81
17. Thomas, J., Jesseit, R., Naab, T., Saglia, R. P., Burkert, A., & Bender, R. 2007, MNRAS, 381, 1672

Gas Orbits in a Spiral Potential

G.C. Gómez[1] and M.A. Martos[2]

[1] Centro de Radioastronomía y Astrofísica, Universidad Nacional Autónoma de México, Apartado Postal 3-72 (Xangari) Morelia, Mich. 58089, México
g.gomez@astrosmo.unam.mx

[2] Instituto de Astronomía – Universidad Nacional Autónoma de México, Apartado Postal 70-264, Ciudad Universitaria, D.F. 04510, México
marco@astroscu.unam.mx

Summary. We performed MHD simulations of the response of a gaseous galactic disk to a spiral perturbation in the background potential. In this poster, as a complement to Martos oral presentation, we present the results of our analysis of the gas flow and its interaction with the resonances expected from stellar orbit theory.

1 Introduction

In a previous work [1] it was shown that the response of a gaseous disk to a rotating two-armed spiral potential representing the potential of the Milky Way is four-armed beyond a certain distance. The axisymmetric part of that model included a disk, a bulge and a DM halo [2]. This is qualitatively consistent with the conventional picture for the spiral arms in our Galaxy, usually traced by HII regions. In a study of the self-consistency of the stellar disk to the imposed spiral perturbation [3], it was found that the stellar component was not in phase with the spiral beyond the 4:1 resonance, as predicted by Contopoulos and Grosbøl [4, 5], using a similar model of non-axisymmetric perturbation plus a background potential (see also references within [3]). In the observational study of Drimmel and Spergel [6], the gas spiral extends up to corotation, again in consistency with the weak case in [4].

In this work, the study presented in [1] is extended to the MHD regime, complementing the analysis with the calculation of the gaseous orbits (in a Lagrangian sense) in order to compare them with the corresponding stellar orbits.

2 The Simulations

The simulations were set up with an isothermal gaseous disk at $8,000\,^{\circ}$K in rotational equilibrium in the axisymmetric background potential [2] taking into account the centrifugal force, (thermal + magnetic) pressure gradient

G. Contopoulos, P.A. Patsis, *Chaos in Astronomy.* Astrophysics and Space Science Proceedings, doi: 10.1007/978-3-540-75826-6,

and magnetic tension. The gas density followed an exponential law with a 15 kpc scale length. The magnetic field was purely azimuthal in the beginning (although it rapidly adopted a spiral geometry), with a strength of 2 μG at the solar circle (at 8.5 kpc) and an exponential decay with a 25 kpc scale length.

This equilibrium was perturbed by a two arm spiral [3] with a pitch angle of 15.5°, imposed as a fixed background potential with an amplitude such that the radial force (averaged over the radial extent) is 2% the axisymmetric background force. This spiral perturbation rotates with a pattern speed $\Omega_P = 25$ km s^{-1} kpc^{-1}, which places the inner Lindblad, 4:1, −4:1 and outer Lindblad resonance radii at 2.42, 5.38, 11.69 and 14.41 kpc, respectively. The corotation radius is at 8.80 kpc. A variety of values for the arm strength, magnetic field intensity, gas temperature, and pattern speed were tested.

The full MHD simulations were performed with the ZEUS code [7, 8] on a 2D grid in cylindrical geometry. The numerical grid extended from 1 kpc through 18 kpc in radius and a full circle in azimuth, with $512 \times 1,024$ grid points. All cases were solved in the pattern reference frame in order to avoid smearing due to numerical diffusion. No self-gravity of the gas was considered, although we plan to add it in a forthcoming paper.

Very rapidly (less than 200 Myr), the gas reacts to the perturbation developing two 14°-pitch spiral arms in the internal region, up to the inner Lindblad resonance (ILR). Just outside this resonance, these arms break (at which point a spur is formed), but can be traced further out (with a 9° pitch angle), until just outside the 4:1 resonance. Also, a second pair of arms develops (with a 13° pitch) spanning from just outside the ILR, across the 4:1 resonance, until just before corotation (although a very weak extension can be traced outside corotation almost to the outer Lindblad resonance). This behavior is somewhat different in several details from that reported by [9] and it will be explored in the near future.

It is important to mention that the simulations reach a grand design quasi-steady state, although the position and pitch angle of the arms oscillate slightly around the above quoted values, and waves travel through the disk reflecting off the resonances and numerical boundaries.

It is quite noticeable the presence of an MHD instability at the position of corotation. Its seeds can already be seen as early as 1.5 Gyr, but it fully develops 2.5 Gyr into the run. After approximately 6 Gyr, the turbulence generated by the instability already spans all but the innermost disk regions. This instability appears to be quite robust; it is always present in the magnetized simulations, but absent in the pure HD simulations. More details about this phenomenon are provided in Martos contribution and in [10].

3 The Gaseous Orbits

Since ZEUS is an Eulerian code, it does not directly provide the actual orbits the gas follows during the evolution of the disk. Nevertheless, they can be reconstructed by performing a Runge–Kutta integration of the velocity data,

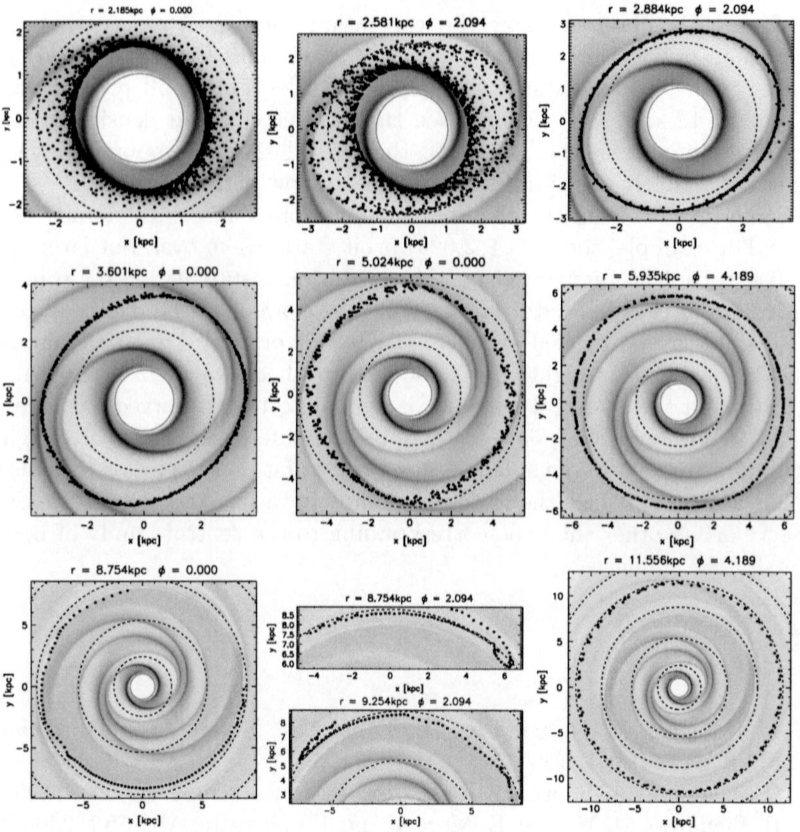

Fig. 1. Orbits of the gas in the simulation. Each panel is labeled with the initial position of the gas element, with angles measured counterclockwise starting from the positive x-axis. The *dashed circles* represent the position of the resonances (see text). The orbits are plotted on top of the density distribution averaged over 3 Gyr

interpolating in time between data dumps. Figure 1 presents several orbits of the gas integrated over 3 Gyr of evolution.

Several families of orbits, very similar to the principal family orbits obtained in stellar dynamics, are readily distinguishable, with transitions from one family to another near the position of the resonances. For example, the orbits integrated from initial position near or internal to the ILR [e.g. $(r/\,\mathrm{kpc}, \phi) = (2.185, 0.000)$] rapidly spiral inwards and settle in the ring/arm+spur structure at $r \approx 1.5$ kpc. Outside the ILR, [e.g. $(r/\,\mathrm{kpc}, \phi) = (2.884, 2.094)$], orbits tightly follow a simple oval, although they show some dispersion. The oval orbit becomes squarish as we move outwards [(3.601, 0.000) orbit], with the dispersion steadily increasing as one pair of arms weakens [(5.024,0.000) orbit], until we reach the 4:1 resonance and oval (but cuspy) orbits are recovered [(5.935, 4.189) orbit]. Banana-shaped orbits are frequently observed near the corotation radius, as expected. Outside corotation, orbits are nearly circular with deviations from the initial radius smaller than 0.5 kpc, even if the spiral perturbation is still present (it dies off at $r = 12$ kpc). Of course, since the gas velocity and density equations are solved simultaneously by ZEUS, the orbital shapes support the gaseous structures, namely the spiral arms, spurs and low density regions.

Sometimes we observe the gas moving from one orbital family to another. For example, the $(2.581, 2.094)$ orbit starts as an oval, but later spirals in. The turbulence generated by the corotation instability makes somewhat unpredictable which orbital family will be followed by the gas near corotation, as it may be pushed away [$(8.754, 0.000)$ orbit], trapped at corotation [$(8.754, 2.094)$], or brought across and then pushed away [$(9.254, 2.094)$].

Although these orbital behavior is similar to that observed in stellar dynamics [4], we are able point out a couple of differences. First, although the simulations are close to a steady state, orbits that do not close on themselves should be allowed since the gas keeps falling inwards. And second, we do not observe orbits other than the corresponding to the central family of periodic orbits of the stellar case.

References

1. M. Martos, X. Hernández, M. Yáñez, E. Moreno and B. Pichardo, MNRAS **350**, L47 (2004)
2. C. Allen and A. Santillán, Rev. Mex. Astron. Astrofis. **22**, 255 (1991)
3. B. Pichardo, M. Martos, E. Moreno and J. Espresate, ApJ **582**, 230 (2003)
4. G. Contopoulos and P. Grosbøl, A&A **155**, 11 (1986)
5. G. Contopoulos and P. Grosbøl, A&A **197**, 83 (1986)
6. R. Drimmel and D. Spergel, ApJ **556**, 181 (2001)
7. J. M. Stone and M. L. Norman, ApJS **80**, 753 (1992)
8. J. M. Stone and M. L. Norman, ApJS **80**, 791 (1992)
9. P. A. Patsis, N. Hiotelis, G. Contopoulos and P. Grosbøl, A&A **286**, 46 (1994)
10. M. Martos, *submitted for publication*

The Structure of the Phase Space in Galactic Potentials of Three Degrees of Freedom

M. Katsanikas[1], P.A. Patsis[1], and L. Zachilas[2]

[1] Research Center for Astronomy, Academy of Athens, Soranou Efessiou 4, GR-11527, Athens, Greece
mkatsan@yahoo.gr, patsis@academyofathens.gr

[2] Department of Economic Studies, University of Thessaly, Volos, Greece
zachilas@uth.gr

Summary. In order to study the structure of the phase space in galactic potentials of three degrees of freedom, we visualize a 4D space of section of the 6D phase space in a 3D Hamiltonian system. The method used is the method of color and rotation [9]. We apply this method to some cases of families of simple periodic orbits in a 3D potential, which describes the potential of the Milky Way [7]. We describe the differences in the orbital behavior in the neighborhood of **S**table **(S)**, Simple **U**nstable **(U)**, **D**ouble **U**nstable **(DU)** and Complex Unstable (Δ) periodic orbits.

Keywords: Chaos and Dynamical Systems, Galactic Dynamics

1 Introduction

The phase space of a conservative system with three degrees of freedom has 6 dimensions, i.e. in Cartesian coordinates, $(x, y, z, \dot{x}, \dot{y}, \dot{z})$. For a given value of the Jacobi constant a trajectory lies on a 5-dimensional manifold. In this manifold the surface of section is 4-dimensional. This does not allow us to visualize directly the surface of section in systems of three degrees of freedom. For this reason we use the method of color and rotation [9]. With this method we first visualize 3D projections and rotate the figures on a computer screen, as if we moved around to see the figure from all its sides. Furthermore we use colors to indicate the 4th dimension. For this purpose we use in this paper the "Mathematica" package and subroutines in "Mathematica" [6]. Each point is colored according to the value of its 4th coordinate in the following way: Firstly we define the surface of section that we will use, e.g. $y = 0$ with $\dot{y} > 0$. Secondly we select a 3D subspace of the surface of section, e.g. $(x, \dot{x}, \dot{z})$. Then we determine the minimum and maximum values of the 4th coordinate z. Finally we normalize the resulting interval $[\min(z), \max(z)]$ into [0,1] from

G. Contopoulos, P.A. Patsis, *Chaos in Astronomy*. Astrophysics and Space Science Proceedings, doi: 10.1007/978-3-540-75826-6,

which interval the color functions of "Mathematica" take values. In our figures we give always the color function of "Mathematica" that we used in a color-coded bar.

The system we used for our applications rotates around the z-axis with angular velocity Ω_b . The Hamiltonian of the system is :

$$H(x,y,z,\dot{x},\dot{y},\dot{z}) = \frac{1}{2}(\dot{x}^2+\dot{y}^2+\dot{z}^2) + \Phi(x,y,z) - \frac{1}{2}\Omega_b^2(x^2+y^2) \quad (1)$$

where $\Phi(x,y,z)$ is the potential which we used for our applications, i.e.:

$$\Phi(x,y,z) = -\frac{GM_1}{(x^2+\frac{y^2}{q_a^2}+[a_1+(\frac{z^2}{q_b^2}+b_1^2)^{1/2}]^2)^{1/2}} - \frac{GM_2}{(x^2+\frac{y^2}{q_a^2}+[a_2+(\frac{z^2}{q_b^2}+b_2^2)^{1/2}]^2)^{1/2}} \quad (2)$$

The potential in its axisymmetric form ($q_a = 1, q_b = 1$) can de considered as representing the potential for the Milky Way [7]. In our units distance R=1 corresponds to 1 kpc and Jacobi constant E_j=1 corresponds to 43950 $(km/sec)^2$. We have used the following values for the parameters: $a_1 = 0\ kpc$, $b_1 = 0.495\ kpc$, $M_1 = 2.05 \times 10^{10}\ M_{\odot}$, $a_2 = 7.258\ kpc$, $b_2 = 0.520\ kpc$, $M_2 = 25.47 \times 10^{10}\ M_{\odot}$, $q_a = 1.2$, $q_b = 0.9$ and $\Omega_b = 60\ km\ s^{-1}\ kpc^{-1}$, which is the value of the angular velocity of the bar of the Milky Way [4]. These values of the parameters secure positive densities everywhere.

At first we calculated the main families of periodic orbits of the system. Then we used the method by Broucke [1] and Hadjidemetriou [5] to calculate the stability of the periodic orbits. We studied the stability of 3D families bifurcated from the basic 2D family x1 [2] for our model. For the names of these families we follow the nomenclature by Skokos et al ([10], [11]). By giving initial conditions very close to the initial conditions of a periodic orbit of a 3D bifurcation of the basic family x1, we calculate a large number (in some cases even 10000) of intersections of the trajectory with the surface of section, since our goal is to understand the dynamical mechanisms in 3D systems.

2 4D Surfaces of Section

We studied the orbital behavior at the neighborhood of **S**table (**S**), Simple **U**nstable (**U**) , **D**ouble **U**nstable (**DU**) and Complex Unstable (Δ) periodic orbits.

Close to **S**table periodic orbits, we find a torus in the 3D projections of the space of section. We find a smooth color variation that starts at the external surface of the torus, but then proceeds continuing on the internal surface and then back again. Details can be found in [6]. The smooth succession of colors on the 3D surface of the torus indicates the distribution of the consequents on a smooth 4D hypersurface in the 4D space.

The orbital behavior close to a Simple **U**nstable periodic orbit is, in general, characterized by the presence of an "8-like", ribbon shaped, surface with succession of colors. The initial conditions of the periodic orbit are located in the middle of this "8-like" surface, in agreement with [9]. We also observed cases with a double "8-like" structure [6].

Close to a **D**ouble **U**nstable periodic orbit we find a cloud of consequents without succession of colors i.e. colors are mixed. This means that consequents visit large regions of the phase space very quickly.

At the neighborhood of a Complex Unstable (Δ) periodic orbit [12] for a few intersections we observed a spiral in the consequents in the $2D$ projections of the $4D$ surface of section, as in [3] and [8].

In Fig. 1 we observe the orbital behavior close to a Complex Unstable (Δ) periodic orbit of the 3D family x1v1 ([10],[11]) at $E_j = -4.62$ for 50 intersections. We use the $(x, \dot{x}, z)$ space for plotting the points and the $\dot{z}$ value to color them. Our point of view in spherical coordinates is given by $(R, \theta, \phi) = (4.730, 32°, 53°)$. We observe a spiral (with three spiral arms) at

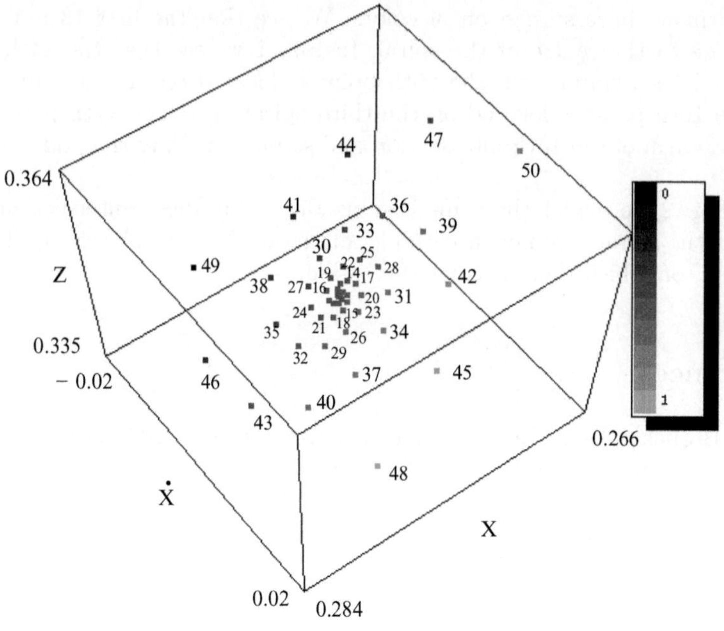

Fig. 1. The orbital behavior close to a Complex Unstable (Δ) periodic orbit (for 50 intersections).

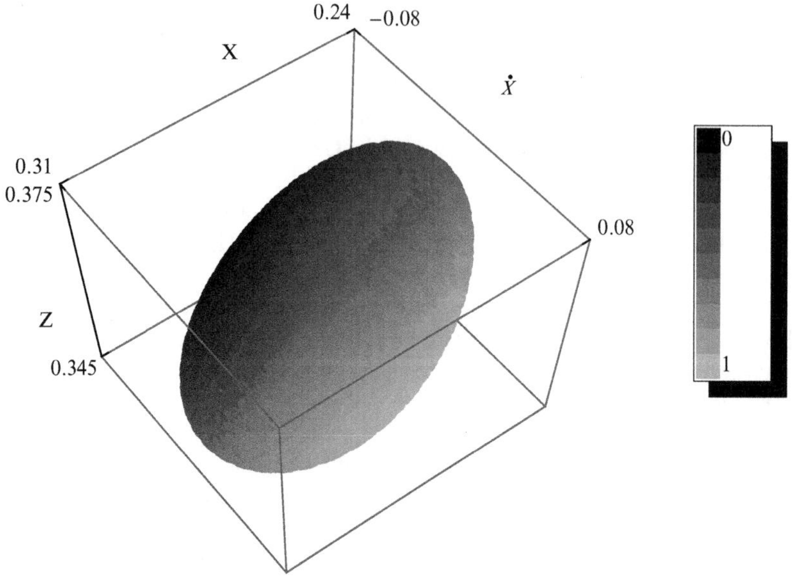

Fig. 2. The same case as Fig.1 for 10000 intersections.

the neghborhood of the Complex Unstable (Δ) periodic orbit. Along every spiral arm we have succession of colors. We see that the first 13 points are very close to the centre of the spiral. In Fig. 1 we see that the 14th point is located on a spiral arm, the 15th point is located on another spiral arm and the 16th point is located on the third spiral arm. The 17th point is on the spiral arm of the 14th intersection and so on. This way the points fill the space.

In Fig. 2 we depict the same case as Fig. 1 for 10000 intersections. We see that the spirals are not discernible any more. Now we observe a disk like structure, on which there is color succession.

References

1. R. Broucke: Amer. Inst. Aeronaut. Astronaut. J. **7**, 1003 (1969).
2. G. Contopoulos: *Order and Chaos in Dynamical Astronomy*, (Springer, Berlin Heidelberg New York 2002) pp 390–404.
3. G. Contopoulos, S.F. Farantos, H. Papadaki and C. Polymilis: Phys. Rev. E **50**, 4399 (1994).
4. P. Englmaier and O.E. Gerhard: Mon. Not. R. Astr. Soc. **304**, 512 (1999).
5. J. Hadjidemetriou: Cel. Mech. **12**, 255 (1975).
6. M. Katsanikas: M.Sc Thesis (in greek), University of Athens (2005).
7. M. Miyamoto and R. Nagai: Publ. Astron. Soc. Japan **27**, 533 (1975).

8. H. Papadaki, G. Contopoulos and C. Polymilis: Complex instability. In: *From Newton to Chaos*, ed by A.E. Roy, B.A. Steves (Plenum Press, New York 1995), pp 485–494.
9. P.A. Patsis and L. Zachilas: Int. J. Bifurc. Chaos **4**, 1399 (1994).
10. Ch. Skokos, P.A. Patsis and E. Athanassoula: Mon. Not. R. Astr. Soc. **333**, 847 (2002).
11. Ch. Skokos, P.A. Patsis and E. Athanassoula: Mon. Not. R. Astr. Soc. **333**, 861 (2002).
12. L.G. Zachilas: Astron. Astrophys. Suppl. **97**, 549 (1993).

Regular and Chaotic Orbits in Narrow 2D Bar Models

D.E. Kaufmann[1] and P.A. Patsis[2]

[1] Department of Space Operations, Southwest Research Institute, 1050 Walnut Street, Suite 300, Boulder, CO 80302-5142, USA
kaufmann@boulder.swri.edu

[2] Research Center for Astronomy, Academy of Athens, Soranou Efessiou 4, Athens, GR-11527, Greece
patsis@academyofathens.gr

Summary. We study orbits in a sequence of narrow 2D $n = 2$ Ferrers bar models with bar axial ratios a/c ranging from 2.5 to 7.5. We find that the central, or x_1, family of periodic orbits is the most important one in models with lower values of a/c. However, in models with $a/c > \sim 6$, we find that a new family of stable orbits having propeller shapes plays the dominant role. In our models this propeller family is in fact a distant relative of the x_1 family. We also find intermediate cases in which both families are important. The dominance of one family over the other may have direct consequences on the morphological properties of the bars that can be constructed from them. Finally, we confirm that the general level of chaos increases with increasing bar axial ratio, ultimately capping the value of the bar axial ratio.

1 Model Description

Our model barred galaxy potential consists of the superposition of two components representing separately a bar and an axisymmetric background. For the axisymmetric background we employ Plummer's spherical model. To model the bar we use the prolate $n = 2$ Ferrers ellipsoid [3, 6]. This particular bar component was chosen not only because it represents many features of observed galactic bars rather well, but also because its dynamics has been studied extensively [1, 2, 5, 7–10]. We take as our physical units the bar mass M_b, the bar semiminor axis length c, and the dynamical timescale along this same axis $\sqrt{c^3/(GM_b)}$. In these units the Newtonian gravitational constant $G = 1$. The bar is assumed to rotate about the z-axis, positively in the sense of the right hand rule, with a constant angular pattern speed Ω_p. In order to determine periodic orbits in this model, we integrated the equations of motion in the frame of reference corotating with the bar using the Bulirsch–Stoer method [11]. We used an iterative Newton method in two dimensions to find the periodic orbits. For calculating the stability of the periodic orbits we used the method described in [4].

G. Contopoulos, P.A. Patsis, *Chaos in Astronomy*. Astrophysics and Space Science Proceedings, doi: 10.1007/978-3-540-75826-6,

Table 1. Parameters of the models

Model	M	b	M_b	a	c	Ω_p
1	10.0	3.00	1.0	2.500	1.0	0.4200
2	10.0	6.00	1.0	5.000	1.0	0.1500
3	10.0	6.36	1.0	5.300	1.0	0.1376
4	10.0	6.75	1.0	5.625	1.0	0.1260
5	10.0	7.50	1.0	6.250	1.0	0.1100
6	10.0	9.00	1.0	7.500	1.0	0.0820

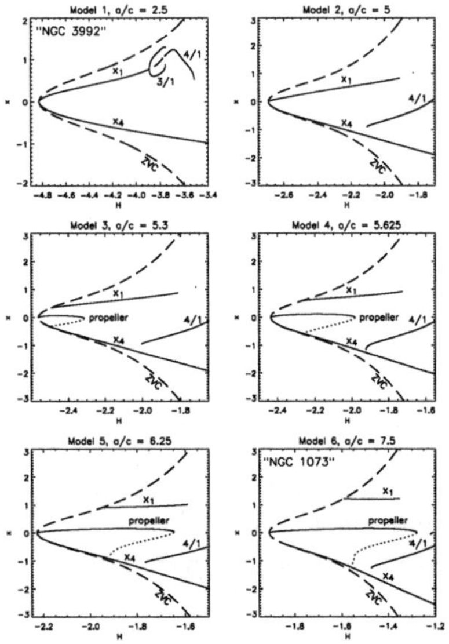

Fig. 1. Orbital characteristic diagrams for all six models

2 Variation of the Bar Axial Ratio a/c

Table 1 summarizes the values that define each model. We set the Plummer sphere mass $M = 10$ in each case, and set its shape parameter $b = 1.2a$. We let the bar semimajor axis length, a, vary from 2.5 to 7.5, and adjusted the bar pattern speed, Ω_p, such that the radial force vanished in the rotating frame at $r = 1.2a$ along the bar major axis. Figure 1 shows the basic orbital characteristic diagrams for the models.

3 Surfaces of Section

Figure 2 shows surface of section plots for six different values of the Hamiltonian, H, from model 6. These plots were generated by sampling the energetically available phase space in the (x, v_x) plane (with $y = 0$) at the different values of H, starting orbits with these initial conditions (choosing $v_y > 0$), and

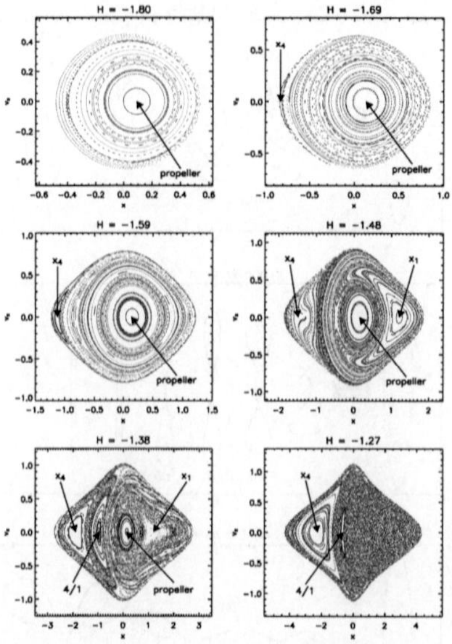

Fig. 2. Surface of section plots for six values of the Hamiltonian in model 6

following them through 200 subsequent upward intersections with the x-axis. Each of these sets of 201 "consequents" were plotted in (x, v_x) space. At the three lower values of H the propeller family of orbits completely dominates the phase space structure of the bar. Only at $H = -1.48$ does the x_1 family emerge as a more or less equal competitor to the propeller family in terms of controlling the prograde portion of the available phase space. As H increases further, the relative importance of the two families remains roughly constant, but the influence of both families decreases in the face of rising stochasticity.

4 A Closer Look at the Propeller Orbit Family

Figure 3 shows the morphologies of orbits from the x_1, propeller, and rectangular 4/1 families in model 6. For the x_1 and 4/1 families, the leftmost orbits are from the low H end, and subsequent orbits trace the respective characteristics toward higher H. All of these orbits are stable. For the propeller family, the leftmost orbit is from the low H end of the unstable branch. Subsequent orbits are taken from the unstable branch toward higher H, then down the stable branch toward lower H. Despite the similarity between the leftmost orbits of the propeller and 4/1 families, the closest direct relationship is between the x_1 and propeller families. Figure 4 shows a more detailed characteristic diagram of the main orbit families in the bar region of model 6. Both the stable x_1 and unstable propeller characteristics terminate on the zero velocity curve at $H \approx -1.591$ and $|x| \approx 1.230$. The leftmost propeller orbit in Fig. 3 is the lowest H periodic orbit *of multiplicity one* on the unstable branch. At

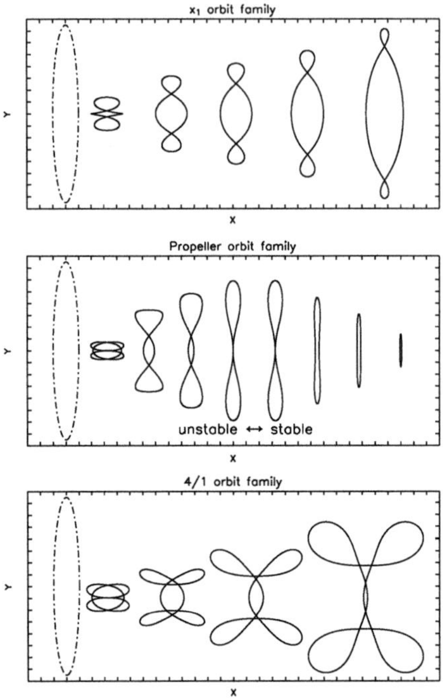

Fig. 3. Shapes of orbits in the x_1, propeller, and rectangular 4/1 orbit families in model 6 in comparison with the bar at left. *Tick mark* separations on both axes represent unit length

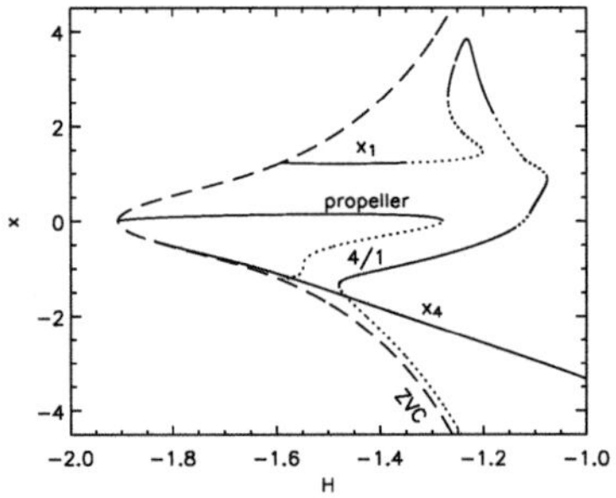

Fig. 4. Detailed characteristic diagram for model 6 showing both stable (*solid*) and unstable (*dotted*) portions of the main orbit families in the bar region

lower values of H the orbit loops interpenetrate and the orbit changes to multiplicity *three*. The propeller family characteristic continues to lower H and lower x and crosses the retrograde x_4 family characteristic, at which point all three loops coincide. Beyond the x_4 crossing, the multiplicity three propeller family continues on and terminates on the zero velocity curve. The propeller

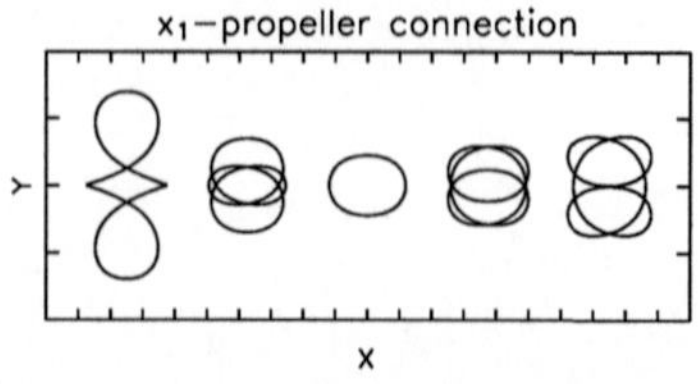

Fig. 5. Shapes of orbits in model 6 along the unstable branch of the propeller family from the zero velocity curve (ZVC) up to the point where the orbits become of multiplicity one

orbit at this point is the same as the x_1 orbit at the termination of the stable x_1 characteristic. Figure 5 shows the morphologies of orbits connecting the leftmost x_1 and propeller family orbits of Fig. 3.

5 Conclusions

The x_1 family does not dominate the structures of all realistic 2D galactic bar potentials. We have found cases in which the propeller family plays the central role. The three orbit families emphasized here, the propeller, the x_1, and the rectangular 4/1 families, can rightly be considered to be part of a larger single family having a complicated characteristic and changing multiplicity at a certain value of H. The importance of propeller orbits in real galactic bars appears to be constrained by the accompanying high level of chaos that seems to prevent real bars from reaching such elongated states.

References

1. E. Athanassoula: MNRAS **259**, 328 (1992)
2. E. Athanassoula, O. Bienaymé, L. Martinet, D. Pfenniger: A&A **127**, 349 (1983)
3. J. Binney, S. Tremaine: *Galactic Dynamics*, 1st edn (Princeton Univ. Press, Princeton 1987)
4. G. Contopoulos, P. Grosbøl: A&A **155**, 11 (1986)
5. G. de Vaucouleurs, K.C. Freeman: Vistas Astron. **14**, 163 (1972)
6. N.M. Ferrers: Quart. J. Pure Appl. Math. **14**, 1 (1877)
7. D.E. Kaufmann, G. Contopoulos: A&A **309**, 381 (1996)
8. T. Papayannopoulos, M. Petrou: A&A **119**, 21 (1983)
9. D. Pfenniger: A&A **134**, 373 (1984)
10. D. Pfenniger: A&A **150**, 112 (1985)
11. W.H. Press, S.A. Teukolsky, W.T. Vetterling, B.P. Flannery: *Numerical Recipes in FORTRAN: The Art of Scientific Computing*, 2nd edn (Cambridge Univ. Press, Cambridge 1992)

The Coalescence of Invariant Manifolds in Barred-Spiral Galaxies

P. Tsoutsis and C. Efthymiopoulos

Research Center for Astronomy, Academy of Athens,
Soranou Efessiou 4, GR-11527, Athens, Greece
ptsoutsi@phys.uoa.gr, cefthim@academyofathens.gr

1 Unstable Periodic Orbits and their Invariant Manifolds

In a previous paper ([7], see also [2, 3]) we explored the role of the *invariant manifolds* of the family of short period unstable periodic orbits which exist around the unstable equilibria L_1 or L_2 in supporting a spiral pattern in barred galaxies. Besides our study, the role of these manifolds has been pointed out in the formation both of rings [4] and spiral arms [5].

Here we consider the N-Body model of Voglis et al. [7, 8] and examine the invariant manifolds of *many different periodic orbits* covering a significant part of the range of Jacobi constants in the histogram of the energies of the particles supporting the spiral pattern (Fig. 1a). Our main finding is that the invariant manifolds of *all* the examined families of unstable periodic orbits produce essentially the same pattern in configuration space, i.e., the same spiral pattern. These manifolds produce a phenomenon of 'stickiness', namely while the phase space is, in general, open to fast chaotic escapes (with escape times of only a few radial periods), the chaotic orbits with initial conditions on or close to an invariant manifold can only escape by following this manifold, and then the escape time increases considerably, i.e., it is of order 10–10^2 periods. During this time the chaotic orbits support the spiral pattern.

Figure 1b shows the characteristic curves of nine different families of periodic orbits with apocenters close to or beyond the corotation region. The characteristic curves correspond to monoparametric functions $r(E_J)$, $\theta(E_J)$ and $p_\theta(E_J)$ yielding the apocentric radius, azimuthal angle and angular momentum for which an orbit with initial conditions $r, \theta, p_\theta, \dot{r} = 0$, and Jacobi constant E_J is periodic. Figure 1b shows only the projection $r(E_J)$ of each characteristic curve. The commensurabilities of Fig. 1b correspond to the ratio of the number of radial oscillations (apocentric passages) of an orbit per azimuthal period. Finally, the black and gray parts correspond to intervals of values of the Jacobi constant at which the corresponding periodic orbit is stable or unstable respectively.

G. Contopoulos, P.A. Patsis, *Chaos in Astronomy.* Astrophysics and Space Science Proceedings, doi: 10.1007/978-3-540-75826-6,

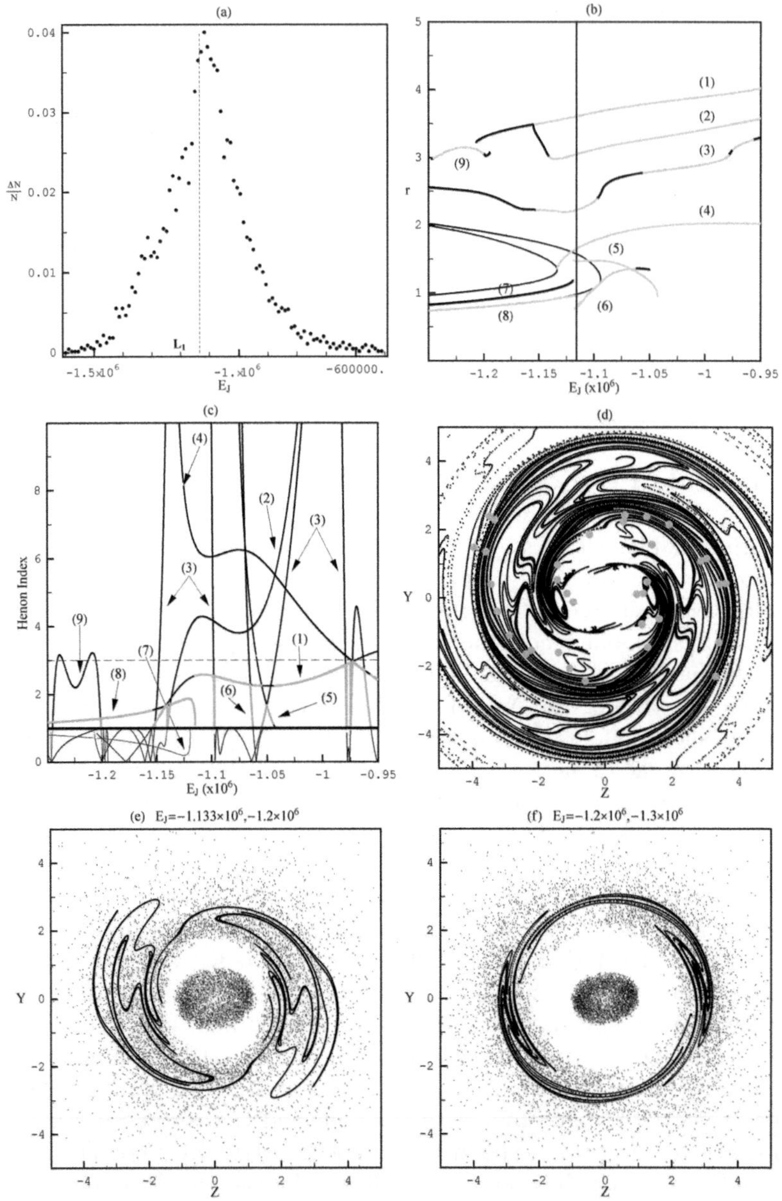

Fig. 1. (**a**) The distribution of the values of the Jacobi constants of the N-body particles supporting the spiral arms in the N-body simulation of Voglis et al.[7]. (**b**) Characteristic *curves* of nine different families of periodic orbits: (**1**) −1:1, (**2**) −1:1 of multiplicity two (bifurcating from the −1:1 family), (**3**) −2:1, (**4**) PL_1, (**5**) −4:1, (**6**) 4:1, (**7**) and (**8**) 3:1, (**9**) −1:1. (**c**) The Hénon index of the periodic families of (**b**). (**d**) The 'coalescence' of invariant manifolds in configuration space for seven different unstable periodic orbits at $E_J = -1.116 \times 10^6$. (**e**), (**f**) The unstable manifolds of the orbits −2:1,−1:1 respectively at the Jacobi constants indicated in the *panels*

Our main observation is that, for most values of the Jacobi constant beyond $E_{J,L1}$, there are more than one low order periodic orbits which are unstable. The vertical line at $E_J = -1.116 \times 10^6$ (in the N-body units, of [8]) shows an example of a value of E_J at which the orbits of seven out of the nine considered families are unstable.

Furthermore, for most considered values of E_J in Fig. 1b we can find at least one unstable periodic orbit with small absolute value of the Hénon stability index. The absolute values of the stability indices $|b|$ of all the orbits of Fig. 1b are shown in Fig. 1c. An orbit is stable if $|b| < 1$ and unstable if $|b| > 1$. The dependence of the stability index of every family on E_J shows intervals at which there is an abrupt rise of $|b|$ to very high values (of order 10^3–10^4), followed by other intervals at which $|b|$ falls to relatively small values (the upper limit of Fig. 1c is taken at $|b| = 10$). The main conclusion is drawn by calculating $b_{min}(E_J)$, i.e., the minimum of the values of $|b|$ of all the periodic orbits which are unstable at a particular value of E_J (gray thick curve of Fig.1c that follows different families for various intervals of E_J). From this curve we see that, for almost any value of E_J in the considered interval, there is one out of the total ensemble of unstable periodic orbits which has a relatively small stability index (the whole gray curve is below $|b| < 3$). This remark is important, because it implies that the invariant manifold of that particular periodic orbit is dynamically important.

Figure 1d shows now the main result. The unstable manifolds of *seven* different families of periodic orbits, which, for $E_J = -1.116 \times 10^6$ are all unstable, are superposed in the same plot of the configuration space in which we show the maxima of the spiral pattern (thick dots) as calculated by the distribution of the N-body particles. We immediately recognize that the unstable manifolds of all the families contribute to the formation of the same spiral pattern. In fact, given that the unstable manifold of one periodic orbit cannot intersect the unstable manifolds of other periodic orbits (see, e.g., [1, pp. 144–157]), the manifolds emanating from two different periodic orbits describe nearly parallel paths in the configuration space. The so-created set of all the manifolds generates a pattern, that we call the *'coalescence'* of invariant manifolds.

For values of $E_J < E_{J,L1}$, there is no communication of the domains inside and outside corotation. In this case we find (Figs. 1e, f) that the outer extensions of the spiral arms are again supported by the invariant manifolds of periodic orbits reaching large apocentric distances from corotation. In fact, as the value of E_J decreases, the patterns generated by the invariant manifolds become more and more axisymmetric, and this tendency of the manifolds marks the end of the spiral pattern. In our case we find this to be the outer 1:1 resonance, i.e., beyond the outer Lindblad resonance.

2 Conclusions

The main conclusions of our work are:

1. In a barred – spiral galaxy the invariant manifolds of many different families of unstable periodic orbits near and beyond corotation support a spiral pattern beyond the bar.
2. The unstable manifolds of all the studied families follow nearly parallel paths in phase space which are reflected in nearly parallel patterns in the configuration space. The overall pattern produced by the invariant manifolds of different families determines a 'coalescence' of invariant manifolds. This occupies a locus in configuration space which is invariant under a Poincaré mapping, i.e., as the chaotic orbits move from one apocentric passage to the next, the successive apocenters remain confined on the same locus.
3. The *real* observed spiral pattern of the system coincides with the spiral pattern produced by the coalescence of invariant manifolds. A dynamical justification of this phenomenon is given elsewhere [6].
4. Different families of unstable periodic orbits play the dominant dynamical role in the above phenomena at different values of the Jacobi constant. The short period unstable family PL_1, plays an important role for values of the Jacobi constant $E_J > E_{J,L1}$, which is however amplified by the extensions of other families near corotation, e.g. the 3:1 and 4:1 families. The main contribution to the spiral pattern is by the manifolds of families which are unstable in the energy range $E_{J,L1} \leq E_J \leq E_{J,L4}$. On the other hand, the manifolds of families which are unstable for $E_J < E_{J,L1}$ are responsible for extensions of the spiral arms at distances well beyond corotation.
5. Beyond the outer Lindblad resonance, the manifolds tend to become more and more axisymmetric, and this tendency marks the end of the spiral arms.

References

1. G. Contopoulos: Order and Chaos in Dynamical Astronomy, Springer, Berlin (2004)
2. G. Contopoulos: Chaos in Astronomy, Springer, Berlin (2008)
3. C. Efthymiopoulos, P. Tsoutsis, C. Kalapotharakos and G. Contopoulos: Chaos in Astronomy, Springer, Berlin (2008)
4. M. Romero-Gomez, J.J. Masdemont, E. Athanassoula and C. Garcia Gomez: Astron. Astrophys. **453**, 39 (2006)
5. M. Romero-Gomez, E. Athanassoula, J.J. Masdemont and C. Garcia-Gomez: Astron. Astrophys. **472**, 63 (2007)

6. P. Tsoutsis, C. Efthymiopoulos and N. Voglis, Mon. Not. R. Astr. Soc. **387**, 1264–1280 (2008)
7. N. Voglis, P. Tsoutsis and C. Efthymiopoulos, Mon. Not. R. Astr. Soc. **373**, 280 (2006)
8. N. Voglis, I. Stavropoulos and C. Kalapotharakos, Mon. Not. R. Astr. Soc. **372**, 901 (2006)

Part II

Celestial Mechanics

Chaotic Dynamics in Planetary Systems

R. Dvorak

Institute of Astronomy, University of Vienna, Türkenschanzstr. 17,
1180 Vienna, Austria
dvorak@astro.univie.ac.at

Summary. The discovery of the first extrasolar planets was a big surprise for astronomers, because the orbits of these planets turned out to be much more eccentric than in our Solar System. In this review we present the dynamical structure of the extrasolar planetary systems with their more than 300 planets known up to now. We shortly report the main techniques to detect extrasolar planets, namely through measurements of the radial velocities or/and photometric observations of the transit of a planet. The investigations of the dynamics of planets in our Solar System showed, that they suffer from chaotic motion, but this does not mean that we expect them to be unstable within the next millions of years. In extrasolar planetary systems with more than one planet on high eccentric orbits we did not understand for a long time how they can be stable. These orbits in mean motion resonances are stable due to a special mechanism which is the apsidal locking. Nevertheless they undergo the typical irregular jumps in the action like variable (in this case the eccentricity) which we know from other dynamical systems. Finally we discuss the satellite mission CoRoT, aimed to find extrasolar planets via transits and show also their first quite interesting results.

1 Introduction

For centuries dynamical studies in planetary systems were just a topic of the motion of planets, asteroids and comets in our own Solar System (SS). The stability of the orbits of the planets (Mercury to Neptune) over gigayears was established during the last decades. With the discovery of chaotic motions in our planetary system the belief in the 'divine' order of the motion of celestial bodies was distressed. This turning point was caused by the non predictability of asteroid collision with the Earth, which is due to the fact that after each 'reflection' by a close approach to a planet the uncertainty in its orbit doubles. This means that multiple close approaches to a planet make it impossible to compute a sufficiently 'precise' orbit to exclude collisions with a planet. But this is only one example of chaotic motion: asteroids in the 3:1 mean motion

G. Contopoulos, P.A. Patsis, *Chaos in Astronomy*. Astrophysics and Space Science Proceedings, doi: 10.1007/978-3-540-75826-6,

resonance (MMR) with Jupiter[1] were found to suffer from another sort of chaotic behaviour, known as separatrix crossing. Finally some 10 years ago, it was discovered that even our planets suffer from chaotic motion on a time scale of some million years which may lead – after billions of years – to an escape of Mercury [7].

In this review we want to discuss the dynamics of extrasolar planetary systems (=EPS), but before that, we shall, in the next chapter, shortly describe the detection methods for planets around other stars. Then we deal with the motion of planets in our own Solar system and speak in the following chapter about the dynamical structure of the known extrasolar planetary systems. We finally discuss some interesting multiplanetary systems with respect to their orbits in MMR and their chaotic signatures. In an appendix first interesting results of the first space mission ever, *CoRoT*, dedicated to detect planets around other suns are presented.

2 Detection Methods

It took astronomers almost 400 years after the invention of the telescope[2] to develop appropriate instruments and techniques to detect extrasolar planets.[3] Nowadays we have in principle four different methods for the detection of a planet around another star:

1. Photometry of the star
2. Radial velocity measurements
3. Astrometric measurements
4. Gravitational lensing

The first method uses basically the transits of a planet in front of the disc of the star[4] The second and third methods can be applied because of dynamical perturbations of the planet on the star which hosts the planet. The fourth method uses an effect of general relativity: a planet close to the lens star can increase or decrease the brightness magnification factor of the lens, which consequently unveils the presence of a planet.[5] Because by far most of the extrasolar planets were detected by the first two methods we will shortly describe these two methods.

---

[1] An asteroid which moves three times around the Sun whereas Jupiter fulfills only one revolution.

[2] In 1610 Galileo Galilei constructed the first telescope after the construction plans of the Dutch Lippershey.

[3] The first planet was discovered around the pulsar PSR 1257+12 in 1990 [15].

[4] In our SS this happens when an inner planet (Mercury or Venus) seen from the Earth passes in front of the Solar disc.

[5] The involved background star (which is photometrically measured), the lens star (with the planet) and the observer on Earth need to be aligned.

2.1 The Photometric (Transit) Method

When a star hosts a planet with an orbit such that the observer, the host star and the planet in between are almost aligned, we can observe a decrease in the measurement of the brightness of the star when the planet passes in front of it. The method described is very effective for large planets orbiting the host star very close. The chances to detect a planet far away from its sun is, unfortunately, very small. Figure 1 shows how this method works in principle. In Fig. 2 we show such a planet transit observed by CoRoT (see Appendix).

Thirty-five planets were observed up to now with this method where all of them are comparable to the size of the gas giants in the SS. They have periods in the order of some days (<6 days with the exception of the planet around HD17156 with $P \approx 20$ days). For a detailed presentation of this method I refer to Rauer and Erikson [13].

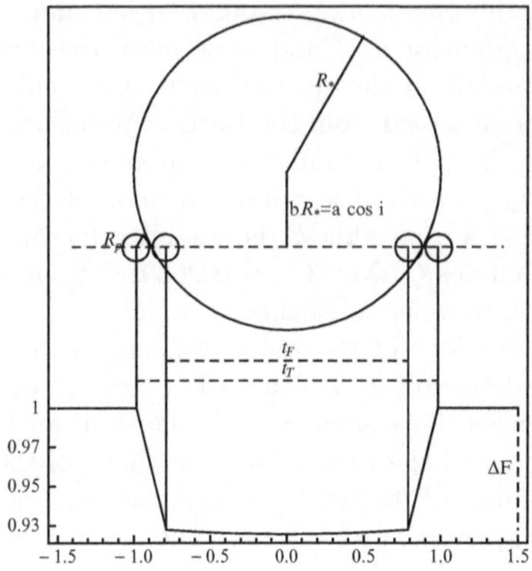

Fig. 1. Schematic view of the detection of a planet when it passes in front of the disc of its host star

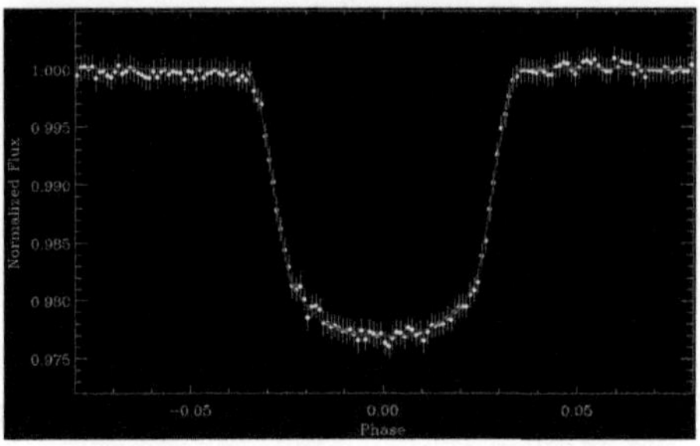

Fig. 2. The first light curve of a transit observed by CoRoT

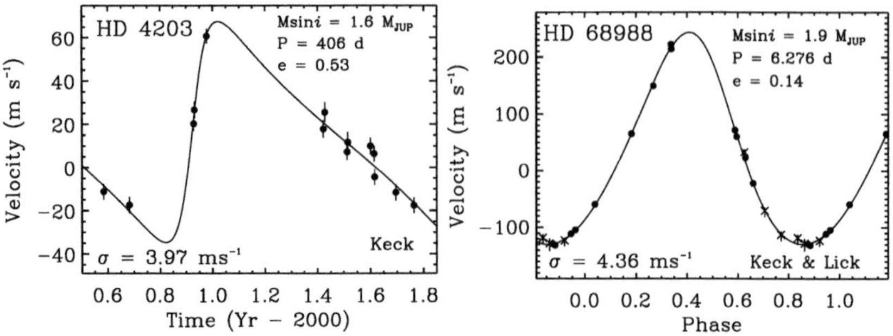

Fig. 3. Radial velocity curves for HD 4203 and HD 68988 [14]

2.2 The Radial Velocity Method

Since the star and the planet move around their common barycenter sometimes the star is approaching us, and sometimes it is moving away from us. The caused Dopplershift in the spectral lines (blue shifted respectively red shifted lines) can be observed from the Earth. From these measurements one can determine e.g. the period (and the semimajor axis of the planet's orbit), the eccentricity and also the minimum mass of the planet (due to the unknown inclinations we only know the quantity '$m \sin i$'). As examples we show in Fig. 3 two curves (HD4203 and HD68988) from Vogt et al. [14]. For a detailed discussion we refer to Beaugé et al. [1].

If we have observations of the radial velocity of the host star via spectroscopy after the detection of a planet via a transit we have quite a good knowledge of all physical parameters of a planet in an EPS. Then one can determine the true mass, the radius from the shape of the light curve (when one knows the diameter of the star) and also the mean density of the planet. Large telescopes may even provide – with the aid of high resolution spectra – most valuable information about the atmosphere of the planet and possibly also of biomarkers.[6]

3 The Structure of EPS

As we already mentioned, the orbital structure of the discovered EPS was a real surprise for astronomers: whereas in our Solar system the planets have small to moderate eccentricities the extrasolar planets have relatively large eccentric orbits. This is well visible from Fig. 4.

In the respective plots we separated planets in single planetary systems and planets in multiplanetary systems (MPS). In Fig. 4 (upper plot) there are

[6] When in the spectrum 0_2 and CH_4 is present this is regarded as a signature of life [6].

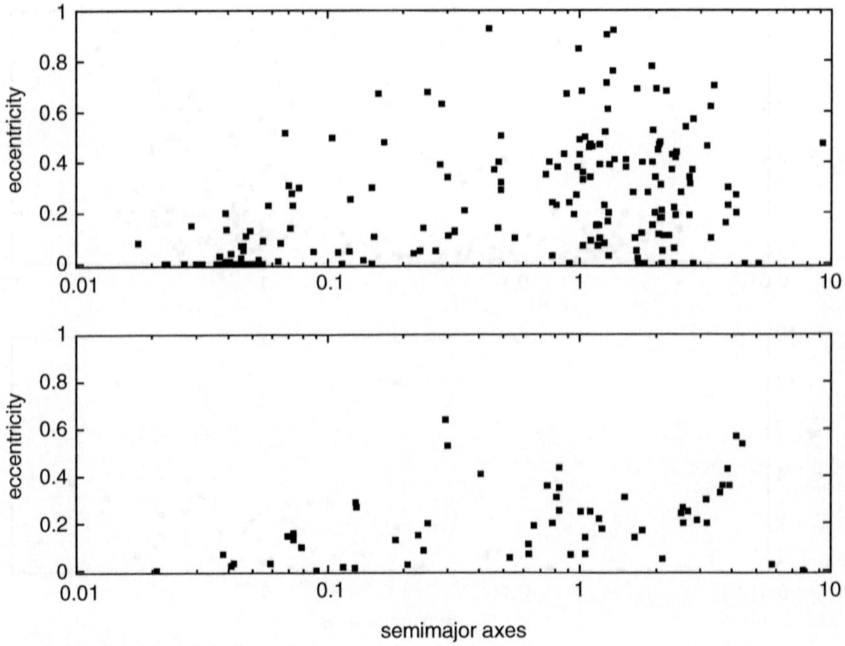

Fig. 4. Distribution of planets in single planetary systems (*upper graph*) and planets in multiplanetary systems (*lower graph*): eccentricities are plotted versus semimajor axes (data from J. Schneider: http://exoplanet.eu)

two properties which are striking, namely the clustering at $e = 0$ and small semimajor axis and the almost uniformly distribution for planets for $1 < a < 5$ AU concerning the eccentricities for single planets. For the MPS, when more planets are present, there is neither a trend nor a clustering to observe in the respective distributions (Fig. 4, lower plot).

Comparing the distribution of the masses of the planets with respect to their semimajor axes (Fig. 5) we can again see the clustering caused by planets with masses in the order of Jupiter's mass (upper graph), which are the so-called hot Jupiters. Because of the lack of hot Jupiters in MPS this clustering is not visible in the same figure (lower graph). It is evident that single planets tend to be more massive than the planets in MPS. We need to say that from the plots with the masses we excluded five planets where the planet is even more massive than ten times $M_{Jupiter}$.

Because from the dynamical point of view our main interests are the systems with more than one planet we will in the following concentrate on them and start with a discussion of our own SS.

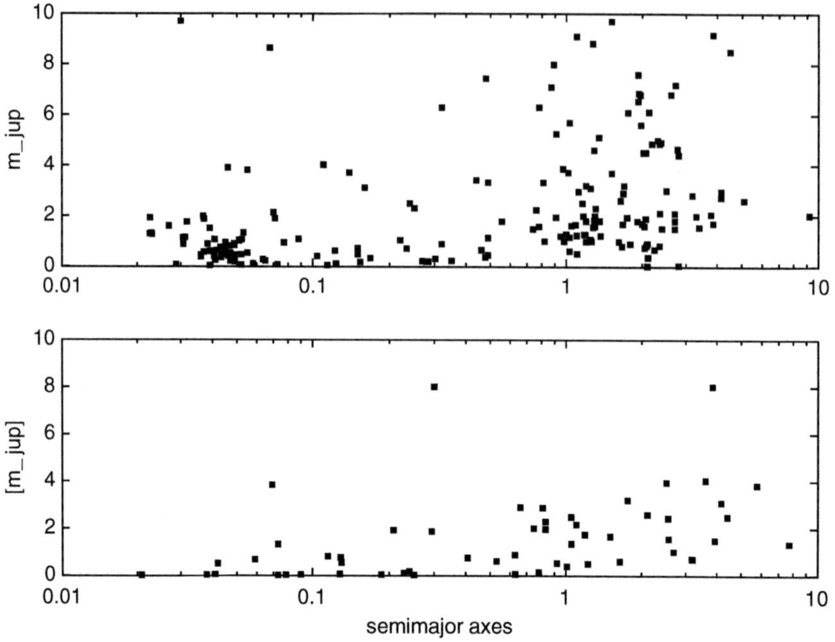

Fig. 5. Distribution of planets in single planetary systems (*upper graph*) and planets in multiplanetary systems (*lower graph*): masses of the planets are plotted versus semimajor axes (data from J. Schneider: http://exoplanet.eu)

4 Our Planetary System as Dynamical Model

Many investigations are devoted to the study of the long-term evolution of our own planetary system (e.g. [5, 9]). We discuss shortly the results of a precise integration of the full equations of motion in a Newtonian dynamical model with all planets (Mercury to Neptune). To derive these results we used the Lie-Integration method [4] with an automatic step size and covered the time span of 1 Gigayear, namely from −500 to +500 Megayear.

There is a natural separation into an inner planetary system with the terrestrial planets Mercury, Venus, Earth and Mars and the outer system with the gas giants Jupiter, Saturn, Uranus and Neptune. And, in fact, when we look at the plots of the maximum values of the eccentricity and the inclinations of the inner and outer planets (Fig. 6) one can see that these values for the inner planets (abscissa 1,2,3 and 4) are in general larger than for the outer planets (abscissa 5, 6, 7 and 8).

Especially the two smallest planets (Mercury and Mars have about one-tenth of the mass of the Earth) suffer from large variations in e and i. In Fig. 7 we show the details of the evolution of the eccentricities e and orbital inclinations i for Mercury and also for Mars for 5 million years of integration.

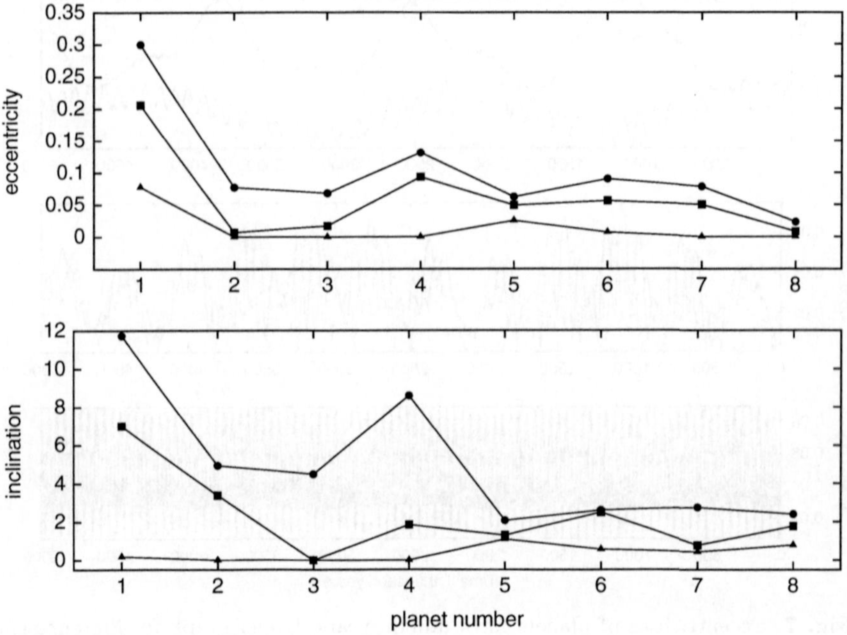

Fig. 6. Maximum (*circles*) – current (*squares*) and minimum values (*triangles*) of the eccentricities of the inner planets (1...Mercury, 2...Venus, 3...Earth, 4...Mars) and of the outer planets(5...Jupiter, 6...Saturn, 7...Uranus, 8...Neptune) are shown on the upper graph from −500 to +500 Megayear and for the inclinations (*lower graph*)

An interesting property, namely the coupled motion of two planets, can be explained in the following way: besides the total energy integral the integral of the angular momentum (L)

$$L = m_1 n_1 a_1^2 \sqrt{1-e_1^2} \cos I_1 + m_2 n_2 a_2^2 \sqrt{1-e_2^2} \cos I_2, \tag{1}$$

determines fully the secular behaviour of planetary systems with two planets. This behaviour depends only on the planetary masses involved and the semi-major axes and links the eccentricities and inclinations of the two planets. The semimajor axes are inversely proportional to the total energy of the planets' motion and they are more or less constant when no close approaches occur. The functions $\sqrt{1-e_k^2}$ are varying monotonically in the interval [0,1] which means that when the eccentricity of the planet increases, $\sqrt{(1-e_k^2)}$ decreases. Because L needs to be constant for two almost coplanar orbits the eccentricity e of the other planet decreases. In addition L constraints the value of the eccentricity of the planets; they cannot grow unlimited and – a fact which is very important – close approaches are avoided. A detailed discussion can be found in [12].

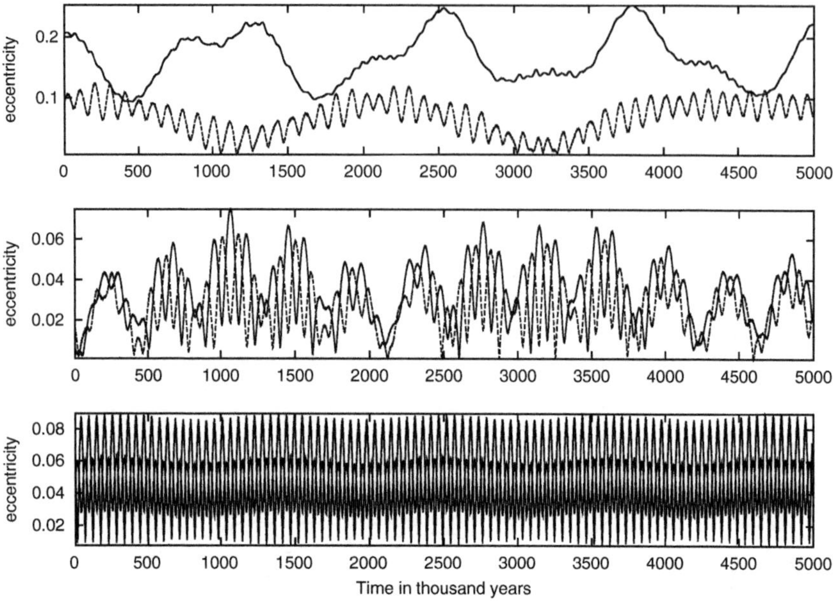

Fig. 7. Eccentricities of planets for 5 million years. Upper graph for Mercury (*full line*) and Mars (*dashed line*), middle graph for Earth (*full line*) and Venus (*dashed line*), lower graph for Jupiter (*full line*) and Saturn (*dashed line*)

The best example for the coupling between planets is the 5:2 MMR between Jupiter and Saturn.[7] This means that the eccentricities of the couple Jupiter and Saturn are in antiphase to each other (Fig. 7, lower graph); the same dynamics is present for the inclinations.

In the corresponding Fig. 7 (middle graph) we can see a similar behaviour for the planets Earth and Venus, which have almost the same masses and therefore the amplitudes of variation are quite similar. Because of the perturbations of the couple Jupiter-Saturn acting on these planets the periods and amplitudes are less regular. These two neighboring planets are in a 13:8 MMR.[8]

By using a semianalytical method (via series expansions, averaging over the fast variables and then integrating these equations of motion),[9] Laskar [8] computed a solution of the planets' motion up to Gigayears. The determination of the periods of the motions of the nodes and of the perihelions of the planets – which are the fundamental frequencies of the secular system – have been undertaken for a time span of 200 million years. These frequencies g_n and s_n refer to the secular motion of Ω_n and ω_n of each planet n.

[7] the respective periods are 11.87 for Jupiter and 29.57 years for Saturn.

[8] whereas the Earth fulfils one revolution around the Sun Venus in 1 year make 13/8 revolutions because of its orbital period of 225 days.

[9] which means to keep the semi-major axis of all planets constant.

All resonances and also their linear combinations were found with the help of a frequency analysis [7] of the numerical solutions of the secular systems. During the time evolution the fundamental frequencies change slightly which leads to a characteristic change from libration to circulation and vice versa from time to time for some combinations of these frequencies.

There are two main secular resonances present among the planets, which showed the typical behaviour for chaotic motion, namely the crossing of the separatrix:

- $\theta = 2(g_4 - g_3) - (s_4 - s_3)$ related to Mars(4) and the Earth(3)
- $\sigma = (g_1 - g_5) - (s_1 - s_2)$ related to Mercury(1), Venus(2) and Jupiter(5)

The qualitative change of the character of the motion is typical for chaotic behaviour: a simple example is the perturbed pendulum with a motion close to the separatrix. A slight change in the initial conditions may lead to a separatrix crossing – and the pendulum is no longer in libration but in circulation and vice versa.

5 Multiplanetary Systems

Up to now 25 systems with more than one planet around the host star are known. These MPS can be classified according to their number of planets:

- 18 systems with 2 planets
- 5 systems with 3 planets
- 1 systems with 4 planets
- 1 systems with 5 planets

In Fig. 8 we plot the eccentricity versus the mass of the host star for all MPS. The three planet systems Gl581 and Gl876 are special: they have very small host stars, the respective masses of the planets are small and the planets are also very close to their sun.[10] Also characteristic are the relatively large eccentricities but, as numerical simulations have shown, they are on stable orbits. From this plot we can see, that even MPS may host planets with relatively large eccentricities (left part of the respective plot).

5.1 HD 82943a and HD 82943b in the 2:1 Mean Motion Resonance

We will discuss as an example of a EPS with two planets in the 2:1 resonance the system HD 82943: they have large eccentricities[11] and it is – at a first

[10] Gl581 has two so-called 'hot jupiters' with $a_1 = 0.04$ and $a_2 = 0.07AU$ and Gl876 has one planet as close as $a = 0.02AU$.

[11] $a_1 = 1.190000AU, a_2 = 0.746000AU, e_1 = 0.2190, e_2 = 0.3590$.

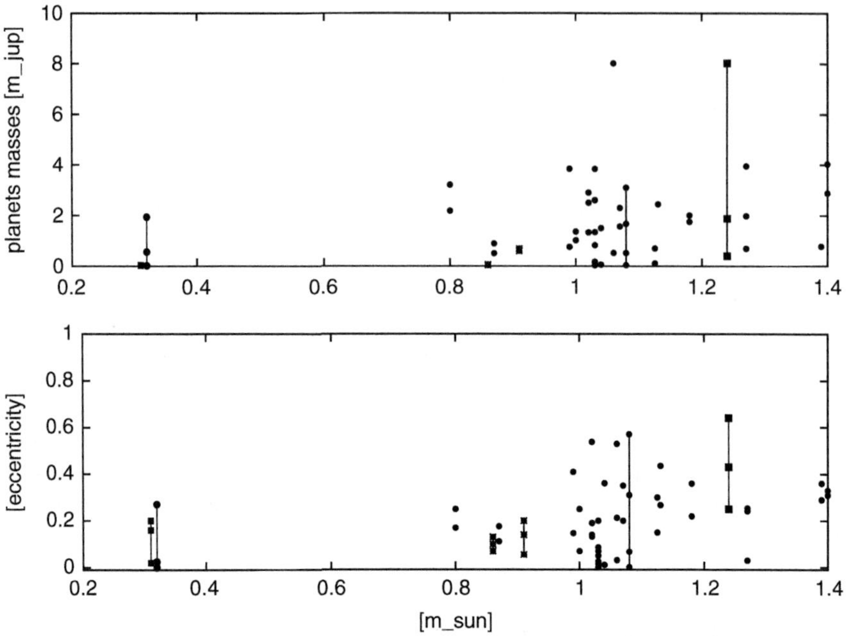

Fig. 8. Masses of the planets (*upper graph*) and eccentricities (*lower graph*) versus mass of the host star

sight unlikely that they may have stable orbits because the periastron distance of the outer planet and the apoastron distance of the inner planet are almost crossing. This crossing is avoided because of the resonance 2:1 and the following reasons: first when the two pericenters are almost equal $\omega_1 \sim \omega_2$, and second both planets pass through their pericenter more or less at the same moment, they stay in stable orbits because they never really approach. In Fig. 9 one can see the orbits at the beginning of the numerical integration (solid line); then both ellipses start to circulate in the direction of motion and at the same time they librate with small amplitudes around their pericenter. This apsidal locking prevents the orbits to be unstable. On the other hand when the two pericenters are different ($\omega_1 - \omega_2 \sim 180°$) the motion is unstable.

5.2 The Five Planets System 55 Cancri

This system with five planets is the most populated one. The host star (Cnc 55a) is a G8 V main sequence star with a mass of 1.08 M_{Sun} with a radius of $0.6 R_{Sun} (\pm 0.3)$. The planet Cnc 55b was detected some 10 years ago [2]; two more planets were discovered later [11], one was the first giant planet with an orbital semimajor axis compared to Jupiter's ('d') and another one ('c') in the 3:1 MMR with the planet 'b'. After 2 years a fourth planet ('e') was found in the data by McArthur et al. [10] with a mass comparable to that of

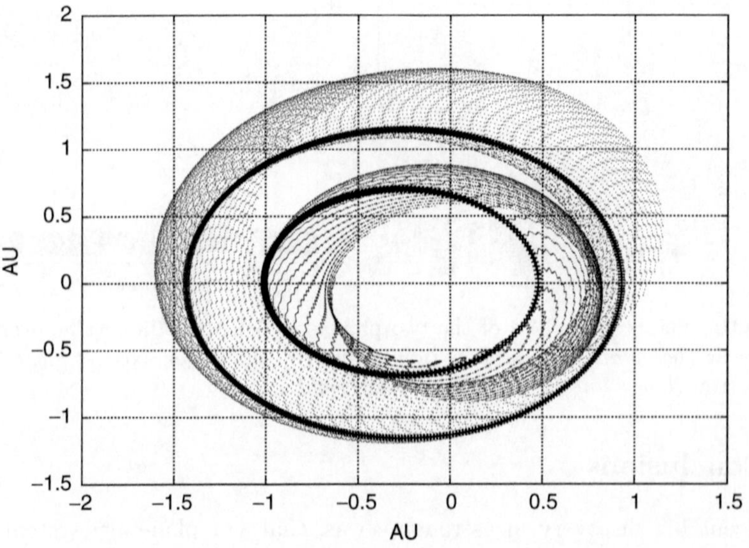

Fig. 9. The orbits of the two planets in HD 82943 in the 2:1 resonance locked in apsidal alignment

Table 1. The system 55 Cnc

Name	Mass (M_{Jup})	a	e	Period (days)
Cnc 55 b	0.824	0.115	0.022	14.65
Cnc 55 c	0.169	0.24	0.138	44.34
Cnc 55 d	3.835	5.77	0.025	5218
Cnc 55 e	0.034	0.038	0.07	2.82
Cnc 55 f	0.144	0.781	0.4	260

Neptune, a ‘hot Neptune’ in the distance of only 1/10 of the one of Mercury in our SS. Finally a recent discovery of a fifth planet ‘f’ was reported based on radial velocity measurements from the Lick observatory [3] which moves in the ‘gap’ between the planets ‘c’ and ‘d’. It seems that the EPS is now ‘fully packed’.

The system itself is stable, and what one can see from the Table 1, the planets ‘b’ and ‘c’ are in a 3:1 MMR. This resonance is known to be ‘dangerous’; e.g. the distribution of the asteroids in the main belt of these bodies, which move between Mars and Jupiter, has a gap there, one of the so-called Kirkwood gaps. In fact the two planets suffer from chaotic motion, which can be seen in Fig. 10. Nevertheless, because of the existence of the angular momentum (see Chap. 3) the two planets never come close; whenever one planet has a large eccentricity, the other one has a small eccentricity and, in addition, there is the constraint on the upper value of the eccentricity.

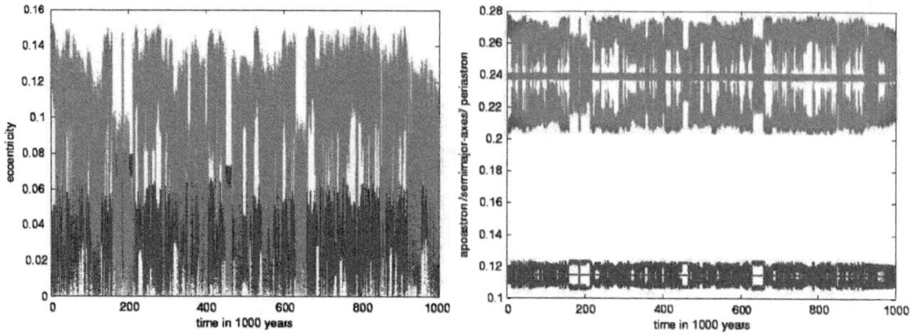

Fig. 10. The eccentricities of the two planets in 3:1 MMR in 55 Cnc over 1 million year (*left graph*). Apoastron (*upper lines*) – semimajor axis (*straight lines*) – periastron (*lower lines*) for 1 million year (*right graph*)

6 Conclusions

A recent big discovery in astronomy was, that our planetary system is not unique in our galaxy, although this was already postulated by Giordano Bruno[12] about 400 years ago. But despite this fact our SS seems[13] very special compared to the EPS concerning their orbital characteristics with appreciable eccentric orbits.[14]

We reviewed in this article what we know after some 15 years after the first discovery of an extrasolar planet about the 'architecture' of the EPS. We have shown shortly how most of the planets were discovered, namely on one hand with the measurements of the radial velocity of stars and on the other hand with the photometric observations of transits of planets in front of the disc of a host star.

Actually in our planetary system chaotic motion is present which does not mean that the planets are on unstable orbits. We have furthermore seen that even in a system like HD 160691 with four planets the orbits are stable despite their relatively large eccentricities. The discussion of MPS unveiled that these large eccentric orbits for relatively close planets need not to destabilize their motion: the apsidal locking makes such orbits stable for hundreds of millions of years. As an example we discussed the most populated MPS 55 Cancri which hosts two planets in a MMR, planets b and c are in the 3:1 resonance. Although they suffer from chaotic motion, they also seem to be in quite safe orbits for very long times, like our own planetary system. We are looking forward to very interesting results for the proximate future because of space missions like the already very successful satellite CoRoT (which we present

[12] 1548–1600.

[13] at least up to now when we just have knowledge of about 300 planets, where most of them are in our Solar neighborhood (up to 60 pcs).

[14] Because Mercury and Mars are an order of a magnitude less massive than Venus and Earth, they suffer more from the perturbations of all the other planets, and, have somewhat more eccentric orbits.

shortly in the appendix). Although we do not know of terrestrial like planets around Solar type stars, it will be only a question of time when we can say: we found a second Earth!

Appendix: CoRoT – The First Exo-Planetary Mission

The CoRoT (**Co**nvection, **Ro**tation & planetary **T**ransits)[15] mission is designed – besides measurements for astroseimology – to detect extrasolar planets with the aid of planetary transits (as explained in Sect. 2). The satellite project was initiated by the French space agency, CNES, and it is supported from the European Space Agency (ESA), Austria, Brazil, Belgium, Germany and Spain. The telescope has a 45 cm main mirror and a field of view of 2.8 by 2.8 square degrees. The four CCDs consist of 2,048 by 2,048 pixels each. The CoRoT instrument is observing the same field of view for 150 consecutive days with up to 12,000 targets, stars brighter than magnitude 15.5. Occultations in the order of some days could – in principle – be used to detect even planets with size of the Earth!

The satellite was launched in late December 2006 into a polar circular orbit of 896 km altitude and is planned to work at least for 2 1/2 years. We have shown the first picture of a transit of a hot Jupiter observed by CoRoT in Fig. 2

The period of this planet orbiting a star similar to our Sun is about 1.5 days and the radius is approximately 1.5–1.8 Jupiter radii. With the aid of additional measurements of the radial velocity its mass could be determined to be 1.3 times that of Jupiter. In addition we show a lightcurve (Fig. 11) derived by CoRoT where the transits repeat regularly. From this graph the advantage of space observations compared to ground observations is quite obvious, namely the continuous measurements without any interruption for long time. A similar project like CoRoT is planned by NASA, the so called Kepler-mission. It will be launched in 2009 and it is – in principle a larger

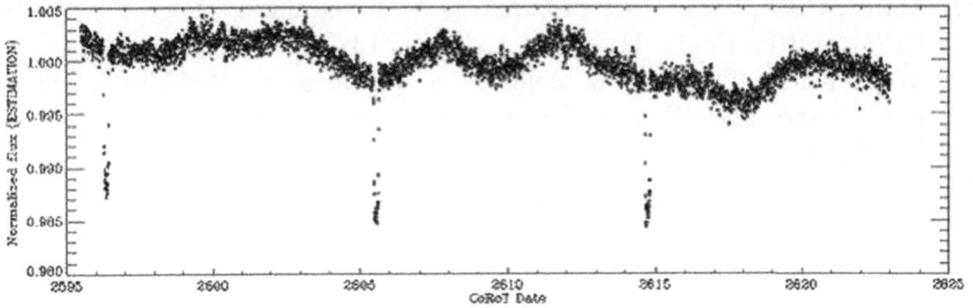

Fig. 11. The measured lightcurve of a transit of a planet with a period of 9 days observed by CoRoT

[15] http://smsc.cnes.fr/COROT/.

CoRoT mission because it will be equipped with a 0.95 m Schmidt telescope. It will also make photometric observations and because of being able to observe many more stars we expect the detection of many Earth-like planets even in the so-called habitable zone.[16]

Acknowledgement

I have to thank the organizers of the meeting Chaos in Astronomy 2007 in Athens for inviting me to give this lecture. I also thank my colleagues from the AstrodynamicsGroup in Vienna A. Bazso, Ch. Lhotka and R. Schwarz to prepare this article.

References

1. Beaugé, C., Ferraz-Mello, S., and Michtchenko, T., 2007, in R. Dvorak (ed.): Extrasolar Planets, Formation, Detection and Dynamics, *Wiley-VCH* 1
2. Butler, R.P., Marcy, G.W., Williams, E., Hauser, H., and Shirts, P., 1997, ApJ, 474, L115
3. Fischer, D.A. and Valenti, J.A., 2005, ApJ, 632, 1102
4. Hanslmeier, A. and Dvorak, R., 1984, A&A 132, 204
5. Ito, T. and Tanikawa, T., 2002, MNRAS, 336, 483
6. Kaltenegger, L. and Selsis, F., 2007, in R. Dvorak (ed.): Extrasolar Planets, Formation, Detection and Dynamics, *Wiley-VCH* 79
7. Laskar, J., 1990, Icarus, 88, 266
8. Laskar, J., 1994, A&A, 287, L9
9. Lecar, M., Franklin, F.A., Holman, M.J., and Murray, N.J., 2001, in Chaos in the Solar System, Ann. Rev. Astron. Astrophys., 39, 581
10. McArthur, B.E., Endl, M., Cochran, W.D., Benedict, G.F., Fischer, D.A., Marcy, G.W., Butler, R.P., Naef, D., Mayor, M., Queloz, D., Udry, S., and Harrison, T.E., 2004, ApJ, 614, L81
11. Marcy, G.W., Butler, R.P., Fischer, D.A., Laughlin, G., Vogt, S.S., Henry, G.W., and Pourbaix, D., 2002, ApJ, 581, 1375
12. Michtchenko, T.A., Ferraz-Mello, S., and Béauge, C., 2007, in R. Dvorak (ed.): Extrasolar Planets, Formation, Detection and Dynamics, *Wiley-VCH* 151
13. Rauer, H. and Erikson A., 2007, in R. Dvorak (ed.): Extrasolar Planets, Formation, Detection and Dynamics, *Wiley-VCH* 207
14. Vogt, S.S., Butler, R.P., Marcy, G.W., Fischer, D.A., Pourbaix, D., Apps, K., and Laughlin, G., 2002, ApJ, 568, 352
15. Wolszczan, A. and Frail, D.A., 1992, Nature, 355, 145

[16] That is the region around a host star where water can exist in liquid form, which we regard as basis for the possible development of life.

Routes to Chaos in Resonant Extrasolar Planetary Systems

J.D. Hadjidemetriou and G. Voyatzis

Department of Physics, University of Thessaloniki, Thessaloniki, Greece
hadjidem@auth.gr, voyatzis@auth.gr

Summary. We study the factors that affect the stability and the long term evolution of a *resonant* planetary system. For the *same* resonance, the long term evolution of a resonant planetary system depends on the relative orientation of the planetary orbits, on the phase of the planets on their orbits and on the proximity to a *periodic* resonant planetary system, either stable or unstable. Chaotic property is not always associated with a disruption of the system and the system may remain bounded for a long time.

1 Introduction

During the past 15 years many planetary systems were observed, called *extrasolar planetary systems*. Some of these systems are very different from our own Solar System, because they have massive planets close to the central star, which we will call the *sun*, of the order of the mass of Jupiter or larger, and large planetary eccentricities. In some of these planetary systems two or more planets were observed. In the case of two planets close to each other, the two planets are in mean motion resonance. Examples are HD 82943 and GLIESE 876, at the 2:1 resonance and 55Cnc at the 3:1 resonance. There are several problems associated with the study of the extrasolar systems, as their formation, their evolution and their dynamical stability. In this paper we will study some aspects of the dynamical stability of extrasolar planetary systems.

There are different approaches to the study of the dynamical evolution of a planetary system and on the mechanisms that stabilize the system, or generate chaotic motion and instability: Beaugé and Michtchenko 2003 [2], Beaugé et al. 2003 [1], Gozdziewski et al. 2003 [7], Michtchenko et al. 2006 [14], Malhotra 2002 [13], Lee 2004 [12], Dvorak et al. 2005 [5]. In these papers different methods have been applied, as the averaging method, direct numerical integrations of orbits, or various numerical methods which provide indicators for the exponential separation of nearby orbits. In this way the regions where stable motion exists have been detected, in the orbital elements space. A special problem in the study of the extrasolar planetary systems is the existence

G. Contopoulos, P.A. Patsis, *Chaos in Astronomy.* Astrophysics and Space Science Proceedings, doi: 10.1007/978-3-540-75826-6,

of large planetary eccentricities. It is believed that the observed planetary systems were not created in their present configuration, but came to their observed, evidently *stable*, state with large eccentricities, following a *migration* process, as the protoplanetary disc in which they were formed slowly dissolved [4, 6]. It is obvious that the planetary system was trapped, at the end of the migration process, to a stable configuration.

We will study the regions of the phase space where we can have stable, bounded, motion of a planetary system, and in particular stable *resonant* motion. This requires the knowledge of the topology of the phase space. The topology of the phase space in any dynamical system is determined critically by the periodic orbits, which are the "backbone" of the phase space, although they are a set of measure zero. Close to a stable periodic orbit we have stable liberations and the motion in phase space takes place on a torus. On the contrary, close to an unstable periodic orbit we may have irregular, chaotic, motion and in many cases the system disrupts into a binary system (the star and one planet) and an escaping planet. We study the evolution from order to chaos as the perturbation on a stable resonant periodic orbit increases. Since an arbitrary resonant planetary system is not always stable, we study the factors that affect the stability of a planetary system at a *fixed* resonance.

2 Resonant Extrasolar Planetary Systems

We consider planetary systems with two planets close to each other, moving in the same plane. Since the gravitational interaction between the planets cannot be ignored, even for very small planetary masses, the model we use is the *general three body problem*, for planar motion. The three bodies are the sun, S, and the two planets, P_1 and P_2. It can be proved [9] that families of periodic orbits in the planar general three body problem exist, in a *rotating* frame xOy, whose x-axis is the line $S-P_1$, with origin at the center of mass of these two bodies, where S is the Sun and P_1 the inner planet. This implies that the *relative* configuration is repeated in the inertial frame. We assume that the center of mass of the whole system is at rest with respect to an inertial frame. We have four degrees of freedom, for planar motion, with generalized variables x_1, x_2, y_2, θ (see Fig. 1a). This is a non uniformly rotating frame, where the first planet, P_1, moves on the x-axis and the second planet, P_2, moves in the xy plane. It turns out that the angle θ is ignorable, so the degrees of freedom are reduced to three and the study of the system can be made in the rotating frame only, with variables x_1, x_2, y_2. Note that the phase space in the rotating frame is six dimensional. The dimensions of the phase space can be reduced further by considering the *Poincaré map* on a surface of section. By the Poincaré map we reduce the dimensions of the phase space, without losing the generality of the problem, and in addition we eliminate the unnecessary details that are not important in the study of the long term evolution of the

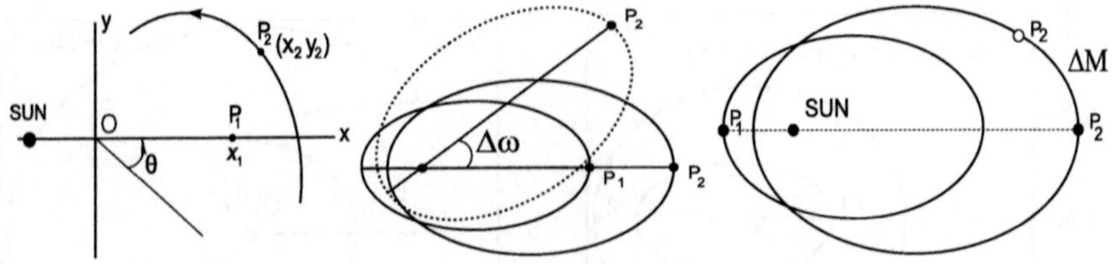

Fig. 1. (**a**) The rotating frame xOy and the coordinates x_1, x_2, y_2 for the position of the three bodies in the rotating frame and θ for the orientation of the rotating frame. (**b**) The perturbation by rotating the orbit of P_2 by $\Delta\omega$. (**c**) The perturbation by shifting the planet P_2 on its orbit by ΔM, where M is the mean anomaly

system. In the present study, in all our computations, we consider the surface of section

$$y_2 = 0 \ (\dot{y}_2 > 0), \quad H = h = \text{constant}.$$

The phase space of the Poincaré map is the four dimensional space x_1, $\dot{x}_1$, x_2, $\dot{x}_2$ ($y_2 = 0$ and $\dot{y}_2$ is obtained from $H = h$, $\dot{y}_2 > 0$) and the periodic orbits are represented as *fixed points* of the map.

There are two types of periodic planetary orbits in the general planar three body problem. These are *non resonant* periodic orbits with nearly circular orbits of the two planets and *resonant* periodic orbits, with nearly elliptic orbits of the two planets. In this latter case the two planets are in *mean motion resonance.* All these orbits belong to *monoparametric families of periodic orbits.* In particular, along the families of the resonant elliptic orbits, the resonance (ratio T_2/T_1 of the planetary periods in the inertial frame) is almost constant, but the planetary eccentricities increase, starting from zero values, and may reach high values. There are resonant families for all values of the resonance T_2/T_1. Some of these families (or parts of them) are stable. The elliptic orbits may be symmetric or asymmetric. In the symmetric orbits, that we shall mainly study, the lines of apsides of the two planetary orbits coincide and the perihelia may be aligned ($\Delta\omega = 0^0$) or antialigned ($\Delta\omega = 180^0$), while the planets are, at $t = 0$, at their perihelia or aphelia. We remark that the above mentioned nearly circular or elliptic planetary orbits refer to the inertial frame, but the motion is *exactly periodic* in the rotating frame, where the shape of the orbit may be quite different (see, for example, Figs. 2b and 9b).

As we mentioned, all the periodic orbits with elliptic planetary orbits are resonant. The periodic orbits appear as *fixed points* on the Poincaré map and it is these fixed points that determine the topology of the phase space. It is close to the stable, resonant, fixed points that a planetary system can be trapped. This makes clear the importance of the resonances in the dynamical study of the extrasolar planetary systems, and explains why there are many observed resonant planetary systems. A systematic study of the dynamics of planetary systems along these lines is presented in Hadjidemetriou 2006 [11].

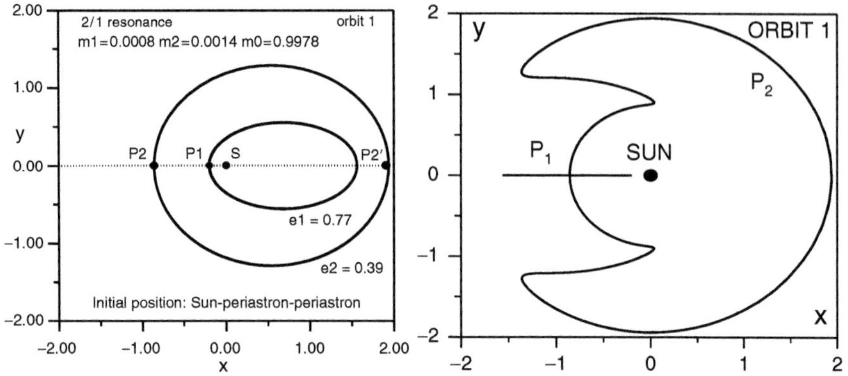

Fig. 2. (**a**) The 2:1 resonant periodic orbit with the two planetary orbits aligned, $\omega_2 - \omega_1 = 0^0$, in the inertial frame. The motion is not periodic in this frame. (**b**) The same periodic orbit in the rotating frame, where it is exactly periodic

Works on the dynamics of planetary systems based on periodic orbits are in [8, 10, 15, 16]

We focus our attention on the stability in the neighborhood of a resonant orbit. There are two types of resonance: The resonant *periodic* orbits, mentioned above, which we shall call *exact resonance*, and the orbits where the semimajor axes of the two planets correspond to a certain resonance, but the eccentricities and the orientation of the planetary orbits are different from those of the exact periodic motion. We start from an exact symmetric resonant periodic orbit and we consider two types of initial perturbation, by destroying the symmetry: (1) We rotate the orbit of the outer planet P_2 (in the inertial frame) by an angle $\Delta\omega$, keeping all other elements fixed and the two planets at their perihelia or aphelia, accordingly (Fig. 1b). (2) We keep the position of the two planetary orbits in their symmetric configuration, but now we shift, at $t = 0$, the second planet P_2 on its orbit by an angle ΔM (Fig. 1c). In both cases the resonance is kept fixed. Thus, by increasing the perturbation $\Delta\omega$, or ΔM, we study the evolution from order to chaos, for *the same resonance.* Close to a *stable* periodic motion we expect a libration of the orbital elements about their exact resonant values, and in particular a libration of the angle $(\omega_2 - \omega_1)$ between the line of apsides of the two planetary orbits. But what happens for larger perturbations? Or what behavior should we expect in the vicinity of an *unstable* periodic orbit? We present four typical examples of resonant periodic orbits, with different behavior as far as their evolution is concerned, as a perturbation increases. Two at the 2:1 resonance, both stable, and two in the 3:1 resonance, one stable and one unstable. As we will see in the following, the resonance alone is not enough to stabilize a planetary system. There are many factors that affect the stability of a resonant system, and mainly the phase and the deviation from symmetry. We study the different routes to chaos, as the perturbation increases, and also the behavior of the angle $(\omega_2-\omega_1)$, from libration, to rotation, to chaos, or to chaotic interchange between libration and rotation, as the perturbation increases.

3 A Stable System at the 2:1 Resonance. $(\omega_2 - \omega_1) = 0^0$

We consider a 2:1 resonant stable periodic orbit, corresponding to a planetary system with masses $m_{SUN} = 0.9978$, $m_1 = 0.0008$, and $m_2 = 0.0014$ (normalized to $m_{SUN} + m_1 + m_2 = 1$), semimajor axes $a_1 = 0.884$ AU, $a_2 = 1.415$ AU ($T_2/T_1 = 2.024$) and eccentricities $e_1 = 0.774$, $e_2 = 0.394$. The orbit is symmetric, with the lines of apsides of the two planets aligned, $(\omega_2-\omega_1) = 0^0$ and the two planets are at perihelion at $t = 0$ (see Fig. 2). This orbit belongs to a family of stable periodic orbits for the masses of the extrasolar planetary system HD82943 [10].

We perturb the periodic motion by rotating the orbit of the second planet by an angle $\Delta\omega$. The semimajor axes and the eccentricities of the two planets are fixed, and also the planets are at their perihelia at $t = 0$. In this way, we destroy the symmetry, but the resonance is always the same, equal to 2:1. We start with a small angle of rotation $\Delta\omega = 15^0$ (Fig. 3) and increase the perturbation to $\Delta\omega = 20^0$ (Fig. 4), $\Delta\omega = 25^0$ (Fig. 5), $\Delta\omega = 27^0$ (Fig. 6) and $\Delta\omega = 30^0$ (Fig. 7). In these figures we present the Poincaré map (a projection on a coordinate plane, as indicated), the evolution of the eccentricities and the evolution of the angle $(\omega_2-\omega_1)$ between the line of apsides. We note that for a small value of the perturbation, $0^0 < \Delta\omega < 25^0$, the Poincaré map is on a well defined torus and the eccentricities librate around their original values (corresponding to the exact periodic motion). The angle $(\omega_2 - \omega_1)$ *librates* around 0^0 with an amplitude that increases as the perturbation increases. When the perturbation increases to $\Delta\omega = 27^0$, the motion is still ordered, but now the angle $(\omega_2-\omega_1)$ *rotates*. For a still larger perturbation, $\Delta\omega \geq 30^0$,

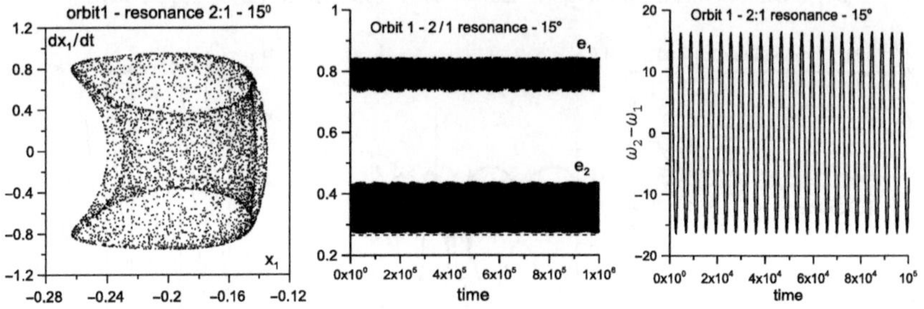

Fig. 3. The evolution of the orbit of Fig. 2, corresponding to the perturbation $\Delta\omega = 15^0$. (**a**) The Poincaré map (projection on the $x_1\,\dot{x}_1$ plane). The motion is near a torus. (**b**) The evolution of the eccentricities, which librate around the values of the periodic motion. (**c**) The libration of $(\omega_2 - \omega_1)$ around 0^0

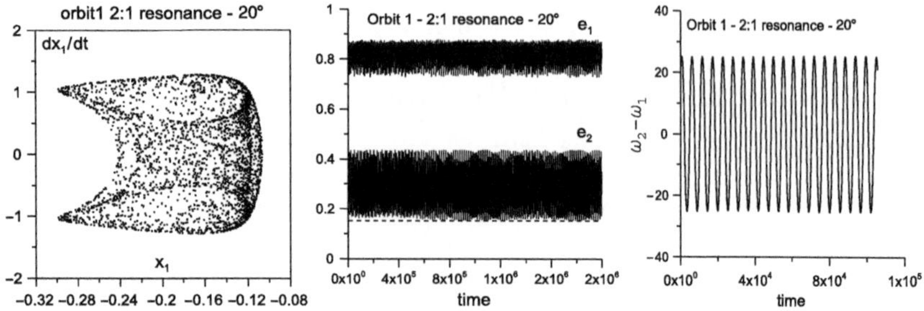

Fig. 4. The same as Fig. 3, $\Delta\omega = 20^0$. The angle $(\omega_2 - \omega_1)$ librates around 0^0

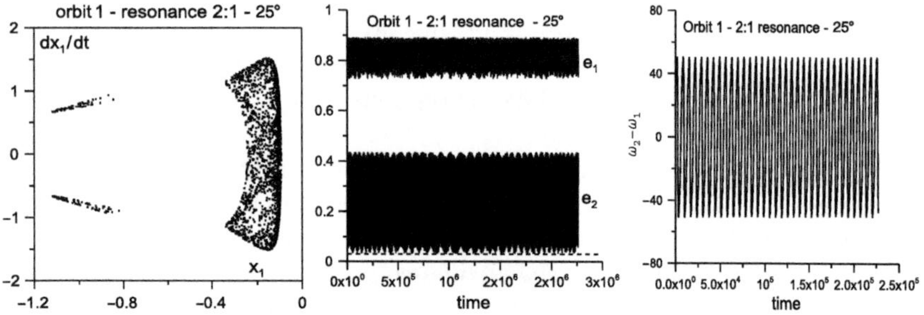

Fig. 5. The same as Fig. 3, $\Delta\omega = 25^0$. The angle $(\omega_2 - \omega_1)$ librates around 0^0

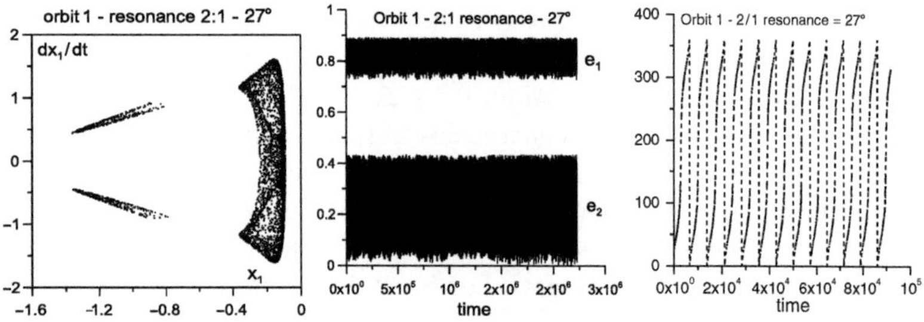

Fig. 6. The same as Fig. 3, $\Delta\omega = 27^0$. The angle $(\omega_2 - \omega_1)$ rotates

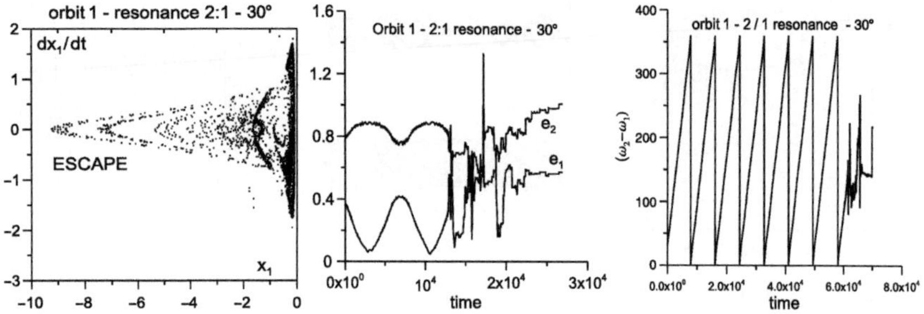

Fig. 7. The evolution of the orbit of Fig. 2, corresponding to the perturbation $\Delta\omega = 30^0$. The evolution is strongly chaotic. (**a**) The Poincaré map (projection on the $x_1\,\dot{x}_1$ plane). The planet P_1 escapes. (**b**) The evolution of the eccentricities. (**c**) The evolution of $(\omega_2 - \omega_1)$

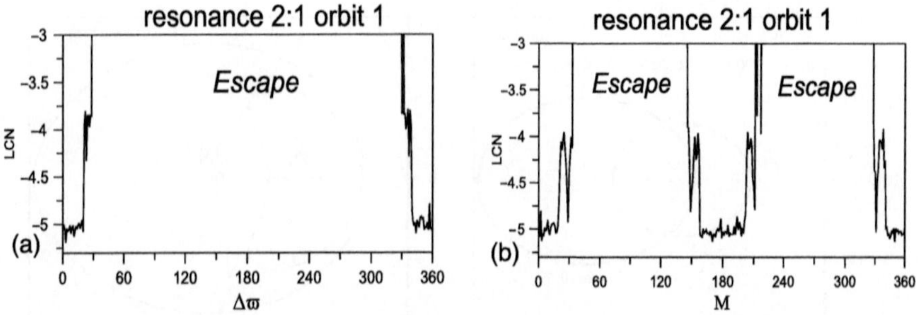

Fig. 8. A global view of the stable and unstable regions by the method of LCN: (**a**) Perturbation in $\Delta\omega$. (**b**) Perturbation in ΔM

the motion becomes chaotic, and eventually the system disrupts to a binary system (the sun and P_2) while the planet P_1 escapes. In all bounded cases the resonance T_2/T_1 remains very close to the 2:1 value. A global view of the behavior presented above with some typical cases, is shown in Fig. 8a, where we present the Lyapunov Characteristic Numbers, LCN [3] (for all values of $\Delta\omega$).

In Fig. 8b we present a global view of the behavior of the resonant system when the perturbation is the shifting of the outer planet P_2 on its orbit by ΔM (Fig. 1c). The results are similar to those for the $\Delta\omega$ perturbation. We note that there is another stable window at $\Delta M = 180^0$, but this is not different from the stable window at $\Delta M = 0^0$. This is so because, due to the 2:1 resonance, after a time interval equal to half the period, the planet P_1 is again at perihelion, but the planet P_2 is at aphelion, $(\omega_2 - \omega_1) = 180^0$. So, these two stable windows represent the same configuration.

4 A Stable System at the 2:1 Resonance. $\omega_2 - \omega_1 = 180^0$

In this section we consider another 2:1 resonant stable periodic orbit, corresponding to a planetary system with masses $m_{SUN} = 0.9978$, $m_1 = 0.0008$, and $m_2 = 0.0014$, semimajor axes $a_1 = 0.782$AU, $a_2 = 1.238$AU ($T_2/T_1 = 1.992$) and eccentricities $e_1 = 0.471$, $e_2 = 0.573$. The orbit is symmetric, with the lines of apsides of the two planets antialigned, $(\omega_2 - \omega_1) = 180^0$, and the two planets are at aphelion at $t = 0$ (see Fig. 9a) and belongs to a family of stable periodic orbits [10]. We work as in the previous case, and the results are presented in Figs. 10 and 11. We note that for $\Delta\omega < 40^0$ the motion is ordered and the angle $(\omega_2 - \omega_1)$ librates around 180^0. For a perturbation $\Delta\omega \geq 42^0$ the motion becomes chaotic, after a long time of motion on a torus. A global view, by making use of LCN is given in Fig. 12a. In this figure we note that there exists another stable window at $\Delta\omega \approx \pm 90^0$. This corresponds to a stable family of non symmetric periodic orbits at the 2:1 resonance (Voyatzis

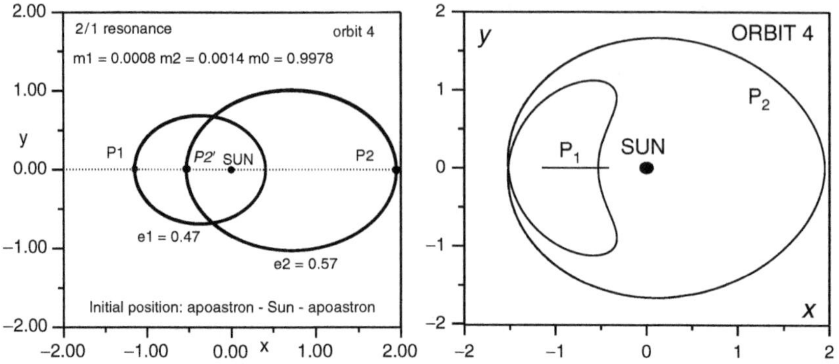

Fig. 9. (**a**) The 2:1 resonant periodic orbit with the two planetary orbits antialigned, $(\omega_2 - \omega_1) = 180^0$, in the inertial frame. In this frame it is not periodic. (**b**) The same orbit in the rotating frame, where it is exactly periodic

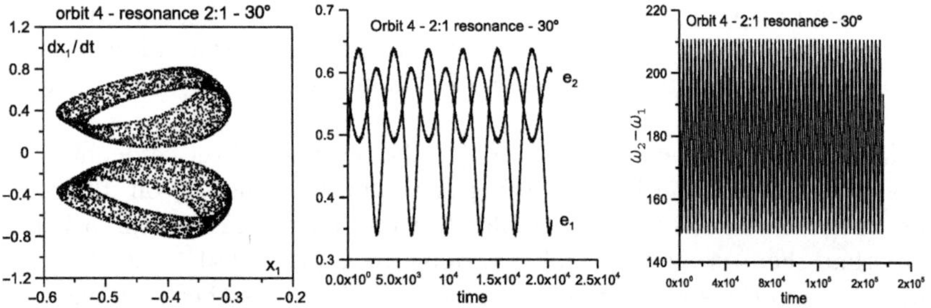

Fig. 10. The evolution of the orbit of Fig. 9, corresponding to the perturbation $\Delta\omega = 30^0$. (**a**) The Poincaré map (projection on the $x_1\,\dot{x}_1$ plane). The motion is on a torus. (**b**) The evolution of the eccentricities, which librate around the values of the periodic motion, with large amplitude. (**c**) The libration of $(\omega_2 - \omega_1)$ around 180^0

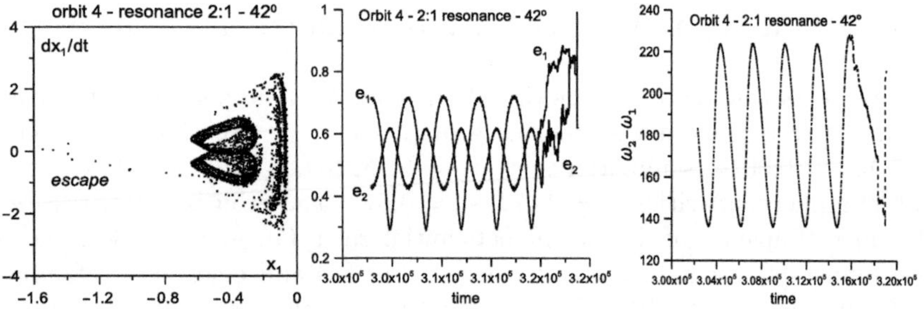

Fig. 11. The evolution of the orbit of Fig. 9, corresponding to the perturbation $\Delta\omega = 42^0$. (**a**) The Poincaré map (projection on the $x_1\,\dot{x}_1$ plane). The motion is chaotic, and the planet P_1 escapes. (**b**) The evolution of the eccentricities, which is chaotic, after a libration for a long time. (**c**) The evolution of $(\omega_2 - \omega_1)$ which is chaotic, after a libration for a long time. In (b) and (c) only the end part of the evolution is shown, after $t > 3 \times 10^5$

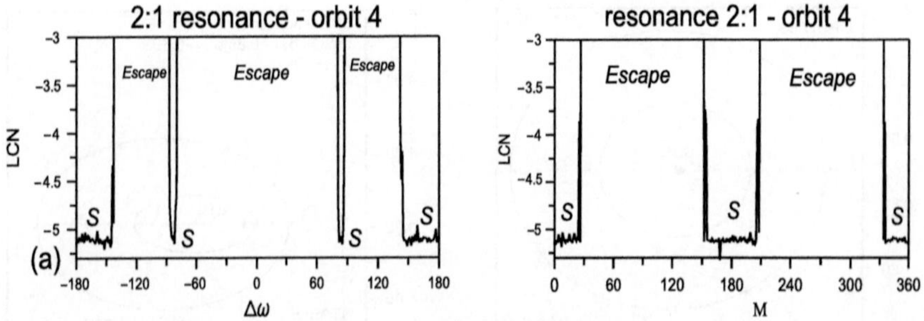

Fig. 12. A global view of the stable and unstable regions by the method of LCN: (**a**) Perturbation in $\Delta\omega$. (**b**) Perturbation in ΔM

and Hadjidemetriou, 2006). In Fig. 12b we present a global view of the stable regions when the perturbation is ΔM. The behavior is similar to the $\Delta\omega$ perturbation.

5 An Unstable System at the 3:1 Resonance. $(\omega_2 - \omega_1) = 0^0$

In this section we consider a 3:1 resonant *unstable* periodic orbit with small eccentricities, corresponding to a planetary system with masses $m_{SUN} = 0.99903$, $m_1 = 0.00078$, and $m_2 = 0.00019$, semimajor axes $a_1 = 1.035$ AU, $a_2 = 2.163$ AU $(T_2/T_1 = 3.019)$ and eccentricities $e_1 = 0.051$, $e_2 = 0.074$. The orbit is symmetric, with the lines of apsides of the two planets aligned, $(\omega_2 - \omega_1) = 0^0$, and the two planets are at perihelion at $t = 0$ (see Fig. 13a) and belongs to a family of unstable periodic orbits [16] for the masses of 55Cnc. We work as in the previous case, and the results are presented in Figs. 14 and 15. We start with a small perturbation $\Delta\omega = 5^0$, close to the unstable periodic orbit (Fig. 14a) and we note that the motion is bounded, as shown by the Poincaré map. (Computations for a much longer time than presented in this figure revealed that the motion is indeed bounded). A further increase of the perturbation, to $\Delta\omega = 45^0$ (Fig. 14b) and to $\Delta\omega = 90^0$ (Fig. 14c) shows that the motion is bounded. The difference however of these latter two cases from the perturbation close to the unstable periodic orbit is that the outer boundary of the Poincaré map close to the unstable periodic orbit seems to be a *fractal*, while the perturbed orbits far from the unstable periodic orbit are well defined *tori*. In Fig. 15a–c we present the evolution of the angle $(\omega_2 - \omega_1)$ for the above three perturbations, respectively. We also note a different behavior close to the unstable periodic orbit and far from it. For the small perturbation $\Delta\omega = 5^0$ there is a chaotic interchange between libration (with large amplitude) and rotation, while for the larger perturbation $\Delta\omega = 45^0$ we have a regular libration and for the still larger perturbation $\Delta\omega = 90^0$ we have a slow increase of $(\omega_2 - \omega_1)$, followed by a rotation for a short time, and this

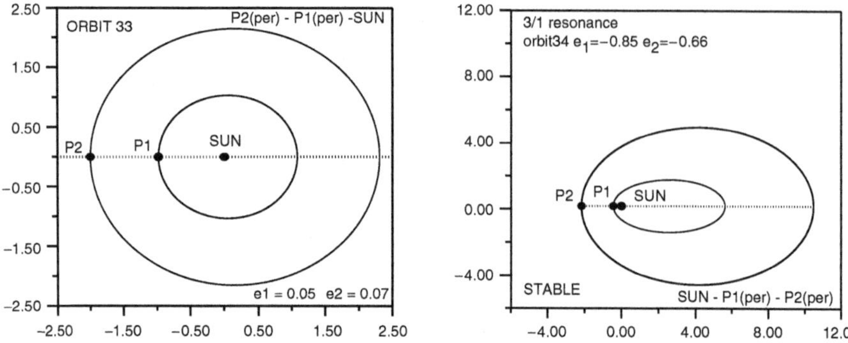

Fig. 13. Two periodic orbits at the 3:1 resonance, at $\omega_2 - \omega_1 = 0^0$: (**a**) An unstable orbit. (**b**) A stable orbit. These orbits are not exactly periodic in this frame

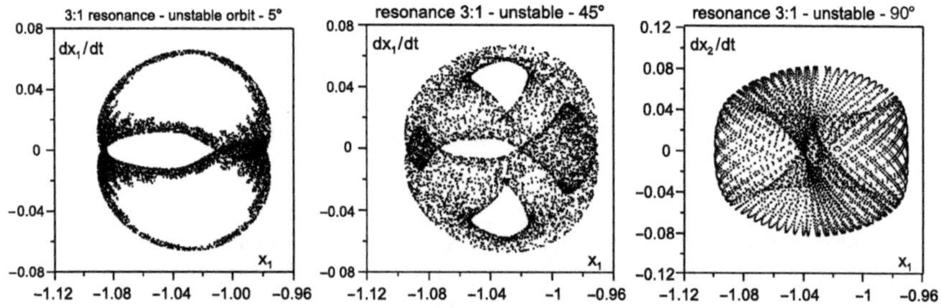

Fig. 14. The Poincaré map at the unstable periodic orbit of Fig. 13a (projection on a plane, as indicated). (**a**) Perturbation $\Delta\omega = 5^0$. The Poincaré map is bounded, but its boundary is a fractal. (**b**) Perturbation $\Delta\omega = 45^0$. The Poincaré map is on a torus. (**c**) Perturbation $\Delta\omega = 90^0$. The Poincaré map is on a torus

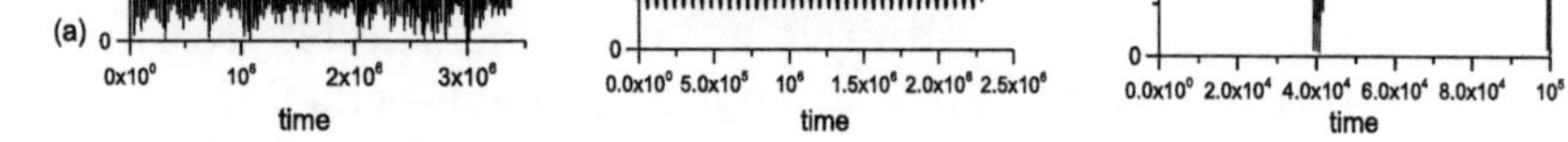

Fig. 15. The evolution of the angle $(\omega_2 - \omega_1)$ at the unstable periodic orbit of Fig. 13a. (**a**) Perturbation $\Delta\omega = 5^0$. There is a chaotic transition between libration with large amplitude and rotation. (**b**) Perturbation $\Delta\omega = 45^0$. The motion is ordered and the angle $(\omega_2 - \omega_1)$ librates. (**c**) Perturbation $\Delta\omega = 90^0$. A slow increase of $(\omega_2 - \omega_1)$ for a long time followed by a rotation for a short time, at regular intervals

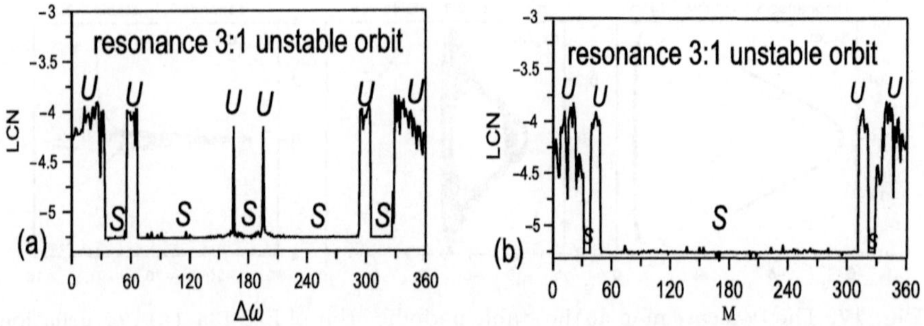

Fig. 16. A global view of the stable and unstable regions by the method of LCN: (**a**) Perturbation in $\Delta\omega$. (**b**) Perturbation in ΔM. Note the windows of stability, far from the unstable periodic orbit, corresponding to asymmetric motion

is repeated in a regular way. This behavior is clearly seen in the global view, by making use of LCN, given in Fig. 16a. In this figure we note that there are stable and unstable windows. The stable windows correspond to asymmetric configurations of the two planetary orbits. The stable motions in Fig. 14b,c are clearly in the stable windows of Fig. 16a.

In Fig. 16b we present a global view of the stable regions when the perturbation is ΔM. The behavior is similar to the $\Delta\omega$ perturbation. Close to the unstable orbit the motion appears as chaotic (but bounded), and far from the unstable periodic orbit ordered motion exists. Note that it this case the deviation from symmetry results to ordered motion, because we come close to families of stable asymmetric periodic orbits at the 3:1 resonance.

6 A Stable System at the 3:1 Resonance. $(\omega_2 - \omega_1) = 0^0$

In this section we consider a 3:1 resonant *stable* periodic orbit, with large eccentricities, corresponding to a planetary system with masses $m_{SUN} = 0.99903$, $m_1 = 0.00078$, and $m_2 = 0.00019$, semimajor axes $a_1 = 3.041$AU, $a_2 = 6.453$AU ($T_2/T_1 = 3.091$) and eccentricities $e_1 = 0.853$, $e_2 = 0.664$. The orbit is symmetric, with the lines of apsides of the two planets aligned, $(\omega_2 - \omega_1) = 0^0$, and the two planets are at perihelion at $t = 0$ (see Fig. 13b) and belongs to the stable part of a family of periodic orbits [16] for the masses of 55Cnc (the same family as in Sect. 5). We work as in the previous case, and the results are presented in Figs. 17 and 18. We start with a small perturbation $\Delta\omega = 15^0$ (Fig. 17a) and increase the perturbation to $\Delta\omega = 18^0$ (Fig. 17b) and to $\Delta\omega = 25^0$ (Fig. 17c). Close to the stable periodic orbit ($\Delta\omega = 15^0$) the Poincaré map is on a well defined torus. The same is true for a further small increase of the perturbation $\Delta\omega = 18^0$, but for another small increase of the perturbation to $\Delta\omega = 25^0$ the motion is chaotic. Note that close to

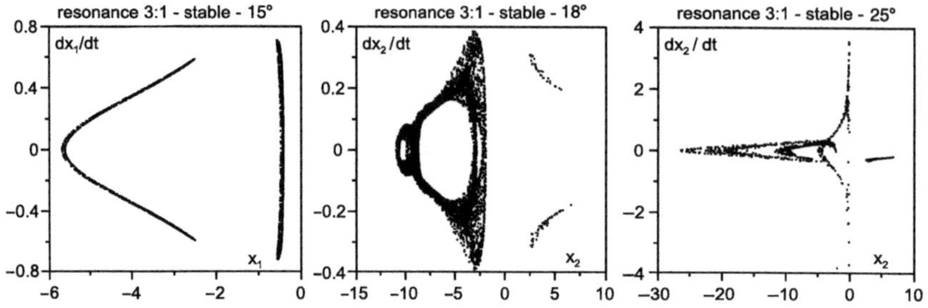

Fig. 17. The Poincaré map at the stable periodic orbit of Fig. 13a. (**a**) Perturbation $\Delta\omega = 15^0$. The Poincaré map is bounded, on a torus. (**b**) Perturbation $\Delta\omega = 18^0$. The Poincaré map is still on a torus. (**c**) Perturbation $\Delta\omega = 25^0$. The Poincaré map is now chaotic and the planet P_2 escapes

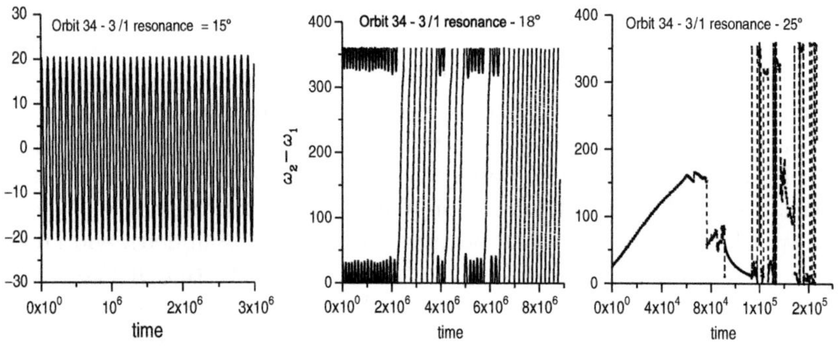

Fig. 18. The evolution of the angle $(\omega_2 - \omega_1)$ at the stable periodic orbit of Fig. 13a. (**a**) Perturbation $\Delta\omega = 15^0$. There is a libration around 0^0. (**b**) Perturbation $\Delta\omega = 18^0$. There is a chaotic transition between libration and rotation, while the motion is bounded. (**c**) Perturbation $\Delta\omega = 25^0$. The evolution is chaotic

the stable periodic orbit we have ordered motion, but the region of stability is small, because in this case the planetary eccentricities are large. A similar behavior appears for the evolution of the angle $(\omega_2 - \omega_1)$, as presented in Fig. 18a–c. For a small perturbation we start with a regular libration, but as the perturbation increases, an intermittent interchange between libration and rotation appears and for a further increase of the perturbation the evolution is chaotic. This behavior is clearly seen in the global view, by making use of LCN, given in Fig. 19a. In this figure we note that the only stable region is close to the stable periodic orbit, with a short range of stability. In Fig. 19b we present a global view when the perturbation is in ΔM. The behavior is similar to the $\Delta\omega$ perturbation.

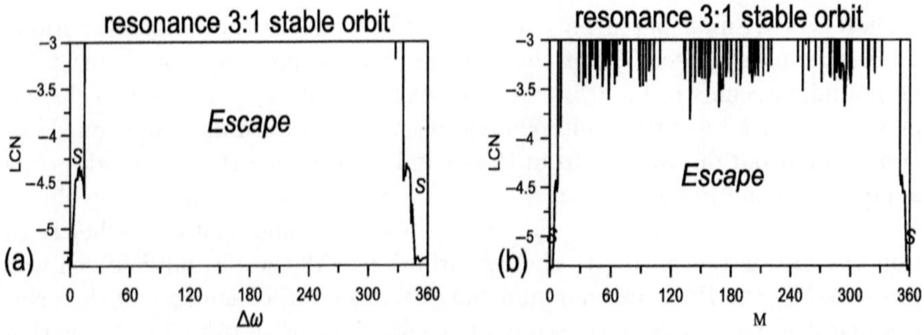

Fig. 19. A global view of the stable and unstable regions by the method of LCN: (**a**) Perturbation in $\Delta\omega$. (**b**) Perturbation in ΔM

7 Discussion

The mean motion resonance of a planetary system with two planets depends only on the ratio of the semimajor axes. So, we may have an infinity of planetary systems, with *the same resonance*, but different eccentricities, different orientations of the planetary orbits, as defined by the angle $(\omega_2 - \omega_1)$ between the line of apsides (see Fig. 1b) and different phases of the two planets, i.e. the position of the second planet on its orbit when the first planet is at perihelion, as defined by the angle ΔM (see Fig. 1c). In this work we studied the factors that affect the stability and the long term evolution of a *resonant* planetary system, keeping the resonance *fixed* and perturbing the system by changing the angles $\Delta\omega$ or ΔM. We presented four typical examples, starting from *symmetric periodic orbits* (Sects. 2–6) and we considered both stable and unstable periodic orbits as starting points. Special emphasis is given on the evolution of the angle $(\omega_2 - \omega_1)$ as the perturbation increases. The study was made by computing the Poincaré map and by computing the LCN for the whole range of perturbations. For a small perturbation close to a *stable* periodic orbit, we find a libration of all the orbital elements around their unperturbed values, and also the Poincaré map was on a well defined torus. The behavior however of a resonant system is not the same for larger deviations from the exact periodic motion, and this is more evident in the evolution of $(\omega_2 - \omega_1)$. In some cases (Sect. 3) we have a smooth transition from libration to rotation, before the system develops chaotic motion for still larger perturbations. In other cases (Sect. 4) we start with a libration of $(\omega_2 - \omega_1)$, but then we go directly to chaotic motion, as the perturbation increases, without passing through rotation. In a third typical case (Sect. 6), we start with libration of $(\omega_2 - \omega_1)$ but then, as the perturbation increases, an intermittent transition between libration and rotation appears, while the system remains bounded. For still larger perturbations we have chaotic motion, as in all previous cases. In all the above cases where we had bounded motion, the Poincaré map is on a well defined torus. The situation is different in the vicinity of an *unstable* periodic motion. For a small perturbation, the motion close to the unstable periodic orbit is bounded (Sect. 5) but the Poincaré map is no longer on a torus, but its boundary seems to be a *fractal.* The evolution of $(\omega_2 - \omega_1)$ is also chaotic, presenting an intermittent interchange between libration and rotation. For a larger perturbation, further from the unstable periodic orbit, ordered motion appears and the Poincaré map is now on a well defined torus. The evolution of $(\omega_2 - \omega_1)$ in this latter case is represented by regular motion, either libration or rotation (according to the perturbation). All these results, which were presented by the Poincaré map and the evolution of the elements of the orbit, were verified by a systematic study of the stability of an orbit by the method of LCN. In this way we also detected regions in the phase space where we have stable resonant motion, but *non symmetric.*

From all the above we see that the resonance alone is not enough to determine the stability at a *fixed* resonance, but there are other factors that determine the stability, as the deviation from symmetry, defined by $\Delta\omega$ and the change of the phase between the planets, defined by ΔM. We also noted that the chaotic property is not necessarily associated with the disruption of the system, and the system may be bounded, even if the LCN (or similar chaotic indicators) show chaos. In this latter case the chaotic property may appear in a different way: Some elements of the orbit, notably the angle $(\omega_2 - \omega_1)$, may have a chaotic evolution and the system may move in a region of the phase space with fractal dimensions.

References

1. Beaugé, C., Ferraz-Mello, S., and Michtchenko, T. (2003): *Extrasolar Planets in Mean-Motion Resonance: Apses Alignment and Asymmetric Stationary Solutions*, ApJ **593**, 1124.
2. Beaugé C. and Michtchenko T. (2003): *Modelling the High-Eccentricity Planetary Three-Body Problem. Application to the GJ876 Planetary System*, MNRAS **341**, 760.
3. Contopoulos, G. and Voglis, N. (1998): *Spectra of Streching Numbers and Helicity Angles*, in Benest, D. and Froeschlé C. (eds.) *Analysis and Modelling of Discrete Dynamical Systems*, Gordon and Breach Scientific Publication, 55–89.
4. Beaugé C., Ferraz-Mello S., and Michtchenko T.A. (2006): *Planetary Migration and Extrasolar Planets in the 2/1 Mean-motion Resonance*, MNRAS **365**, 1160–1170.
5. Dvorak, R., Freistetter, F., Kurths, J. (eds). (2005): *Chaos and Stability in Planetary Systems*, Lecture Notes in Physics, Springer, Berlin Heidelberg New York.
6. Ferraz-Mello, S., Beaugé, C., and Michtchenko T. (2003): *Evolution of Migrating Planet Pairs in Resonance*, Cel. Mech. Dyn. Astr. **87**, 99–112.

7. Gozdziewski, K., Bois, E., and Maciejewski A. (2002): *Global Dynamics of the Gliese 876 Planetary System*, MNRAS **332**, 839.
8. Hadjidemetriou, J.D. (1976): *Families of Periodic Planetary Type Orbits in the Three-Body Problem and their Stability*, Astrophys. Sp. Sci. **40**, 201–224.
9. Hadjidemetriou, J.D. (1975): *The Continuation of Periodic Orbits from the Restricted to the General Three-Body Problem*, Cel. Mech. **12**, 155–174.
10. Hadjidemetriou, J.D. (2002): *Resonant Periodic Motion and the Stability of Extrasolar Planetary Systems*, Cel. Mech. Dyn. Astr. **83**, 141–154.
11. Hadjidemetriou, J.D. (2006): *Symmetric and Asymmetric Librations in Extrasolar Planetary Systems: A global view*, Cel. Mech. Dyn. Astron. **95**, 225–244.
12. Lee, M.H. (2004): *Diversity and Origin of 2:1 Orbital Resonance in Extrasolar Planetary Systems*, ApJ **611**, 517.
13. Malhotra, R. (2002): *A Dynamical Mechanism for Establishing Apsidal Resonance*, ApJ **575**, L33–36.
14. Michtchenko T.A., Beaugé C., and Ferraz-Mello S. (2006): *Stationary Orbits in Resonant Extrasolar Planetary Systems*, Cel. Mech. Dyn. Astron. **94**, 381–397.
15. Psychoyos, D. and Hadjidemetriou, J.D. (2005): *Dynamics of 2/1 Resonant Extrasolar Systems. Application to HD82943 and Gliese876*, Cel. Mech. Dyn. Astr. **92**, 135–156.
16. Voyatzis, G. and Hadjidemetriou, J.D. (2006): *Symmetric and Asymmetric 3:1 Resonant Periodic Orbits: An Application to the 55Cnc Extra-Solar System*, Cel. Mech. Dyn. Astr. **95**, 259–271.

Prometheus and Pandora, the Champions of Dynamical Chaos

I.I. Shevchenko

Pulkovo Observatory of the Russian Academy of Sciences,
Pulkovskoje ave. 65/1, St. Petersburg 196140, Russia
iis@gao.spb.ru

Summary. Chaos in the orbital dynamics of Prometheus and Pandora, the 16th and 17th satellites of Saturn, as well as hypothetical chaos in their rotational dynamics, are considered and discussed. The Lyapunov time estimates, calculated analytically by means of a method based on the separatrix map theory, are presented. It is shown that the Lyapunov times of both chaotic orbital and hypothetically chaotic rotational dynamics of the satellites are the smallest ones amongst all known Lyapunov times in the Solar system; thus these two satellites are real "champions" of dynamical chaos.

Chaotic regimes played significant role in the long term orbital evolution of planetary satellites in the Solar system. The Miranda–Umbriel system (U5 and U2) is perhaps the best studied one in this respect; see [21, 23, 43] and references therein. Besides Miranda and Umbriel, chaotic orbital states in past epochs of orbital evolution and their effect on the present orbital states were studied in the orbital dynamics of the Enceladus–Dione system (S2 and S4) [5, 13, 18], Miranda–Ariel system (U5 and U1) [43], Ariel–Umbriel system (U1 and U2) [42, 43], Titan–Iapetus system (S6 and S8) [27], and Galilean system (J1, J2, J3 and J4) [27, 28, 41].

However, chaotic states in the dynamics of satellite systems took place not only in the past. Champenois and Vienne [7, 8] considered dynamics of Mimas (S1) and Tethys (S3), residing at present in the 4:2 orbital inclination-type mean motion resonance, and showed that the dynamics of this system might be chaotic, due to large amplitudes of libration in the resonance. Later on, estimates of the Lyapunov time (characteristic time of predictable dynamics) were made in [25], both analytically and in numerical experiments, by means of direct integration of the equations of motion. They turned out to be 300–600 years in various models. Thus chaos in this system does not manifest itself on timescales which allow observational verification.

Only one example of observable orbital chaos in planetary satellite systems can be given at present; this is the Prometheus–Pandora system. The Saturnian satellites Prometheus (S16) and Pandora (S17) are the shepherds of the

G. Contopoulos, P.A. Patsis, *Chaos in Astronomy.* Astrophysics and Space Science Proceedings, doi: 10.1007/978-3-540-75826-6,

F ring. Both satellites were discovered by the Voyager spacecraft in 1980–1981 [38]. The Hubble Space Telescope observations in 1995 revealed that the mean longitudes of Prometheus and Pandora were different at that time by some 20° from the values predicted by the Voyager ephemerides [4, 24, 26]. Prometheus lagged its predicted longitude, and Pandora led by the same amount. What is more, abrupt changes in the mean motions ("kinks" in the mean longitudes) were observed [14]. Goldreich and Rappaport [15] interpreted these deviations as a signature of dynamical chaos due to the mutual gravitational interactions of these almost coorbital satellites. In a numerical experiment involving numerical integration of the equations of motion they found that the system gradually deviated from the Voyager ephemerides due to chaotic diffusion in the phase space. Goldreich and Rappaport related the "kinks" in the mean longitudes to the increase of gravitational interactions during apse antialignment that occurs every 6.2 years.

Generally, the reason for emergence of chaos in the dynamics of satellite systems, as well as in the dynamics of asteroids and planets, lies, as a rule, not in the interaction of separate mean motion resonances, but in the interaction of subresonances in a multiplet corresponding to mean motion resonance (see, e.g., [21, 22]). The orientation of orbits is subject to variations including secular precession of pericentres and nodes, and this precession is just the reason of splitting of orbital resonances into subresonances.

The cause of chaos in the Prometheus–Pandora system lies in interaction of subresonances in the mean motion resonance multiplet corresponding to the 121:118 mean motion commensurability [16, 29]. This was identified by surveying the rates of change of all possible critical arguments for the Prometheus–Pandora system [16, Fig. 1]. Chaos manifests itself in numerical simulations of system's dynamics in various settings of the problem [11, 15–17, 30], with or without including perturbations from other Saturnian satellites.

Saturn's oblateness causes rapid precession of the orbits of Prometheus and Pandora. Therefore the effect of gravitational interactions between the satellites on the apsidal angles and orbital eccentricities can be neglected, and one can analyze solely changes in mean motions, or equivalently in the resonant angle $\psi \equiv 121\lambda' - 118\lambda$, where λ and λ' are the mean longitudes of Prometheus and Pandora, respectively. As shown by Goldreich and Rappaport [16], in this way the equations of orbital motion for the Prometheus–Pandora system are reduced to the equation of the pendulum with periodic perturbations. The dynamics of the Prometheus–Pandora system in this approximation was studied in detail by Farmer and Goldreich [12] in terms of adiabatic invariants and separatrix crossing phenomena.

The separatrix map theory [32, 34–36] is readily applied to the perturbed pendulum motion not only in the case of "fast" chaos, but in the case of slow, or adiabatic, chaos as well. In particular, one can obtain analytical estimates of the Lyapunov time [34, 36] and the width [35] of the chaotic layer.

The Hamiltonian for the general perturbed pendulum model of nonlinear resonance (see, e.g., [32]) in the form relevant to our problem is

$$H = \frac{\mathcal{G}p^2}{2} - \mathcal{F}\cos\varphi + a\cos(\varphi - \tau) + b\cos(\varphi + \tau). \tag{1}$$

The first two terms in (1) represent the Hamiltonian H_0 of the unperturbed pendulum; φ is the pendulum angle (the resonance phase angle), p is the momentum. The periodic perturbations are given by the last two terms. The quantity τ is the phase angle of perturbation: $\tau = \Omega t + \tau_0$, where Ω is the perturbation frequency, and τ_0 is the initial phase of the perturbation. The quantities $\mathcal{F}$, $\mathcal{G}$, a, b are constants; $\mathcal{F} > 0$, $\mathcal{G} > 0$. In what follows, we use the notations $\varepsilon_1 = a/\mathcal{F}$ and $\varepsilon_2 = b/\mathcal{F}$ for the relative amplitudes of perturbation. The so-called adiabaticity parameter is defined as $\lambda = \Omega/\omega_0$, where $\omega_0 = (\mathcal{F}\mathcal{G})^{1/2}$ is the frequency of small-amplitude oscillations of the resonance phase angle.

A formula for the maximum Lyapunov exponent for the slowly chaotic triad of interacting resonances was given in [34], but without detailed derivation; the latter is presented in [36]. A theory for estimating the width of the chaotic layer in the adiabatic case was developed in [35]. It was shown rigorously in [36] that the Prometheus–Pandora system moves in adiabatic regime, and the elements of the chaotic layer theory [35, 36], comprising estimation of the maximum Lyapunov exponent and the width of the chaotic layer, were applied to the analysis of the orbital dynamics of this system. The Lyapunov time was calculated analytically, and an analytical estimate of the width of the chaotic layer was made as well. The ranges of chaotic diffusion in the mean motion were shown to be almost twice as big compared to previous estimates for both satellites. Here we consider the dynamics of Prometheus and Pandora in a more general framework.

The adiabaticity parameter λ, the relative perturbation strengths ε_1 and ε_2, and the perturbation period $T_{\text{pert}} = 2\pi/\Omega$ for the Prometheus–Pandora system in the perturbed pendulum model of the resonance are easily evaluated from the observational data on the system, compiled in [36]: $\lambda = 0.40$, $\varepsilon_1 = 0.52$, $\varepsilon_2 = 0.65$, $\varepsilon = \frac{1}{2}(\varepsilon_1 + \varepsilon_2) = 0.58$, $T_{\text{pert}} = 6.24$ years. The condition for adiabatic chaos is $\lambda < 1/2$ [35]; thus the dynamics of the Prometheus–Pandora system represents the case of adiabatic chaos (for details see [36]).

According to [34, 36], the formula for the Lyapunov time for the "slowly chaotic triplet" resonance type (i.e., $\lambda < 1/2$, $a = b$) in model (1) is

$$T_L \approx \frac{T_{\text{pert}}}{2\pi} \ln \left| \frac{16}{\lambda\varepsilon} \sin\left(\frac{\lambda}{2} \ln \frac{4}{\lambda|\varepsilon|} \right) \right| . \tag{2}$$

This formula has specific limits of applicability, namely, one of the separatrix map parameters, c (approximately equal to $\lambda \ln \frac{4}{\lambda|\varepsilon|}$ in the adiabatic case), should not be close to 0 mod 2π. The general theory for estimating Lyapunov times, incorporating fast chaos and the case of asymmetric perturbation, can be found in [34].

Substituting $\lambda = 0.40$, $\varepsilon = 0.58$, $T_{\text{pert}} = 6.24$ years in formula (2), one gets the Lyapunov time for the Prometheus–Pandora system: $T_L = 3.6$ years.

The parameter $c \approx 1.2$, so the formula is applicable. The uncertainty in the estimate can be appraised in the following way: taking the relative amplitude of perturbation in the range from ε_1 to ε_2, one gets the analytical estimate of the Lyapunov time by formula (2) in the range of 3.5–3.7 years, i.e., $T_L = 3.6 \pm 0.1$ years. The detailed calculation is given in [36].

Goldreich and Rappaport [15, 16] obtained numerical-experimental estimates for the Lyapunov time of the Prometheus–Pandora system in a model setting restricted to mutual gravitational interactions between Prometheus and Pandora. According to their results, the Lyapunov time is $\approx$3.3 years. Later on, the Lyapunov time was estimated by Cooper and Murray [11] by numerical integration in a full problem with taking into account all important perturbations from other satellites, and by Farmer and Goldreich [12] by numerical integrations in the pendulum model. All these numerical-experimental estimates are quite close to each other. In summary, they are: $T_L = 3.0$–4.0 years [15, Fig. 7], [16, Fig. 4]; $T_L \approx 3.3$ years [11, Fig. 10]; $T_L = 3.3$–4.2 years [12, Fig. 9]. Our analytical estimate, $T_L = 3.6$ years, is within the range of these numerical-experimental values.

It is just the small value of the Lyapunov time that makes chaos directly observable. The observed chaotic regime in the motion of the Prometheus–Pandora system can play an essential role in the long term orbital evolution of the system. The parametric proximity of a slowly chaotic system to a low order secondary resonance drastically modifies global properties of the chaotic layer, i.e., its global structure, relative measure of inner regular component, and the value of the maximum Lyapunov exponent. According to [36], the system is not close at present to any low order secondary resonances.

The orbital dynamics of Prometheus and Pandora is hitherto the only example of observable orbital chaos in the Solar system, just as the rotation of the 7th satellite of Saturn, Hyperion, is the only known example of observable chaos [2, 19, 46, 47] in rotational dynamics amongst the Solar system bodies. It is interesting that chaos in the orbital motion of Pandora was theoretically envisaged by Borderies et al. [3] as early as in 1984, in the same year when chaos in the rotation of Hyperion was predicted by Wisdom et al. [47]. However, contrary to the Hyperion case, chaos in the Prometheus–Pandora system was not especially sought for in observations.

The Halley comet dynamics might be considered as another example of observed orbital chaos. However, though "observed", this chaos is not directly observable, but was discovered [9, 10, 44] by analyzing data of historical chronicles. On the contrary, chaos in the motion of Prometheus and Pandora manifests itself on the time scale of current observations.

In what concerns orbital dynamics, the Prometheus–Pandora system has the lowest value of the Lyapunov time amongst objects with known values of the Lyapunov time in the Solar system. The only contenders might be some asteroids and comets, exhibiting close encounters with planets, in particular, some near-Earth objects. However, none of them is known to have Lyapunov times smaller than 10 years; see results of relevant integrations in [45, Fig. 2] and [39, 40].

So, Prometheus and Pandora might be called the "champions" of orbital chaos. What about their rotational dynamics? As follows from our recent theoretical analysis [20], Prometheus and Pandora may rotate chaotically. The survey for possible chaotic dynamics was performed amongst all satellites with known inertial and orbital parameters. The major criterion was the instability of the synchronous state with respect to tilting the axis of rotation. This testifies the dynamical condition for a satellite to rotate chaotically in its final dynamical state of tidal evolution. Apart from Hyperion, already known to rotate chaotically, only Prometheus and Pandora were found to satisfy or to be close to satisfying this criterion. Besides, a simple calculation of the times of despinning due to tidal evolution demonstrated that the despinning times for these satellites were short in comparison with the Solar system age, and the spins of these satellites could have indeed evolved up to entering the chaotic zone near the synchronous state. So, these two satellites were found to be the main candidates, after Hyperion, for being observed in chaotic rotation.

Contrary to the case of Hyperion [46, 47], chaos in rotation of these two satellites is due to fine tuning of the dynamical and physical parameters rather than simply to a large extent of the chaotic domain in the rotation phase space. It is worth noting that the theoretical inferences on chaos in rotation of these satellites are completely independent from the fact of presence of chaos in their orbital motion.

Let us estimate analytically the Lyapunov times of the rotation of Prometheus and Pandora, if it were chaotic. This is possible in the model of planar rotation, i.e., when the axis of rotation of a satellite is set to coincide with that of its maximum moment of inertia and to be orthogonal to the orbit plane. The motion is then described by the Beletsky equation [1] for the planar librations/rotations of a satellite in an elliptic orbit. In the case of small orbital eccentricity, the equations of motion are given by Hamiltonian (1), which is derived neglecting all powers in the eccentricity higher than unity (see, e. g., [6, 47]). According to [31], one has in the pendulum paradigm: $\varepsilon_1 = -7e/2$, $\varepsilon_2 = e/2$; $\omega_0 = (3(B-A)/C)^{1/2}$, $\lambda = 1/\omega_0$. The quantities $A < B < C$ are the principal central moments of inertia of the satellite; e is the eccentricity of the orbit.

Since $|\varepsilon_1| \gg |\varepsilon_2|$, we consider the limit case of ε_2 equal to zero. It means that we set that one of the two perturbing resonances does not exist, and instead of the resonance triad we have a duad. In what follows, we denote ε_1 by just ε. The adiabaticity parameter λ, the relative perturbation strength ε, and the perturbation period $T_{\mathrm{pert}} = 2\pi/\Omega$ for each satellite can be evaluated from the observational data, compiled in [37]. This gives $\lambda = 0.85$, $\varepsilon = -0.014$, $T_{\mathrm{pert}} = T_{\mathrm{orb}} = 0.61$ days for Prometheus; and $\lambda = 1.07$, $\varepsilon = -0.014$, $T_{\mathrm{pert}} = T_{\mathrm{orb}} = 0.63$ days for Pandora. The quantity T_{orb} is the orbital period.

Since $\lambda > 1/2$ in the both cases, the dynamics represents the case of "fast" chaos, according to the condition derived in [35]. The formula for the Lyapunov time for this kind of interacting resonances, i.e., for the fastly chaotic resonance duad, as derived in [34], is

$$T_L \approx \frac{5T_{\mathrm{pert}}}{2\pi} \cdot \left(\frac{4f(2\lambda)}{g(2\lambda, W(\lambda))} + \frac{f(\lambda)}{g(\lambda, W(\lambda))} \right)^{-1}, \tag{3}$$

where

$$W(\lambda) = 4\pi\varepsilon \frac{\lambda^2}{\sinh \frac{\pi\lambda}{2}} \tag{4}$$

and

$$f(x) = C_h \frac{2x}{1+2x}, \quad g(x,y) = x \ln \frac{32e}{x|y|}, \tag{5}$$

where $C_h \approx 0.80$ is Chirikov's constant [33], e is the base of natural logarithms, x and y are "dummy" variables.

From (3) one has $T_L = 1.6$ days for Prometheus and $T_L = 1.9$ days for Pandora. If we compare these theoretical estimates with the numerical-experimental ones, namely, $T_L = 1.5$ days for Prometheus and $T_L = 1.8$ days for Pandora, obtained in the case of planar chaotic rotation in [37], we see that the results are in perfect agreement. In the case of spatial chaotic rotation, the numerical-experimental values computed in [37] are somewhat less: $T_L = 0.8$ days for Prometheus and $T_L = 1.1$ days for Pandora. The lower predictability in the spatial case is understandable, since one has 3 and 1/2 degrees of freedom instead of 1 and 1/2. Still the accuracy of the theoretical predictions are better than by the order of magnitude.

So, the calculated Lyapunov times of the hypothetical chaotic rotation of Prometheus and Pandora are about 1 day, i. e., about 30 times less than that of Hyperion's rotation. This means that the chaotic nature of rotation of these two satellites, if their rotation is indeed chaotic, can be recovered on very short time intervals of observation, much shorter than those needed in the case of Hyperion.

If the rotational chaos, in addition to the orbital one, were confirmed, this would mean that Prometheus and Pandora are real "champions" of dynamical chaos in the Solar system, and not only in what concerns the fact of double (orbital and rotational) chaos itself, but also in the degree of unpredictability of their motion. As we have seen, the Lyapunov times of both chaotic orbital dynamics and hypothetically chaotic rotational dynamics of these satellites are the smallest ones amongst the known Lyapunov times in the Solar system; thus their motion is the most unpredictable. Of course, further studies might reveal new objects with even smaller Lyapunov times of orbital or rotational chaos, but such objects might be much smaller in their physical size.

Acknowledgments

This work was supported by the Russian Foundation for Basic Research (project # 05-02-17555) and by the Programme of Fundamental Research of the Russian Academy of Sciences "Fundamental Problems in Nonlinear Dynamics".

References

1. V.V. Beletsky: *The Motion of an Artificial Satellite about Its Mass Center* (Nauka, Moscow 1965) (in Russian)
2. G.J. Black, P.D. Nicholson, P.C. Thomas: Icarus **117**, 149 (1995)
3. N. Borderies, P. Goldreich, S. Tremaine: Unsolved problems in planetary ring dynamics. In: *Planetary Rings*, ed by R. Greenberg, A. Brahic (University of Arizona Press, Tucson 1984) pp 713–734
4. A.S. Bosh, A.S. Rivkin: Science **272**, 518 (1996)
5. N. Callegari Jr., T. Yokoyama: Celest. Mech. Dyn. Astron. **98**, 5 (2007)
6. A. Celletti: Z. Angew. Math. Phys. **41**, 174 (1990)
7. S. Champenois, A. Vienne: Icarus **140**, 106 (1999)
8. S. Champenois, A. Vienne: Celest. Mech. Dyn. Astron. **74**, 111 (1999)
9. B.V. Chirikov, V.V. Vecheslavov: Inst. Nucl. Phys. Preprint, No. 86–184 (1986)
10. B.V. Chirikov, V.V. Vecheslavov: Astron. Astrophys. **221**, 146 (1989)
11. N.J. Cooper, C.D. Murray: Astron. J. **127**, 1204 (2004)
12. A.J. Farmer, P. Goldreich: Icarus **180**, 403 (2006)
13. S. Ferraz-Mello, R. Dvorak: Astron. Astrophys. **179**, 304 (1987)
14. R.G. French, C.A. McGhee, L. Dones, J.J. Lissauer: Icarus **162**, 143 (2003)
15. P. Goldreich, N. Rappaport: Icarus **162**, 391 (2003)
16. P. Goldreich, N. Rappaport: Icarus **166**, 320 (2003)
17. R.A. Jacobson, R.G. French: Icarus **172**, 382 (2004)
18. M. Karch, R. Dvorak: Celest. Mech. Dyn. Astron. **43**, 361 (1988)
19. J.J. Klavetter: Astron. J. **98**, 1855 (1989)
20. V.V. Kouprianov, I.I. Shevchenko: Icarus **176**, 224 (2005)
21. R. Malhotra: Physica D **77**, 289 (1994)
22. R. Malhotra: Orbital resonances and chaos in the Solar system. In: *Solar System Formation and Evolution*, ed by D. Lazzaro et al., ASP Conf. Series **149**, 37 (1998)
23. R. Malhotra, S.F. Dermott: Icarus **85**, 444 (1990)
24. C.A. McGhee, P.D. Nicholson, R.G. French, K.J. Hall: Icarus **152**, 282 (2001)
25. A.V. Melnikov, I.I. Shevchenko: Astronomicheskii Vestnik **39**, 364 (2005) [Solar Syst. Res. **39**, 322]

26. P.D. Nicholson, M.R. Showalter, L. Dones, R.G. French, S.M. Larson, J.J. Lissauer, C.A. McGhee, P. Seitzer, B. Sicardy, G.E. Danielson: Science **272**, 509 (1996)
27. B. Noyelles, A. Vienne: The k:k+4 resonances in planetary systems. In: *Dynamics of Populations of Planetary Systems* (Proc. IAU Coll. 197), ed by Z. Knežević, A. Milani (Cambridge University Press, Cambridge 2005) 453–458
28. B. Noyelles, A. Vienne: Icarus **190**, 594 (2007)
29. S. Renner, B. Sicardy: BAAS **35**, 1044 (2003) (abstract)
30. S.S. Renner, B. Sicardy, R.G. French: Icarus **174**, 230 (2005)
31. I.I. Shevchenko: Celest. Mech. Dyn. Astron. **73**, 259 (1999)
32. I.I. Shevchenko: Zh. Eksp. Teor. Fiz. **118**, 707 (2000) [Sov. Phys. JETP **91**, 615]
33. I.I. Shevchenko: Pis'ma Zh. Eksp. Teor. Fiz. **79**, 651 (2004) [JETP Lett. **79**, 523]
34. I.I. Shevchenko: On the Lyapunov exponents of the asteroidal motion subject to resonances and encounters. In: *Near Earth Objects, our Celestial Neighbors: Opportunity and Risk* (Proc. IAU Symp. 236), ed by A. Milani, G.B. Valsecchi, D. Vokrouhlický (Cambridge University Press, Cambridge 2007) 15–29
35. I.I. Shevchenko: Physics Letters A **372**, 808 (2008)
36. I.I. Shevchenko: MNRAS **384**, 1211 (2008)
37. I.I. Shevchenko, V.V. Kouprianov: Astron. Astrophys. **394**, 663 (2002)
38. S.P. Synnott, R.J. Terrile, R.A. Jacobson, B.A. Smith: Icarus **53**, 156 (1983)
39. G. Tancredi: Astron. Astrophys. **299**, 288 (1995)
40. G. Tancredi: Celest. Mech. Dyn. Astron. **70**, 181 (1999)
41. W.C. Tittemore: Science **250**, 263 (1990)
42. W. Tittemore, J. Wisdom: Icarus **74**, 172 (1988)
43. W. Tittemore, J. Wisdom: Icarus **85**, 394 (1990)
44. V.V. Vecheslavov, B.V. Chirikov: Pis'ma Astron. Zh. **14**, 357 (1988) [Sov. Astron. Letters **14**, 151]
45. A.L. Whipple: Icarus **115**, 347 (1995)
46. J. Wisdom: Astron. J. **94**, 1350 (1987)
47. J. Wisdom, S.J. Peale, F. Mignard: Icarus **58**, 137 (1984)

Planets in Multiple Star Systems: A Symplectic Approach

P.E. Verrier and N.W. Evans

Institute of Astronomy, University of Cambridge, Madingley Road, Cambridge CB3 0HA, UK
pverrier@ast.cam.ac.uk, nwe@ast.cam.ac.uk

Summary. In recent years exoplanets have been discovered in both binary and triple star systems, and in some cases the stars can significantly affect the dynamics of the planetary system. A study of the Gamma Cephei system, an eccentric giant planet in a relatively close binary, reveals that the stability of small bodies in the system is extremely complex. Developing a symplectic integration scheme specifically for planetary orbits in hierarchical triple stellar systems permits the dynamics of particles to be investigated in these more complicated environments. A general study of the circumbinary region shows that, although often well modelled by the overlay of the effects of two decoupled binary systems, there is a regime when the stars are relatively close and eccentric where the stability is far more complex, as the combined effect of all three stars acts to destabilise test particles.

1 Introduction

The subject of this paper is the symplectic integration of planets in multiple star systems, specifically hierarchical triples, defined here as a close inner binary and a more distant third companion making up a wider outer binary with the close pair. There are three main types of planetary orbits in such systems: circumstellar (around one star), circumbinary (around the inner binary pair) and circumtriple (around all three stars).

Circumstellar exoplanets are thought to exist in about 27 binaries and 3 triples [4], as shown in Fig. 1, some of them fairly close stellar systems such as HD 41004, γ Cephei and Gliese 86. The diagram shows the characteristics of the planetary orbits as a function of stellar mass and semimajor axis. For the three relatively close binaries mentioned above we can note that the planetary orbits themselves are not that small. In addition to these systems circumbinary debris discs are also known in the quadruple systems GG Tau [5] and HD 98800 [8].

The discovery of new systems such as these, so very different from traditional views of planetary systems, requires new numerical and analytical techniques to be developed to study their complex dynamics. An example of

G. Contopoulos, P.A. Patsis, *Chaos in Astronomy*. Astrophysics and Space Science Proceedings, doi: 10.1007/978-3-540-75826-6,

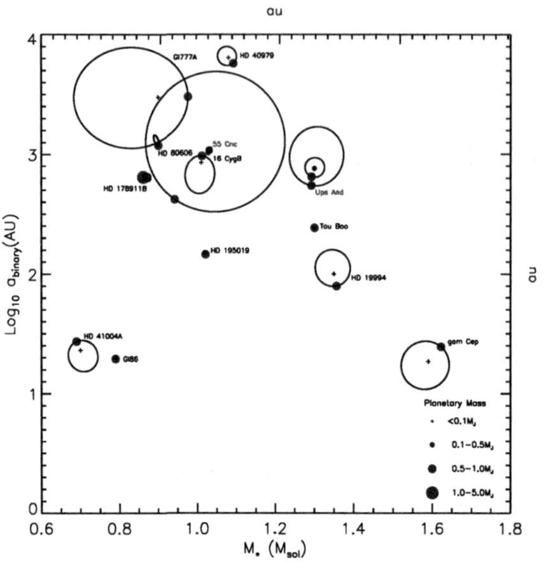

Fig. 1. The distribution of planets in binary systems, plotted as a function of the primary star mass and the binary's semimajor axis. Systems with more than one planet are labelled in *grey*. The position of each planet at periastron is also shown, with the size of the plot point indicating mass. The scales on the *top* and *left hand side* indicate the size of each orbit, with one major tick mark equating to 1 AU. Plotted using data from [4]

such dynamics is given by the binary system γ Cephei [13]. This is an eccentric giant planet in a circumstellar orbit with eccentricity 0.12 and semimajor axis 2.13 AU around a 1.6 $M_\odot$ K-type primary of a binary stellar system, the secondary of which is an M dwarf, about 0.4 $M_\odot$, at 18.5 AU in a 0.36 eccentricity orbit [6]. Numerically investigating the dynamics using the standard Bulirsch-Stoer integrator in the MERCURY software package [2] revealed additional zones of stability for small particles in the system. In the circumbinary region there is a possibility that a Kuiper-Belt analog structure could exist outwards of about 65 AU. For the circumstellar cases, around the secondary there is a small region out to about 2.5 AU where even habitable planets could remain stable, although they are not presently detectable. Around the primary it was found that an asteroid belt-like structure could exist, the detailed dynamics and stability of which were not straightforward, as shown in Fig. 2. Here the stability is shown as a function of inclination, for both a coplanar and inclined planet. Overlaid are the positions of mean motion resonances with the planet, and it is clear that the region is sculpted by these in a similar manner to the asteroid belt in our own solar system.

The dynamics of planets in multiple systems like this are clearly very interesting. Most numerical methods available to integrate the equations of motion are developed for single star systems and cannot be used. Methods

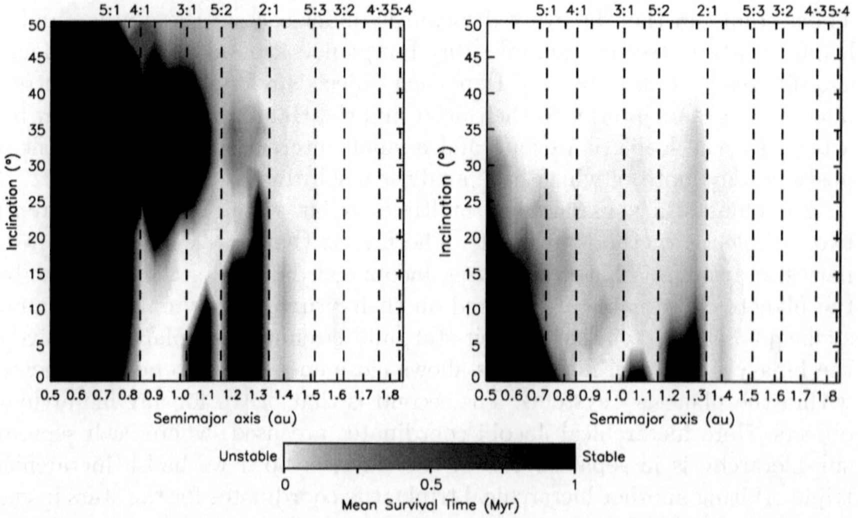

Fig. 2. Stability map for circumstellar test particles interior to the planet in the γ Cephei system as a function of inclination. In the *left-hand panel* the planet and stars are coplanar but the test particles inclined, in the *right-hand panel* the planet shares the particle's inclination. Stability is indicated by the test particle's survival time averaged over longitudes and normalised to 1 Myr. *Darker colours* indicate more stable regions, while *lighter colours* show less stable regions. Over-plotted are the nominal locations of the mean motion resonances with the planet, up to the 5:4 case

like Runge-Kutta or Bulirsch-Stoer – as used for γ Cephei – can be implemented but have disadvantages: they are very slow and have poor long term energy conservation. Hence a symplectic method is desirable, as for systems with low eccentricity planetary orbits these are fast and long term accurate. The development of such a method for hierarchical triples is described in the next section. The results of an application of this integrator to the general planetary stability of a particular orbital type are then described in Sect. 3 and conclusions given in Sect. 4.

2 Method

Symplectic integration was first introduced for planetary systems by [14]. This simple yet elegant method has turned out to be extremely powerful in the field of planetary dynamics. In a symplectic integration scheme the energy error of the system remains bounded, and this gives them a long term accuracy and also permits large time-steps to be taken, making them a fast method as well. Simply, the scheme relies on choosing a new coordinate system in which the system's Hamiltonian is separable into one dominant term and other smaller perturbation terms, each of which is by itself analytically integrable.

The system can then be evolved forwards under each of these in a standard leapfrog pattern (see for example [10]). For planets around a single star, Jacobi coordinates were used by [14]. Here each object is referenced to the centre of mass of all others interior to their orbit and the Hamiltonian can be split into a term for the Keplerian orbits and a small interaction term dependent on position only, both of which have analytical solutions.

For planets in a hierarchical multiple stellar system there are currently two symplectic methods available. The first is that of [3], for planets in binary star systems. This method uses Jacobi coordinates for the two stars, but the planets coordinates are centred on their primary body only. So, circumstellar planets are relative to their star, and circumbinary planets relative to the binary's centre of mass. This allows close encounters to be implemented within the planetary systems. The second is that of [1], for any hierarchy of objects. Here hierarchical Jacobi coordinates are used, where each separate sub-hierarchy is in separate Jacobi coordinates. So if we had a hierarchical triple orbiting another hierarchical triple, the coordinates for the stars in each triple would be relative to the centre of mass of all objects interior to their orbit, and then the centre of mass of one triple would be referenced to the centre of mass of the other. This allows for general hierarchies but is not specific to planetary systems and does not permit close encounters to be implemented.

To derive a symplectic integration scheme specifically for planets in hierarchical triples these ideas can be extended. To do this hierarchical Jacobi coordinates are used for the stars, but the planet's coordinates are still relative only to the centre of their orbit, as shown in Fig. 3. Here the stars in the inner binary have been labelled as A and B and the more distant companion as C. Circumstellar planets are relative to their star (coordinates $\mathbf{X}_{A_i}$,$\mathbf{X}_{B_i}$ and $\mathbf{X}_{C_i}$ in the diagram), circumbinary planets relative to the barycentre of the inner binary ($\mathbf{X}_{SP_i}$) and circumtriple planets relative to the centre of mass of all three stars and the other planetary systems ($\mathbf{X}_{P_i}$). The centre of mass of star

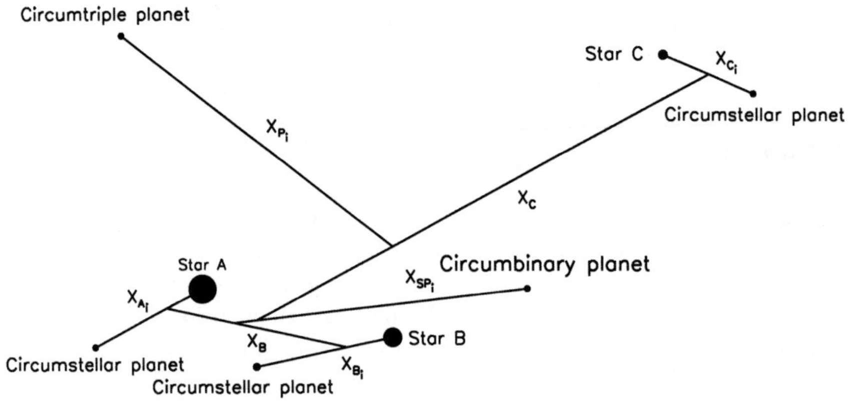

Fig. 3. The coordinate system used for the symplectic integration scheme, as described in the text

B and its planets is then taken relative to the centre of mass of star A and its planets ($\mathbf{X}_B$) and the centre of mass of star C and its planets relative to the inner binary and all associated planets ($\mathbf{X}_C$). These coordinates are advantageous as they are specific to the system structure to be studied and allow the integration scheme to be easily extended to include close encounters between planets, variable time-steps for each stellar orbit and additional forces such as general relativity or tidal effects. The system's Hamiltonian now splits into the dominant Keplerian terms and two interaction Hamiltonians dependent solely on position or momentum (see [12]) and the system can be integrated using the leapfrog method. This has all been implemented in a C++ code, MOIRAI, separately for each individual planetary orbital type and thoroughly tested on a range of cases to ensure its results are accurate and reliable.

3 Application and Results

As a first application of the method the general stability of circumbinary planets in hierarchical triples was investigated. To do this grids of several thousand test particles were set up and followed for 1 Myr in a variety of stellar systems and their survival to the end of the simulation used as an indicator of stability. The particles started on circular orbits spaced by 0.1 AU in semimajor axis, 45° in longitude and were kept coplanar with either the inner or outer binary. Similar studies in binary systems have found that a critical radius for test particle stability exists and is controlled by the binary orbital parameters [7], so it was expected that the outer stellar orbit would control an outer stability edge of the test particles and the inner stellar orbit an inner edge, that is, the system would behave as two decoupled binaries. Because of this the orbits of the inner and outer binary were varied separately, within the observed limits of known hierarchical triples, taken from the Multiple Star Catalog [11], as shown in Table 1.

As expected, the test particles did evolve to form a stable ring with an outer edge primarily controlled by the outer stars orbit and an inner edge controlled by the inner binary. The size of the stable region also scaled with

Table 1. The range of stellar orbital parameters used in the simulations. The inner mass ratio is $q_{in} = m_A/m_B$ (defined to be always greater than unity) and the outer mass ratio is $q_{out} = m_C/(m_A + m_B)$. Note these are different quantities to the ratio denoted by μ

Parameter	Inner orbit	Outer orbit
a (AU)	1–5	20–100
q	1–2	0.1–3
e	0.0–0.6	0.0–0.6
i (°)	0–30	0–30

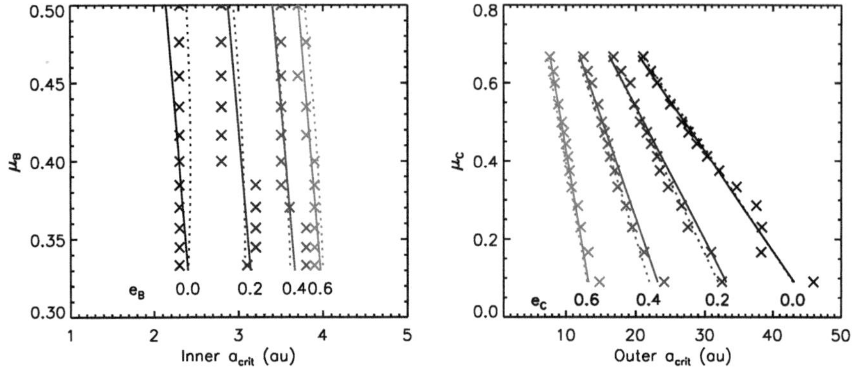

Fig. 4. Test particle inner and outer stability boundaries as a function of mass ratio and eccentricity, in the coplanar case with $a_C = 100\,\text{AU}$ and $a_B = 1\,\text{AU}$. The *left panel* shows the inner edge and the *right panel* the outer edge. *Crosses* indicate the simulation results, the *solid lines* are the fits to the data and the *dotted lines* the locations of the stability boundaries as determined from the empirical formulae of [7] when modelling the triple as two decoupled binary stars. The stability boundary is taken as the location of the first or last completely stable semimajor axis

the stellar semimajor axis, again as expected, for all but the smallest stellar orbital separations and highest mass ratios and eccentricities. The results for the completely coplanar case are shown in Fig. 4. Here the location of the inner and outer stability radii are plotted as a function of stellar eccentricity and mass ratio, defined as $\mu_B = m_B/(m_A+m_B)$ and $\mu_C = m_C/(m_A+m_B+m_C)$. A fit to the data as a polynomial in mass ratio and eccentricity is also shown and compared to the binary results, and can be seen to be in good agreement.

There is, however, a deviation from these well defined boundaries as the distance between the inner and outer stellar orbits decreases and the stability zone shrinks. This is shown in Fig. 5 for the outer edge in the case $e_C = 0.6$. Here the stability radii start to deviate from the predictions of the binary results, and islands of instability begin to appear in the middle of the stable test particle zone, more prominent as the mass ratio and eccentricity increase. The system can no longer be viewed as two decoupled binaries and the combined effect of all three stars causes additional instability, even though the stellar orbits themselves remain constant. Thus, the empirical fits to the data must be viewed as an upper limit to the stability.

There are two limitations to this study so far. Firstly, these results are only for completely coplanar systems. However, an investigation of the effects of inclination of either stellar orbit up to 30° reveals little change to the stability boundaries. Secondly, to limit the parameter space, when one stellar orbit has been varied the other has been kept as circular. Rerunning the simulations for a range of eccentricities of both stars also results in little difference in the location of the stability boundaries.

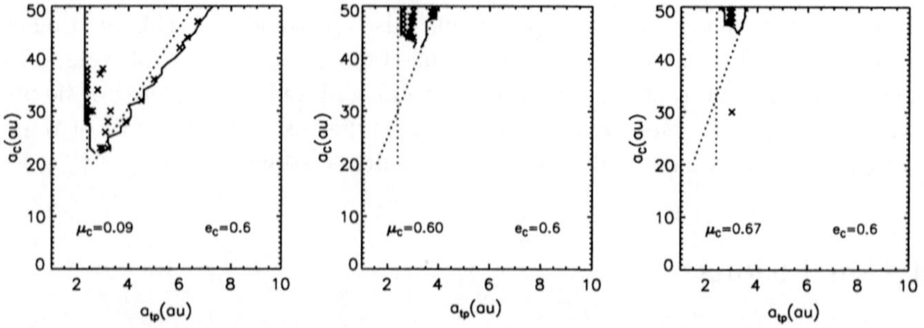

Fig. 5. Test particle stability as a function of test particle initial semimajor axis, for an eccentricity of 0.6 and increasing mass ratio and semimajor axis of the outer binary. The *solid lines* are the inner and outer stability boundaries, defined as the first and last radius where all test particles are stable for the length of the simulation, *crosses* are unstable locations within these, and *dotted lines* the fit of [7] applied to two decoupled binary stars representing the triple

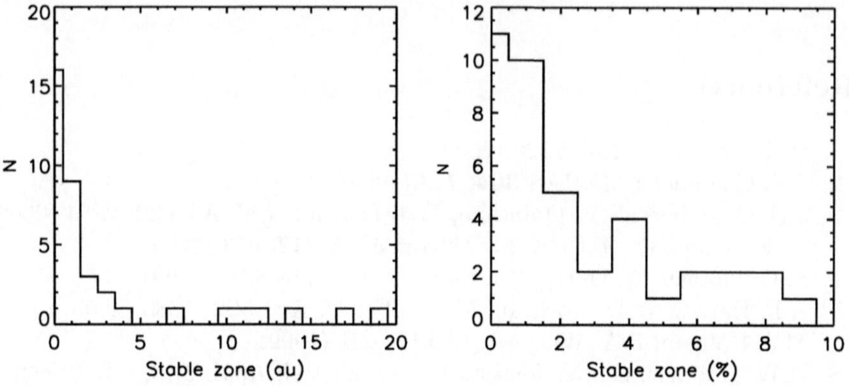

Fig. 6. The calculated widths of the circumbinary stability zones for known triple systems, in AU and as a percentage of the area between the inner and outer binary orbits

The fits to the stability radii as a function of mass ratio and eccentricity are therefore applicable to most hierarchical triples. As such, they can be applied to the known hierarchical triples in the Multiple Star Catalog [11] to characterise the size of their circumbinary stability zones, as shown in Fig. 6. There are 54 systems with well determined orbits, and of these 13 are completely unstable to circumbinary planets, that is, the inner stability radius is larger than the outer stability radius, and 16 have stable zones less than an AU wide, which implies little stability in these cases too. It is therefore likely that circumbinary planets in known triples are fairly uncommon. Whether this is an intrinsic feature of triples or an observational bias in the catalog is undeterminable.

There are many other applications also possible for this symplectic method. The inclusion of variable time-steps for different objects, close encounters and interacting planetary systems will add a great deal of flexibility. Future applications then include such studies as stellar fly-bys of binary star-planetary systems or debris disks in multiple stellar systems.

4 Conclusions

A symplectic method has been derived to study planets in hierarchical triple star systems. Using this method a general study of stability zones has revealed that the circumbinary stability is in most cases well approximated by representing the system as two coplanar decoupled binaries. Deviations from this are seen only when the stars are very close, massive and eccentric. Using these results it is possible to determine that most known triples are unlikely to have such planets.

References

1. H. Beust: A&A **400**, 1129 (2003)
2. J.E. Chambers: MNRAS **304**, 793 (1999)
3. J.E. Chambers, E.V. Quintana, M.J. Duncan et al: AJ **123**, 2884 (2002)
4. A. Eggenberger, S. Udry, M. Mayor: A&A **417**, 353 (2004)
5. S. Guilloteau, A. Dutrey, M. Simon: A&A **348**, 570 (1999)
6. A.P. Hatzes, W.D. Cochran, M. Endl et al: ApJ **599**, 1383 (2003)
7. M.J. Holman, P.A. Wiegert: AJ **117**, 621 (1999)
8. D.W. Koerner, E.L.N. Jensen, K.L. Cruz et al: ApJL **533**, L37 (2000)
9. L.R. Mudryk, Y. Wu: ApJ **639**, 423 (2006)
10. P. Saha, S. Tremaine, AJ **104**, 1633 (1992)
11. A.A. Tokovinin: A&AS **124**, 75 (1997)
12. P.E. Verrier, N.W. Evans: MNRAS **382**, 1432 (2007)
13. P.E. Verrier, N.W. Evans: MNRAS **368**, 1599 (2006)
14. J. Wisdom, M. Holman: AJ **102**, 1528 (1991)

Stabilized Chaos in the Sitnikov Problem

A.R. Dzhanoev[1], A. Loskutov[2], J.E. Howard[3], and M.A.F. Sánju[1]

[1] Departamento de Fisica, Universidad Rey Juan Carlos, 28933 Mostoles, Madrid, Spain
arsen.dzhanoev@urjc.es
[2] Moscow State University, 119992 Moscow, Russia
[3] Laboratory for Atmospheric and Space Physics and Center for Integrated Plasma Studies, University of Colorado, Boulder, CO 80309, USA

1 Formulation of the Problem

The Sitnikov problem consists of two equal masses M (called primaries) moving in circular or elliptic orbits about their common center of mass and a third, test mass μ moving along the straight line passing through the center of mass normal to the orbital plane of the primaries. This problem has attracted the attention of many other authors (see for instance [1–9]). The equation of motion can be written in scaled coordinates and time as

$$\ddot{z} + \frac{z}{[\rho(t)^2 + z^2]^{3/2}} = 0, \tag{1}$$

where z denotes the position of the particle μ along the z-axis and $\rho(t) = 1 + e\cos(t) + O(e^2)$ is the distance of one primary body from the center of mass. Here we see that the system (1) depends only on the eccentricity, e, which we shall assume to be small. The linear approach to this system with assumption that (1) possesses moderate eccentricity and small amplitudes was carried out in [7]. We first consider the circular Sitnikov problem i.e. when $e = 0$, for which $H = \frac{1}{2}p^2 - \frac{1}{\sqrt{1+z^2}}, p = \dot{z}$. The level curves $H = h$, where $h \in [-2, +\infty)$, partition the phase space (p, z) into qualitatively different types of orbits. We are interested in solutions that correspond to the level curves $H = 0$, namely two parabolic orbits that separate elliptic and hyperbolic orbits and can be considered as a separatrix between these two classes of behavior. To make clear how this problem is related to homoclinic orbits, let us employ the non-canonical transformation [9] $z = \tan u, p = \dot{z}, u \in \left[-\frac{\pi}{2}, \frac{\pi}{2}\right], v \in R$. Then the Hamiltonian for (1) in the new variables (u,p) has the form $H(u, p) = \frac{1}{2}p^2 - \frac{1}{\rho(t)^2 + \tan^2 u} = H_0(u, p) + eH_1(u, p, t, e)$, where $H_0(u, p) = \frac{1}{2}p^2 - \cos u$. One can see that when $e = 0$ the form of the Hamiltonian that obtained after

G. Contopoulos, P.A. Patsis, *Chaos in Astronomy.* Astrophysics and Space Science Proceedings, doi: 10.1007/978-3-540-75826-6,

non-canonical transformation exhibits the pendulum character of motion. In this work we consider only small values of e. Hence due to the KAM-theorem, since our system has 3/2 degrees of freedom the invariant tori bound the phase space and chaotic motion is finite and takes place in a small vicinity of a separatrix layer. Our analysis is directed to the stabilization of this chaotic behavior in the elliptic Sitnikov problem.

2 Stabilization of Chaotic Behavior in the Extended Sitnikov Problem

The idea that chaos may be suppressed goes back to the publications. [10–14] We consider the problem of stabilization of chaotic behavior in systems with separatrix contours that can be described by (2.1) $\dot{\mathbf{x}} = \mathbf{f}_0(\mathbf{x}) + \varepsilon\mathbf{f}_1(\mathbf{x}, t)$, where $\mathbf{f}_0(\mathbf{x}) = (f_{01}(\mathbf{x}), f_{02}(\mathbf{x}))$, $\mathbf{f}_1(\mathbf{x}, t) = (f_{11}(\mathbf{x}, t), f_{21}(\mathbf{x}, t))$. For this equation the Melnikov distance, which "measures" (in the first order of ε) the distance between stable and unstable manifolds (Fig. 1), $D(t_0)$ is given by $D(t_0) = -\int\limits_{-\infty}^{\infty} \mathbf{f}_0 \wedge \mathbf{f}_1 dt \equiv I[g(t_0)]$. We assume that $D(t_0)$ changes its sign. To suppress chaos we should get *a function of stabilization* $\mathbf{f}^*(\omega, t)$ that leads to a situation when separatrices are not intersected: (2.2) $\dot{\mathbf{x}} = \mathbf{f}_0(\mathbf{x}) + \varepsilon\left[\mathbf{f}_1(\mathbf{x}, t) + \mathbf{f}^*(\omega, t)\right]$, where $\mathbf{f}^*(\omega, t) = (f_1^*(\omega, t), f_2^*(\omega, t))$. Suppose $D(t_0) \in [s_1, s_2]$ and $s_1 < 0 < s_2$. After the stabilizing perturbation $\mathbf{f}^*(\omega, t)$ is applied we have two cases: $D^*(t_0) > s_2$ or $D^*(t_0) < s_1$, where $D^*(t_0)$ – Melnikov distance for system (2.2). We consider the first case (analysis for the second one is similar). Then $I[g(t_0)] + I[g^*(\omega, t_0)] > s_2$, where $I[g^*(\omega, t_0)] = -\int\limits_{-\infty}^{+\infty} \mathbf{f}_0 \wedge \mathbf{f}^* dt$. This expression is true for all left hand

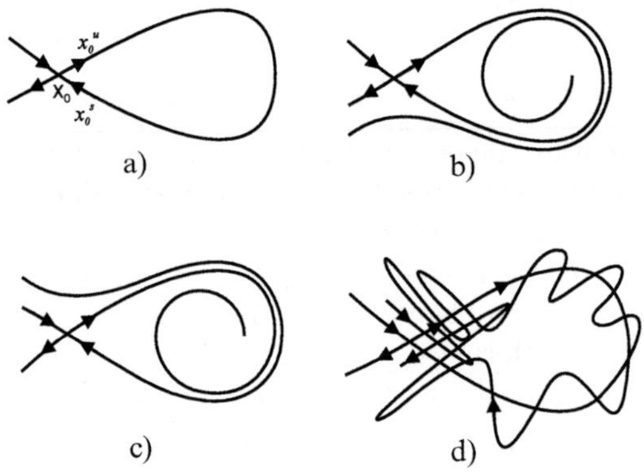

Fig. 1. Poincaré section t = const (mod T) of the system (2.1) for $\varepsilon = 0$ (**a**) and $\varepsilon \neq 0$ (**b–d**). Only in case of (**d**) we have homoclinical chaos

side values of inequality that is greater than s_2. It is derived that $I[g(t_0)] + I[g^*(\omega, t_0)] = s_2 + \chi = const$, where $\chi, s_2 \in \mathbb{R}^+$. Therefore $I[g^*(\omega, t_0)] = \text{const} - I[g(t_0)]$. On the other hand, $I[g^*(\omega, t_0))] = -\int_{-\infty}^{\infty} \mathbf{f}_0 \wedge \mathbf{f}^* dt$. We choose $\mathbf{f}^*(\omega, t)$ from the class of functions that are absolutely integrable on an infinite interval such that they can be represented in Fourier integral form. Then $\mathbf{f}^*(\omega, t) = \text{Re}\{\hat{A}(t)e^{-i\omega t}\}$. Here we suppose that $\hat{A}(t) = (A(t), A(t))$ i.e. the regularizing perturbations applied to both components of (2.2) are identical. Therefore $-\int_{-\infty}^{\infty} \mathbf{f}_0 \wedge \left\{\hat{A}(t)e^{-i\omega t}\right\} dt = \text{const} - I[g(t_0)]$. The inverse Fourier transform yields: $\mathbf{f}_0 \wedge \hat{A}(t) = \int_{-\infty}^{\infty} (I[g(t_0)] - \text{const})\, e^{i\omega t} d\omega$. Hence, $A(t) = \frac{1}{f_{01}(x) - f_{02}(x)} \int_{-\infty}^{\infty} (I[g(t_0)] - \text{const})\, e^{i\omega t} d\omega$. Here $A(t)$ can be interpreted as the amplitude of the "stabilizing" perturbation. Thus, for system (2.1) the external stabilizing perturbation has the form: $f^*(\omega, t) = \text{Re}\left[\frac{e^{-i\omega t}}{f_{01}(x) - f_{02}(x)} \int_{-\infty}^{\infty} (I[g(t_0)] - \text{const})\, e^{i\omega t} d\omega\right]$. In conservative case const=0. From the physical point of view the dynamics of the chaotic system are stabilized by a series of "kicks". The orbit that was chaotic and became regular under influence of the external perturbation we call the *stabilized orbit*.

Let us now consider two bodies of mass m that are placed in the vicinity of the stable triangular Lagrange points of the Sitnikov problem (as shown in Fig. 2). Here we treat only the hierarchical case : $\mu \ll m \ll M$. In the new

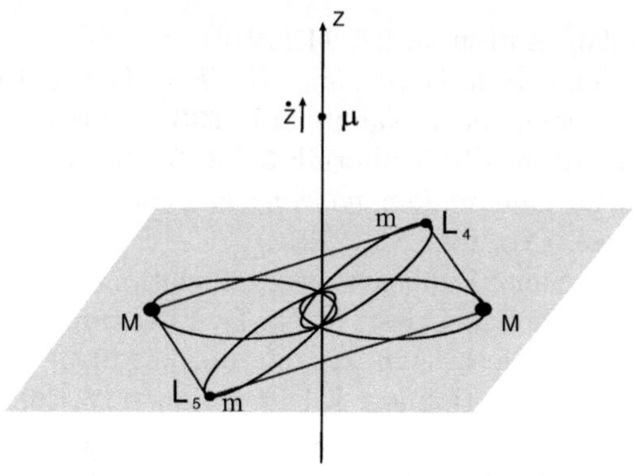

Fig. 2. Geometry of the Extended Sitnikov problem

configuration we can achieve the situation when the influence of bodies placed close to the triangular Lagrange points to the particle μ can be presented as a series of periodic impulses. Taking into account the new configuration (Fig. 2) we may say that there is a connection between the extended elliptical Sitnikov problem and the motion of the chaotic nonlinear pendulum with an external impulse-like perturbation. The Hamiltonian of such system changes to

$$H(u,p) = H_0(u,p) + e\left[H_1(u,p,t,e) + \sum_{n=-\infty}^{+\infty} \delta(t - n\tau)\right], \tag{2}$$

where τ is the duration of the impulsive forces that the particle μ experiences from bodies in the vicinity of L_4 and L_5. Now taking into account the result from the first part of this section we conclude that the forces which the particle experiences from bodies in the neighborhood of L_4 and L_5 act on the chaotic behavior of μ as an external stabilizing perturbation and the system (2) represents the system with stabilized chaotic behavior that corresponds to the stabilized orbits in the extended Sitnikov problem. The extension of the analysis carried out above to the corrections of higher order in ε of the (1) and numerical verification of the obtained results could be found in [16].

In summary, on the basis of the elliptic Sitnikov problem we constructed a configuration of five bodies which we called the extended Sitnikov problem and analytically showed that in this configuration along with chaotic and regular orbits a new type of orbit (stabilized) could be realized. We thank Carles Simó and David Farrelly for valuable discussions. A. Dzhanoev acknowledges that this work is supported by the Spanish Ministry of Education and Science under the project number SB2005-0049. Financial support from project number FIS2006-08525 (MEC-Spain) is also acknowledged. This work was supported in part by the Cassini project.

References

1. W.D. MacMillan: Astron. J. **27**, 11 (1913)
2. K. Stumpff: *Himmelsmechanik Band II VEB*, (Berlin, 1965) pp 73–79
3. K.A. Sitnikov: Dokl. Akad. Nauk USSR **133**, 2, 303 (1960)
4. V.M. Alexeev: Math. USSR Sbornik **5**, 73; **6**, 505; **7**, 1 (1969)
5. J. Moser: *Stable and random motions in dynamical systems*, (Prinston University Press, Prinston, N.J., 1973)
6. L. Llibre and C. Simó: Publicaciones Matemàtiques U.A.B. **18**, 49 (1980)
7. J. Hagel and C. Lhotka: Celest. Mech. Dyn. Astron. **93**, 201 (2005)
8. R. Dvorak: Celest. Mech. Dyn. Astron. **56**, 71 (1993)
9. H. Dankowicz and Ph. Holmes: J. Differ. Equ. **116**, 468 (1995)
10. V.V. Alexeev and A. Loskutov: Sov. Phys.-Dokl. **32**, 270 (1987)
11. A. Loskutov and A.I. Shishmarev: Chaos **4**, 351 (1994)
12. E. Ott, C. Grebogi, and J.A. Yorke: Phys. Rev. Lett. **64**, 1196 (1990)

13. A. Loskutov: Discr. Continuous Dyn. Syst., Ser. B **6**, 1157 (2006)
14. T. Schwalger, A.R. Dzhanoev, and A. Loskutov: Chaos **16**, 2, 023109 (2006)
15. A.R. Dzhanoev, A. Loskutov et al: Discr. Continuous Dyn. Syst., Ser. B **7**, 275 (2007)
16. A.R. Dzhanoev, A. Loskutov et al: submitted to Phys. Rev. E, (2008)

Part III

Fundamental Concepts and Methods

Nonextensive Statistical Mechanics – An Approach to Complexity

C. Tsallis

Centro Brasileiro de Pesquisas Físicas, Rua Xavier Sigaud 150 2 2290-180,
Rio de Janeiro, Brazil
tsallis@santafe.edu

Summary. Central concepts within extensive statistical mechanics are briefly reviewed, namely the extensivity of the nonadditive entropy S_q, the finite entropy production per unit time for weakly chaotic systems, and its connections with q-generalized central limit theorems. Bibliography for current applications in astrophysical systems is provided as well.

1 Introduction

Boltzmann-Gibbs (BG) statistical mechanics is essentially based on a specific connection between the Clausius thermodynamic entropy S and the set of probabilities $\{p_i\}$ $(i = 1, 2, \dots, W)$ of the microscopic configurations of the system. This connection is provided by the Boltzmann-Gibbs (or Boltzmann-Gibbs-von Neumann-Shannon) entropy S_{BG} defined (for the discrete case) through

$$S_{BG} = -k \sum_{i=1}^{W} p_i \ln p_i \,, \tag{1}$$

where k is a positive constant, the most usual choices being either the Boltzmann constant k_B, or just unity ($k = 1$).

This entropic functional and its associated statistical mechanics cover an impressive amount of interesting systems, and have exhibited brilliant successes along 130 years. However, very many natural, artificial and even social complex systems escape from the domain of applicability of the BG theory. As an attempt to improve the situation, a generalization of (1) was proposed in 1988 [1]. See Table 1, where the *q-logarithm* is defined as

$$\ln_q x \equiv \frac{x^{1-q} - 1}{1-q} \quad (x > 0;\, q \in \mathcal{R};\, \ln_1 x = \ln x)\,, \tag{2}$$

the inverse function, the *q-exponential*, being

G. Contopoulos, P.A. Patsis, *Chaos in Astronomy.* Astrophysics and Space Science Proceedings, doi: 10.1007/978-3-540-75826-6,

Table 1. The nonadditive entropy S_q $(q \neq 1)$, and its particular, additive, case S_{BG}

ENTROPY	Equal probabilities	Arbitrary probabilities
	$(p_i = 1/W\,, \forall i)$	$(\sum_{i=1}^{W} p_i = 1)$
$q = 1$	$S_{BG} = k \ln W$	$S_{BG} = -k \sum_{i=1}^{W} p_i \ln p_i$
		$= k \sum_{i=1}^{W} p_i \ln \frac{1}{p_i}$
$\forall q$	$S_q = k \ln_q W$	$S_q = k \frac{1-\sum_{i=1}^{W} p_i^q}{q-1}$
		$= k \sum_{i=1}^{W} p_i \ln_q \frac{1}{p_i}$

$$e_q^x \equiv [1 + (1-q)\,x]_+^{\frac{1}{1-q}} \quad (q \in \mathcal{R};\; e_1^x = e^x)\,, \tag{3}$$

with $[x]_+ = x$ if $x \geq 0$, and zero otherwise.

If we consider a system composed by two probabilistically *independent* subsystems A and B, i.e., $p_{ij}^{A+B} = p_i^A p_j^B \;\forall (ij)$, we straightforwardly verify that

$$\frac{S_q(A+B)}{k} = \frac{S_q(A)}{k} + \frac{S_q(B)}{k} + (1-q)\,\frac{S_q(A)}{k}\,\frac{S_q(B)}{k}\,. \tag{4}$$

Therefore, S_{BG} is *additive* [2], and S_q $(q \neq 1)$ is *nonadditive.* The entropy S_q leads to a natural generalization of BG statistical mechanics, currently referred to as *nonextensive statistical mechanics* [1, 3, 4]. The nonextensive theory focuses on nonequilibrium stationary (or quasi-stationary) states in the same way the BG theory focuses on thermal equilibrium states. Recent reviews of the theory and its various applications can be found in [5, 6], and a regularly updated bibliography is available at [7].

2 Nonextensivity of the Nonadditive Entropy S_q

From its very definition, additivity depends *only* on the functional form of the entropy. Extensivity *also* depends on the system. Indeed, the entropy $S(N)$ of a given system composed by N (independent or correlated) elements is said *extensive* if $\lim_{N\to\infty} S(N)/N$ is finite, i.e., if $S(N)$ scales like N for large N, hence matching the classic thermodynamic behavior. Consequently, for systems whose elements are independent or weakly independent, S_{BG} is extensive, whereas S_q is nonextensive for $q \neq 1$. There are, in contrast, systems whose elements are strongly correlated, having as a consequence that S_q is extensive only for a special value of q, noted q_{ent}, with $q_{ent} \neq 1$. In other words, S_q is nonextensive for any value of $q \neq q_{ent}$. In particular, for such anomalous systems, S_{BG} is nonextensive, hence inadequate for thermodynamical purposes at the limit $N \to \infty$ (*thermodynamic limit*). There are finally systems that are even more complex, and for which there is no value of q such that $S_q(N)$ is extensive. Such strongly anomalous systems remain outside the realm of nonextensive statistical mechanical concepts, and are not addressed here.

Abstract probabilistic models with either $q_{ent} = 1$ or $q_{ent} \neq 1$ are exhibited in [8]. Physical realizations are exhibited in [9] for strongly entangled systems. For example, the block entropy of magnetic chains in the presence of a transverse magnetic field at its critical value and at $T = 0$, belonging to the universality class characterized by the *central charge* c [10], yield [9]

$$q_{ent} = \frac{\sqrt{9+c^2}-3}{c} \quad (c > 0)\,. \tag{5}$$

Therefore, for the one-dimensional Ising model with short-range interactions (hence $c = 1/2$), we have $q_{ent} = \sqrt{37} - 6 \simeq 0.08$, and for the one-dimensional isotropic XY model with short-range interactions (hence $c = 1$), we have $q_{ent} = \sqrt{10} - 3 \simeq 0.16$. At the $c \to \infty$ limit, we recover the BG value $q_{ent} = 1$ (see [11] for $c = 26$).

3 Entropy Production per Unit Time

Let us illustrate on one-dimensional unimodal nonlinear dynamical systems $x_{t+1} = f(x_t)$ some interesting concepts, namely that of *sensitivity to the initial conditions* and that of *entropy production per unit time.* A system is said *strongly chaotic* (or just *chaotic*) if its Lyapunov exponent λ_1 is positive, and *weakly chaotic* if it vanishes.[1] The *sensitivity to the initial conditions* ξ is defined as

$$\xi(t) \equiv \lim_{\Delta x(0)\to 0} \frac{\Delta x(t)}{\Delta x(0)}\,. \tag{6}$$

If $\lambda_1 > 0$, we typically have

$$\xi(t) = e^{\lambda_1 t}\,. \tag{7}$$

The *entropy production per unit time* K_1 in such a system is defined as

$$K_1 \equiv \lim_{t\to\infty} \lim_{W\to\infty} \lim_{M\to\infty} \frac{S_{BG}(t)}{t}\,, \tag{8}$$

[1] In order to avoid notation ambiguities, let us warn that, in somewhat different contexts, the expression *weakly chaotic* is used to refer to systems whose maximal Lyapunov exponent is positive but relatively small in value. The case of vanishing maximal Lyapunov exponent is then refereed to as *ordered.*

where M is the number of initial conditions that have been randomly chosen within one of the W intervals in which the allowed phase space of the system has been partitioned. K_1 typically is positive and satisfies a Pesin-like equality, i.e.,

$$K_1 = \lambda_1 \,. \tag{9}$$

For a large class of nonlinear functions $f_a(x)$, where a is a control parameter, the system has one or more values of a for which it is at the *edge of chaos*, with $\lambda_1 = 0$. As anticipated, such systems are called weakly chaotic. $\xi(t)$ may present a complex behavior which typically admits an upper bound. This upper bound has been shown, in many cases, to be given by [12–14]

$$\xi(t) = e_{q_{sen}}^{\lambda_{q_{sen}}\, t} \,, \tag{10}$$

where *sen* stands for *sensitivity*, $q_{sen} < 1$, and $\lambda_{q_{sen}} > 0$. Such systems satisfy [12, 15], again as an upper bound,

$$K_{q_{sen}} \equiv \lim_{t\to\infty} \lim_{W\to\infty} \lim_{M\to\infty} \frac{S_{q_{sen}}(t)}{t} \,. \tag{11}$$

Although no proof is available, it has been conjectured [16] that

$$q_{sen} = q_{ent} \,. \tag{12}$$

4 q-Generalization of the Standard and the Lévy-Gnedenko Central Limit Theorems

The Central Limit Theorem (CLT) constitutes a mathematical building block of BG statistical mechanics. For example, under simple and appropriate constraints, the optimization of S_{BG} yields a Gaussian (e.g., the Maxwellian distribution of velocities), which precisely is the attractor within the standard CLT. A slight extension yields Lévy distributions as well, whose second moment diverges. But all these distributions are obtained within the basic hypothesis of independence or quasi-independence, in the sense that the correlations do *not* resist up to large values of N. The Fokker-Planck counterpart of these facts can be focused on as follows. Consider the equation

$$\frac{\partial p(x,t)}{\partial t} = D\frac{\partial^\alpha p(x,t)}{\partial |x|^\alpha} \quad (D > 0;\, 0 < \alpha \le 2) \,. \tag{13}$$

Assume also that the initial distribution is a Dirac delta, i.e., $p(x,0) = \delta(x)$. The exact solution, $\forall(x,t)$, is a Gaussian if $\alpha = 2$ (x scaling like $t^{1/2}$), and a Lévy distribution if $0 < \alpha < 2$ (x scaling like $t^{1/\alpha}$). The equation being *linear*, these two results precisely correspond to the standard and the Lévy-Gnedenko

CLT's (finite and diverging second moment, respectively, of the distribution that is being summed up to N random variables).

Let us now generalize (13) into the following *nonlinear* Fokker-Planck equation:

$$\frac{\partial p(x,t)}{\partial t} = D_q \frac{\partial^\alpha [p(x,t)]^{2-q}}{\partial |x|^\alpha} \quad (D_q \in \mathcal{R};\, 0 < \alpha \le 2;\, q < 3)\,. \tag{14}$$

Assume once again that the initial distribution is a Dirac delta. The exact solutions are now [17, 18] *q-Gaussians* if $\alpha = 2$ (x scaling like $t^{1/(3-q)}$), the q-Gaussians being the distributions which optimize S_q under appropriate constraints. A more complex type of distributions, named *(q,α)-stable distributions*, emerge if $0 < \alpha < 2$ (the scaling between x and t going through a crossover between two different regimes). The corresponding CLT's are the natural q-generalizations of the usual theorems, the main new ingredient in the hypothesis being now the fact that the N random variables are strongly correlated, in a manner from now referred to as *q-independence*. See [19–25] for details. A synopsis is presented in Fig. 1.

The existence of these theorems might explain the ubiquity of q-Gaussians in natural, artificial and social systems. See, for instance [26–32].

$N^{1/[\alpha(2-q)]}$-scaled attractor $\mathbb{F}(x)$ when summing $N \to \infty$ q-independent identical random variables with symmetric distribution $f(x)$ with $\sigma_Q = \int dx\, x^2 [f(x)]^Q / \int dx\, [f(x)]^Q$ $\left(Q = 2q-1,\ q_1 = \frac{1+q}{3-q}\right)$

	$q=1$ [*independent*]	$q \ne 1$ (*i.e.*, $Q = 2q-1 \ne 1$) [*globally correlated*]
$\sigma_Q < \infty$ ($\alpha = 2$)	$\mathbb{F}(x) =$ *Gaussian* $G(x)$, *with same* σ_1 *of* $f(x)$ Classic CLT	$\mathbb{F}(x) = G_q(x) = G_{(3q_1-1)/(1+q_1)}(x)$, *with same* σ_Q *of* $f(x)$ $G_q(x) \sim \begin{cases} G(x) & \text{if } \lvert x\rvert \ll x_c(q,2) \\ f(x) \sim C_q / \lvert x\rvert^{2/(q-1)} & \text{if } \lvert x\rvert \gg x_c(q,2) \end{cases}$ *with* $\lim_{q\to 1} x_c(q,2) = \infty$ S. Umarov, C. T. and S. Steinberg (2006) [cond-mat/0603593]
$\sigma_Q \to \infty$ ($0 < \alpha < 2$)	$\mathbb{F}(x) =$ *Levy distribution* $L_\alpha(x)$, *with same* $\lvert x\rvert \to \infty$ *behavior* $L_\alpha(x) \sim \begin{cases} G(x) & \text{if } \lvert x\rvert \ll x_c(1,\alpha) \\ f(x) \sim C_\alpha / \lvert x\rvert^{1+\alpha} & \text{if } \lvert x\rvert \gg x_c(1,\alpha) \end{cases}$ *with* $\lim_{\alpha\to 2} x_c(1,\alpha) = \infty$ Levy-Gnedenko CLT	$\mathbb{F}(x) = L_{q,\alpha}$, *with same* $\lvert x\rvert \to \infty$ *asymptotic behavior* $L_{q,\alpha} \sim \begin{cases} G_{\frac{2(1-q)-\alpha(1+q)}{2(1-q)-\alpha(3-q)},\alpha}(x) \sim C^*_{q,\alpha} / \lvert x\rvert^{\frac{2(1-q)-\alpha(3-q)}{2(1-q)}} \\ \text{(intermediate regime)} \\ G_{\frac{2\alpha q-\alpha+3}{\alpha+1},2}(x) \sim C^L_{q,\alpha} / \lvert x\rvert^{(1+\alpha)/(1+\alpha q-\alpha)} \\ \text{(distant regime)} \end{cases}$ S. Umarov, C. T., M. Gell-Mann and S. Steinberg (2006) [cond-mat/0606038] and [cond-mat/0606040]

Fig. 1. Standard and Lévy-Gnedenko central limit theorems and their q-generalizations. See details in [19–25]

5 Current Applications of Nonextensive Statistical Mechanics to Astrophysical Systems

Finally, let us briefly mention here some among the many current applications of the present theory to astrophysical systems. Consistent evidence of nonextensivity has been found in the solar wind in the heliosphere and the heliosheath [33–41]. The non-Gaussian distribution of fluctuations of the temperatures of the universe in the cosmic background radiation appears to be q-Gaussian [42]. The largest part of the celebrated solar neutrinos defect can be explained through neutrino oscillations. A remaining part could be explained through non-Boltzmannian statistics [43–46]. The flux of cosmic rays verifies [47], over very many decades, a distribution of the type emerging within the present theory. Various cosmological applications are already available in the literature [48–65]. A slightly non-Planckian black-body distribution has been conjectured some years ago [66]. The velocities of systems such as the Pleiads have been interpreted in nonextensive terms [67–69]. And a great variety of approaches to self-gravitational systems is available as well [70–95]. For further applications see [7].

Acknowledgments

I am very grateful for warm and delightful hospitality at the Academy of Athens by G. Contopoulos and P. Patsis. Partial financial support by CNPq and Faperj (Brazilian agencies) is acknowledged as well.

References

1. C. Tsallis, J. Stat. Phys. **52**, 479 (1988).
2. O. Penrose, *Foundations of Statistical Mechanics: A Deductive Treatment* (Pergamon, Oxford, 1970), page 167.
3. E.M.F. Curado and C. Tsallis, J. Phys. A **24**, L69 (1991); Corrigenda: **24**, 3187 (1991) and **25**, 1019 (1992).
4. C. Tsallis, R.S. Mendes and A.R. Plastino, Physica A **261**, 534 (1998).
5. M. Gell-Mann and C. Tsallis, eds., *Nonextensive Entropy – Interdisciplinary Applications* (Oxford University Press, New York, 2004).
6. J.P. Boon and C. Tsallis, eds., *Nonextensive Statistical Mechanics: New Trends, New Perspectives*, Europhysics News **36** (6) (2005).
7. http://tsallis.cat.cbpf.br/biblio.htm this is a long list which contains, at the date of 9 July 2008, 2485 items (co)authored by 1897 scientists from 62 countries.
8. C. Tsallis, M. Gell-Mann and Y. Sato, Proc. Nat. Acad. Sci. (USA) **102**, 15377 (2005).

9. F. Caruso and C. Tsallis, in *Complexity, Metastability and Nonextensivity,* eds. S. Abe, H.J. Herrmann, P. Quarati, A. Rapisarda and C. Tsallis, American Institute of Physics Conference Proceedings **965**, 51 (New York, 2007); F. Caruso and C. Tsallis, Phys. Rev. E **78** (2008), in press.
10. P. Calabrese and J. Cardy, J. Stat. Mech.: Theor. Exp. P06002 (2004).
11. P. Ginsparg and G. Moore, *Lectures on 2D String Theory* (Cambridge Univ. Press, Cambridge, 1993).
12. C. Tsallis, A.R. Plastino and W.-M. Zheng, Chaos, Solitons and Fractals **8**, 885 (1997).
13. F. Baldovin and A. Robledo, Europhys. Lett. **60**, 518 (2002).
14. F. Baldovin and A. Robledo, Phys. Rev. E **66**, R045104 (2002).
15. F. Baldovin and A. Robledo, Phys. Rev. E **69**, 045202(R) (2004).
16. C. Tsallis, M. Gell-Mann and Y. Sato, in *Nonextensive Statistical Mechanics: New Trends, New perspectives*, eds. J.P. Boon and C. Tsallis, Europhysics News **36**, 186 (European Physical Society, 2005).
17. A.R. Plastino and A. Plastino, Physica A **222**, 347 (1995).
18. C. Tsallis and D.J. Bukman, Phys. Rev. E **54**, R2197 (1996).
19. S. Umarov, C. Tsallis and S. Steinberg, Milan J. Math. 76 (2008) [DOI 10.1007/s00032-008-0087-y].
20. S. Umarov, C. Tsallis, M. Gell-Mann and S. Steinberg, cond-mat/0606038 (2006).
21. S. Umarov, C. Tsallis, M. Gell-Mann and S. Steinberg, cond-mat/0606040 (2006).
22. S. Umarov and C. Tsallis, in *Complexity, Metastability and Nonextensivity,* eds. S. Abe, H.J. Herrmann, P. Quarati, A. Rapisarda and C. Tsallis, American Institute of Physics Conference Proceedings **965**, 34 (New York, 2007).
23. C. Tsallis and S.M.D. Queiros, in *Complexity, Metastability and Nonextensivity,* eds. S. Abe, H.J. Herrmann, P. Quarati, A. Rapisarda and C. Tsallis, American Institute of Physics Conference Proceedings **965**, 8 (New York, 2007).
24. S.M.D. Queiros and C. Tsallis, in *Complexity, Metastability and Nonextensivity,* eds. S. Abe, H.J. Herrmann, P. Quarati, A. Rapisarda and C. Tsallis, American Institute of Physics Conference Proceedings **965**, 21 (New York, 2007).
25. C. Vignat and A. Plastino, J. Phys. A **40**, F969 (2007).
26. A. Upadhyaya, J.-P. Rieu, J.A. Glazier and Y. Sawada, Physica A **293**, 549 (2001).
27. K.E. Daniels, C. Beck and E. Bodenschatz, Physica D **193**, 208 (2004).
28. P. Douglas, S. Bergamini and F. Renzoni, Phys. Rev. Lett. **96**, 110601 (2006).
29. U. Tirnakli, C. Beck and C. Tsallis, Phys. Rev. E **75**, 040106(R) (2007).
30. A. Pluchino, A. Rapisarda and C. Tsallis, Europhys. Lett. **80**, 26002 (2007).
31. R. Arevalo, A. Garcimartin and D. Maza, Eur. Phys. J. E **23**, 191 (2007).

32. R. Arevalo, A. Garcimartin and D. Maza, in *Complex Systems – New Trends and Expectations*, eds. H.S. Wio, M.A. Rodriguez and L. Pesquera, Eur. Phys. J.-Special Topics **143** (2007).
33. L.F. Burlaga and A.F. Vinas, Geophys. Res. Lett. **31**, L16807 (2004).
34. L.F. Burlaga and A.F. Vinas, J. Geophys. Res. – Space Phys. **109**, A12107 (2004).
35. A.F. Vinas, R.L. Mace and R.F. Benson, J. Geophys. Research **110**, A06202 (2005).
36. L.F. Burlaga and A.F. Vinas, Physica A **356**, 375 (2005).
37. L.F. Burlaga and A.F. Vinas, J. Geophys. Res. – Space Phys. **110**, A07110 (2005).
38. L.F. Burlaga, N.F. Ness and M.H. Acuna, J. Geophysical Res.-Space Physics **111**, A09112 (2006).
39. L.F. Burlaga, A.F. Vinas and C. Wang, J. Geophysical Res.-Space Physics **112**, A07206 (2007).
40. L.F. Burlaga, N.F. Ness, M.H. Acuna, Astrophys. J. **668**, 1246 (2007).
41. L.F. Burlaga and A.F. Vinas, in *Complexity, Metastability and Nonextensivity*, eds. S. Abe, H.J. Herrmann, P. Quarati, A. Rapisarda and C. Tsallis, American Institute of Physics Conference Proceedings **965**, 259 (New York, 2007).
42. A. Bernui, C. Tsallis and T. Villela, Europhys. Lett. **78**, 19001 (2007).
43. G. Kaniadakis, A. Lavagno and P. Quarati, Phys. Lett. B **369**, 308 (1996).
44. P. Quarati, A. Carbone, G. Gervino, G. Kaniadakis, A. Lavagno and E. Miraldi, Nucl. Phys. A **621**, 345c (1997).
45. G. Kaniadakis, A. Lavagno and P. Quarati, Astrophysics and space science **258**, 145 (1998).
46. G. Kaniadakis, A. Lavagno, M. Lissia and P. Quarati, in *Relativistic Aspects of Nuclear Physics*, eds. T. Kodama, C.E. Aguiar, S.B. Duarte, Y. Hama, G. Odyniec and H. Strobele (World Scientific, Singapore, 1998), p. 193.
47. C. Tsallis, J.C. Anjos and E.P. Borges, Phys. Lett. A **310**, 372 (2003).
48. V.H. Hamity and D.E. Barraco, Phys. Rev. Lett. **76**, 4664 (1996).
49. H.P. de Oliveira, S.L. Sautu, I.D. Soares and E.V. Tonini, Phys. Rev. D **60**, 121301 (1999).
50. H.P. de Oliveira, I.D. Soares and E.V. Tonini, Physica A **295**, 348 (2001).
51. M.E. Pessah, D.F. Torres and H. Vucetich, Physica A **297**, 164 (2001).
52. M.E. Pessah and D.F. Torres, Physica A **297**, 201 (2001).
53. C. Hanyu and A. Habe, Astrophys. J. **554**, 1268 (2001).
54. H.P. de Oliveira, I.D. Soares and E.V. Tonini, Phys. Rev. D **67**, 063506 (2003).
55. C. Castro, Chaos, Solitons and Fractals **15**, 797 (2003).
56. H.P. de Oliveira, I.D. Soares and E.V. Tonini, Phys. Rev. D **70**, 084012 (2004).
57. C. Beck, Physica A **340**, 459 (2004).

58. A.R. Plastino, A. Plastino, H.G. Miller and H. Uys, Proc. XIth United Nations/European Space Agency Workshop on Basic Space Science, eds. H. Haubold and M. Rabolli, Astrophysics and Space Science **290**, 275 (Kluwer, Dordecht, 2004).
59. T. Matos, D. Nunez and R.A. Sussman, Class. Quantum Grav. **21**, 5275 (2004).
60. H.P. Oliveira and I.D. Soares, Phys. Rev. D **71**, 124034 (2005).
61. R. Silva and J.A.S. Lima, Phys. Rev. E **72**, 057101 (2005).
62. M.B. Cantcheff and J.A.C. Nogales, Int. J. Mod. Phys. A **21**, 3127–3131 (2006).
63. N.W. Evans and J.H. An, Phys. Rev. D **73**, 023524 (2006).
64. D. Nunez, R.A. Sussman, J. Zavala, L.G. Cabral-Rosetti and T. Matos, American Institute of Physics Conference Proceedings, **857** A, 316 (2006).
65. J. Zavala, D. Nunez, R.A. Sussman, L.G. Cabral-Rosetti and T. Matos, J. Cosmol. Astropart. Phys. **6**, 008 (2006).
66. C. Tsallis, F.C. Sa Barreto and E.D. Loh, Phys. Rev. E **52**, 1447 (1995).
67. B.B. Soares, J.C. Carvalho, J.D. do Nascimento Jr. and J.R De Medeiros, Physica A **364**, 413 (2005).
68. J.C. Carvalho, B.B. Soares, B.L. Canto Martins, J.D. do Nascimento Jr. and J.R De Medeiros, Physica A **384**, 507 (2007).
69. A. Lavagno, G. Kaniadakis, M. Rego-Monteiro, P. Quarati and C. Tsallis, Astrophysical Letters and Communications **35**, 449 (1998).
70. A.R. Plastino and A. Plastino, Phys. Lett. A **174**, 384 (1993).
71. K.S. Fa and E.K. Lenzi, J. Math. Phys. **42**, 1148 (2001) [Erratum: **43**, 1127 (2002)].
72. A. Taruya and M. Sakagami, Physica A **307**, 185 (2002).
73. K.S. Fa, Phys. Lett. A **299**, 345 (2002).
74. P.-H. Chavanis, Astro. Astrophys. **386**, 732 (2002).
75. P.H. Chavanis, Astro. Astrophys. **401**, 15 (2003).
76. W.H. Siekman, Chaos, Solitons and Fractals **16**, 119 (2003).
77. A. Taruya and M. Sakagami, Physica A **318**, 387 (2003).
78. A. Taruya and M. Sakagami, Physica A **322**, 285 (2003).
79. A. Taruya and M. Sakagami, Phys. Rev. Lett. **90**, 181101 (2003).
80. R. Silva and J.S. Alcaniz, Physica A **341**, 208 (2004).
81. A. Taruya and M. Sakagami, Physica A **340**, 453 (2004).
82. M. Sakagami and A. Taruya, Physica A **340**, 444 (2004).
83. M. Sakagami and A. Taruya, in *Nonadditive entropy and nonextensive statistical mechanics*, ed. M. Sugiyama, Contin. Mech. Thermodyn. **16**, 279 (Springer, Heidelberg, 2004).
84. J.L. Du, Phys. Lett. A **320**, 347 (2004).
85. J.L. Du, Europhys. Lett. **67**, 893 (2004).
86. A. Taruya and M. Sakagami, Mon. Not. R. Astron. Soc. **364**, 990–1010 (2005).
87. P.H. Chavanis, J. Vatteville and F. Bouchet, Eur. Phys. J. B **46**, 61–99 (2005).

88. S.H. Hansen, D. Egli, L. Hollenstein and C. Salzmann, New Astro. **10**, 379 (2005).
89. A. Taruya and M.A. Sakagami, in *Complexity and Nonextensivity: New Trends in Statistical Mechanics*, eds. M. Sakagami, N. Suzuki and S. Abe, Prog. Theor. Phys. Suppl. **162**, 53 (2006).
90. S.H. Hansen and B. Moore, New Astro. **11**, 333 (2006).
91. P.-H. Chavanis, Int. J. Mod. Phys. B **20**, 3113–3198 (2006).
92. P.H. Chavanis, Eur. Phys. J. B **53**, 487–501 (2006).
93. J. Du, Astrophys. Space Sci. **305**, 247 (2006).
94. M. Razeira, B.E.J. Bodmann and C.A. Zen Vasconcellos, Internat. J. Mod. Phys. D **16**, 365 (2007).
95. J.L. Du, Astrophys. Space Sci. **312**, 47 (2007).

Absolute Versus Relative Motion in Mechanics

D. Lynden-Bell

The Institute of Astronomy, Madingley Road, Cambridge CB3 OHA, UK
and Clare College Cambridge, CB3 OHA, UK
dlb@ast.cam.ac.uk

Summary. A thought experiment demonstrates that the inertial frame is influenced by swirling gravitational waves, even in the flat space into which the waves do not reach. Since such waves have no energy tensor it is deduced that Mach's principle can not involve only the energy tensor as the source of inertia. Within Newtonian mechanics there are no gravitational waves and a Relative Mechanics that gives the Newtonian results when applied to any non-rotating universe is given. Such a mechanics has neither absolute space nor absolute rotation.

The gravitational force between two bodies depends on the product of their masses and inversely as the square of the distance between them. The principle of equivalence insists that gravitational and inertial forces are fundamentally the same. If this were fully true, then inertia must depend, not merely on the mass of a body, but also on the masses of all other bodies in the universe.

In a 1913 letter to Mach, Einstein [7] explained that in his new theory, an untouched body inside an upward accelerated heavy spherical shell of mass M and radius a would be weakly accelerated upwards in sympathy. The strength of the effect being smaller by a factor $\sim GM/(ac^2)$. He also noted that if the same shell was rotating, then the inertial frame inside it would, like Foucault's pendulum, rotate too with a similar reduction factor. Later, in fully fledged general relativity, Thirring [9] found the rotational effect which is now called Lense-Thirring [13] precession. These thought experiments demonstrated that, in General Relativity, inertial frames were not determined by Newton's Absolute Space but were affected by the motion and distribution of neighbouring matter. Einstein extrapolating from this said "The relativity of inertia is not fulfilled in the statement that local masses affect inertia, they must entirely cause it".

He also said "There is no inertia of mass against space, but only inertia of mass against mass".

Mach's principle, that the inertial frame at any point is determined by some average of the positions and motions of the other masses in the universe, was one of the driving influences on Einstein's development of General

G. Contopoulos, P.A. Patsis, *Chaos in Astronomy.* Astrophysics and Space Science Proceedings, doi: 10.1007/978-3-540-75826-6,

Relativity. Nevertheless General Relativity admits solutions that do not obey Mach's principle, such as Minkowski space, in which the inertia of test particles can be traced to the boundary conditions at infinity. Einstein thought it very objectionable that part of the inertial influence should come from neighbouring masses and part from a boundary condition. It was his attempt to solve this problem that led to his "abolition of infinity" by closing the universe. He hoped that, in such a closed space, the metric would be essentially determined by the stress energy tensor $T_{\mu v}$. However, in his final years he was very aware that Mach's Principle had never been shown to be fully compatible with General Relativity.

In work done this summer with Bičak and Katz [22], we have demonstrated that Mach's Principle can not be formulated in terms of the stress-energy-momentum tensor $T_{\mu v}$. Gravitational waves carry energy momentum and angular momentum but have NO $T_{\mu v}$. To get waves that carry angular-momentum, we must have wave vectors with a component around an axis. The non-linear Einstein-Rosen waves [8] are fully cylindrically symmetric with the wave vectors radial to the axis, so these carry no angular momentum. We therefore start with linearised gravitational waves which generate a space-time with one Killing vector $\partial/\partial z$ which is hypersurface orthogonal. The work of Ashtekar et al. [1] shows that such metrics can be put in the form

$$ds^2 = e^{2\psi} d\bar{s}^2 + e^{-2\psi} dz^2 \ , \tag{1}$$

where $d\bar{s}^2$ is a 2+1 metric $g_{ab}dx^a dx^b$ with $a, \ b = 0, \ 1, \ 2$

$$\text{and} \qquad \bar{R}_{ab} = -2\partial_a \psi \partial_b \psi \ , \tag{2}$$

This is the Ricci tensor of the $2+1$ metric, where $g^{ab}\partial_a \partial_b \psi = 0$.

We take cylindrical polar coordinates so the linearised version of this last equation is the cylindrical wave equation

$$-\frac{1}{c^2}\frac{\partial^2 \psi}{\partial t^2} + \frac{1}{R}\frac{\partial}{\partial R}\left(R\frac{\partial \psi}{\partial R}\right) + \frac{1}{R^2}\frac{\partial^2 \psi}{\partial \phi^2} = 0 \ , \tag{3}$$

whose solution with $c = 1$, angular frequency ω, in Bessel functions is

$$\psi = A(\omega) e^{\pm i\omega t \pm im\phi} J_m(\omega R) \ . \tag{4}$$

We take a linear superposition of such waves all rotating positively about the axis

$$\psi = \mathrm{Re}\left[2B\int_0^\infty (\omega a)^m e^{im\phi - i\omega t - a\omega} J_m(\omega R)\, a d\omega\right] \ . \tag{5}$$

This gives a pulse of gravitational waves of duration $\sim a$ which comes in from infinity swirling around the axis in the form of a leading spiral. At $t = 0$ it straightens up to cartwheel form with m spokes still rotating and thereafter it retreats as a trailing spiral. Because of the rotation, the wave amplitude near the origin remains very small at all times, being almost zero for $R < 0.4a$ even at $t = 0$. This follows because $J_m(\omega R) \sim (\omega R)^m$ for small R.

Having derived the linear wave, we now look for its gravitational effects by second order perturbation theory. We solve (2) for the departure of the $g_{0\phi}$ from its flat space value, 0, with the RHS now known from (5). We find the inertial frame at the axis rotates at the rate

$$\omega_0 = 2^{1-2m}(2m)!\,(B^2/a)\,\frac{1+m(1+T^2)}{(1+T^2)^2}\,, \tag{6}$$

where $T = t/a$. Thus the angular momentum contained in the gravitational wave (which has no $T_{\mu v}$) causes a rotation of the inertial frame just as the angular momentum of matter does within a rotating shell.

As the inertial frame is certainly affected by gravitational wave angular momentum, such effects must be accounted for in any theory of the origin of inertia [22].

As the localisation of gravitational wave energy, angular momentum and momentum continues to be a problem that vexes relativists, it is not clear how to account for their inertial influence.

Although Mach's Principle within General Relativity remains elusive, we showed, some years ago [16–18, 20] that it can readily be achieved by a revision within the context of Newtonian Mechanics of point particles. Such a mechanics is often derived from the Lagrangian

$$\mathrm{L} = \mathrm{T} - \mathrm{V}$$

where

$$\mathrm{V} = \mathrm{G}\sum\sum_{\mathrm{I<J}} m_{\mathrm{I}} m_{\mathrm{J}}/|\mathbf{r}_{\mathrm{J}} - \mathbf{r}_{\mathrm{I}}|\,,$$

and

$$\mathrm{T} = \sum_{\mathrm{I}} \tfrac{1}{2} m_{\mathrm{I}} \left(d\mathbf{r}_{\mathrm{I}}/dt\right)^2\,.$$

V is independent of the axes with respect to which $\mathbf{r}_{\mathrm{I}}$ and $\mathbf{r}_{\mathrm{J}}$ are measured. It is a true property of the system independent of the observer's axes. However, the same can not be said of T, so the two constituents of the Lagrangian have very different transformation properties. In particular, with respect to axes moving with velocity $\mathbf{u}(t)$

$$\mathrm{T}' = \sum_{\mathrm{I}} \tfrac{1}{2} m_{\mathrm{I}} \left[\frac{d\mathbf{r}_{\mathrm{I}}}{dt} - \mathbf{u}(t)\right]^2 = \mathrm{T} - \mathbf{u}(t).\,\mathbf{P} + \tfrac{1}{2} M\,[\mathbf{u}(t)]^2\,,$$

where M is the total mass.

A way of making T independent of the axes chosen is to minimise T' over all choices of $\mathbf{u}(t)$. This yields $\mathbf{u} = \mathbf{P}/M$ and with this choice we find $\mathrm{T}'_{\mathrm{min}}$ can be put in the form

$$\mathrm{T}'_{\mathrm{min}} = \frac{1}{2M}\sum\sum_{I<J} m_{\mathrm{I}} m_{\mathrm{J}} \left(\frac{d\mathbf{r}_{\mathrm{J}}}{dt} - \frac{d\mathbf{r}_{\mathrm{I}}}{dt}\right)^2\,,$$

which is a sum over relative velocities and a double sum like V, so this looks a much better building block for a Langrangian independent of the observer. If, in place of L, we start mechanics from the Lagrangian

$$\mathrm{L}' = \mathrm{T}'_{\mathrm{min}} - \mathrm{V} ,$$

we find that L′ is totally independent of the motion of the centre of mass of the whole system (i.e. the whole universe). There is, therefore, no equation of motion governing its motion. Indeed we may take it to move in any way we please, like a fly's trajectory around a room, for instance! Despite this the relative motions of all the bodies are correctly described and are just as they would be in Newton's Mechanics. Were we to choose the centre of mass to be fixed or to move uniformly in a straight line then, relative to such axes, every particle would behave just as it would in Newton's Mechanics. However, despite this $\mathrm{T}_{\mathrm{min}}$ is not truly independent of the observer. It still involves the concept of fixed directions for all time, so our new Lagrangian still involves absolute rotation. Using Newtonian terminology, the velocity $\dot{\mathbf{r}}$ with respect to axes rotating with angular velocity $\boldsymbol{\Omega}(t)$ is related to the velocity in non-rotating axes $d\mathbf{r}/dt$, through the equation

$$\frac{d\mathbf{r}}{dt} = \dot{\mathbf{r}} + \boldsymbol{\Omega}(t) \times \mathbf{r} .$$

An observer who does not know about absolute rotation will use, as the kinetic energy, T'_{m} in his/her system.

$$\begin{aligned} \mathrm{T}^* &= \frac{1}{2M} \sum\sum_{I<J} m_{\mathrm{I}} m_{\mathrm{J}} \left(\dot{\mathbf{r}}_{\mathrm{I}} - \dot{\mathbf{r}}_{\mathrm{J}}\right)^2 \\ &= \frac{1}{2M} \sum\sum_{I<J} m_{\mathrm{I}} m_{\mathrm{J}} \left[\frac{d\mathbf{r}_{\mathrm{I}}}{dt} - \frac{d\mathbf{r}_{\mathrm{J}}}{dt} - \boldsymbol{\Omega}(t) \times \left(\mathbf{r}_{\mathrm{I}} - \mathbf{r}_{\mathrm{J}}\right)\right]^2 . \end{aligned}$$

To get a kinetic energy that is truly intrinsic to the system itself and independent of the idea of fixed directions in space, we minimise $\mathbf{T}^*$ over all choices of $\boldsymbol{\Omega}(t)$. This leads to a Lagrangian from which the idea of absolute rotation has been eliminated and to which only relative rotations contribute. Does such a Lagrangian give us a mechanics of the relative motions equivalent to that which we deduce from Newton's Mechanics?

In general the answer is NO! However, when Newton's mechanics is applied in a universe with no angular momentum of the whole (in Newtonians language), then the relative mechanics gives exactly the same relative motions as those derived from Newton's laws.

What then does the relative mechanics give when applied to a universe which has a total angular momentum relative to Newton's Absolute Space?

If axes are chosen at the centre of mass with such a rotation that, relative to them, the universe has initially no angular momentum $\sum_{\mathrm{I}} m_{\mathrm{I}} \left(\mathbf{r}_{\mathrm{I}} \times \dot{\mathbf{r}}_{\mathrm{I}}\right)$

then the relative mechanics gives the same result as Newton's would give if those axes were (wrongly) assumed to be the inertial axes. The new mechanics propagates the motions forward in these axes as though there were no Coriolis or Centrifugal forces needed in these axes.

Further details of this relative mechanics can be found in [16–18, 20].

Over recent years we have worked out the consequences of a number of thought experiments [4–6, 11, 19] based on General Relativity. These have, to some extent, enhanced the understanding of the classical results in the field which culminated in the fine paper of Lindblom and Brill [14], which was the first to demonstrate the "apparently instantaneous" rotation of the inertial frame inside a collapsing rotating spherical shell. Our result with swirling gravitational waves also shows this effect. The inertial frame's rotation rate on the axis (6) is a maximum **exactly** when the waves (which never reach the axis) are closest. It is not delayed by the light travel time from the wave to the axis. However, an observer on the axis is in flat space and can not tell that he is rotating without signals from the external world that take time to propagate. The apparently instantaneous rotation of the inertial frame can not be locally measured and can not be used to send signals faster than light.

References

1. Ashtekar A, Bičak J, Schmidt BG (1997) Phys. Rev D 55: 669; 30: 1
2. Besak D. (2007) Leptogenesis in two different models of the Universe. Diplom Thesis, University of Dortmund
3. Bičak J, Lynden-Bell D, Katz J (2004) Phys. Rev D, 69: 064011
4. Bičak J, Lynden-Bell D, Katz J (2004) Phys. Rev D 69: 064012
5. Bičak J, Katz J, Lynden-Bell D (2007) Phys. Rev D 76: 063501
6. Bičak J, Katz J, Lynden-Bell D (2008) Classical and Quantum Gravity 25
7. Einstein A (1993) Collected Papers. 5(448): 340, Princeton University Press
8. Einstein A, Rosen N (1937) J. Franklin Inst. 223: 43
9. Thirring H (1918) Physikalische Zeitschrift 19: 33; 22: 29
10. Kartavtsev A (2007) Leptogenesis in the superstring inspires E6 model. Dissertation, University of Dortmund
11. Katz J, Lynden-Bell D, Bičak J (1998) Classical and Quantum Gravity 15: 3177
12. Kolb EW, Turner MS (1990) The Early Universe. Addison Wesley, Reading, MA, USA
13. Lense J, Thirring H (1918) Physikalische Zeitschrift 19: 156
14. Lindblom L, Brill D (1974) Phys. Rev D. 10: 3151
15. Luty MA (1992) Phys. Rev. D45: 455
16. Lynden-Bell D (1992) Variable Stars and Galaxies, in honour of M. W. Feast on his retirement, ed. Warner, B. ASP Conference Series

17. Lynden-Bell D (1993) Mach's Principle from Newton's Bucket to Quantum Gravity, Einstein Studies vol 6, 172. Birkhauser Boston, eds Barbour, J., Pfister, H.
18. Lynden-Bell D, Katz J (1995) Phys. Rev D 52: 7322
19. Lynden-Bell D, Katz J, Bičak J (1995) MNRAS, 272: 150
20. Lynden-Bell D (1996) Gravitational Dynamics, eds. Lahav O., Terlevich, E., Terlevich, R.J., Cambridge University Press, Cambridge
21. Lynden-Bell D, Bičak J, Katz J. (1999) Annals of Physics 271: 1
22. Lynden-Bell D, Bičak J, Katz, J (2008) Classical and Quantum Gravity 25

Far Fields, from Electrodynamics to Gravitation, and the Dark Matter Problem

A. Carati[1], S.L. Cacciatori[2], and L. Galgani[1]

[1] Department of Mathematics, Milan University, Via Saldini 50, I–20133 Milano, Italy
andrea.carati@unimi.it, luigi.galgani@unimi.it
[2] Department of Physical and Mathematical Sciences, Insubria University, Via Valleggio 11, I-22100 Como, Italy
sergio.cacciatori@uninsubria.it

Summary. We describe the steps we followed in obtaining a recent result, according to which it may not be necessary to make recourse to dark matter in order to explain the observed velocity dispersion in the Coma cluster. The main idea is that the missing gravitational action may be due to the external far galaxies. In particular, we describe how relevant for us was the analogy between the retarded far fields of electrodynamics and those of gravitation theory, in order to understand the dominant role of the far galaxies.

1 Introduction

Very recently it occurred to us to write a paper (see [1]) in which it was pointed out that the existence of a dark matter may not be necessary in order to explain the phenomenology, at least in the application of the viral theorem to the Coma cluster, which is the first case where the existence of a dark matter was conceived (see [2]). The point we made is that one should take into account the gravitational force due to the external galaxies, which usually are not even mentioned at all. We gave an estimate for such a force, showing that it seems to be of the right order of magnitude. Three points play a fundamental role in our argumentation. Two of them (Hubble's law and the fractal nature of the Universe) are of a cosmological character. Preliminarily, there is however a point of a kinematic–dynamic character. This concerns the role played in the theory of gravitation by the far fields, inasmuch as they are the ones that give the dominant contribution to the force on a test particle (especially if Hubble's law is taken into account). Now, it is true that the far fields are a standard topic in the theory of gravitation, in connection with the problem of the gravitational waves. However, the idea that they may be relevant for the dynamics of the galaxies seems to be new. In the present

G. Contopoulos, P.A. Patsis, *Chaos in Astronomy*. Astrophysics and Space Science Proceedings, doi: 10.1007/978-3-540-75826-6,

paper we describe first how it occurred to us to come to such a conception by analogy with the case of electrodynamics, and then how we finally came to our result.

The recognition of the relevance of the analogy between far fields in electrodynamics and in gravitation actually had a rather occasional origin. During a conference in honour of Claude Froeschlé, held in Spoleto at the end of June 2007, one of us had the opportunity to illustrate some recent results concerning the classical microscopic theory of matter–radiation interaction for a system of dipoles located at the sites of an infinite lattice [3, 4]. Particular emphasis was given to two features, namely, the retarded character of the forces and the global character of the interactions of the infinitely many dipoles composing the system. Such features had indeed played an essential role in the deduction of the main result, i.e., a proof of the Wheeler–Feynman identity. It then naturally occurred that, in a meeting mostly devoted to celestial mechanics, the question would be raised whether such features concerning retardation, the occurring of far fields and the global character of the interactions may perhaps play some role also in the theory of gravitation. Many discussions in this connection took place with George Contopoulos and Christos Efthymiopoulos, and these were particularly stimulating for us.

In the meantime, it had occurred to us to read, in the occasion of the thesis of a student (see [5]) the papers in which Zwicky was applying the virial theorem to the Coma cluster, while at the same time we were lecturing on the classical application of such a theorem to gases by Clausius. The comparison between the two applications then came up very naturally, with the realization of the lack of any justification for the neglect of the external forces in the case of Zwicky. So the circle was closed, and the idea was formed that one should take into account the gravitational force exerted by the external galaxies. In particular, one had to understand how it may happen that such a force can have the character of a pressure, thus playing a role analogous to that of the walls in the case of gases. Here, it will be described how, step by step, we came to our final estimate.

2 Far Fields in Electrodynamics, and the Wheeler Feynman Identity

The recognition of the global role of the far fields in electrodynamics came about as follows. In the paper [6] it was pointed out that there is a deep inconsistency in the way Planck was dealing, in his classical papers, with the dynamical aspects of the black–body problem. Indeed, he was studying the interaction of a "material resonator" with the electromagnetic field (which in some approximation could be reduced to just the electric field $\mathbf{E}$). He had previously understood (possibly being the first one to do so) that the "self-interaction" of the material resonator with the field should be taken into account through an effective self-force proportional to the third derivative of its position. However, for the sake of simplicity Planck introduced the approximation in which he replaced such a self-force with a damping force proportional to the velocity. So for the resonator he wrote down an equation of the type

$$m(\ddot{\mathbf{q}} + \omega_0^2 \mathbf{q} + \omega_0^2 \tau \dot{\mathbf{q}}) = \mathbf{E}(t)$$

where $\mathbf{q}$ is the position vector of the resonator and $\mathbf{E}(t)$ the external electric field acting on it; moreover, ω_0 a characteristic angular frequency of the resonator, while $\tau = (2/3)e^2/(mc^3)$ is the familiar characteristic time of the classical electron theory, c being the speed of light and e and m the charge and the mass of the resonator. An equation of such a type is discussed by Planck for example in the celebrated paper where his "second theory" was first advanced. One can check how Planck introduced some generic assumptions for the function $\mathbf{E}(t)$, without taking into account the fact that such a field is the one produced by the other resonators composing the considered system. This should enforce one to deal with a many–body problem, whereas Planck was explicitly assuming (see the quotation in [6, 7]) that all the resonators were acting "*independently of one another*".

The mutual retarded interactions among all resonators was taken into account in the paper [3], where an extremely simple many–body model was introduced. This is a system of infinitely many dipoles each of which can oscillate about a site of an infinite regular chain (and along the direction of the chain), being attracted to its equilibrium position by a linear restoring force. Moreover, each dipole is subject to a retarded interaction with all the other ones through an electric force which is a solution of the Maxwell equations having all the other dipoles as sources. The system is then linearized, inasmuch as the distance between the dipoles is approximated by the distance between the corresponding equilibrium points. Moreover, for what concerns the self-force, no approximation is made, and its standard form, proportional to the third derivative, is used. If q_j denotes the displacement of the jth dipole from its equilibrium position, then one has the infinite system of equations with delay given by

$$m(\ddot{q}_j + \omega_0^2 q_j - \tau \dddot{q}_j) = 2e^2 \sum_{k \neq j} \left[\frac{q_k(t - r_{jk}/c)}{r_{jk}^3} + \frac{1}{c} \frac{\dot{q}_k(t - r_{jk}/c)}{r_{jk}^2} \right], \tag{1}$$

where $r_{jk} = a|j - k|$ is the distance between the equilibrium positions of dipoles j and k, a being the lattice step.

In the paper [3] the rather astonishing result was proven that there exist normal-mode solutions of the complete system, namely, solutions of the type $q_j(t) = A \exp(\kappa a j - \omega t)$ with a suitable dispersion relation $\omega = \omega(\kappa)$. Thus, in such a solution there occurs for each dipole an exact compensation between the energy it emits and the energy it receives from all the other ones. Such an exact compensation occurs in virtue of an identity which had already been conceived by Oseen [8] in the year 1916, and is essentially equivalent to the more familiar one that was amply discussed by Wheeler and Feynman [9] in the year 1945. An essential point concerning such an identity is its global character, inasmuch as it makes reference to the whole system of all dipoles.

This feature becomes even more evident if one considers the analogous three-dimensional model (see [4]), because in such a case the main contribution to the force acting on a single dipole is the one coming from the far ones. This is due to the fact that the four-potential acting on a dipole is the retarded solution of the wave equation having each of the other dipoles as a source. Thus, concerning the force "created" by any other dipole, in addition to the Coulomb term there appear both the "near" term proportional to the velocity of the source, which decays as $1/r^2$, and the "far" one proportional to the acceleration of the source (and normal to its velocity), which decays as $1/r$. The latter term, which is the dominating one, was actually lacking in the one-dimensional model, due to the accidental fact the dipoles were constrained to oscillate along the direction of the chain. So we have described how we came to understand the role of far fields in electrodynamics.

3 The Virial Theorem in the Theory of Gases and in Astrophysics, and the Role of the External Forces

Let us now come to the problem of gravitation, and start up recalling how the virial theorem was applied by Zwicky to the Coma cluster, in contrast to the classical application to gases made by Clausius. One considers a system S composed of n points. For Clausius S is a gas enclosed in a box, for Zwicky it is the Coma cluster, whose "points" are galaxies, immersed in the Universe. One considers Newton's equations of motion $\ddot{\mathbf{x}}_i = \mathbf{F}_i/m_i \equiv \mathbf{f}_i$, $i = 1, \ldots, n$ (the dot denoting time derivative) where $\mathbf{x}_i$ is the position vector of the ith particle with respect to the center of mass of the system S, while $\mathbf{F}_i$ is the force acting on the ith particle (of mass m_i) and $\mathbf{f}_i = \mathbf{F}_i/m_i$ is the corresponding force per unit mass, i.e., the corresponding acceleration. Then one takes the dot product with $\mathbf{x}_i$, and adds over i. Performing a time average, under the hypothesis that the system remains confined one immediately obtains

$$\overline{\sigma_v^2} = -\overline{\mathcal{V}}/n \,, \tag{2}$$

where $\sigma_v^2 = (1/n)\sum_i v_i^2$ is the variance of the velocity distribution of the galaxies of the cluster, whereas $\mathcal{V} = \sum_i \mathbf{f}_i \cdot \mathbf{x}_i$ is called the virial of the forces (per unit mass), and overline denotes time-average. This is the form of the theorem suited to the case of gravitation (which involves forces per unit mass), whereas the analogous theorem for the case of general forces relates twice the kinetic energy to the virial of the forces.

Notice that one has the decomposition $\mathbf{F}_i = \mathbf{F}_i^{int} + \mathbf{F}_i^{ext}$ of the force acting on the ith particle as the sum of an internal force and an external one (that exerted by the walls of the box confining the gas in the case of Clausius, and by the galaxies external to Coma in the case of Zwicky). In the case of Clausius one has $\mathbf{F}_i^{int} = 0$ for a perfect gas, and in any case the virial of the internal forces is considered to be negligible with respect to the external one. In turn, the virial of the external forces is related to the pressure, and so from the virial theorem one obtains the thermodynamic interpretation of the translational kinetic energy as proportional to temperature. In the case of Coma, instead, Zwicky doesn't make any mention of the external forces at all, and considers only the internal ones, given by Newton's gravitational law. He thus finds, analogously to the case of gases, that $|\overline{\mathcal{V}^{int}}|/n$ is just a negligible fraction of σ_v^2, and so he is led to the conjecture that some non visible mass exists, whose contribution to the virial may restore the balance with the observed velocity variance.

Now, why should the external forces be neglected at all? The first idea underlying our work was that in astrophysics, just as in the case of gases, the virial of the external forces may actually be what is needed in order to restore the balance in the virial theorem.

4 Relevance of the Far Matter if Hubble's Law is Taken into Account

So we had the problem of estimating the gravitational field of force due to the external galaxies. From the point of view of general relativity, in the weak-field approximation this amounts to writing down the equations for the geodesic motion when the metric tensor $g_{\mu\nu}$ is a solution of the Einstein equation with the external galaxies as sources. It is well known that, for small velocities of the test particle, the equation for the geodesic motion is the same as for a point particle with a Lagrangian

$$L = g_{ik}\dot{x}^i\dot{x}^k + c\, g_{0k}\dot{x}^k + c^2 g_{00} \ , \tag{3}$$

where c is the speed of light (the summations over the spatial indices i and k from 1 to 3 being understood). So at first sight, neglecting the corrections due to the kinetic energy, the forces per unit mass appear to be the same as if the test particle were in presence of an electromagnetic field having g_{00} as scalar potential and g_{0k} as vector potential (although relevant differences exist between the two cases, as particularly emphasized by Zeldovich and Novikov [10]). On the other hand, it is very well known that in the weak-field approximation, writing the metric tensor as a perturbation of the Lorentzian background $\eta_{\mu\nu}$, namely, as $g_{\mu\nu} = \eta_{\mu\nu} + h_{\mu\nu}$, the perturbation $h_{\mu\nu}$ turns out to be a solution of the wave equation, so that its components are the familiar retarded potentials of electrodynamics. In fact one finds

$$h_{\mu\nu} = \frac{-2G}{c^4} \sum_{j=1}^{N} M_j \frac{1}{\gamma_j} \left. \frac{2\dot{q}_\mu^{(j)}\dot{q}_\nu^{(j)} - c^2\eta_{\mu\nu}}{|\mathbf{x} - \mathbf{q}^{(j)}|} \right|_{t=t_{ret}} \ , \tag{4}$$

where G is the gravitational constant, while M_j, $\mathbf{q}^{(j)}$ and γ_j, $j = 1, \ldots, N$, are the mass, the position vector and the Lorentz factor of the jth source galaxy, dealt with as a point particle, and the dot denotes derivative with respect to proper time along the worldline of the source. The notation $\mathbf{q}^{(j)}$ in place of $\mathbf{q}_j$ for the position vector of the jth external galaxy was introduced just in this formula, in order to avoid confusion with the tensorial indices.

Formula (4) implies first of all that, according to general relativity in the weak-field approximation, the gravitational field of force presents (in analogy with electrodynamics) both a "classical" Newtonian near-field term, decaying as $1/r^2$, and a far-field one, decaying as $1/r$. This in turn implies that the contribution to the gravitational field of force due to the far matter is in principle dominant with respect to that due to the near matter. In other terms, the problem of estimating the external gravitational field of force acting on a localized system such as the Coma cluster, immediately presents itself as a problem of a cosmological character, and this compels one to introduce a cosmological model.

To this end we introduced an oversimplified model. Let us make reference to a local chart with Lorentzian coordinates, having as origin the center of mass of the considered localized system (the Coma cluster). As $h_{\mu\nu}$ depends on each source not only through its position $\mathbf{q}_j$, but also through its velocity, the latter has to be assigned in order that the model be defined. In our simple model, this is obtained by introducing Hubble's law as a phenomenological prescription, namely, by requiring that for the external galaxies one has

$$\dot{\mathbf{q}}_j = \gamma_j^{-1} H_0 \mathbf{q}_j \ , \quad j = 1, \ldots, N. \tag{5}$$

(the dot denoting now derivative with respect to the background Lorentzian time). Here, H_0 is the Hubble constant which, in our extremely simplified model, we take fixed to its present value.

Notice that the Hubble assumption (5) has an essential impact on the size of the gravitational field of force. Indeed, in the field of force one has a term (decreasing as $1/r^2$) proportional to the velocity of the source, and a term (decreasing as $1/r$) proportional to the acceleration of the source. Thus, as Hubble's law (5), implies that also the acceleration has a contribution proportional to the distance, it turns out that the term proportional to the acceleration actually doesn't depend on distance at all. This is the main reason why the far matter gives the dominant contribution to the gravitational field of force.

The force per unit mass at the origin corresponding to such a dominant term (which we denote by $\mathbf{f}$) turns out to have the form

$$\mathbf{f} = \frac{4GH_0^2 M}{c^2} \mathbf{u}, \tag{6}$$

where we have introduced the vector $\mathbf{u}$, depending on the number N of external galaxies, defined by

$$\mathbf{u}(N) = \sum_{j=1}^{N} \frac{\mathbf{q}_j}{|\mathbf{q}_j|} . \tag{7}$$

Here, the masses of the sources were all put equal to a common value M, and the Lorentz factors γ_j were approximated to 1. As will be shown later, this approximation is justified for the aims of our estimate. Notice the extremely simple nature of this force per unit mass: apart from a multiplicative factor, it is just the sum of all the unit vectors pointing to each of the N external galaxies.

Concerning the main procedure which was followed in obtaining the force exerted by the external matter, the following comment may be in order. The idea of taking into account the role of the external matter was discussed by Einstein, in his Princeton lectures of the year 1921 (see [11]), in connection with the Mach principle, where he pointed out that the perturbation $h_{\mu\nu}$ to the metric (that he was denoting by $\gamma_{\mu\nu}$) can be obtained "*by the method. familiar in electrodynamics, of retarded potentials*" [see his (101), p. 87]. The only difference is that at those times he did non yet had available Hubble's law, and on the other hand he had in mind the application to astronomy. So he wrote (p. 88) "*The previous developments are valid however rapidly the masses which generate the field may move relatively to our chosen system of quasi-Galilean coordinates. But in astronomy we have to do with masses whose velocities, relatively to the coordinate system employed, are always small compared to the velocity of light, ... We therefore get an approximation which is sufficient for nearly all practical purposes if in (101) we replace the retarded potential by the ordinary (non-retarded) potential ...*". This is the way it happened that only the Newtonian, fast decaying, potential was considered, and consequently only the near matter, and not the far one, did play a role in connection with Mach's principle (see [11], p. 100).

5 Estimate of the Gravitational Forces. Role of the Discreteness of the Sources, and of the Fractal Nature of the Universe

So we had to estimate the gravitational force per unit mass exerted by the external galaxies, namely essentially the sum (7), and we now briefly describe the steps we followed. The first result was a negative one: One shows that the external force exactly vanishes if the external matter is described as a continuum, with a radially symmetric density.

As a further step, we took into account the fact that the matter actually is a discrete system of point particles. In order to make an estimate with concrete sets of positions for the galaxies, we took the probabilistic point of view that Chandrasekhar and von Neumann (see the review [12]) had taken in order to estimate the vector sum of the Newtonian forces exerted on a star by the near ones. Following such authors, we started assuming that the positions $\mathbf{q}_j$ of the N galaxies are independent random variables, uniformly distributed with respect to the Lebesgue measure. Then the sum (7) is found to grow as $\sqrt{N}$, just as a consequence of the central limit theorem. Correspondingly, apart from a constant factor, the force per unit mass (6) behaves, with increasing N, as the fraction $MN/\sqrt{N}$, where the total mass MN of the galaxies was put in evidence in the numerator. One easily checks (see later) that such a force is completely negligible.

So we modified the previous assumption, and considered the case in which the distribution of mass is assumed to be fractal [13] (see also [14–17]). This means first of all that the positions of the galaxies are no more independently distributed, and this has the relevant consequence that the sum (7) is no more constrained to grow as $\sqrt{N}$, and can instead have a faster growth, as required by the observations. The removal of the assumption of independence of the positions of the galaxies has however the consequence that the analytical discussion becomes now much more difficult than in the case of Chandrasekhar and von Neumann. So we were forced, at least provisionally, to investigate the problem by numerical methods.

We proceeded as follows. We actually considered the component of the force $\mathbf{f}$ (or of the corresponding vector $\mathbf{u}$) along a given direction. Such a component of $\mathbf{f}$ was simply denoted by f and, analogously, the corresponding component of $\mathbf{u}$ by u. To estimate the sum, the positions of the galaxies were extracted (with the method described in [13]) in such a way that the mass distribution has a fractal dimension, precisely the fractal dimension 2. The quantity u was thus dealt with as a random variable, and its probability distribution was investigated by considering 10,000 samples, with N ranging from 1,000 to 512,000, the density being kept constant. This means that the positions of the N points were taken to lie inside a cutoff sphere whose volume was made to increase as N. For the values of N investigated, the corresponding radius turns out to be so small with respect to the present horizon, that the Lorentz factors γ could altogether be put equal to 1.

The mean of u was found to practically vanish for all N, while its variance σ_u^2 was found to grow as N^2 (actually, as $0.2\,N^2$), rather than as N, as occurs in the uniform case (see [1], Fig. 1). We thus could conclude that the standard deviation σ_f of the component of the force per unit mass along a direction is proportional to N, being given by

$$\sigma_f \simeq \sqrt{0.2}\,\frac{4GH_0^2}{c^2}\,MN \;=\sqrt{0.2}\,\frac{4G}{R_0^2}\,MN\,, \tag{8}$$

where $R_0 = c/H_0$ is the present horizon. We then took such a result, which was obtained for extremely small values of N, and extrapolated it up to the present horizon $R_0 = c/H_0$, i.e., we inserted in (8) the actual value of N, so that the quantity MN could be identified with the total visible mass of the Universe.

Concerning the total visible mass MN of the Universe, one can write

$$MN = \frac{4}{3}\pi \, \rho_{\text{eff}} \, R_0^3 \,,$$

with a suitable effective density ρ_{eff}. In the paper [1] it was shown that a natural consistency condition for the model gives

$$\rho_{eff} = \frac{1}{4}\frac{3H_0^2}{8\pi G} \simeq 5\rho_0$$

where $\rho_0 = \Omega_0 \, (3H_0^2)/(8\pi G)$, with $\Omega_0 \simeq 0.05$, is the actual density. This gives $\sigma_f \simeq 0.2 \, cH_0$. On the other hand, if a random variable f has zero mean and a finite variance σ_f^2, with great probability it will take on values very near to its standard deviation σ_f. In such a sense we may say we have found for the force per unit mass along a direction, exerted by the external galaxies, the typical value

$$|f| \simeq 0.2 \, cH_0 \,. \tag{9}$$

6 Application to the Virial for a Cluster of Galaxies

We finally applied such a result to estimating the virial for a cluster of galaxies. This could be done at a very heuristic level. First of all, one has to assume that locally, in the region of interest, the field of force has somehow a central character, because otherwise the cluster itself could not exist at the considered place. In other words, we are assuming that, for a given realization of the positions of the external galaxies, on the average (over the positions of the internal galaxies) locally the field of force is directed towards a center, i.e., is acting as a pressure. However, apart from such a correlation, the intensity of the field of force should not be thought of as a smooth function, being for the rest uncorrelated with respect to the position.

Let us consider a cluster composed of n galaxies. We have to estimate the time-average of the virial of the forces (per unit mass) due to the N external galaxies, namely, the quantity

$$\overline{\mathcal{V}} = \sum_{i=1}^{n} \overline{\mathbf{f}_i \cdot \mathbf{x}_i} \,, \tag{10}$$

where $\mathbf{f}_i$ is the force on the ith internal galaxy due to the external ones. We can assume all terms of (10) to be equal, and given by $\overline{\mathbf{f}_i \cdot \mathbf{x}_i} \simeq -f \, \overline{|\mathbf{x}_i|}$ with f given by (9), while taking $\overline{|\mathbf{x}_i|} \simeq L/4$, where L is the diameter of the cluster. So one finds $|\overline{\mathcal{V}}| \simeq nfL/4$. Inserting for f the expression (9), one thus obtains the result that, according to the virial theorem, if the external force due to the far galaxies is taken into account, the velocity variance of a cluster should obey the law

$$\overline{\sigma_v^2} \simeq 0.2\,\frac{cH_0L}{4}\,, \tag{11}$$

where L is the linear dimension of the cluster. In the case of Coma, for the velocity variance one thus finds a value $\simeq 6 \cdot 10^5\,\mathrm{km}^2/\mathrm{s}^2$, which is very near to the value $5 \cdot 10^5\,\mathrm{km}^2/\mathrm{s}^2$ reported by Zwicky.

Notice the linear dependence on L in (11), which seems to be in agreement with the observations (see [18], Fig. 2, p. 539). This property is also confirmed by a dimensional analysis. Indeed, with the parameters entering the problem, the square of a velocity can be formed only as c^2, or as cH_0L or as $(H_0L)^2$. But the first term is by far too large, the last term by far too small, while the term linear in L is indeed about of the correct order of magnitude.

7 Conclusion

So we have described how, starting from the analogy with electrodynamics, we were led to conceive that the gravitational action of the far galaxies may be a substitute for the dark matter, if Hubble's law is taken into account. We also emphasized the role that the fractal nature of the Universe has in allowing for the corresponding force per unit mass to have the right order of magnitude, namely, about cH_0.

Acknowledgement

This paper is dedicated to the memory of Nikos Voglis.

References

1. Carati, A., Cacciatori, S.L., Galgani, L.: Europh. Lett. (2008), in press.
2. Zwicky, F.: Helv. Phys. Acta **5**, 110 (1933); A. Phys. J. **86**, 217 (1937).
3. Carati, A., Galgani, L.: Nuovo Cim. B **118**, 839 (2003), arXiv:physics /0312075v1[physics.optics].
4. Marino, M, Carati, A., Galgani, L.: Annals of Physics **322**, 799 (2007).
5. Giaz, A.: Thesis in Physics, University of Milan (2007).
6. Carati, A., Galgani, L.: Int. Journ. of Mod. Phys. **B 18**, 549 (2004).
7. Planck, M.: *The theory of heat radiation* (Dover, New York 1959).
8. Oseen, C.W.: Physik. Zeitschr. **17**, 341 (1916).
9. Wheeler, J.A., Feynman, R.: Rev. Mod. Phys. **17**, 157 (1945).
10. Zeldovich, Ya.B., Novikov, I.D.: *Stars and relativity* (Dover, New York 1971).
11. Einstein, A.: *The meaning of relativity* (Princeton U.P., Princeton 1922).
12. Chandrasekhar S.: Rev. Mod. Phys. **15**, 1 (1943).

13. Mandelbrot, B.: *The fractal geometry of nature* (Freeman, New York 1977).
14. Pietronero, L.: Physica A **144**, 257 (1987).
15. Ruffini, R., Song, D.J., Taraglio, S.: Astron. Astrophys. **190**, 1 (1988).
16. Sylos–Labini, F., Montuori, M., Pietronero, L.: Phys. Rep. **293**, 61 (1998).
17. Combes, F.: Astrophysical fractals: interstellar medium and galaxies. In *The chaotic Universe*. ed. by Gurzadyan, V.G., Ruffini, R. (World Scientific, Singapore 2000).
18. Kazanas, D., Mannhein, P.D.: Dark matter or new physics? In *After the first three minutes, AIP Conference Proceedings 222*, ed. by Holt, S., Bennett, C.L., Trimble, B.V. (American Institute of Physics, New York 1991).

Distribution Functions for Galaxies using Quadratic Programming

V. Dury[1], S. De Rijcke[1], V. Debattista[2], and H. Dejonghe[1]

[1] Sterrenkundig Observatorium, Universiteit Gent, Krijgslaan 281, B 9000 Gent, Belgium
Herwig.Dejonghe@UGent.be

[2] Center for Astrophysics, University of Central Lancashire, Preston, PR1 2HE, UK
vpdebattista@uclan.ac.uk

1 Introduction

One way to characterize the amount of chaos in real stellar systems is to construct distribution functions for them that depend on as many integrals as possible. While it may be difficult to quantify the amount of chaos in any formal way, the above statement is certainly qualitatively correct: stellar systems with distribution functions that depend on many integrals will leave little room to their stars for chaotic behaviour. The challenges of this approach to chaos in stellar systems are at least twofold: firstly on must be able to identify integrals of the motion, and secondly one must be able to fit distributions functions to data. The former requirement sets the framework, and in a sense defines the kind of chaos one is after, the latter requirement leads to a goodness of fit, that then may give a measure of the amount of chaos present. In the first part of this contribution, a program based on quadratic programming is reviewed that is designed to be as flexible as possible in the task of producing distribution functions for real stellar systems. In the second part some recent theoretical modeling work with possible applications in the quadratic programming modeling method is presented.

2 The Quadratic Programming Modeling Method

2.1 The Principle

In very general terms, an observable quantity μ^{ob} of a stellar system is a linear operator $\mathcal{L}|_A \{F\}$ of the distribution function F, with A additional constraints that come from the details of the observational setup. These operators range from simple bins to complicated integrals over multi-connected

G. Contopoulos, P.A. Patsis, *Chaos in Astronomy.* Astrophysics and Space Science Proceedings, doi: 10.1007/978-3-540-75826-6,

multi-dimensional domains. If there are L observables, a model F is successful to the degree that the following set of equations is satisfied:

$$\begin{cases} \mu_1^{\text{ob}} = \mathcal{L}_1|_A\{F\} \\ \quad\vdots \\ \mu_l^{\text{ob}} = \mathcal{L}_l|_A\{F\} \\ \quad\vdots \\ \mu_L^{\text{ob}} = \mathcal{L}_L|_A\{F\} \end{cases}$$

We now define basis functions f^i and write

$$F(x) = \sum_i c_i\, f^i(x)$$

These basis functions give rise to the known functions

$$\mu_l^i = \mathcal{L}_l|_A\{f^i\}$$

and we thus need to satisfy as accurately as possible

$$\begin{cases} \mu_1^{\text{ob}} = \sum_i c_i\, \mu_1^i \\ \quad\vdots \\ \mu_l^{\text{ob}} = \sum_i c_i\, \mu_l^i \\ \quad\vdots \\ \mu_L^{\text{ob}} = \sum_i c_i\, \mu_L^i \end{cases}$$

This system is not very useful, since it requires as many basis functions as observables, and there is moreover no guarantee that F is everywhere possible. Therefore we resort to the weaker version: minimize

$$\chi^2 = \sum_l w^l \left(\mu_l^{\text{ob}} - \sum_i c_i\, \mu_l^i \right)^2 ,$$

with the w^l (arbitrary) weights, together with the linear constraints

$$F(x_m) = \sum_i c_i\, f^i(x_m) \geq 0, \quad m = 1, \ldots, M$$

over a grid in integral space. This formulation is by definition a quadratic programming problem. In this form, it is (1) data driven, (2) quite open to, or independent of, most strategies concerning dynamical analysis, at least in principle. Moreover it has a few attractive features. (1) There is a unique solution (or no solution). (2) There is a statistically significant fit possible if the data allow for it. In that case, one can obtain error bars on the derived quantities, such as the distribution function. (3) Tuning is flexible, either by playing with the weights, or by adding additional constraints. (4) The data can be very heterogeneous. (5) The method is well known computationally, and is well documented.

The code that is in use at the "Sterrenkundig Observatorium" has been growing over the years. It counts more than 1,000 functions, more than 1,500 subroutines and about 600 additional entry points. The code totals over 110,000 lines of FORTRAN code. It is documented by an almost 200 page manual. The code has been used for various modeling jobs. It was originally conceived for making equilibrium models for clusters of galaxies [4], but it has been used extensively, in a long series of papers, for modeling elliptical galaxies, including population synthesis [2], and dwarf elliptical galaxies dynamics [3]. It has also been used to model star samples, either with radial velocities as dynamical input, such as in [8, 9], or using proper motions as the dynamical information [11].

2.2 Basis Functions

The choice of the basis functions is probably the most important aspect in any application of the method. The most obvious basis functions are analytical ones, either in phase space or in configuration space. The latter can be used to deproject a given projected light distribution, in order to calculate a potential. Numerical basis functions are an option, but they are much harder to deal with. Pixel functions lead generally to unstable results. Schwarzschild orbital densities can also be used, but it is notoriously difficult to do it nicely. A new and alternative option is the use of modes (see next section). In this way, also non-equilibrium models can be envisaged, that can be included in a fit.

Basis functions can also carry an astrophysical tag. An obvious option is to set up basis functions that point to different stellar populations. When the observables are spectra, the basis functions can refer to some parts of the spectra of representative stars. In this way, a dynamical analysis and a population synthesis can be performed simultaneously [2].

3 Self-consistent Instabilities

3.1 Method

We have developed a computer code that searches for self-consistent perturbations in razor-thin disc galaxies. We only consider the stellar component of the disc and neglect the dynamical influence of gas and dust. The equilibrium configuration of the stellar disc is characterised completely by the global potential $V_0(r)$ and the distribution function $f_0(E, J)$, with binding energy E and angular momentum J. The stellar disc is embedded in an axisymmetric or spherical dark matter halo that is considered dynamically too hot to develop any instabilities and thus only enters the calculations by its contribution to $V_0(r)$. An instability is described as the superposition of a time-independent axisymmetric equilibrium configuration and a infinitesimal perturbation. The total potential can be written as:

$$V(r, \theta, t) = V_0(r) + \epsilon V_1(r, \theta, t), \tag{1}$$

with a perturbing potential that can be expanded in a series of normal modes of the form:

$$V_1(r, \theta, t) = V'(r)e^{i(m\theta - \omega t)} \tag{2}$$

with a pattern speed $\mathrm{Re}(\omega)/m$ and a growth rate $\mathrm{Im}(\omega)$. Analogously, we can write the response of the distribution function to a perturbation as:

$$f(r, \theta, v_r, v_\theta, t) = f_0(r, v_r, v_\theta) + \epsilon f'(r, \theta, v_r, v_\theta, t). \tag{3}$$

The evolution of the perturbed part of the distribution function can be calculated using the linearised collisionless Boltzmann equation. We can write (using Poisson brackets):

$$\begin{aligned} \frac{\partial f'}{\partial t} - [f', E] &= [f_0, V'] \\ &= \frac{\partial f_0}{\partial E}\frac{\mathrm{dV}'}{\mathrm{dt}} + i(\omega\frac{\partial f_0}{\partial E} - m\frac{\partial f_0}{\partial J})V'. \end{aligned} \tag{4}$$

Note that potentials are defined as binding energies. Because E is the unperturbed part of the energy, the left hand side of this last equation is the total time derivative along an unperturbed orbit.

Along an unperturbed orbit, the radial coordinate r is a periodic function of time with angular frequency ω_r, as are also v_r and v_θ. Because the mean value of v_θ can be different from zero, θ will be the superposition of a periodic function $\theta_p(t)$ and a uniform drift velocity ω_θ:

$$\theta = \omega_\theta t + \theta_p(t). \tag{5}$$

If we integrate (4) along the unperturbed orbit and separate the part of the integrand that is periodic with frequency ω_r from the aperiodic part and expand it in a Fourier series, we can write the response of the distribution function to the perturbation as:

$$f'(\mathbf{r}_0, \mathbf{v}_0; t_0) = e^{i(m\theta_0 - \omega t_0)} \sum_{l=-\infty}^{\infty} I_l \frac{(l\omega_r + m\omega_\theta)\frac{\partial f_0}{\partial E} - m\frac{\partial f_0}{\partial J}}{l\omega_r + m\omega_\theta - \omega}. \tag{6}$$

With the perturbed distribution function we can calculate the perturbed density:

$$\rho'(r, \theta, t) = \int\int f'(r, \theta, v_r, v_\theta, t)\, \mathrm{d}v_r \mathrm{d}v_\theta. \tag{7}$$

For self-consistent perturbations, the gravitational potential produced by this response density should equal the original perturbing potential. We search these instabilities with a matrix method [7] and use the same basis set of potential-density couples (V'_k, ρ'_k) as in [10]. The perturbing potential $V'(r)$ becomes:

$$V'(r) = \sum_i a_i V'_i(r). \tag{8}$$

Each response density generated by one of the basic potentials can be expanded in terms of the basic density distributions:

$$\rho'_i(r) = \sum_j c_{ij}(\omega)\rho'_j(r) \tag{9}$$

and thus

$$\begin{aligned} \rho'(r) &= \sum_i a_i \sum_j c_{ij}(\omega)\rho'_j(r) \\ &= \sum_j a_j\ \rho'_j(r). \end{aligned} \tag{10}$$

This gives rise to a matrix $C(\omega)$ that contains the coefficients of the basic expansions. Self-consistent perturbations have a pattern speed and growth rate ω for which $C(\omega)$ has an eigenvalue $\lambda = 1$. The corresponding eigenvector contains the coefficients of the expansion of the response density in terms of the basis set of density distributions. Thus, for a self-consistent perturbation of order m we are left with the numerical search within the complex plane for a value of ω for which $C(\omega)$ has a unity eigenvalue. For a more detailed description of the method, we refer the reader to [5, 10].

3.2 Models

We will analyse the stability of two unperturbed models with the same unperturbed potential and unperturbed density but different distribution functions. The unperturbed potential in the plane of the disc is given by:

$$V_0(r) = \frac{GM}{\sqrt{1+r^2}} + \frac{GM}{\sqrt{1+(\frac{r}{4.4})^2}}, \tag{11}$$

with $M = 10^{10} M_\odot$ and r expressed in kpc. This is a global potential and thus includes the contribution from the disc and the halo. For the unperturbed density profile we choose an approximately exponential profile:

$$\rho_0 = \alpha e^{-1.3\sqrt{0.2+r^2}}. \tag{12}$$

The mass of the disc, and thus the mass of the halo, is determined by α. We impose an outer limit at $r_{\max} = 6\,\text{kpc}$. We have chosen α so that the proportion H/D of the total mass inside the outer radius $r_{\max}$ for the halo and the disc is about 2.5.

The distribution function $f_0(E, J)$ is written as a linear combination of basic distributions:

$$f_0(E, J) = \sum_t c_t f_{0,t}(E, J). \tag{13}$$

The expansion coefficients c_t are determined by a least square fit of the corresponding mass density to the proposed exponential form. The disc is truncated by demanding that the distribution function $f_0(E, J)$ is zero for orbits that venture outside this outer limit.

Model 1

The first model is strongly tangentially anisotropic and is known to develop strong spiral arms [10]. The distribution function is shown in Fig. 1. An unperturbed orbit of a star is determined by the value of its radius at apocentre r_+ and at pericentre r_-. This determines the energy E and orbital moment J, and thus $f_0(E, J)$. Most stars in this model populate nearly circular orbits that rotate in one direction. The strongest self-consistent instabilities for this model are a strong rotating two-armed spiral (Fig. 1c) and a weak rotating one-armed spiral (Fig. 1b).

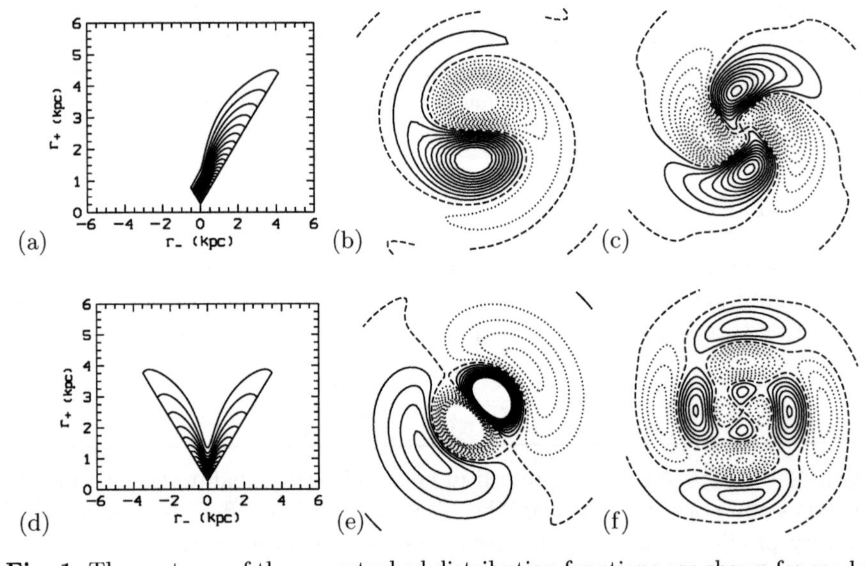

Fig. 1. The contours of the unperturbed distribution functions are shown for model 1 (**a**) and model 2 (**d**). The perturbed density is shown for the strongest self-consistent instabilities in model 1: a weak rotating one-armed spiral (**b**) and a strong rotating two-armed spiral (**c**), and in model 2: a strong non-rotating lopsided mode (**e**) and a weaker pair of rotating bars, one of which is presented in (**f**). (overdensities: *solid line*; underdensities: *dotted line*)

Model 2

For the second model the disc is built up with the same disc as in the first model and an equally massive counterrotating disc (Fig. 1d). The strongest self-consistent instabilities for this model are a strong non-rotating lopsided mode (Fig. 1e) and a weaker pair of counterrotating bars (Fig. 1f).

3.3 Physical Interpretation

Binney and Tremaine used linear perturbation theory in combination with the epicyclic approximation to calculate the response of near-circular orbits to a general m-armed perturbation ([1], Sect. 3.3, 3(a)). They calculated the evolution of the radial coordinate as:

$$R(t) = R_0 + C_1 \cos(\kappa t + \psi) + C_2 \cos(m(\Omega_0 - \Omega_p)t)), \tag{14}$$

with

$$C_2(r) = \frac{1}{\kappa^2 - m^2(\Omega - \Omega_p)^2} \left(\frac{\mathrm{d}V'(r)}{\mathrm{d}r} + 2\frac{\Omega V'(r)}{r(\Omega - \Omega_p)}) \right). \tag{15}$$

Just like them we will now study the orbits of stars for the two distinct $m = 1$ perturbations. We calculated the orbits of stars that move in the global perturbed potential:

$$V(r, \theta, t) = V_0(r) + \epsilon V'(r) e^{i(m\theta - \omega t)}. \tag{16}$$

In order to simplify the interpretation, we keep the amplitude of the perturbation fixed, i.e. we set $\mathrm{Im}(\omega) = 0$, and only consider its pattern speed, $\mathrm{Re}(\omega)/m$. The prefactor ϵ is determined by requiring that the maximum difference between the perturbed and the unperturbed density nowhere exceeds 10% of the unperturbed density. With a maximum density contrast of 10%, the orbits are not per se in the linear regime. Here we wish to illustrate how the orbits are affected by the perturbation and which mechanism induces the instability to grow.

Model 1

For the rotating lopsided mode we have a resonance at the corotation radius and one at the outer Lindblad resonance. Orbits inside the corotation radius are shifted into the direction of the overdensity ($C_2 > 0$) (Fig. 2b). The corotation radius of this mode is at $r_c = 0.8$ kpc, which coincides with the position of the one-armed spiral. Outside corotation, the sign of C_2 changes and orbits are shifted into the other direction (Fig. 2c), thus the sign of C_2 traces that of the orbit displacement and as a consequence, also roughly the sign of the density perturbation (Fig. 2a).

At corotation, C_2 becomes infinite and the orbits become banana orbits (Fig. 3). The orbits all start at the corotation radius $r = r_c$ with phases $\theta = 45°$

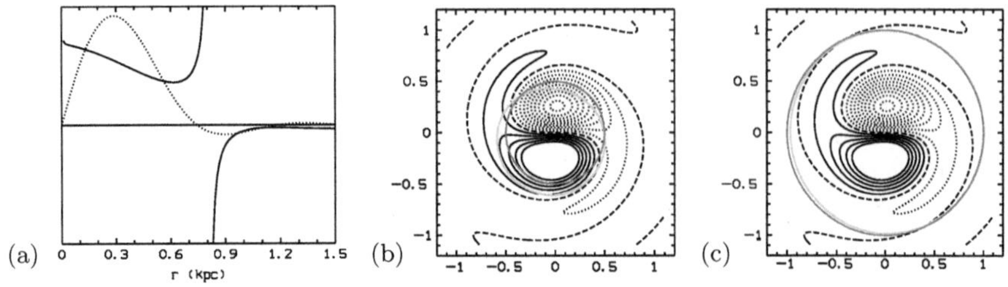

Fig. 2. (**a**) The sign of the quantity $C_2(r)$ (*solid line*) roughly traces that of the density perturbation (*dotted line*). Perturbed orbits (*gray*) are displaced from the unperturbed orbit (*black orbit*). Orbits with $C_2 > 0$ are shifted into the direction of the overdensity (**b**), those with $C_2 < 0$ are shifted into the opposite direction (**c**)

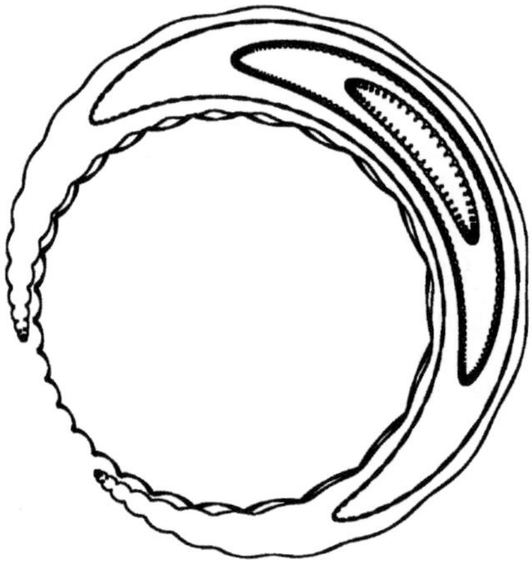

Fig. 3. A few banana orbits in a frame rotating with the lopsided mode. All orbits start from co-rotation with the same energy and angular momentum, but at a different angle θ

(inner orbit), $90°, \ldots 225°$ (circular orbit). All banana orbits evolve around a stable Lagrange point of the system which is located at the local maximum of the effective potential at a phase angle θ_L. In the corotating reference frame, a star at position (r_c, θ_L) will remain there forever. Orbits started at corotation but with different phase angles will trace out a banana-shaped curve. The further away from the equilibrium point (r_c, θ_L) an orbit is started, the more elongated the banana becomes. The orbit started at $(r_c, 180° + \theta_L)$ is circular in the corotating reference frame.

Unlike in the next case, where aligned loop orbits are almost solely responsible for creating the instability, here the two orbit families fulfil different tasks. The aligned loop orbits support the inner lobes of the lopsided structure whereas the banana orbits make up the outer one-armed spiral.

Model 2

The perturbation has no Lindblad resonances, thus (15) becomes:

$$C_2(r) = \frac{1}{\kappa^2 - \Omega^2}\left(\frac{\mathrm{d}V'(r)}{\mathrm{d}r} + 2\frac{V'(r)}{r})\right). \tag{17}$$

Tangential orbits with $C_2 > 0$ are shifted into the direction of the overdensity (Fig. 4b), those with $C_2 < 0$ are shifted into the opposite direction (Fig. 4c). The condition that the sign of the quantity $C_2(r)$ roughly traces that of the density perturbation is fulfilled everywhere within the stellar disc (Fig. 4a) and the perturbed near-circular orbits will all strengthen the $m = 1$ mode.

In Fig. 5, we present some perturbed stellar orbits that occupy the region in which the density perturbation is maximal. We can easily recognise two general orbit families: butterflies (Fig. 5c) and loops (Fig. 5a, b). Both orbit families are also found by [6] for a disc galaxy model with a lopsided potential that is of Stäckel form in elliptic coordinates and with two separate strong density cusps. They also found two other orbit families, nucleophilic bananas and horseshoe orbits. It is clear from their formulation that these two orbit families are associated with the cusps having diverging central densities which is why we do not find them in our models.

An infinitesimal $m = 1$ perturbation will cause near-circular orbits to become somewhat more elliptic and to shift towards the slight overdensity

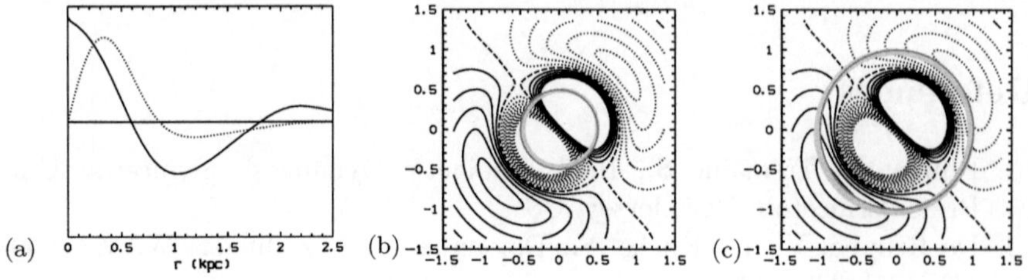

Fig. 4. (**a**) The sign of the quantity $C_2(r)$ (*solid line*) traces roughly that of the density perturbation (*dotted line*). Perturbed orbits (*gray*) are displaced from the unperturbed orbit (*black orbit*): orbits with $C_2 > 0$ are shifted into the direction of the overdensity (**b**), those with $C_2 < 0$ are shifted into the opposite direction (**c**)

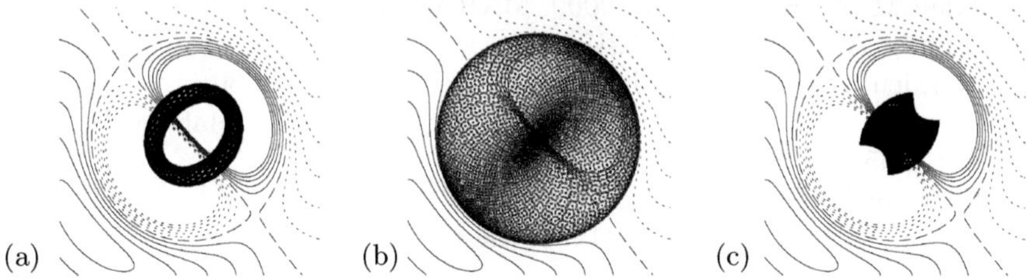

Fig. 5. A few perturbed non-circular orbits in the region where the density perturbation is maximal

(Fig. 5a), thus adding to this over-density, which in turn causes other orbits to become more elongated and to shift, and so on. Radial orbits also become loop orbits, but don't support the lopsided structure. They are more likely to form a bar (Fig. 5b). Figure 5c reveals a feature that is not captured by our linear mode-analysis. Once the perturbation is strong enough, radial orbits are deformed into the new family of butterfly orbits whose centres of gravity are also shifted in the direction of the over-density. Thus, this orbit family may potentially contribute to the $m = 1$ perturbation but only after the amplitude of the perturbation has become large enough.

4 Conclusions and Future Prospects

Quadratic programming is a tool that is well suited for modeling galaxy data of all sorts. In the version that has been implemented at the "Sterrenkundig Observatorium", care has been taken to make it a tool that is very flexible. It can also handle large data sets, which is an asset for future work. Since the goodness of fit can be quantified, the amount of chaos in a stellar system can be assessed when components are used that imply the existence of many integrals of the motion. The (numerical) components presented in Sect. 2 will take the program beyond equilibrium models. Since these components can be interpreted and physically understood, the same will be true for a model to data that is a linear combination of these components.

References

1. Binney J., Tremaine S., 1987, "Galactic Dynamics", Princeton Univ. Press, Princeton, New Jersey, US
2. De Bruyne, V., De Rijcke, S., Dejonghe, H. & Zeilinger, W. W., 2004, MNRAS, 349, 461
3. De Rijcke, S., Prugniel, P., Simien, F., Dejonghe, H., 2006, MNRAS, 369, 1321
4. Dejonghe, H., 1989, ApJ, 343, 113
5. Dury V., De Rijcke S., Debattista V. P., Dejonghe H., MNRAS, submitted
6. Jalali M. A., Rafiee A. R., 2000, MNRAS, 320, 379
7. Kalnajs A. J., 1977, ApJ, 212, 637
8. Mathieu, A., Dejonghe, H., Hui, X., 1996, A&A, 309, 30
9. Sevenster, M.N., Dejonghe, H., Van Caelenberg, K & Habing, H., 2000, A&A, 355, 537
10. Vauterin P., Dejonghe H., 1996, A&A, 313, 465
11. Wybo, M. & Dejonghe, H., 1996, A&A, 312, 649

Chaos Analysis Using the Patterns Method

I. Sideris

Institute for Theoretical Physics, University of Zürich, Winterthurerstrasse 190, Zürich, CH-8057, Switzerland
sideris@physik.unizh.ch

Summary. We discuss quantification of chaos using the patterns method in non-periodic flows in a time-independent hamiltonian and in a time-dependent hamiltonian regime. In the first case, emphasis is given on how sticky zones are identified. In the second case, orbits can experience transient chaos, i.e. they can experience both chaotic and regular epochs throughout their evolution (transient chaos), and the main focus is given on identification of these transitions. Patterns method is a new measure of chaos which treats orbits in a local level (epoch by epoch), rather than globally (as one entity).

1 Introduction

The realization – not so long ago – of the important role chaos plays in generic dynamical systems, was the ignition point for a strong effort to comprehend how exactly this new concept behaves. In reality the idea was not so new; Poincaré himself became aware of the existence of chaos, but lacking the power of computers he was unable to pursue its depths. When computers became available in the fifties and later, new concepts and ideas emerged, and unsurprisingly a demand for new tools followed. In the heart of these tools was the distinction between regular and chaotic orbits.

A number of measures were based on the very idea of chaotic instability: if we choose an orbit initially located infinitesimally close to a chaotic orbit, and then evolve these two orbits simultaneously they will diverge from each other exponentially fast in time. This exponential divergence can define an exponent which is called the largest Lyapunov exponent [1–3] and is typically used to measure how chaotic an orbit is. On the other hand if one follows the same process for two regular orbits the divergence is much slower (a power law) and the largest Lyapunov exponent owes to be zero. This basic theoretical expectation was enhanced and translated into several algorithmic recipes, which have been implemented successfully over the past [4–7]. A comparison of these methods can be found in [8].

G. Contopoulos, P.A. Patsis, *Chaos in Astronomy*. Astrophysics and Space Science Proceedings, doi: 10.1007/978-3-540-75826-6,

Another interesting way to distinguish between chaos and regularity is by use of Fourier spectra analyses. On the very basis of these techniques is the recognition that regular orbits are characterized by a set of discrete fundamental frequencies – which are as many as the degrees of freedom – and their harmonics. On the other hand, chaotic orbits are characterized by broad continuous spectra [9, 10]. This means that in practice even using a finite set of recorded data, a chaotic orbit will appear to have a large number of frequencies with significant power, commonly much larger than a regular one. (It has been shown at least for one model that the number of frequencies with significant power in the power spectrum of orbits correlate linearly with their largest Lyapunov exponents [11].) However, the emphasis of the most sophisticated spectral analysis schemes has been mainly given to the precise identification of the fundamental frequencies associated with motion, as well as the categorization of the orbital shapes [12–14].

Looking more carefully at the problem of quantification of chaos one may notice that there is some important information associated with the evolution of chaotic orbits moving in a divided (both regular and chaotic regions) phase space. A chaotic orbit, if left to evolve for long enough timescales, will experience two kinds of morphologically and to some extent physically different behaviours. Specifically, when the orbit moves inside the broad chaotic sea it will experience strong chaos in the sense that it will become extremely unpredictable. A number of different frequencies with significant power characterize the orbit in this kind of motion. Now, occasionally the orbit will be entrapped inside regions surrounding the regular islands of the phase space. The mechanism of this entrapment is associated with the existence of cantori surrounding the regular islands, which function as porous obstacles in the potential and do not allow orbits to easily escape the region after they get entrapped. When this happens orbits get "sticky", meaning that they stay constrained around the regular island for long timescales [8, 15, 16]. Physically these orbits try to mimic regularity; their frequencies approach the discrete frequencies picture regular orbits have, but never quite become discrete. Eventually these orbits get unstuck and experience again strong chaos. In time they will get trapped again and the two behaviours will alternate throughout the evolution ad infinitum.

A second interesting scenario is that of time-dependent Hamiltonian systems. There is a significant differentiation in the behaviour of orbits in time-dependent and time-independent Hamiltonian systems. In a time-independent hamiltonian regime energy is conserved, so orbits maintain their regular or chaotic nature forever; orbits which are initially regular, stay always regular, while orbits starting chaotic stay always chaotic. Now, in time-dependent regimes the energy is not conserved so the nature of an orbit may change. A time-dependent hamiltonian orbit may experience both chaos and regularity, a behaviour which has been coined "transient chaos" [17] and we can call these orbits "mixed".

2 The Patterns Method

The established measures of chaos [2, 5, 7, 13, 18] may be limited to analyze adequately either of the two aforementioned behaviours, weak versus strong chaos, and evolution in time-dependent regimes where alternation from regularity to chaos and vice versa exists. The main reason for this limitation is that these measures treat orbits as monolithic entities, emphasizing into distinguishing whether the whole orbit is either regular or chaotic. In order to analyze stickiness and/or time-dependent chaos, however, it is more useful to treat the orbit as a set of sequential epochs, with every epoch having no predefined size.

A new measure of chaos which attempts to do exactly that appeared recently [19]. It treats the orbit locally, piece by piece, without presupposing the time-length of weakly and strongly chaotic, and regular epochs. It has been constructed by design to recognize the different epochs in the evolution of an orbit. The new measure is called "patterns method" and attempts to identify repetitive patterns in the evolution of the orbit.

The method is simple but powerful. In a signal (for example $x(t)$) associated with an orbit one can easily identify the extrema. The whole method is based on the observation that if the extrema get connected in the right step, let's say extremum i with extrema $i-2n$, $i-n$, $i+n$, $i+2n$ etc., smooth curves may emerge (one can see that in Fig. 2 in [19]). For every extremum the algorithm tries all possible steps starting from $n = 3$ and increasing it to the maximum possible step available from the specific extremum. Using a simple interpolation scheme it assigns a smoothness value to everyone of these curves and then identifies the step n which provides the smoothest possible curve.

Assume for example that we have a regular orbit and we use step n to connect extremum i, happening at time t_i with extrema $i-2n$ (time $= t_{i-2n}$), $i-n$ (time $= t_{i-n}$), $i+n$ (time $= t_{i+n}$), and $i+2n$ (time $= t_{i+2n}$). We can use the extrema $i-2n$, $i-n$, $i+n$, and $i+2n$ to compute where the extremum i should be located. Assume we compute that its position should be at x_{iC}. We can now compare this position with the real position x_i (recorded in our data) where the extremum i is found. The relative error is $s_i = |x_i - x_{iC}|/|x_i|$. This error can serve as a measure of chaos, or otherwise chaos strength of that extremum. If the curve for step n is very smooth then this error obviously becomes very small since our interpolation scheme can predict accurately where x_i should be. For chaotic orbits this error cannot be very small because the curves are never very smooth, but at the sticky epochs the curves are smoother than in the wildly chaotic epochs. We repeat again here that the program tries all possible n and chooses to keep the smallest possible error found. Although this may sound computationally expensive one should remember that the program operates on the extrema of the signal, and not on all recorded points. Therefore every period in the orbit is represented only by 2 extrema which makes at least galactic orbits (typically of the order of 100 dynamical times) easy to treat.

The basic assumption in the patterns method is that the smoother the connecting curves between extrema are the more regular the particular epoch is. Since the above method is applied to every extremum of the signal, eventually every extremum has been assigned a chaos strength value. When an orbit is regular all extrema are assigned typically very low chaos strengths. Chaotic orbits on the other hand take an even better advantage of the method; since every extremum is assigned a chaos strength, sticky parts of the orbit will have typically lower value of chaos strength while wildly chaotic parts will have high values of chaotic strength. This scheme was employed successfully in a number of time-independent potentials, all chosen because they admit a large number of both regular and chaotic orbits. The sticky areas were identified efficiently and it was shown how the stickiness recesses as an orbit increases its distance itself from the regular island.

In Fig. 1 one can see the results for a the time-independent case of Hénon–Heiles potential [20]:

$$V(x, y) = \frac{1}{2}(x^2 + y^2) + x^2 y + \frac{1}{3}y^3 \tag{1}$$

for energy $E = 0.125$. For this particular picture, 20000 orbits where integrated and for each one we computed the points on the Poincaré plot

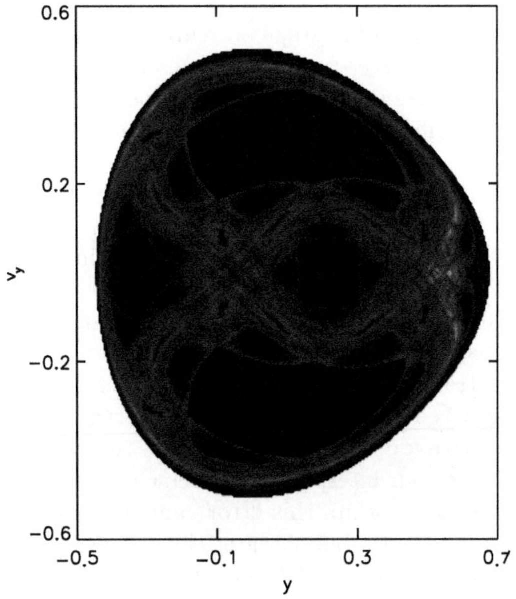

Fig. 1. The phase space structure of the Hénon–Heiles potential for energy $E = 0.125$. In *black* are the regions of regularity. The chaotic regions are marked with *shades of gray* where deeper gray corresponds to weaker chaos. Notice that the very *dark gray* regions are located around the regular islands. Also notice the intricate structure of the chaotic zones located inside the chaotic sea

for $v_x = 0$. Using the patterns method we assigned a strength of chaos to each point. Then we binned the strengths, finding the average for very small squared regions in the plot. Regular areas had very small average strengths (≤ 0.007) so they appear black in the contour plot. The value 0.007 is a distinction criterion computed and used in [19] by taking advantage of features of the distribution of the chaos strengths. Regions with chaotic strengths bigger than 0.007 appear in different scales of strength. The shades become lighter for larger strength of chaos. One can easily notice that deep gray areas (stickiness) are located around the regular islands. But even more exciting is the high resolution of the intricate nature of phase space inside the chaotic sea. It is obvious that there are paths of weak chaos right in the middle of it, reminiscent of the location of the separatrices in smaller energies of the same potential, where the whole or almost all phase space was regular.

The next important step was to apply the method to time-dependent Hamiltonian systems. Two main events happen in these systems, which although not necessarily disconnected from each other, is better to distinguish here so we make the picture more visually comprehensible. Firstly, the structure of the phase space is changing for every energy value. If one chooses a specific value for an energy and plots the phase space he/she will find that this phase space is changing with time, new regular or chaotic regions appear at the positions occupied previously by chaotic or regular areas. Secondly, the energy per orbit is not conserved, so every orbit will experience several different phase spaces. So every orbit during its evolution is moving on an ever changing phase space. What the orbit will be, chaotic or regular, depends exclusively on not only what energy it will experience but also on what time it will experience this particular energy.

Understandably this is a very challenging regime to analyze by any chaotic measure. For example if an orbit is originally chaotic then becomes regular for a very short number of orbital periods and then becomes chaotic again, it is very difficult to identify the short regularity epoch. It is well hidden inside the two chaotic epochs. However, if the regularity insists for a sufficient time it may be possible to identify this epoch. The patterns method attempts to analyze these phenomena. For a system that changes with medium or slow speed, orbits may experience regularity for a sufficiently long period of time so repetitive patterns, and therefore smooth curves can be identified. The patterns method does this distinction automatically, working on the time-dependent orbit without the need to choose any predefined size windows or any sensitive parameters.

Using the technique as described earlier it connects extrema i, $i+n$, $i+2n$, etc. trying, in practice, to identify the step n which gives the smoothest possible curve. The smoother the curve the smallest the strength of chaos is for the extrema involved in the curve. The smoothness of the curve is computed using a simple interpolation scheme. This is how it assigns a strength of chaos for every extremum of the signal of the orbit. Different extrema have different strengths. The smaller the strength of chaos the bigger the regularity is.

The patterns method applied to a set of orbits integrated into a time-dependent generalization of the Hénon–Heiles system [21]:

$$V(x, y) = (m + 1)\left[\frac{1}{2}(x^2 + y^2) + x^2 y + \frac{1}{3}y^3\right] \tag{2}$$

with $m = m_0 e^{-at}$, where the values for the parameters were $m_0 = 0.8$ and $a = 0.003$. (For these values of m_0 and a, m decreases to almost zero at physical time $t = 2000$.

In Fig. 2 one can see the results of this analysis. Several energies are involved in the evolution but we want to see how the phase space structure is evolving for one energy. Therefore one energy value $E \simeq 0.143$ was chosen and several orbits found to have that energy in that particular time were

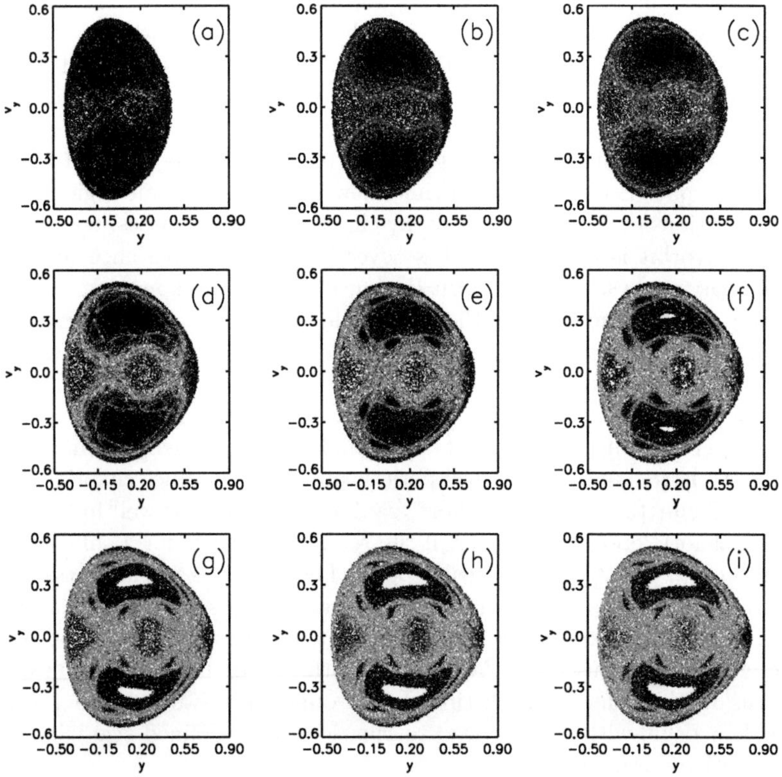

Fig. 2. The *panels* show nine consecutive times in the evolution of the phase space structure for a unique energy $E \simeq 0.143$ for the time-dependent generalization of the Hénon–Heiles potential. *Black color* signifies regularity, *dark gray* signifies weak chaos and *light gray* signifies strong chaos. For early times the phase space structure is almost completely regular, but throughout the evolution significant chaos emerges. (**a**) $t \simeq 25$, (**b**) $t \simeq 175$, (**c**) $t \simeq 325$, (**d**) $t \simeq 475$, (**e**) $t \simeq 625$, (**f**) $t \simeq 775$, (**g**) $t \simeq 1{,}125$, (**h**) $t \simeq 1{,}475$, (**i**) $t \simeq 1{,}925$

used to make the plots. One can observe that while the phase space is almost regular initially (black color points), as time evolves, more and more gray areas appear (dark gray signifies weak chaos while light gray signifies strong chaos), and the phase space is changing into heavily chaotic. It is interesting to observe how small chains of islands appear and then tend to disappear throughout the evolution. In some sense these pictures encompass a good part of chaotic dynamics for Hamiltonian systems since they show all the effects a time-dependent perturbation has on a system.

The results were compared to the results associated with freezing the potential in specific times [21]. When we freeze a time-dependent potential at some time t, in practice we generate a time-independent potential and then we can explore the structure of these potential as we do usual for any time-independent regime. This technique has been used before (for example in the context of N-body simulations [22, 23]) and it is in general very time consuming, by far more consuming than analyzing the time-dependent orbit.

The comparison of the results between the frozen potentials method and the patterns method was excellent. However, there are cases where the only option is to freeze the potential. In particular when a system is changing very fast, for example in a violent relaxation regime in galactic dynamics, one should be unwilling to use any chaotic measure that analyzes the time-dependent orbit. The time-dependent orbit may pass from regular to chaotic and vice versa several times in very short timescales (a couple of dynamical times). In these timescales any analysis by any existing measure of chaos may be at least dangerous. If one freezes the potential, however, then this analysis becomes possible but to have a complete picture about what happens with the phase space structure of the system he/she needs to freeze the potential in many time snapshots and to integrate in the frozen potentials a sufficient number or orbits for a sufficient number of energies. Usually this is annoyingly or even formidably expensive for available computational resources.

On the other hand, if the system happens to change in medium speeds one may choose to apply first the patterns method to the involved orbits and investigate if there seem to be any interesting epochs (in time and energy) associated with the evolution. That can be seen as a very first, quick filtering of the information. Precise information can then been given for some chosen epochs by freezing the potential.

3 Conclusions

The patterns method is a new technique for quantification of chaos. Usually chaotic measure aim to distinguish between regularity and chaos. The patterns method attempts to do something more; its goal is to assign a chaos strength to every small segment of the orbit. So while it provides distinction between regularity and chaos like all the other measures, it also provides information about when a chaotic orbit experiences stickiness during its evolution. To achieve this it analyzes the orbit not as one entity but segment by segment.

Because of its design it can directly, without alterations, apply to time-dependent regimes. In these regimes an orbit may experience both chaotic and regular behaviour, an effect directly related to the non-conservation of energy. When the system does not change violently, the patterns method may be able to provide at least one first picture of the evolution in a number of different energies.

Patterns method is efficient in several ways. It does not involve sensitive parameters because it tries all possible combinations of correlations between extrema. It is very fast (since it works only with the extrema points) for orbits evolved for less than 500 orbital periods. (For longer orbits one can divide an orbit into several sizeable pieces). It treats output data of an integration, which saves the time of simultaneous integration of extra equations during the evolution like the ones for computation of the Lyapunov exponents. And at least in the potentials treated till now it has shown that it is able to characterize correctly a significant number of orbit as regular or chaotic even when the integration time is as limited as 20–30 orbital periods.

References

1. A.M. Lyapunov, Ann. Math. Stud. **17**, 1947 (1907)
2. G. Benettin, L. Galgani, A. Giorgilli, and J.-M. Strelcyn, Meccanica **15**, 9 and 21 (1980)
3. A.J. Lichtenberg and M.A. Lieberman, Regular and Chaotic Dynamics, Springer, New York (1992)
4. N. Voglis and G. Contopoulos, J. Phys. A **27**, 4899 (1994)
5. C. Froeschle C. Froeschle, and E. Lohinger, Cel. Mech. Dyn. Astron. **56**, 307 (1993)
6. C. Froeschle, E. Lega, and R. Gonczi, Cel. Mech. Dyn. Astron. **67**, 41 (1997)
7. C. Skokos, J. Phys. A**34**, 10029 (2001)
8. G. Contopoulos, Order and Chaos in Dynamical Astronomy, Springer, Berlin (2002)
9. M. Tabor, Chaos and Integrability in Nonlinear Dynamics: An Introduction, Wiley, New York (1989)
10. H.E. Kandrup, B.L. Eckstein, and B.O. Bradley, Astron. Astroph. **320**, 65 (1997)
11. C.L. Bohn and I.V. Sideris, Phys. Rev. ST-AB **6**, 034203-1 (2003)
12. J. Binney and D. Spergel, Astroph. J. **252**, 308 (1982)
13. J. Laskar, Icarus **88**, 266 (1990)
14. D.D. Carpintero and L.A. Aguilar, Mon. Not. R. Astron. Soc., **298**, 1 (1998)

15. R.B. Shirts and W.P. Reinhardt, J. Chem. Phys. **77**, 5204 (1982)
16. G. Contopoulos, Astron. J. **76**, 147 (1971)
17. H.E. Kandrup, I.M. Vass, and I.V. Sideris, Mon. Not. R. Astron. Soc. **341**, 927 (2003)
18. G. Contopoulos and N. Voglis, Cel. Mech. Dyn. Astron. **64**, 1 (1996)
19. I.V. Sideris, Phys. Rev. E, **73**, 066217, (2006)
20. M. Hénon and C. Heiles, Astron. J. **69**, 73 (1964)
21. I.V. Sideris, Phys. Rev. E, submitted, (2007)
22. C. Kalapotharakos, N. Voglis, and G. Contopoulos, Astron. Astrophys. **428**, 905 (2004)
23. E. Athanassoula, Ann. N.Y. Acad. of Sci. **1045**, 168 (2005)

On the Topology of Regions of 3-D Particle Motions in Annular Configurations of n Bodies with a Central Post-Newtonian Potential

T. Kalvouridis

Department of Mechanics, National Technical University of Athens, Athens, Greece
tkalvouridis@gmail.com

Summary. We study the topology of regions where three-dimensional motions of a small body can take place in a force field created by $\nu = \eta - 1$ big bodies with equal masses m located at the vertices of a regular polygon centered in the $(\nu + 1)$th body with a mass m_0 that is generally different from the previous ones and is characterized by a post Newtonian potential.

1 Introduction

In 1859 Maxwell used a regular polygon configuration with Newtonian potentials in order to describe the rings of Saturn. Mioc and Stavinschi [5, 6] used post-Newtonian potentials and studied the dynamics of this configuration. Here, we study the dynamics of a small body in such a configuration by assuming that the force field created by each peripheral mass is Newtonian, while that of the central mass is a post-Newtonian one, either:

- A Manev-type potential of the form $A/r_0 + B/r_0^2$ or
- A Schwarzschild-type one of the form $A/r_0 + B/r_0^3$

where r_0 is the distance of the small body from the central primary. These potentials model several situations belonging mainly to Astronomy, such as the case of a spheroid central primary [1], as well as various relativistic fields like those proposed by Fock [4] and Reissner–Nordstrom [6]. The system is characterized by the number of the peripheral primaries ν, the mass parameter $\beta = m_0/m$ and the two coefficients A and B.

2 The Three-Dimensional Ring Problem of $(n + 1)$ Bodies and the Normalized Equations of Motion

The motion of the small body S in a synodic system is described by the equations [2],

G. Contopoulos, P.A. Patsis, *Chaos in Astronomy*. Astrophysics and Space Science Proceedings, doi: 10.1007/978-3-540-75826-6,

$$\ddot{x} - 2\dot{y} = \frac{\partial U}{\partial x}$$
$$\ddot{y} + 2\dot{x} = \frac{\partial U}{\partial y}$$
$$\ddot{z} = \frac{\partial U}{\partial z}$$

The potential function U has the following expression in various post-Newtonian potentials of the central primary.

- **Case 1.** Manev-type potential

$$U(x,y,z) = \frac{1}{2}(x^2+y^2) + \frac{1}{\Delta}\left[\beta\left(\frac{A}{r_0}+\frac{B}{r_0^2}\right)+\sum_{i=1}^{\nu}\frac{1}{r_i}\right]$$

- **Case 2.** Schwarzschild-type potential

$$U(x,y,z) = \frac{1}{2}(x^2+y^2) + \frac{1}{\Delta}\left[\beta\left(\frac{A}{r_0}+\frac{B}{r_0^3}\right)+\sum_{i=1}^{\nu}\frac{1}{r_i}\right]$$

where r_0, r_i $i = 1, 2, \ldots$ are the distances of the particle from the primaries. In any case,

$$\Delta = 2\sin^3\theta\left(\sum_{i=2}^{\nu}\frac{1}{\sin(i-1)\theta}+\beta\right) \qquad \theta = \pi/\nu.$$

There is a Jacobian-type integral of motion

$$C = 2U(x,y,z) - (\dot{x}^2+\dot{y}^2+\dot{z}^2).$$

3 Equilibrium Zones and Evolution of the Zero-Velocity Surfaces in 3-D Motion

The equilibrium positions are distributed on imaginary circles that are called equilibrium zones [2] (Fig. 1b). When we consider Newtonian potentials in all primaries, the existing equilibrium zones are five or three depending on the mass parameter value. Bifurcations in the topology of the zero-velocity surfaces occur at values $C = C_J$, where $C_J = C_{A1}, C_{A2}, B, C_{C2}, C_{C1}$. The way that the zvs evolve is directly related to the sequence of the Jacobian constants of the existing equilibrium zones. There is generally a critical value l_ν of the mass parameter β, where the five equilibrium zones reduce to three. Table 1 shows the critical values and the existing equilibrium zones when we consider that the potential of the central primary is either a Manev-type or a Schwarzschild-type one.

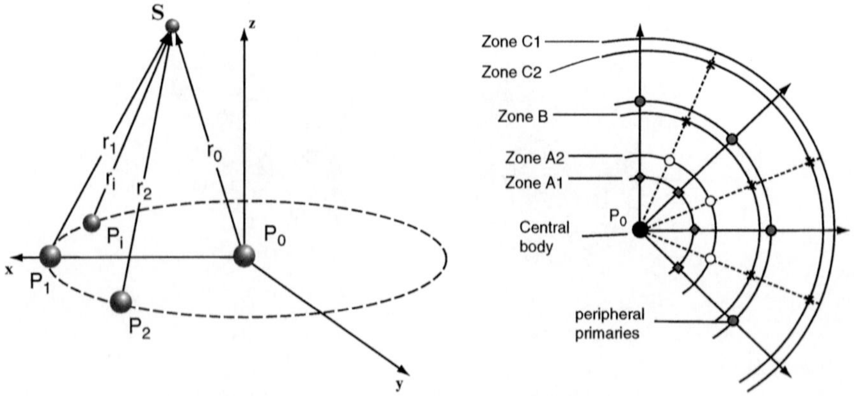

(a) (b)

Fig. 1. a The ring configuration of $N = \nu + 1$ primaries and the small body S. **b** The five equilibrium zones

Table 1. Critical values l_ν for the case $\nu = 7$, $A = 1$ and a post-Newtonian potential

Potential	B	l_ν	Existing equilibrium	Zones
			$\beta < l_\nu$	$\beta > l_\nu$
Manev	−0.14	5.2	All five equilibrium zones	A_1, B, C_1
	0.1	1.9		A_1, C_2, C_1
	0.3	1.25		A_1, C_2, C_1
Schwarzschild	0.1	0.65	All five equilibrium zones	
	0.3	0.7		A_1, C_2, C_1

4 Applications

(i) Manev-type potential of the central primary ($\beta = 2$ ($\beta < l_\nu$), $A = 1$, $B = -0.14$, 5 equilibrium zones). When C is relatively high, the particle can move either inside the closed surfaces that surround the peripheral primaries or outside the external surface (Fig. 2a). At $C = C_B$ the internal closed surfaces around the peripheral primaries touch each other. Then, with a further decrease C, ν channels of intercommunication are created between these surfaces. At $C = C_{C_1}$ the internal closed formation touches the external surface at the points of zone C_1. Then, as C further decreases, ν openings are created at the external surface and the particle can escape outside it (Fig. 2b). When $C < C_{A_2}$ the zero-velocity surface splits in two parts that shrink as C further decreases (Figs. 2c, d) until they finally vanish.

(ii) Schwarzschild-type potential of the central primary ($\beta = 2$ ($\beta < l_\nu$), $A = 1$, $B = -0.14$, 3 equilibrium zones). When $C > C_B$ the particle can move either inside the closed surfaces that surround the peripheral primaries or outside the external surface (Fig. 2e). At $C = C_B$ the closed internal surfaces

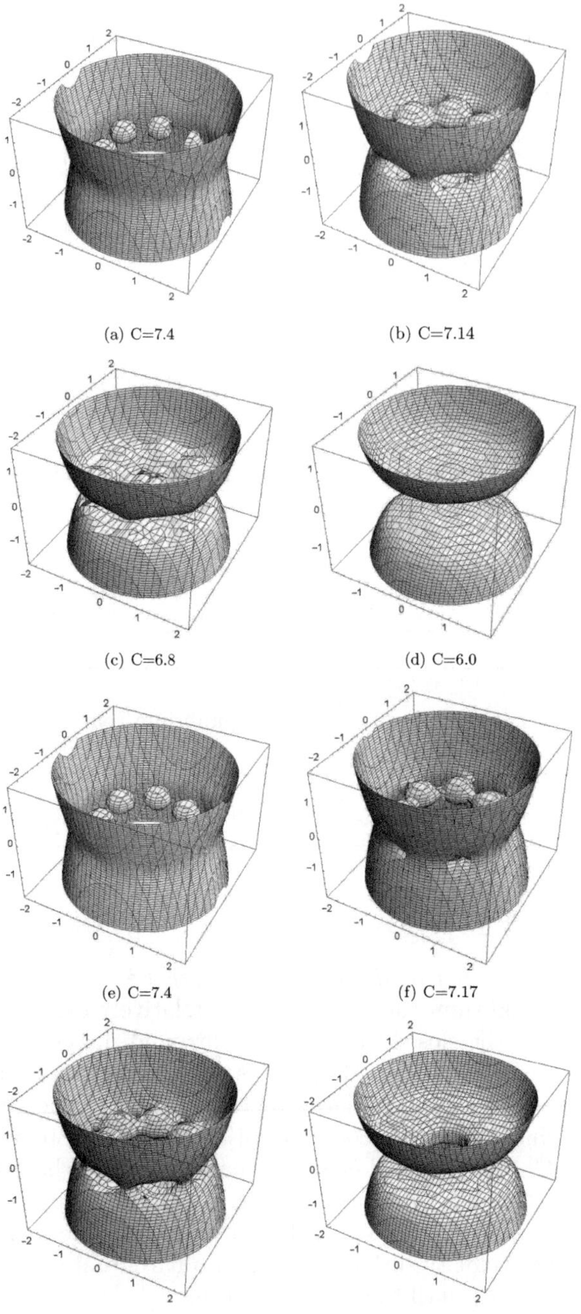

(a) C=7.4 (b) C=7.14

(c) C=6.8 (d) C=6.0

(e) C=7.4 (f) C=7.17

(g) C=7.0 (h) C=6.5

Fig. 2. a–d Evolution of the zero-velocity surfaces with a central Manev-type potential when $\nu = 7$, $\beta = 2$, $A = 1$, $B = -0.14$. **e–h** Evolution of the zero-velocity surfaces with a central Schwarzshild-type potential when $\nu = 7$, $\beta = 2$, $A = 1$, $B = -0.14$

touch each other at the equilibrium positions of zone B. With a reduction of C, ν small channels are created between these surfaces (Fig. 2f). At $C = C_{C_1}$ the internal surface touches the external one at the points of zone C_1. Then ν small windows are created that enlarge as C decreases even more (Fig. 2g). When $C > C_{C_2}$ the zero-velocity surface splits in two parts (Fig. 2h) that shrink as C decreases until they vanish.

References

1. Arribas, M., Elipe, A.: 2004. *Bifurcations and equilibria in the extended N-body problem.* Mech. Res. Commun. 31, 1–8
2. Kalvouridis, T.J.: 1999. *A planar case of the $n+1$ body problem: the 'ring' problem.* Astroph. Space Sci., 260 (3), 309–325
3. Maxwell, J.C.: 1890. *On the stability of the motion of Saturn's rings.* Scientific Papers of James Clerk Maxwell, Cambridge University Press, Vol. 1, p. 228
4. Mioc, V.: 1994. *Elliptic type motion in Fock's gravitational field.* Astron. Nachr., 315, 175–179
5. Mioc, V., Stavinschi, M.: 1998. *On the Schwarzschild-type polygonal* $(n+1)$-*body problem and on the associated restricted problem.* Baltic Astron., 7, 637–651
6. Mioc, V., Stavinschi, M.: 1999. *On Maxwell's* $(n+1)$-*body problem in the Manev-type field and on the associated restricted problem.* Phys. Scrip. 60, 483–490

The Average Power-Law Growth of Deviation Vector and Tsallis Entropy

G. Lukes-Gerakopoulos, N. Voglis*, and C. Efthymiopoulos

Research Center for Astronomy, Academy of Athens, Soranou Efesiou 4, GR-11527, Athens, Greece
`gglukes@phys.uoa.gr`, `cefthim@academyofathens.gr`

1 Introduction

The usefulness of the so-called 'non-extensive q-entropy' [1] in characterizing the statistical mechanical properties of nonlinear dynamical systems has so far been demonstrated in a number of instructive examples in the literature (see [2] for a comprehensive review). Here we focus on one particular property of the Tsallis q-entropy, first reported in [3, 4], and further explored in [5–7]. These authors demonstrated that in the so-called 'edge of chaos' of dissipative systems the rate of increase of the q-entropy remains constant for a quite long time interval. In this time interval the chaotic spreading of the orbits exhibits a power-law rather than exponential sensitivity on the initial conditions [3, 8]. The same phenomenon was found numerically in low-dimensional mappings [9–11], and it was called 'metastable state' [9]. Here we consider the question of how can the 'metastable behavior' associated with a constant rate of production of the Tsallis q-entropy be justified theoretically by an analysis of the behavior of the variational equations in the limit of weak chaos. Our method leads to the calculation of the q-exponent along a flow of chaotic nearby orbits directly from the variational equations of motion. The same method can be used as a 'chaotic indicator' distinguishing weakly chaotic orbits from nearby regular orbits. We call this indicator APLE (average power-law exponent) and give examples of its application in the standard map.

2 Q-Entropy and the Growth of Deviation Vectors

Consider a partitioning of the n-dimensional phase space $\mathcal{M}$ of a conservative system into a large number of volume elements of size δ^n for some small $\delta > 0$. Let $\mathbf{x}(0)$ be the initial condition of an orbit located in a particular volume element $V_0 = \xi_{01}\xi_{02}...\xi_{0n}$, where ξ_{0k}, $k = 1, ..., n$ are the linear dimensions

* Deceased.

G. Contopoulos, P.A. Patsis, *Chaos in Astronomy.* Astrophysics and Space Science Proceedings, doi: 10.1007/978-3-540-75826-6,

of V_0 in a locally orthogonal set of coordinates in the neighborhood of $\mathbf{x}(0)$. Without loss of generality, we set all the initial values ξ_{0k} equal, i.e., $\delta \equiv V_0^{1/n} = \xi_{0k}$, $\forall k$. Because of the volume preservation, the orbital flow defines a mapping $V_0 \rightarrow V(t)$ of the volume V_0 to an equal volume $V(t)$ at the time t. Let $\boldsymbol{\xi}(t) = D_t \boldsymbol{\xi}_0$ be the solution of the variational equations for an initial deviation vector $\boldsymbol{\xi}_0$ acted upon by a linear evolution operator D_t determined solely by the orbit $x(t)$. Let $\boldsymbol{\xi}'_k(t)$ be the images of $\boldsymbol{\xi}_{0k}$, $k = 1, 2, ..., n$ under the action of the operator D_t. The vectors $\{\boldsymbol{\xi}'_k(t)\}$ form a complete basis of the tangent space to $\mathcal{M}$ at the point $\mathbf{x}(t)$ iff $\{\boldsymbol{\xi}_{0k}\}$ form a complete basis of the tangent space to $\mathcal{M}$ at the point $\mathbf{x}(0)$ and $rank(D_t) = n$. It is possible to obtain an orthogonal basis $\{\boldsymbol{\xi}_k(t)\}$ starting from $\{\boldsymbol{\xi}'_k(t)\}$ via the Gramm-Shmidt procedure [12]. The volume $V(t)$ is given by $V(t) = \xi_1(t)\xi_2(t)...\xi_n(t)$. We reorder this basis by decreasing the length of the vectors $\boldsymbol{\xi}_k(t)$. Let $V_c(t)$ be a coarse-grained volume equal to the total volume of all the cells visited by the orbits in $V(t)$. $V_c(t)$ is determined by only those vectors with lengths greater or equal to δ, that is $V_c(t) = \xi_1(t)\xi_2(t)...\xi_m(t)\delta^{n-m}$ where m is defined by the condition $m = \sup\{m' : \xi_k(t) \geq \delta \text{ for all } k = 1, \ldots, m'\}$. It follows that $m = n$ when $t = 0$ and $V_c(0) = V(0) = \delta^n$. Let $W(t) = \frac{\xi_1(t)\xi_2(t)...\xi_m(t)}{\delta^m}$ be the number of cells (or 'micro-states') occupied by $V_c(t)$. The time evolution of the q-entropy [1] for an ensemble of orbits with initial conditions within the volume V_0 is given by $S_q(t) = \frac{W(t)^{1-q}-1}{1-q}$. If t_1 is a transient initial time of evolution of the orbits, the mean rate of evolution of S_q is given by

$$\frac{S_q}{t/t_1} = \frac{1}{(t/t_1)(1-q)}[(\frac{\xi_1\xi_2\xi_m}{\delta^m})^{1-q} - 1] \ . \tag{1}$$

For every ξ_k, $k = 1, 2, ...m$, we define an *Average Power Law Exponent* (APLE) p_k according to

$$\xi_k(t) = \xi_k(t_1)(\frac{t}{t_1})^{p_k}, \qquad k = 1, 2, ...m \ . \tag{2}$$

In conservative systems we have $p_1 + p_2 + ... + p_m \geq 0$, since, by the preservation of volumes, the components $\xi_k(t)$ cannot be all decreasing functions of the time. Equation (1) now takes the form $\frac{S_q}{t/t_1} = \frac{\left(\left(\frac{t}{t_1}\right)^{(p_1+\cdots p_m)(1-q)} - 1\right)}{(t/t_1)(1-q)}$. In the limit $t \rightarrow \infty$ the quantity $\frac{S_q}{t/t_1}$ tends to a non-zero finite value only if a) the p_is take constant limiting values, and b) the entropic index q satisfies the relation $(p_1 + p_2 + ...p_m)(1 - q) = 1$. In all other cases, $\frac{S_q}{t/t_1}$ tends either to zero or to infinity. If the deviations $\xi_k(t)$ grow asymptotically as a power law, then condition (a) is satisfied and the mean rate of Tsallis entropy $\frac{S_q}{t/t_1}$ tends to the sum of the positive APLEs

$$\lim_{t\rightarrow\infty} \frac{S_q}{t/t_1} = p_1 + p_2 + ...p_m \tag{3}$$

for the value of q given by $q = 1 - \frac{1}{p_1+p_2+\dots p_m}$. In practice, the maximum of the APLEs can be used as a lower bound of the limit of $\frac{S_q}{t/t_1}$. Furthermore we can use (3) for a long but finite time t in order to estimate the average value of the q-exponent in the interval from t_1 and t. The APLE is related to the limit of the mean rate of growth of Tsallis entropy via $q = 1 - 1/p$.

3 Numerical Examples

We give numerical examples that reveal a number of interesting properties of the APLE $p = \frac{\ln \xi(t)}{\ln t}$ ($\xi(t_0) = 1$ for $t_0 = 1$). In order to avoid a singular value of p when $t_0 = 1$, we set $t_1 = 10$ and we calculate p for $t \geq t_1$, with $p = 0$ when $t = t_1$. We use the standard map in the form $y_{n+1} = y_n + \frac{k}{2\pi}\sin(2\pi x_n), x_{n+1} = x_n + y_{n+1}$, with x_n, y_n endowed with mod(1) in the intervals [0,1) and [−0.5,0.5), respectively.

In Fig. 1a the dotted line gives the time evolution of the APLE along a regular orbit with initial conditions $(x_0, y_0) = (0.0, 0.07)$ for $k = 10^{-4}$, forming a librational torus close to the stable periodic orbit at (0,0.5). The initial deviation vector $\boldsymbol{\xi}(0)$ is perpendicular to this torus. The solid line in Fig. 1a gives the behavior of the APLE in the case of a weakly chaotic orbit for $k = 10^{-4}$ with initial conditions on the unstable eigendirection, i.e. on the unstable asymptotic curve emanating from the unstable periodic orbit at $(0, 0)$ at a small distance (10^{-8}) from $(0, 0)$. The initial deviation vector $\boldsymbol{\xi}(0)$ is perpendicular to this eigendirection.

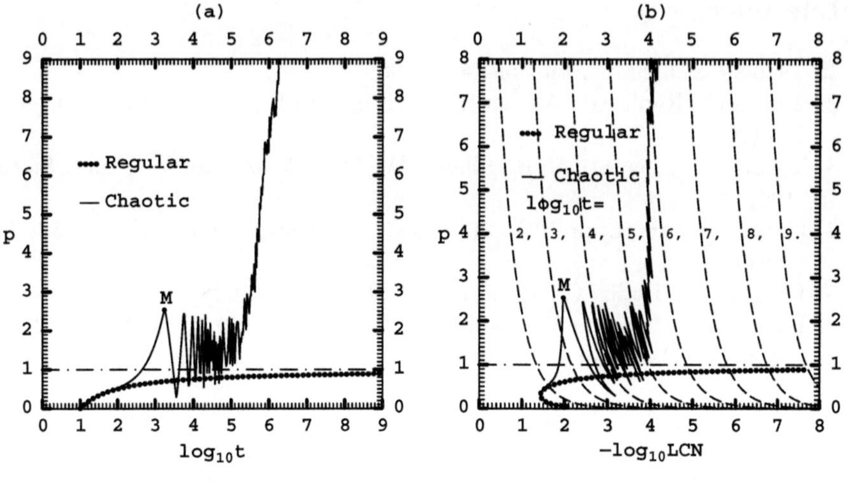

Fig. 1. The APLE p for a regular orbit (*dotted line*) and a weakly chaotic orbit (*solid line*) (**a**) vs. time t and (**b**) vs. the LCN. The isochrone curves are plotted by *dashed lines* for the iteration times $t = 10^2, \dots, 10^9$

In the case of the regular orbit (dotted line in Fig. 1a) the APLE grows slowly tending asymptotically to 1 from smaller values. In the case of the chaotic orbit (solid line in Fig. 1a) APLE grows rapidly, crosses the value 1 at a short time (at about $t = n = 400$ iterations after the first $t_1 = 10$ iterations) and reaches a first local maximum value when the orbit passes near the first homoclinic point M. Then the APLE describes a number of oscillations around a mean value that remains systematically larger than unity and for times longer than the Lyapunov time $t_L = 1/LCN$ increases tending exponentially to infinity.

For a better comparison of the evolution of the APLE with that of the finite time Lyapunov characteristic number $LCN = \frac{\ln \xi(t)}{t}$ Fig. 1b shows the evolution of the two orbits on the plane $-\log_{10} LCN$ vs APLE p The isochrone curves are also plotted in the same figure by dashed lines for the iteration times $t = 10^2, \ldots, 10^9$. In the case of the regular orbit (dotted line), as $-\log LCN$ tends to 0, the APLE tends to the value 1 from smaller values. In the case of the chaotic orbit, for a period t_L of the order $t_L = 10^4$ iterations the quantity $-\log LCN$ increases on average, while p oscillates around a mean value larger than 1. Later on, when $-\log_{10} LCN$ tends to a constant value of about $LCN = 10^{-4}$, p tends to infinity.

Acknowledgments

G. Lukes-Gerakopoulos is supported by the Greek Foundation of State Scholarships (IKY) and by the Research Committee of the Academy of Athens.

References

1. Tsallis C., *J. Stat. Phys.* **52**, 479 (1988)
2. Tsallis C., Rapisarda A., Latora V., Baldovin F., *Lect. Notes Phys.* **602**, 140 (2002)
3. Tsallis C., Plastino A. R., Zheng W.-M., *Chaos Solitons Fract.* **8**, 885 (1997)
4. Costa U. M. S., Lyra M. L., Tsallis C., Plastino A., *Phys. Rev. E* **56**, 245 (1997)
5. Lyra M. L., Tsallis C., *Phys. Rev. Lett.* **80**, 53 (1998)
6. Baranger M., Latora V., Rapisarda A., *arXiv:cond-mat/0007302* (2000)
7. Latora V., Baranger M., Rapisarda A., Tsallis C., *Phys. Lett. A* **273**, 97 (2000)
8. Grassberger P., Scheunert M., *J. Stat. Phys.* **26**, 697 (1981)
9. Baldovin F., Tsallis C., Schulze B., *PhysicaA* **320**, 184 (2003)
10. Baldovin F., Brigatti E., Tsallis C., *Phys. Lett. A* **320**, 254 (2004)
11. Añaños G. F. J, Baldovin F., Tsallis C., *Eur. Phys. J. B* **46**, 409 (2005)
12. Benettin G., Galgani L., Giorgilli A., Strelcyn J.-M., *Meccanica* **15**, 9 (1980)

Global Dynamics of Coupled Standard Maps

T. Manos[1,2], Ch. Skokos[3], and T. Bountis[1]

[1] Department of Mathematics, Center for Research and Applications of Nonlinear Systems (CRANS), University of Patras, GR–26500, Greece
thanosm@master.math.upatras.gr, bountis@math.upatras.gr
[2] LAM, OAMP, 38, rue Frederic Joliot-Curie, 13388 Marseille cedex 13, France
[3] Astronomie et Systèmes Dynamiques, IMCCE, Observatoire de Paris, 77 Av. Denfert–Rochereau, F–75014, Paris, France
hskokos@imcce.fr

Summary. Understanding the dynamics of multi-dimensional conservative dynamical systems (Hamiltonian flows or symplectic maps) is a fundamental issue of nonlinear science. The Generalized ALignment Index (GALI), which was recently introduced and applied successfully for the distinction between regular and chaotic motion in Hamiltonian systems [1], is an ideal tool for this purpose. In the present paper we make a first step towards the dynamical study of multi-dimensional maps, by obtaining some interesting results for a 4-dimensional (4D) symplectic map consisting of $N = 2$ coupled standard maps [2]. In particular, using the new GALI_3 and GALI_4 indices, we compute the percentages of regular and chaotic motion of the map equally reliably but much faster than previously used indices, like GALI_2 (known in the literature as SALI).

1 Definition and Behavior of GALI

Let us first briefly recall the definition of Generalized Alignment Index (GALI) and its behavior for regular and chaotic motion, adjusting the results obtained in [1] to symplectic maps. Considering of a $2N$-dimensional map, we follow the evolution of an orbit (using the equations of the map) together with k initially linearly independent deviation vectors of this orbit $\vec{\nu}_1, \vec{\nu}_2, ..., \vec{\nu}_k$ with $2 \leq k \leq 2N$ (using the equations of the tangent map). The GALI of order k is defined as the norm of the wedge or exterior product of the k unit deviation vectors:

$$GALI_k(i) = \| \hat{\nu}_1(i) \wedge \hat{\nu}_2(i) \wedge ... \wedge \hat{\nu}_k(i) \| \qquad (1)$$

and corresponds to the volume of the generalized parallelepiped, whose edges are these k vectors. We note that the hat (ˆ) over a vector denotes that it is of unit magnitude and that i is the discrete time.

G. Contopoulos, P.A. Patsis, *Chaos in Astronomy.* Astrophysics and Space Science Proceedings, doi: 10.1007/978-3-540-75826-6,

In the case of a chaotic orbit all deviation vectors tend to become linearly dependent, aligning in the direction of the eigenvector which corresponds to the maximal Lyapunov exponent and GALI_k tends to zero following an exponential law $\sim e^{-[(\sigma_1-\sigma_2)+(\sigma_1-\sigma_3)+\ldots+(\sigma_1-\sigma_k)]i}$, where $\sigma_1,\ldots,\sigma_k$ are approximations of the first k largest Lyapunov exponents. In the case of regular motion on the other hand, all deviation vectors tend to fall on the N-dimensional tangent space of the torus on which the motion lies. Thus, if we start with $k \leq N$ general deviation vectors they will remain linearly independent on the N-dimensional tangent space of the torus, since there is no particular reason for them to become aligned. As a consequence GALI_k remains practically constant for $k \leq N$. On the other hand, GALI_k tends to zero for $k > N$, since some deviation vectors will eventually become linearly dependent, following a particular power law, i. e. $GALI_k(i) \sim i^{2(N-k)}$.

2 Dynamical Study of a 4D Standard Map

As a model for our study we consider the 4D symplectic map:

$$\begin{array}{ll} x_1' = x_1 + x_2' & x_2' = x_2 + \frac{K}{2\pi}\sin(2\pi x_1) - \frac{B}{2\pi}\sin[2\pi(x_3 - x_1)] \\ x_3' = x_3 + x_4' & x_4' = x_4 + \frac{K}{2\pi}\sin(2\pi x_3) - \frac{B}{2\pi}\sin[2\pi(x_1 - x_3)] \end{array} (mod\, 1), \quad (2)$$

which consists of two coupled standard maps [2] and is a typical nonlinear system, in which regions of chaotic and regular dynamics are found to coexist. In our study we fix the parameters of the map (2) to $K = 0.5$ and $B = 0.05$. In Fig. 1, we show the behavior of GALIs for two different orbits: a regular orbit R with initial conditions $(x_1, x_2, x_3, x_4) = (0.55, 0.10, 0.54, 0.01)$ (Fig. 1a), and a

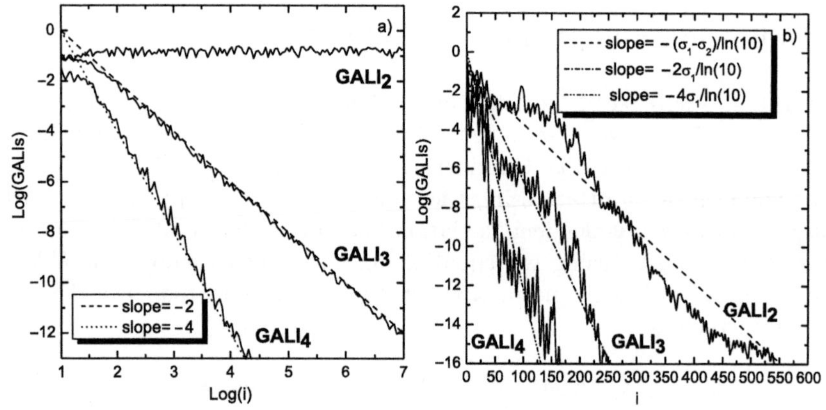

Fig. 1. The evolution of GALI_k, $k = 2, 3, 4$, with respect to the number of iterations i for (**a**) the regular orbit R and (**b**) the chaotic orbit C. The plotted lines correspond to functions proportional to n^{-2}, n^{-4} in (**a**) and to $e^{-(\sigma_1-\sigma_2)i}$, $e^{-2\sigma_1 i}$, $e^{-4\sigma_1 i}$ for $\sigma_1 = 0.070$, $\sigma_2 = 0.008$ in (**b**)

chaotic orbit C with initial conditions $(x_1, x_2, x_3, x_4) = (0.55, 0.10, 0.005, 0.01)$ (Fig. 1b). The positive Lyapunov exponents of orbit C were found to be $\sigma_1 \approx 0.070$, $\sigma_2 \approx 0.008$. From the results of Fig. 1 we see that the evolution of GALIs is described very well by the theoretically obtained approximations presented in Sect. 1.

Let us now turn our attention to the study of the global dynamics of map (2). From the results Fig. 1 we conclude that in the case of 4D maps, $GALI_2$ has different behavior for regular and chaotic orbits. In particular, $GALI_2$ tends exponentially to zero for chaotic orbits ($GALI_2 \sim e^{-(\sigma_1-\sigma_2)i}$) while it fluctuates around non-zero values for regular orbits. This difference in the behavior of the index can be used to obtain a clear distinction between regular and chaotic orbits. Let us illustrate this by following up to $i = 4{,}000$ iterations, all orbits whose initial conditions lie on a 2D grid of 500×500 equally spaced points on the subspace $x_3 = 0.54$, $x_4 = 0.01$, of the 4-dimensional phase space of the map (2), attributing to each grid point a color according to the value of $GALI_2$ at the end of the evolution. If $GALI_2$ of an orbit becomes less than 10^{-10} for $i < 4{,}000$ the evolution of the orbit is stopped, its $GALI_2$ value is registered and the orbit is characterized as chaotic. The outcome of this experiment is presented in the left panel of Fig. 2.

But also $GALI_4$ can be used for discriminating regular and chaotic motion. From the theoretical predictions for the evolution of $GALI_4$, we see that after $i = 1{,}000$ iterations the value of $GALI_4$ of a regular orbit should become of the order of 10^{-16}, since $GALI_4 \sim t^{-4}$, although the results of Fig. 1 show that more iterations are needed for this threshold to be reached, due to an initial transient time where $GALI_4$ does not decrease significantly. On the other hand, for a chaotic orbit $GALI_4$ has already reached extremely small

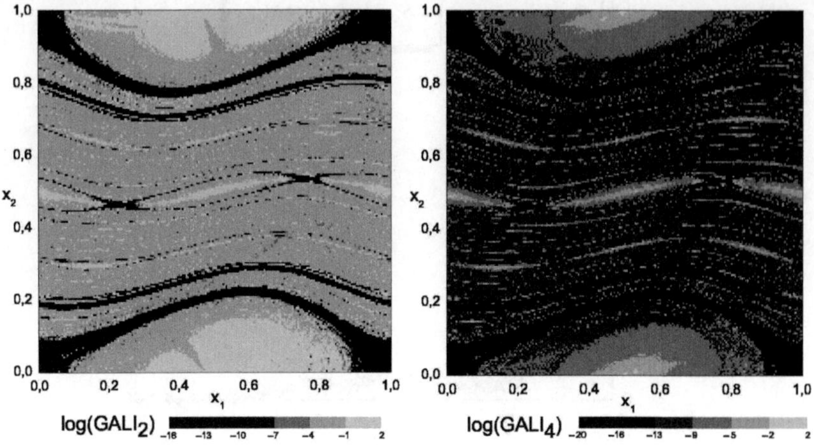

Fig. 2. Regions of different values of the $GALI_2$ (*left panel*) and $GALI_4$ (*right panel*) for a grid of 500×500 initial conditions on the subspace $x_3 = 0.54$, $x_4 = 0.01$ of map (2) for $K = 0.5$ and $B = 0.05$

values at $i = 1{,}000$ due to its exponential decay ($\text{GALI}_4 \sim e^{-4\sigma_1 i}$). Thus, the global dynamics of the system can be revealed as follows: we follow the evolution of the same orbits as in the case of GALI_2 and register for each orbit the value of GALI_4 after $i = 1000$ iterations. All orbits having values of GALI_4 significantly smaller than 10^{-16} are characterized as chaotic, while all others are considered as non-chaotic. In the right panel of Fig. 2 we present the outcome of this procedure.

From the results of Fig. 2, we see that both procedures, using GALI_2 or GALI_4 as a chaos indicator, give the same result for the global dynamics of the system, since in both cases 16% of the orbits are characterized as chaotic. These orbits correspond to the black colored areas n both panels of Fig. 2. One important difference between the two procedures is their computational efficiency. Even though GALI_4 requires the computation of four deviation vectors, instead of only two that are needed for the evaluation of GALI_2, using GALI_4 we were able to get a clear dynamical 'chart', not only for less iterations of the map (1,000 instead of 4,000 needed for GALI_2), but also in less CPU time. In particular, for the computation of the data of the left panel of Fig. 2 (using GALI_2) we needed 1 h of CPU time on an Athlon 64 bit, 3.2 GHz PC, while for the data of the left panel of the same figure (using GALI_4) only 14 min of CPU time were needed.

Using the above-described method, both for GALI_2 and GALI_4, we were able to compute very fast and accurately the percentages of regular motion for several values of parameter B. In Fig. 3 we plot the percentage of regular

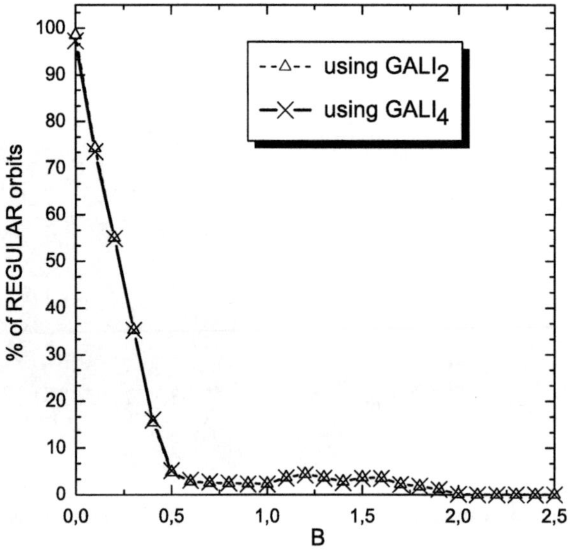

Fig. 3. Percentages of regular orbits on the subspace $x_3 = 0.54$, $x_4 = 0.01$ of map (2) for $K = 0.5$, as a function of the parameter $B \in [0, 2.5]$

orbits for $B \in [0, 2.5]$ where B varies with a step $\delta B = 0.1$. We see that the two curves practically coincide, but using $GALI_2$ we needed almost four times more CPU time. So, it becomes evident that a well-tailored application of $GALI_k$, with $2 < k$, can significantly diminish the CPU time required for the detailed 'charting' of phase space regions, compared with that for $GALI_2$.

Acknowledgments

T. Manos was supported by the "Karatheodory" graduate student fellowship No B395 of the University of Patras, the program "Pythagoras II" and the Marie Curie fellowship No HPMT-CT-2001-00338. Ch. Skokos was supported by the Marie Curie Intra–European Fellowship No MEIF–CT–2006–025678.

References

1. Ch. Skokos, T. Bountis and Ch. Antonopoulos, *Physica D*, **231**, 30, (2007)
2. H. Kantz and P. Grassberger, *J. Phys. A: Math. Gen*, **21** L127, (1988)

Part IV

Other Astronomical Systems

The Chaotic Light Curves of Accreting Black Holes

D. Kazanas

Astrophysics Science Division, Code 663, NASA/GSFC, Greenbelt, MD 20771, USA
Demos.Kazanas@nasa.gov

Summary. We present a brief overview of the temporal and spectral properties of the light curves of accreting black holes. Particular attention is paid on their apparently chaotic character which, as argued, are in fact stochastic. A brief review of the processes that can provide variability power spectral densities similar to those observed is also presented. Considering that the process of spectrum formation is the Comptonization of soft photons by hot electrons, we focus on the analysis of the cross spectra between different photon energies and their corresponding lags. We argue that the observed lag dependence on the Fourier frequency can be easiest accounted for if the process of Comptonization takes place in inhomogeneous plasma clouds that extend over several decades in radius. It is argued that the lag dependence on the Fourier frequency is a direct manifestation of the density profile of the Comptonizing corona and that the majority of data sets argue for density profiles consistent with ADIOS rather than ADAF type of flows.

1 Introduction

The study of the physics of accretion powered sources, whether on galactic (X-ray binaries, neutron stard or Black Hole Candidates; BHC) or extragalactic systems (AGN), involves length scales much too small to be resolved by current technology. As such, this study is conducted mainly through the theoretical interpretation of their spectral and temporal properties. Until recently studies of this class of objects has been focused, for technical mainly reasons, on their spectra. The latter comprise two generic components:(a) A broad quasi-thermal component thought to represent emission by a geometrically thin, optically thick accretion disk that extends from the Innermost Stable Circular Orbit (ISCO) to several hundred Schwarzschild radii emitting the locally dissipated energy in black body form; its peak emission frequency depends on the mass of the accreting compact object, producing the sub keV Multicolor Black Body (MBB) component in accreting binaries or the O-UV Big Blue Bump (BBB) in AGN. (b) A high energy component that extends to $\gtrsim$100 keV both in accreting binaries and AGN. The spectrum of this latter

G. Contopoulos, P.A. Patsis, *Chaos in Astronomy.* Astrophysics and Space Science Proceedings, doi: 10.1007/978-3-540-75826-6,

component is fit very well by the spectra of Comptonization of soft photons by hot ($T_e \lesssim 10^9$ K) electrons, a process explored in great depth over the past twenty or so years [18]. The physical location of the electrons required to produce this component is in a hot corona overlying the geometrically thin, much cooler accretion disk. Because the electron temperatures of the matter accreting onto a black hole are expected to be similar to those necessary to produce the observed 2–100 keV spectra, it has been considered that detailed spectral fits of these sources would lead to insights on the dynamics of accretion onto the compact object.

However, the determination of accretion dynamics requires the knowledge of the density and size of the emitting region, neither of which is provided by radiative transfer and spectral fitting considerations especially those of the Comptonization (the equations of radiative transfer that determine the Comptonization spectra involve the optical depth as the independent variable). Indeed, as shown explicitly in [6] and [5], plasmas of very different radial extent and density profiles can yield identical Comptonization spectra. The degeneracy of this situation can lifted with the additional information provided by timing observations.

However, the timing properties of BHC lead to a rather confused picture: The MBB or BBB components representing emission from the smallest scales of these systems seem to be impervious to the general variability of these objects [8–10]; they appear to respond on time scales much longer than those of the hard, power law component due to the Comptonization by the electrons of the hot corona, despite the presumed proximity of these components.

The X-ray variability of accreting compact objects is by and large dominated by the variations of their high energy (2–100 keV) power law components. Inspection of these light curves in the time domain indicates their variations to be largely random. Greater insights on their variability properties is gained through Fourier analysis. Such analysis has shown that the variability properties of X-rays are generally inconsistent with the notion that the corresponding luminosity is emitted by gas plunging into the black hole from a region of a few Schwarzschild radii, R_S: The Power Spectral Densities (PSD) of BHC exhibit most of their power at scales ~1 s, far removed from the characteristic time scales associated with the fluid dynamics of the vicinity of the black hole horizon, $R_S/c \sim 10^{-3}$ s (the variability of neutron stars is to a certain extent an exception, as they present most of their variability power at frequencies $\nu \sim 10^3$ Hz, a fact that was proposed as a discriminant between neutron stars and black holes [11]; however, there still a need to account for their variability at lower frequencies as well as for the QPO features in their PSD). Until recently rather little attention has been paid to this time scale discrepancy; rather, more attention was paid to the flicker noise–type ($\propto f^{-1}$) of their PSDs. This PSD functional form is of interest because it contains roughly equal power per logarithmic frequency interval, indicating the absence of a characteristic time scale associated with the variations of the source luminosity.

2 From Light Curves to Dynamics

The apparent random character of the light curves of accreting black holes has raised the issue of whether these are completely stochastic or just chaotic, whereby the last term implying the existence of a low dimensionality strange attractor that is responsible for their apparent random character. It this indeed were the case it would severely limit and constrain dynamical models the dynamics of accretion.

To answer this question, [12] employed the well developed procedures of correlation integrals to search for a low dimensionality attractor in the light curves. The method relies in using the time delays to reconstruct the phase space profile of the underlying dynamics and then determine whether the trajectories in phase space fill up its entire volume or are restricted to a submanifold of it; this is decided by constructing the correlation integral by determining the dimension of this submanifold.

Operationally, given the light curve of a source represented by the discrete function $f(t_i)$ one has to proceed as follows: (a) Create the n-dimensional vectors V composed of the value of the flux

$$V_n = \{f(t),\ f(t+\tau),\ \ldots,\ f(t+[n-1]\tau)\} \tag{1}$$

(b) Form the Correlation Integral $C(r)$ defined by

$$C(r) \equiv \frac{1}{M^2}\sum_{i,j}^{M}\theta(r - |V_i - V_j|) \tag{2}$$

as a function of the norm r of the vectors V, where M is the total number of points in the time series. (c) Search for a value of the *embedding dimension* D defined by

$$D = \frac{dlogC(r)}{dlogr} \tag{3}$$

above which the correlation integral slope does not change. The embedding dimension D is in general non-integer and represents the dimensionality of the underlying (strange) dynamical attractor of the corresponding dynamical Hamiltonian.

The method and its application on the HEAO 1 X-ray data is described in detail in [12]. The analysis described in this reference has shown that the underlying dimensionality of an attractor (if indeed there is one) is greater than 12, suggesting that the corresponding light curves are effectively stochastic.

3 The Power Density Spectra

Despite the apparent stochasticity of the light curves, their Power Spectral Densities (PSD) are very different from white noise, the PSD one would naively associate with a truly random process. Rather, the Power Spectral

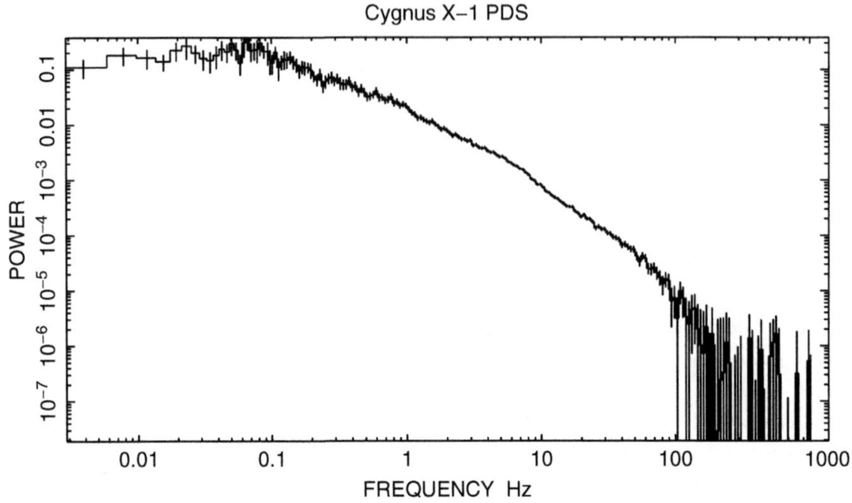

Fig. 1. The PSD of Cyg X-1 in its hard state

Densities were found to be in general power laws of flicker noise–type ($\propto f^{-1}$), as shown in Fig. 1, i.e. variability with roughly equal power per logarithmic frequency interval. This was thought to be rather amazing fact, implying the presence of interesting underlying physics which, given that most of the luminosity in accretion powered sources is derived at the smallest radii, determine the way that the black holes is fed.

Various attempts have been made to reproduce this behavior. The most obvious way to do so is to invoke an ensemble of exponential shots with a distribution of decay times $\tau(f)$; given that the PSD of a single shot is proportional to $1/[f^2+\tau^2(f)]$, a judicious choice of the distribution of $\tau(f)$ can indeed provide power spectra wi In fact, this particular behavior apparently depends on the spectral state $\propto f^{-1}$, and in fact of any index less than 2. Unfortunately, such models are purely kinematic without any relevance to the dynamics of accretion.

A different approach was that of [14], who showed that the scale free variability PSD of AGN (very similar to those of BHC) in a frequency band that covers roughly two decades (10^{-3}–10^{-5} Hz) in frequency can be attributed to emission by an ensemble of hot spots in Keplerian orbits; these authors showed that the observed variability can be attributed entirely to the Keplerian motion of these spots in the relativistic potential of a black hole and the ensuing Doppler beaming.

Finally, an account of the flicker noise of the power spectra of accreting compact objects in terms of the dynamics of accretion disks was provided by Lyubarskii [13], who showed that local fluctuations in the disk parameter α can produce variations in the accretion rate onto the compact object with frequency dependence consistent with the observed PSD. Interestingly, he showed that such a behavior is possible both for the standard geometrically thin, optically thick accretion disk and advection dominated disks (ADAF). It is worth noticing that the frequency of highest variability power of fig. 1, i.e $\sim$0.1 Hz, in the case of ADAF is associated with radii much larger (by a factor of $\sim 10^3$) than the compact object radius.

4 The Lags

Novel insights into the variability properties were introduced by a novel type of analysis introduced by Miyamoto [15, 16]. This author, motivated by the fact that the $\epsilon \gtrsim 1$ keV spectra of galactic BHC are thought to be produced by Comptonization, sought to observe the lags between two different energy bands induced by this process (in Comptonization, the energy of a photon increases exponentially with the scattering time $\tau_s = 1/n\sigma_T c$ in the plasma, where n is the plasma density, σ_T the Thompson cross section, c the speed of light; hence variations at an energy ϵ_1 should lag those at a lower energy ϵ_2 by $\Delta\tau \simeq \tau_s \, \text{Log}(\epsilon_1/\epsilon_2)$). To do so he formed the cross spectrum between two different energy bands including the Fourier phases ϕ_1, ϕ_2 and plotted their difference $\Delta\phi$, or equivalently the corresponding time lag $\Delta\tau = \Delta\phi/f$ as a function of the Fourier frequency f or the Fourier period $P = 1/f$. The results were sufficiently surprising to lead the author to question Comptonization as the process responsible for the formation of the X-ray spectrum: The time lags were found to increase in proportion to the Fourier period P, being as long as $\sim$0.1 s at a Fourier period of $\sim$10 sec. An example of this dependence is given in Fig. 2; open circles indicate positive lags between the 1–3 and 5-10 keV bands in seconds, filled circles negative lags between the same bands. The discrepancy between observation and expectation can be appreciated by comparing the data with the lags expected from a hot quasi-uniform corona of size a few time the Schwarzschild radius of a 10 solar mass black hole, or $R \simeq 10^7$ cm, depicted by the dotted lines; while the expected lags are of similar magnitude at high frequencies with those observed the linear increase of those measured with the Fourier period P is in stark contrast with the constant lags expected in the generally accepted models.

These discrepancies between the expected and the observed variability of BHC led [6], [3], [5] and [7] to propose that, contrary to the prevailing notions, the size of the scattering region responsible for the X-ray emission is not $R \simeq 3 - 10\ R_S$ but rather $R \gtrsim 10^3 R_S$, as implied by the PSDs and lag observations. Furthermore, the scattering medium (corona) is *inhomogeneous*, with the electron density following the law,

$$n(r) = \begin{cases} n_1 & \text{for } r \leq r_1 \\ n_1 (r_1/r)^p & \text{for } r_2 > r > r_1 \end{cases} \tag{4}$$

where r is the radial distance from the center of the corona (assumed to be spherical) and r_1, r_2 are its inner and outer radii, respectively. The index

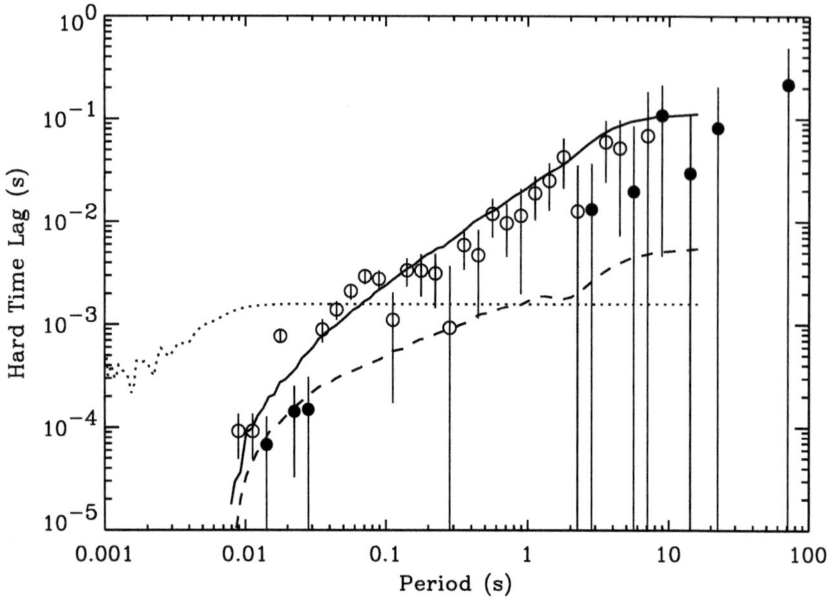

Fig. 2. The time lags of Cyg X-1 as a function of Fourier period P; open circles positive lags, filled circles negative lags. Lines represent lags produced by scattering in configurations of different radial density profiles. Solid line $p = 1$, dashed line $p = 3/2$, dotted line $p = 0$

$p > 0$ is a free parameter whose value depends on the specific dynamical model that determines the electron density. For example, the ADAF of [17] suggest $p = 3/2$ and $T_e \lesssim 10^9$ at radii as large as $r \simeq (m_p/m_e)R_S$, while models combining inflow and outflow [2], [1] (ADIOS) allow a broader range of density profiles, including those with $p \simeq 1$. As pointed out in [6] and [5] most of the data analyzed to date suggest $p = 1$ with $p = 3/2$ also acceptable in certain cases.

It is intuitively easy to see that scattering in the extended configuration of (4) can produce time lags over a range of Fourier periods similar to the range of radii span by the hot corona: Scattering at a given radius R increases the X-ray energy and introduces a lag $\Delta\tau \lesssim R/c$ between the scattered and unscattered photons, the lag appearing at a Fourier period $P = R/c$ [5]. To compare this to the lags given in [15, 16] (which are the average over all such photon pairs scattered in a given decade in R, as a function of R) one has to multiply $\Delta\tau$ by the probability of scattering the photons at a given radius, $\mathcal{P}(R) \simeq \tau(R)$, with $\tau(R)$ the scattering depth over the radius R. For the density profile of (4), $\tau(R) \propto R^{-p+1}$ and since the Fourier period $P \propto R$, $\langle\Delta\tau\rangle \propto R^{-p+2} \propto P^{-p+2}$.

Figure 2 presents the lags of the Cyg X-1 data as a function of the Fourier period P along with fits of Monte Carlo simulations [5] corresponding to scattering configurations of density profiles with different values of the parameter p: Solid line $p = 1$, dashed line $p = 3/2$, dotted line $p = 0$. These indicate that the scaling arguments and analytical considerations discussed above are essentially correct . In addition, these simulations showed that the configuration of (4) produces light curves of high coherence [3, 20] over the range $[r_2/c, r_1/c]$ of the Fourier period, provided that the soft photons are injected near its center. Finally, one should note that while the majority of this type of analysis was performed on the light curves of galactic BHC, similar analysis of the light curve of the AGN NGC 7469 [19] has produced similar results.

One should conclude this particular section by noting that the spectra corresponding to each of these different density profiles are independent of the density profile of the scattering corona. This is not surprising considering that the resulting spectra depend only on the product $\int d\tau(kT/m_e c^2)$ along the observer's line of sight. So while the spectrum provides a constraint on the integral of the plasma column, the lags provide constraints on the differential distribution of column with radius $d\tau/dR$.

5 Modeling the X-ray Light Curves

Following the above analysis, one can now produce model light curves that conform both to the lag constraints, the power spectrum and the spectra of a given source. To this end we construct light curves that are produced by an injection soft photon shots with a Poisson distribution at times $t_i = t_{i-1} + \Delta t_i$ where $\Delta t_i = \bar{t}\ |lnR_i|$ near the center of a configuration with density profile given by Eq. (4), where $\bar{t}$ is the mean time between shots and R_i a random number between 0 and 1. Each such event produces at the outer edge of the scattering configuration an X-ray pulse with time profile that of the response function of this configuration which is of the form $g(t) \propto t^{\alpha-1}\, e^{-t/\beta}$ $(0 < \alpha < 1,\ \beta > 0)$. The final light curve will then be of the form

$$F(t) = \sum_{i=1}^{N} Q_i \theta \left(t - \sum_{i=1}^{N} t_i \right) g(t - t_i)\ . \tag{5}$$

where $\theta(t - t_0)$ is the step function. The values of α and β depend on the photon energy providing responses which are slightly flatter and extend to slightly longer times with energy given that higher energy photons take longer to escape the Comptonizing medium. Given that the final light curve is the convolution of the response function with a Poisson process, the final power spectrum is the product of the corresponding power spectra. The Fourier transform of $g(t)$ is $G(\omega) = \Gamma(\alpha)\beta^{\alpha}(1+\beta^2\omega^2)^{\alpha/2} e^{i\alpha\theta}/\sqrt{2\pi}$, where $\omega = 2\pi\nu = 2\pi/P$, $\Gamma(\alpha)$ is the Gamma function, and $tan\theta = \beta\omega$, while the Poisson PSD is simply a constant. This then implies that the corresponding PSD is $|G(\omega)|^2 \propto (1+\beta^2\omega^2)^{\alpha}$, i.e. a power law with a low frequency break to a Poisson spectrum, as observed, while all the phase information is contained in the term $e^{i\alpha\theta}$.

The dependence of the lags on the Fourier period P can be assessed from the form of the slight difference in the form of response function with energy: The higher the photon energy the larger is the value of β (harder photons stay longer in the scattering cloud and as such they escape at later times) and the larger the value of α (the profile is flatter because the photons are spread over longer time scales). Then, considering the parameters (α_1 ,β_1 and α_2, β_2 of the light curves corresponding to two different energies $\epsilon_1 < \epsilon_2$ we have the following expression for the time lag

$$\delta t = \frac{1}{\omega}\left[\alpha_2 \arctan(\beta_2\omega) - \alpha_1 \arctan(\beta_1\omega)\right] \ . \tag{6}$$

For small P or $\beta\omega \gg 1$ the above formula gives $\delta t \simeq (\alpha_2 - \alpha_1)P/4$, in agreement with the arguments given in the previous section; for long times (large P, $\beta\omega \ll 1$ the lag saturates to the value $\alpha_2; \beta_2 - \alpha_1\beta_1$. In fact, an analysis of the light curves of Cyg X-1 along the lines just described gave the results (open and filled circles) of Fig. 2, along with fits for a Monte Carlo calculation of extended ($p = 1$, solid curve; $p = 3/2$, dashed curve) and compact uniform ($p = 0$, dotted curve) plasma coronae. On the other hand, for a uniform density profile, $\alpha_1 = \alpha_2 = 1$ and the response function has an exponential form; expanding the above expression for the lag to to higher order in $\beta\omega$ we obtain $\delta t \simeq \alpha(1/\beta_1 - 1/\beta_2)P^2$ saturating at a value $\propto \beta_2 - \beta_1$ with power spectra proportional to $1/(1+\beta^2\omega^2)$ as discussed earlier.

Figure 3 shows the light curve that results from the above procedure. Its appearance, after a turning on time of $t \simeq \beta$ is very similar to that of Cyg X-1, its spectrum consistent with observations and the PSD has also the proper

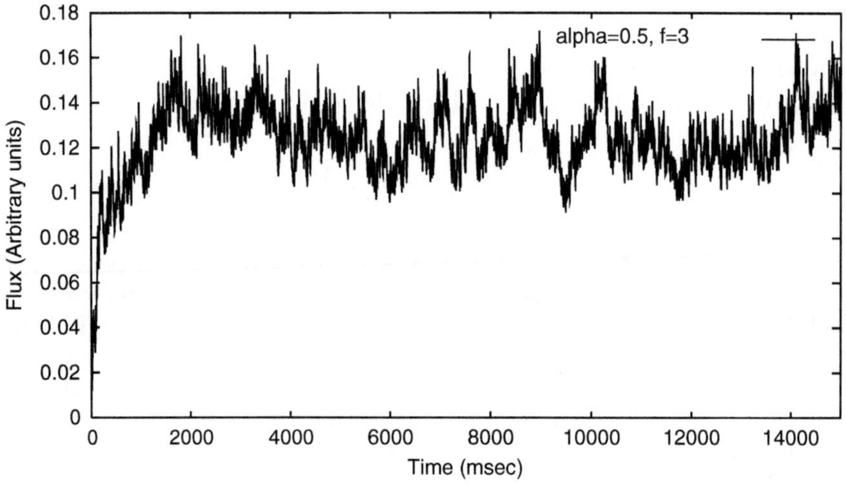

Fig. 3. A model light curve produced as discussed in the text using a scattering medium response function with $\alpha = 0.5$, $\beta = 1.5$ s and $\bar{t} = 0.003$ s, simulating the corresponding light curves of accreting galactic black holes candidates

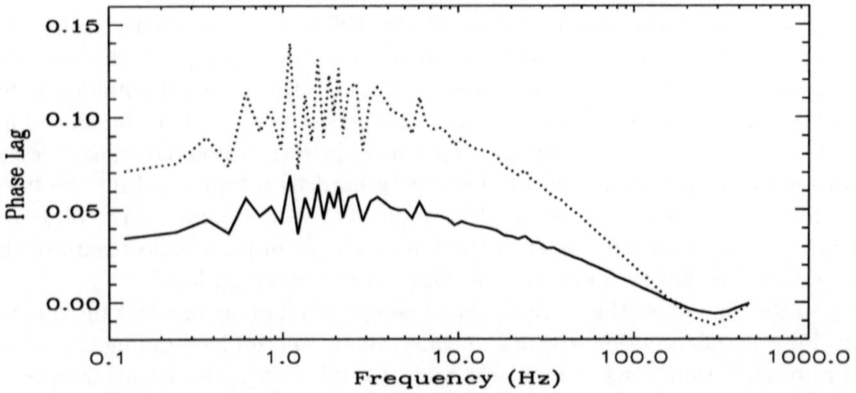

Fig. 4. Phase lags as a function of Fourier frequency associated with model light curves of the type shown in Fig. 3. Three light curves were produced corresponding to $\alpha = 0.5$, 0.55, 0.6 and $\beta = 16$ s. The dotted curve corresponds to the lags between the first and third light curve while the solid curve to the lags between the first and the second one

form, in agreement with observations. The remaining issue is that of the resulting lags. We have produced a similar light curve corresponding to three different energy bands with different corresponding parameters $\alpha = 0.5, 0.55, 0.6$ and $\beta = 16$ sec for all curves. The corresponding phase lags are shown in Fig. 4 between the first and second energy bands (solid line) and between the first and third (dotted line). The phase lags are almost constant across four decades of Fourier period, indicating that the time lags are proportional to the Fourier period as observed.

6 Conclusions, Discussion

We have discussed above that the light curves of accreting compact objects present a number of challenges in the understanding of the underlying dynamics. It was noticed early on that they appear to be random, consisting of shots fit reasonably well by exponentials; however, their corresponding PSD are far from those of a single exponential; kinematic models that fit the PSD as a superposition of such shots provide little insight on the underlying dynamics.

The deeper understanding and development of non-linear dynamics and strange attractors that can provide chaotic trajectories in systems with a small number of degrees of freedom raised the issue that the underlying dynamics of these systems may indeed be simple; this led the explorations discussed in §2 which were useful in pointing out that the number of degrees of freedom are larger than 12, implying that quite likely the dynamics of X-ray emission is by and large stochastic.

The issue of the interpretation of the PSDs was advanced significantly by the work of [13], who managed to related it in a simple fashion to the accretion dynamics that power these sources. In the personal opinion of this author, the possibility of applying this treatment equally well in thin and thick disks provides a great advantage, especially since the observed characteristic time scales of breaks in the PSD are very hard to comprehend on the basis of thin disks. However, these models, while they can account for the observed PSD, because they emit most of the luminosity from the smallest radii of the accretion flow fail to produce an account of the observed lags.

In the opinion of this author, the cross spectra and lag analysis introduced in [15] presents a major advance in constraining and understanding the structure of accretion flows onto black holes. Coupled with the resulting spectra they present a comprehensive spectral – temporal approach which, because it uses all available information, is the approach of choice in the study of accretion onto black holes. This type of analysis provided a novel picture of the structure of these flows and it is the only type of analysis that provides in a straight forward way constraints on the density structure of these flows. In this respect, the preponderance of flows with density profile characterized by $p = 1$ for the density parameter is rather astounding; within the confines of the model that accounts for these lags, their form indicates that underlying flows similar to ADIOS, with density profile $n(r) \propto r^{-1}$, are preferred to those of ADAF with $n(r) \propto r^{-3/2}$.

A simple model for producing light curves consistent with all these constraints was presented in §5, based on the notions of scattering in the extended non-uniform coronae of §4. This model is very economical in its assumptions and produces stochastic light curves simply because the injection of soft photons at the base of the flow is totally stochastic and because the Comptonization process that acts on these photons is also a stochastic one. However, the reprocessing of these photons through the filter of Comptonization in the non-uniform corona, yields light curves with the correct power spectra, lags and energy spectra. Inspection of these light curves does indicate shots of varying durations, as the real data do. However, the shots that result from reprocessing of a delta function injection of soft photons near $r = 0$ through the non-uniform corona, is a shot of power law form rather than an exponential shot. While this is hard to discern in the light curve that is the superposition of these shots (they can be also fit as exponential ones over limited span in time), this makes all the difference in the formation of the PSD and also with the overall appearance of the light curves.

One should caution the casual reader that the calculations of [3, 5, 6] were based on coronae with constant temperature profiles. It is expected however that for large radii the electron temperature is likely to decrease. If the density profile remains the same (e.g. $n(r) \propto r^{-1}$), while the magnitude of the lags will still be determined by the arguments given earlier and therefore preserve their form as a function of the Fourier period, they may actually become *negative* in the sense that hard X-ray precede the soft one, which are now the result of *downscattering* of hot photons on colder electrons. This effect may already have been observed (see e.g. Fig. 2 at $P \gtrsim 10$ sec.), albeit with very large errors.

Finally, one should bear in mind that this way of producing the PSD of accreting sources is not at odds with the work of [13]; one could in fact inject soft photons at the base of the flow at the rates discussed in this reference. As discussed above, the resulting spectrum will be the product of the spectra of the two processes. This makes these models more complicated and perhaps this is the reason they have not pursued so far in earnest.

Acknowledgments

I would like to thank the organizing committee for the invitation to present the above material at the meeting as well as for their financial support. Additional support for this work was provided by INTEGRAL and Chandra GO grants.

References

1. R. D. Blandford & M. C. Begelman: MNRAS **303**, L1 (1999)
2. J. Contopoulos, & R. V. E. Lovelace: ApJ, **429**, 139 (1994)
3. X.-M. Hua, D. Kazanas, & L. G. Titarchuk: ApJ, **482**, L57, (1997)
4. J. E. Grove, *et al.*: ApJ, **502**, L45, (1998)
5. X.-M. Hua, D. Kazanas, W. Cui: ApJ, **512**, 793, (1999)
6. D. Kazanas, X.-M. Hua, & L. G. Titarchuk: ApJ, **480**, 735, (1997)
7. D. Kazanas, X.-M. Hua: ApJ, **519**, 750, (1999)
8. A. J. Berkley, D. Kazanas, J. Ozik: ApJ, **535**, 712 (2000)
9. P. Reig, I. E. Papadakis, C. R. Shrader, D. Kazanas: ApJ, **644**, 424 (2006)
10. C. R. Shrader, P. Reig & D. Kazanas: ApJ, **667**, 1063 (2007)
11. R. Sunyaev & M. Revnivtsev: A&A, **358**, 617 (2000)
12. J. C. Lochner, J. H. Swank, A. E. Szymkowiak: ApJ, **337**, 823 (1989)
13. Yu. E. Lyubarskii: MNRAS, **292**, 679 (1997)
14. M. A. Abramowicz et al.: A&A, **245**, 454 (1991)
15. S. Miyamoto et al.: Nature, **336**, 450, (1988)
16. S. Miyamoto et al.: ApJL, **391**, L21,(1992)
17. R. Narayan & I. Yi: ApJ, **428**, L13, (1994)
18. R. A. Sunyaev & L. G. Titarchuk: A&A, **86**, 121, (1980)
19. I. E. Papadakis, K. Nandra & D. Kazanas: ApJ, **554**, L133 (2001)
20. B. A. Vaughan, & M. A. Nowak: ApJ, **474**, L43, (1997)

3D Accretion Discs Dynamics: Numerical Simulations

D.V. Bisikalo

Institute of Astronomy of the Russian Academy of Sciences, Moscow, Russia
bisikalo@inasan.ru

Summary. In this report the main attention is paid at physics of accretion discs in close binary systems and particularly at formation of waves in discs. The characteristic features and possible observational manifestations of the "hot line"shock wave formed due to interaction between the circumdisc halo and the stream from the inner Lagrangian point, two arms of the tidal shock, the precessional spiral density wave in cold discs, and the spiral-vortex structure in hot discs are discussed. The analysis of the numerical model and comparison of computational results with observational data allow us to define more exactly the physical processes resulting in formation of structures in accretion discs in binary stars.

1 Introduction

The close binaries (CBs) belong to the class of interacting stars, where one of components fills its critical surface, which causes mass transfer between the components of the system. Under the standard assumptions of circular orbits and synchronous rotation of stars the critical surface coincides with the internal surface (the Roche lobe) in a restricted problem of three bodies, and mass transfer between the components of the system occurs through the vicinity of the inner Lagrangian point L_1, where the pressure gradient is not counterbalanced by gravitation.

Let us consider the gas behavior after passing the L_1 point. Under action of the gravitational force of the companion star (hereafter designated as accretor) the stream velocity increases and soon becomes essentially supersonic (see, e.g., [1]). In addition, under action of the Coriolis force the stream deviates from the straight line connecting centers of the stars. If angular momentum of the fluid element is conserved as it takes place in an axially symmetric disc with negligible viscosity, the matter could reach only the so called circularization radius $r_{min} = L^2/2GM_{acc}$. Here L is the vertical component of angular momentum $\mathbf{r} \times \mathbf{v}$, M_{acc} is the mass of central object, G is the gravitational constant.

G. Contopoulos, P.A. Patsis, *Chaos in Astronomy.* Astrophysics and Space Science Proceedings, doi: 10.1007/978-3-540-75826-6,

For binaries where the radius of accretor is larger than R_{min} ($R_{acc} > R_{min}$), the gas stream strikes the surface of the star directly. Alternatively, when $R_{acc} < R_{min}$, the gas stream rotates around the star and eventually crosses itself at a certain point. Let us consider the angular momentum of a particle after passing through the point where the trajectory crosses itself (hereafter the crossing point). Taking into account that the crossing point is located rather close to accretor and, hence, the gravitational influence of donor star can be neglected, we assume that the angular momentum of a gas particle relative to accretor is constant. For a particle rotating around a point mass with constant angular momentum, the energy is minimized at a circular orbit, therefore we should expect that after passing the crossing point a gas particle will move along a circular orbit with radius R_h around the accretor. The azimuthal velocity v_ϕ of a gas element on a circular orbit is determined (under the assumption of neglect of gas pressure) by the balance between the centrifugal force v_ϕ^2/r and the gravitational attraction of accretor that is equal to GM_{acc}/r^2. Equating these forces we can obtain characteristics of the forming gas ring. It is easy to see that this gas ring obeys the Keplerian law of rotation, i.e., $v_K = \sqrt{GM_{acc}/r}$. The angular momentum in this Keplerian ring increases outward as $\sim \sqrt{r}$. The angular velocity of the gas $\Omega_K = v_K/r$ implies that gas particles rotate in the ring differentially, i.e., the gas flow contains non-zero shearing stresses. Dissipation processes try to equalize the angular velocities of gas particles across the disc, and causes the rotation of gas particles to become slower along the internal orbit but faster along the external orbit. As a result, the angular momentum redistributes in the ring which expands in the accretion disc.

Starting from the study of Struve [2] who first conceived the idea of a gas stream appearing between components in β Lyrae to explain the peculiar behavior of the spectrum at eclipse, the effects of mass transfer resulting in formation of gas flows, streams, discs, circumbinary envelopes and other structures were observed in a number of CBs (see, e.g., [3, 4]). The study of the flow structure in CBs is of great importance, and the results can be used both for consideration of the evolution status and for the interpretation of observational data. In this paper we summarize the results of 3D numerical simulation of mass transfer in CBs that were mainly obtained by Bisikalo et al. (1998–2007) in [4–18].

2 Model

To provide the correct description of the gas flow structure the consistent solution of gas dynamic and radiation transfer equations is required. Unfortunately, present-day computing facilities are not sufficient to solve this problem correctly, so the gas dynamic description is carried out with various simplifications of energy losses.

Let us consider a semidetached binary system with masses of accretor and mass-donating star M_{acc} and M_d respectively, separation A, and angular velocity of orbital rotation Ω. Gaseous flows in the system can be described by the Euler equations with the addition of an ideal gas equation of state. The shape of donor-star is defined by the Roche lobe geometry so there are only few dimensionless parameters determining the flow structure: mass ratio $q = M_d/M_{acc}$, the Lubow–Shu parameter $\epsilon = c(L_1)/A\Omega$ [1] where $c(L_1)$ is sound speed in L_1, the relative size of accretor R_{acc}/A, and the adiabatic index γ. We also should expand this list by adding the value of viscosity (expressed as dimensionless parameter α introduced in [19, 20]) since in approximate solutions some numerical viscosity is present. The model implies that the magnetic and radiation fields do not influence the gas motion in the system. To solve the system of gas dynamic equations we use the Roe–Osher TVD scheme of a high approximation order with Einfeldt modification. This numerical method allows one to study flows with a significant density contrast, and, hence, to investigate the morphology of gaseous flows in binaries and consider the impact of the forming circumbinary envelope on the flow patterns. The details of the model used are presented in [4, 18].

The analysis of numerical results shows a sort of stability of the obtained solutions against perturbations of the input parameters when their values lie in intervals pertinent to cataclysmic variables (CVs) and low mass X-ray binaries (LMXB). The calculations prove the qualitative similarity in the structure of flow in the studied LMXB and CV systems. On the other hand, there are some differences in the solutions for discs with high and low temperatures. In particular, in solutions with hot and cold accretion discs there are spiral waves of different origins. It means, in turn, that the accretion disc temperature range in semidetached binaries is principal for numerical modelling. Analysis conducted in [12] has shown that for realistic values of the parameters ($\dot{M} \simeq 10^{-12}$–$10^{-7} M_{\odot}\,\mathrm{yr}^{-1}$ and $\alpha \simeq 10^{-1}$–10^{-2}) the gas temperature in the outer parts of the disc is between $\sim 10^4$ and $\sim 2 \times 10^5$ K. This implies that both cold and hot accretion discs can form in close binaries.

In this paper, we consider the morphology of steady-state gaseous flows in a system with both hot and cold accretion discs.

3 Shock Waves in Accretion Discs

The structure of gaseous flows in a binary system with hot disc can be evaluated from Fig. 1. In this figure the distribution of density over the equatorial plane and velocity vectors are presented. We also put a gas dynamic trajectory of a particle moving from L_1 to the accretor (a white line with circles) and a gas dynamic trajectory passing through the shock wave along the stream edge (a black line with squares).

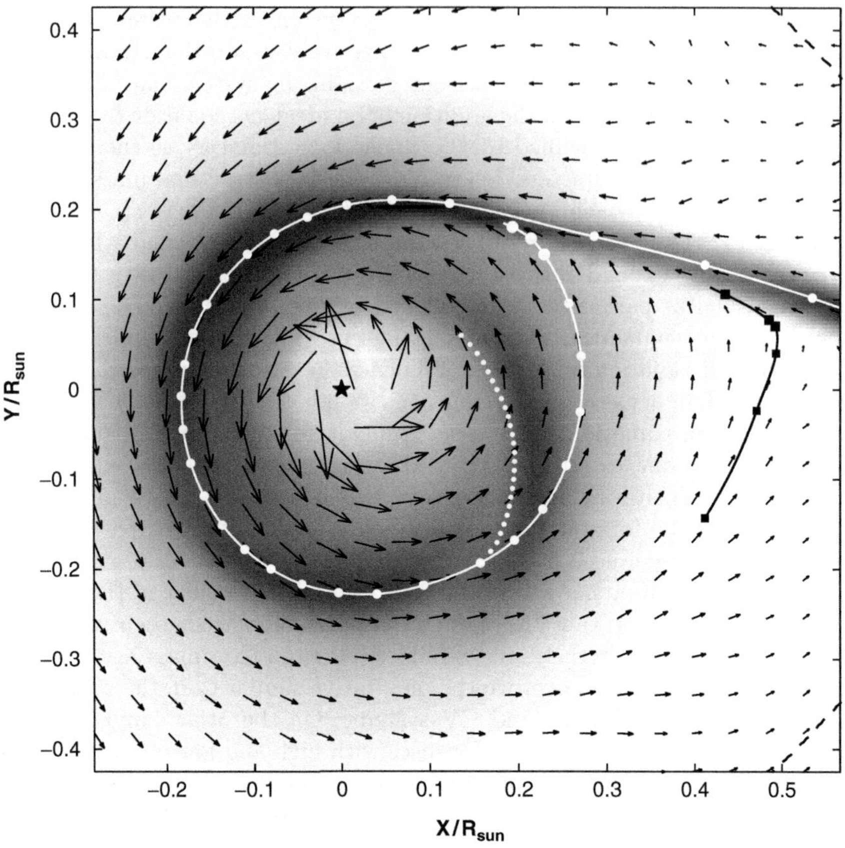

Fig. 1. The distribution of density over the equatorial plane. *Arrows* are the velocity vectors. The *black asterisk* is the accretor. The *dashed line* bounds the Roche lobe. The *dotted line* is the tidally induced spiral shock. Gas dynamic trajectory of a particle moving from L_1 to the accretor is shown by a white line with circles. Another gas dynamic trajectory is shown by a *black line* with *squares*. Symbols of larger size correspond to passing of trajectory through shock wave located on the edge of the stream

The morphology of gaseous flows in semidetached binaries is governed by the stream of matter from L_1, quasi-elliptical accretion disc, circumdisc halo and circumbinary envelope. This classification of the main constituents is based on their physical properties: (a) if the motion of a gas particle is not determined by the gravitational field of the accretor then this particle belongs to the circumbinary envelope filling the space between the components of binary; (b) if a gas particle revolves around the accretor and after that mixes with the matter of the stream then it doesn't belong to the accretion disc but forms the circumdisc halo; (c) the accretion disc is formed by the matter of the stream which is gravitationally captured by the accretor and hereinafter doesn't interact with the stream but moves to the accretor losing the angular momentum. The interaction of matter of circumdisc halo and circumbinary envelope with the stream results in the formation of the shock located along the edge of the stream. This shock is referred as the "hot line." The tidal action of mass-losing star results in the formation of a spiral shock. Our 3D gas dynamic simulations for the "hot" case have shown only one-armed spiral shock while in the place where the second arm should be the flow structure is determined by the stream from L_1.

Let us consider the morphology of gaseous flows in a system with a cold accretion disc (Fig. 2). In the model used the temperature decreases to 13,600 K over the entire computation domain due to the radiative cooling. Analysis of results shows that in the "cold" case the interaction between the circumdisc halo and the stream displays all typical features of an oblique collision of two

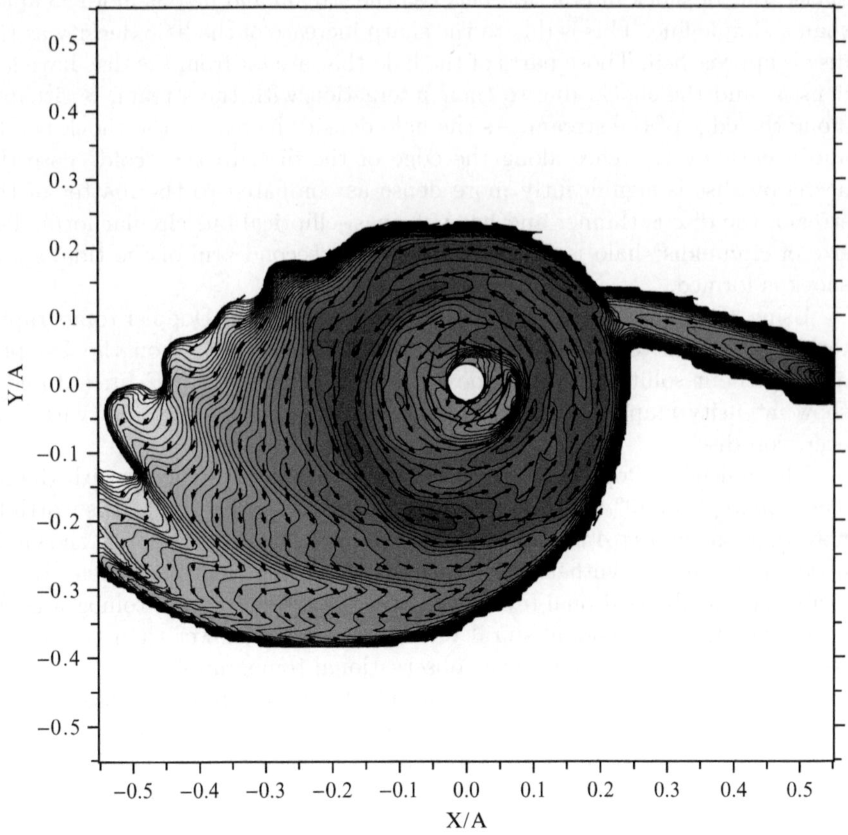

Fig. 2. Contours of constant density in the equatorial plane of the system with cold accretion disc

streams. We can clearly see the "hot line" consisting of two shock waves and a tangential discontinuity between them. The gases forming the halo and stream pass through the shocks corresponding to their flows, mix, and move along the tangential discontinuity between the two shocks. Further, this material forms the disc itself, the halo, and the envelope.

Let us consider the changes occurring during the transition from the hot accretion disc to the cold one. Our 3D gas dynamic simulations have shown that for the "cold" case when the radiative cooling decreases the gas temperature to $\sim 10^4$ K the solution has the same qualitative features as that for the "hot" case, namely: the interaction between the stream and disc is shockless, the region of enhanced energy release is formed due to the interaction between the circumdisc halo and the stream and is located beyond the disc. The resulting shock – the "hot line" is fairly extended, that is particularly important for explaining the observations. However, unlike the solution with a high temperature in the outer regions of the disc, in the "cold" case, the shape of the zone of shock interaction between the stream and halo is more complex than a simple line. This is due to the sharp increase of the halo density as the disc is approached. Those parts of the halo that are far from the disc have low density, and the shock, due to their interaction with the stream, is situated along the edge of the stream. As the halo density increases, the shock bends, and eventually stretches along the edge of the disc. In the "cold" case the accretion disc is significantly more dense as compared to the matter of the stream, the disc is thinner and has not quasi-elliptical but circular form. The size of circumdisc halo is smaller as well. The second arm of the tidal spiral shock is formed.

Using 3D gas dynamic simulations alongside with the Doppler tomography technique allows to identify the main features of the flow on the Doppler maps without solution of an ill-posed inverse problem [21]. Figures 3 and 4 show intensity map and synthetic Doppler tomograms for system with cold accretion disc.

The synthetic Doppler map shown in Fig. 4 was constructed with the inclination angle of 40° (close to the angle estimated for SS Cyg). It is worth to note that on the Fig. 4 the snapshot of the intensity distribution is shown. It means that on the synthetic map one would see the elements which are not visible on the observational (evaluated over the orbital period) tomograms. In particular, the precessional spiral wave, that rotates slower than the orbital period, should not be seen on the observational tomograms.

To make the comparison more convenient we reproduce the observational tomograms of SS Cyg in the Fig. 5 with the same markers as on the synthetic map. The region of the circumbinary envelope near the bow shock (zones A, B, C) makes significant contribution into the tomograms. The contribution of these zones leads to the asymmetry of the tomograms. It is also clear seen that the "hot line" shock waves and two arms of tidal shock make a large input into the tomogram and can be clearly identified from the observations.

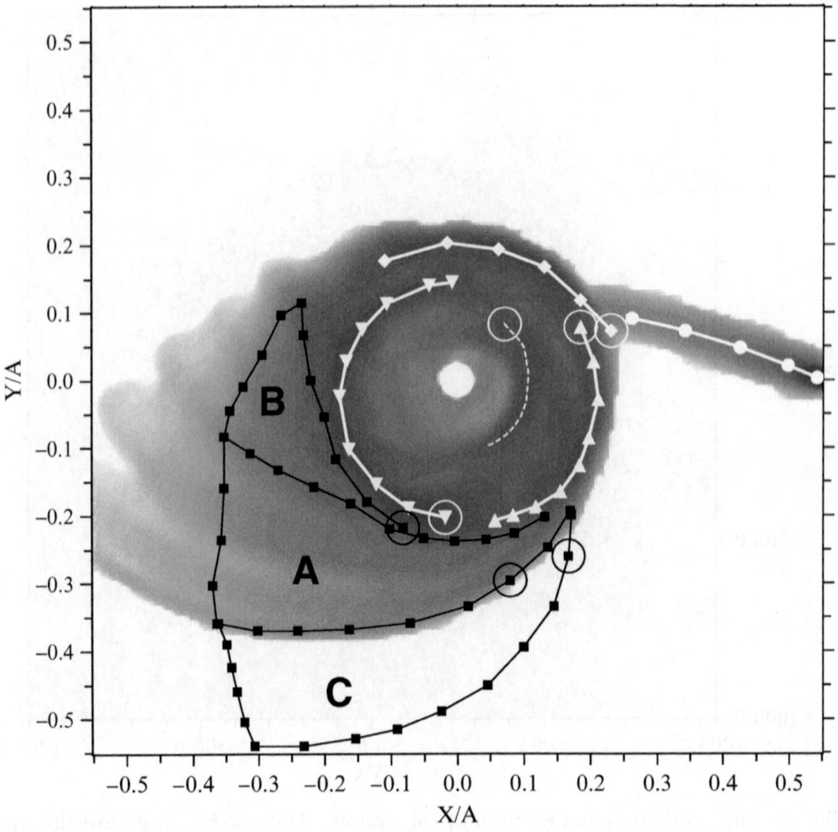

Fig. 3. Intensity map in the equatorial plane of the system with cold accretion disc. The stream from L_1 point is marked by *solid white line* with circles; the "hot line" shock wave – by the *solid white line* with *diamonds*; the two arms of tidal shock by the *solid white line* with *triangles*; the precessional spiral wave – by *dashed line*. The beginning of each line is denoted by the corresponding symbol bounded by the circle. In the figure three zones of circumbinary envelope located close to the bow shock (**A**,**B**,**C**) are also marked by *solid black line* with *squares*

4 The "Precessional" Spiral Density Wave in the Cold Accretion Disc

Taking into account that the stream influences the dense inner part of the disc weakly as well as that all the shocks ("hot line" and two arms of tidal wave) are located in the outer part of the disc we can introduce a new element of flow structure for the "cold" case: the inner region of accretion disc where the influence of gas dynamic perturbations mentioned above is negligible. The formation of a non-perturbed region in the inner part of the disc allows us to consider the latter as a slightly elliptical disc with the typical size of

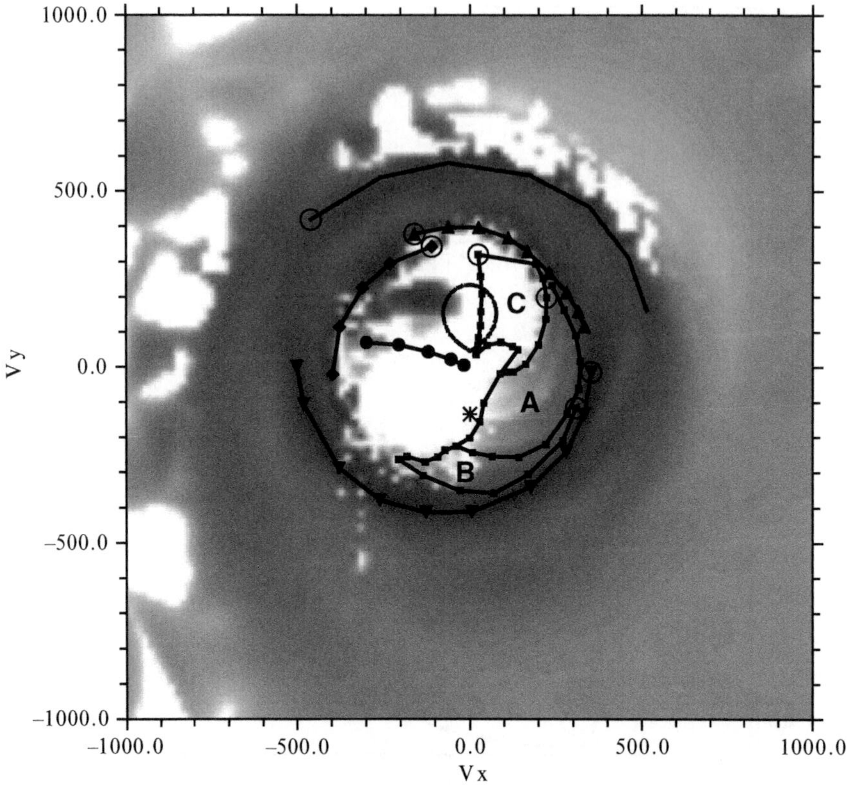

Fig. 4. The synthetic Doppler map of the system. The *markers* used are the same as in Fig. 3

∼0.2–0.3A embedded in the gravitational field of the binary. It is known (see, e.g., [3]), that the influence of companion star results in the precession of orbits of particles rotating around of the binary's component. The precession is retrograde and its period increases with approaching the accretor.

For the accretion disc the orbits must be replaced by flow lines. If the orbits precess such that the precession of distant flow lines tends to be faster, these distant flow lines will constantly overtake those with smaller semimajor axes. Since the flow lines in a disc cannot intersect, an equilibrium solution is established with time and all the flow lines begin to precess with the same angular velocity, i.e., to display rigid-body rotation. Let us consider a solution with the semimajor axes of the flow lines misaligned with respect to some chosen direction by an angle (turn angle) that is proportional to the semimajor axis of the orbit. It is obvious that such a solution should contain spiral structures. In particular, due to the nonuniformity of the motion along the flow line and the formation of a maximum density at apastron, the curve connecting the apastrons will form a spiral density wave (see left panel of

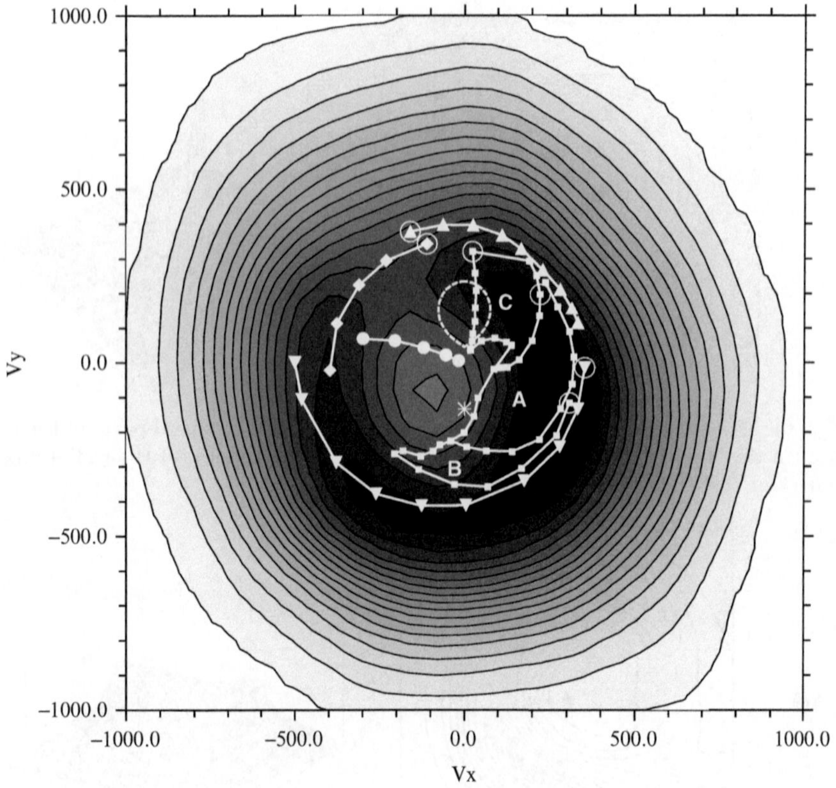

Fig. 5. The observational Doppler tomogram of SS Cyg in and H_γ lines [22]. The same *markers* as in Fig. 3, and 4 are also shown

Fig. 6). Numerical results show that the "precessional" wave moves as a single entity, and its period is 3–7% longer than the orbital one. The presence of the density wave together with the fact that the velocity of the particles increases after passing the apastron leads to an increase in the radial component of the mass flux $F_{rad} = \rho v_r$ due to the increase of both the density ρ and the radial velocity component v_r. The increase in the radial component of the mass flux behind the wave increases the accretion rate by more than a factor of ten in the region where the precessional wave approaches the accretor (see right panel of Fig. 6).

Increasing of the radial flux of matter after passing the density wave results in growth of the accretion rate and formation of a compact zone of energy release on the accretor surface. This zone can be seen as a periodic increase of brightness on light curves of semidetached binaries. The features of the "precessional" spiral wave allow us to propose the new mechanism explaining superoutbursts in binaries of SU UMa type [14]. This mechanism explains both the energy release during the outburst and all its observational manifestations.

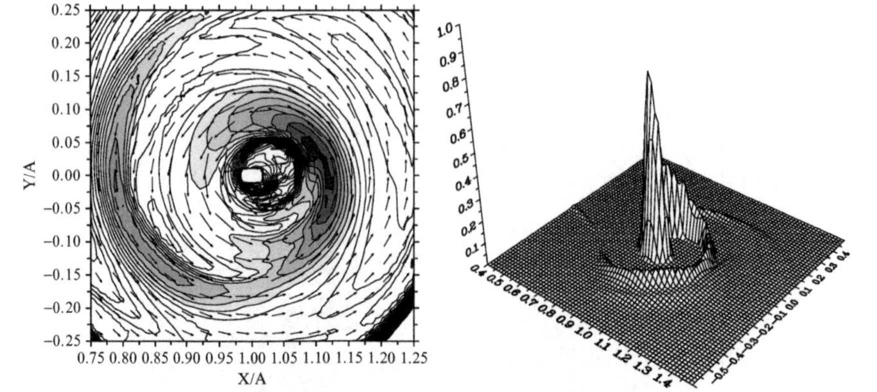

Fig. 6. *Left*: contours of density and velocity vectors in the central parts of the cold disc (in the equatorial plane of the binary); *right*: bird-eye view of the radial flux of matter in the central parts of the disc

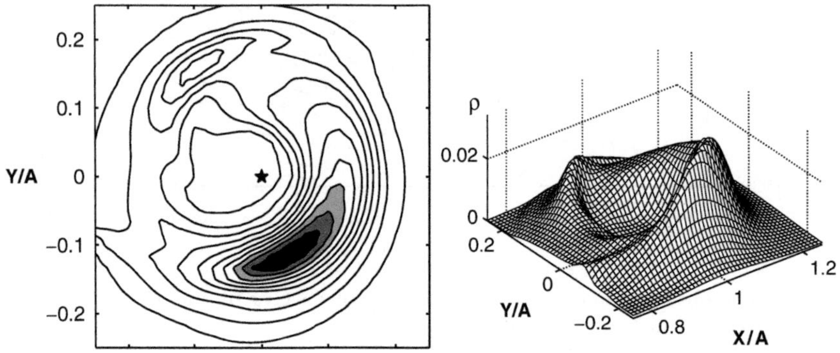

Fig. 7. Density distribution in the equatorial plane for the solution with hot disc

The distinctive characteristic of a superoutburst in a SU UMa type star is the appearance of the superhump on the light curve. The proposed model reproduces well the formation of the superhump as well as its observational features, such as the period that is 3–7% longer than the orbital one and the detectability of superhumps regardless of the binary inclination.

5 The Spiral-Vortex Structure in the Hot Accretion Disc

Results of 3D numerical simulations of CBs with hot accretion disc show that a variation of the mass transfer rate leads to the formation of a dense "blob" (see Fig. 7) moving in the disc with period 5-6 times less than the orbital one [11]. As it was shown in [23] this structure is a density wave containing a one-armed spiral and an anticyclonic vortex with the center at a corotation radius, where the azimuthal phase velocity of the wave coincides with the angular rotation velocity of the disc. The formation of the structure is caused by a hydrodynamical instability rather than the tidal influence of a companion star. The instability results in a turbulence of the accretion disc. A value of the turbulent viscosity coefficient is in accordance with observations.

Analysis of light curve variations for the non-eclipsed part of IP Peg shows that the period of the density wave rotation is in agreement with the typical periods of light curve variations observed in cataclysmic binaries.

6 Conclusions

Results of gas dynamic modelling allow us to realize the existence of different waves in the accretion discs of close binaries. In systems with stationary accretion discs there are shock waves: the "hot line" formed due to interaction between the circumdisc halo and the stream from inner Lagrangian point, and two arms of the tidal shock. Analysis of the results of a 3D gas dynamic simulation fully confirms the possibility of the generation of a spiral wave in the inner parts of a cold disc. The agreement between the qualitative analysis and the computational results makes us confident that the wave has a precessional origin. In systems with hot accretion discs a hydrodynamical instability results in formation of the spiral-vortex structure. Processing of observational data proves the existence of these structures in accretion discs of close binaries.

Acknowledgments

The work was supported by the Russian Foundation for Basic Research (project no. 06-02-16097), the Scientific Schools Program (grant no. NSh-4820.2006.2), and the basic research programs "Mathematical modeling and intellectual systems" and "Origin and evolution of stars and galaxies" of the RAS.

References

1. S.H. Lubow, F.H. Shu: Astrophys. J. **198**, 383 (1975)
2. O. Struve: Astrophys. J. **93**, 93 (1941)
3. B. Warner: *Cataclysmic Variable Stars*, (Cambridge University Press, Cambridge, 1995).
4. A.A.Boyarchuk, D.V.Bisikalo, O.A.Kuznetsov, V.M.Chechetkin: *Mass transfer in close binary stars*, (Taylor and Francis, London, 2002)
5. D.V Bisikalo, A.A. Boyarchuk, V.M. Chechetkin, O.A. Kuznetsov, D. Molteni: Monthly Notices Roy. Astron. Soc., **300**, 39 (1998)

6. D.V Bisikalo, A.A. Boyarchuk, O.A. Kuznetsov, T.S. Khruzina, A.M. Cherepashchuk: Astron. Rep. **42**, 33, (1998)
7. D.V Bisikalo, A.A. Boyarchuk, O.A. Kuznetsov, V.M. Chechetkin: Astron. Rep. **43**, 229, (1999)
8. D.V Bisikalo, A.A. Boyarchuk, V.M. Chechetkin, O.A. Kuznetsov, D. Molteni: Astron. Rep. **43**, 797, (1999)
9. D.V Bisikalo, P. Harmanec, A.A. Boyarchuk, O.A. Kuznetsov, P. Hadrava: Astron. Astrophys. **353**, 1009, (2000)
10. D.V Bisikalo, A.A. Boyarchuk, O.A. Kuznetsov, V.M. Chechetkin: Astron. Rep. **44**, 26, (2000)
11. D.V Bisikalo, A.A. Boyarchuk, A.A. Kilpio, O.A. Kuznetsov, V.M. Chechetkin: Astron. Rep. **45**, 611, (2001)
12. D.V Bisikalo, A.A. Boyarchuk, P.V. Kaigorodov, O.A. Kuznetsov: Astron. Rep. **47**, 809, (2003)
13. D.V Bisikalo, A.A. Boyarchuk, P.V. Kaigorodov, O.A. Kuznetsov, T.Matsuda: Astron. Rep. **48**, 449, (2004)
14. D.V Bisikalo, A.A. Boyarchuk, P.V. Kaigorodov, O.A. Kuznetsov, T.Matsuda: Astron. Rep. **48**, 588, (2004)
15. D.V. Bisikalo: Astrophys. Space Sci. **296**, 391, (2005)
16. D.V Bisikalo, P.V. Kaigorodov, A.A. Boyarchuk, O.A. Kuznetsov: Astron. Rep. **49**, 701, (2005)
17. D.V Bisikalo, A.A. Boyarchuk, P.V. Kaigorodov, O.A. Kuznetsov, T.Matsuda: Chin. J. Astron. Astrophys. **6**, 159, (2006)
18. A.Yu. Sytov, P.V. Kaigorodov, D.V Bisikalo, A.A. Boyarchuk, O.A. Kuznetsov: Astron. Rep. **51**, 836, (2007)
19. N.I. Shakura: Soviet Astron. **16**, 756 (1972)
20. N.I. Shakura, R.A. Sunyaev: Astron. Astrophys. **24**, 337 (1973)
21. O.A. Kuznetsov, D.V Bisikalo, A.A. Boyarchuk, T.S. Khruzina, A.M. Cherepashchuk: Astron. Rep. **45**, 872, (2001)
22. D.V Bisikalo, D.A. Kononov, F.G. Zhilkin, A.A. Boyarchuk: Astron. Rep., *in press* (2008)
23. A.M. Fridman, A.A. Boyarchuk, D.V Bisikalo, O.A. Kuznetsov, O.V. Khoruzhii, Yu.M. Torgashin, A.A. Kilpio: Phys. Lett. A **317**, 181 (1973)

Growth of Density Fluctuations at the Time of Leptogenesis

E.A. Paschos

Institute of Physics, University of Dortmund, 44221 Dortmund, Germany
paschos.e@gmail.com

In the early universe structures are formed through the growth of small density inhomogeneities that will eventually become galaxies or clusters of galaxies. At some epoch in this development an excess of baryons and/or leptons must also be formed in order to produce the material world. An attractive scenario for the formation of a lepton asymmetry is leptogenesis where the decays of heavy Majorana neutrinos produce an excess of leptons over antileptons. The excess survives after the decay of the lightest Majorana particle. The value of mass or energy where this phenomenon occurs still carries a large degree of uncertainty. In all cases it must be larger than the electroweak symmetry breaking scale. At this high energy the universe was radiation dominated and all the authors solve the Boltzmann equations for the development of the asymmetry in a universe with uniform density. At the time of Leptogenesis, however, there were already small density fluctuations which were created earlier. It is now appropriate to ask how do the density differences change during the transient period of Leptogenesis ? I discussed this and similar topics with Nikos Voglis and the good memories of those occasions motivate me to write this progress report. The new property in this case is the flow of Majorana neutrinos into the high density spots.

One approach for studying such effects is to consider the fluctuations as originating in the metric and then solve the Boltzmann equations for the development of the Majorana density and the lepton asymmetry by including perturbations in the metric. These are adiabatic fluctuations where the Majoranas and the photons fluctuate together. The phenomenon has been studied by two of my students [1, 2] who found small effects. The second possibility is to consider the existence of high-density spots, where there is an upward fluctuation in the density of particles, and study the change of densities during the transient time of Leptogenesis.

As the universe expands and the temperature approaches the mass of a Majorana particle the following phenomena take place.

G. Contopoulos, P.A. Patsis, *Chaos in Astronomy.* Astrophysics and Space Science Proceedings, doi: 10.1007/978-3-540-75826-6,

1. The ith Majorana particle N_i becomes non-relativistic and moves very slowly.
2. The reaction cross sections with Majorana particles in the initial state become very small. For example, the s-channel process [3,4]

$$N_i + L \to h \to Q + u$$

with h the higgs particle, L a lepton and Q,u quarks, has the reduced cross section [3,4]

$$\hat{\sigma}_h(s) = \frac{m_Q^2 (g^\dagger g)_{ii}}{2\pi v^2} \left[\frac{s - M_i^2}{s} \right]^2 . \tag{1}$$

Similarly the exchange of a higgs particle in the t-channel produces the cross section [3, 4]

$$\hat{\sigma}(s) = \frac{m_Q^2 (g^\dagger g)_{ii}}{\pi v^2} \left[\frac{s - M_i^2}{s} + \frac{M_i^2}{s} \ln \frac{s - M_i^2 + M_h^2}{M_h^2} \right] \tag{2}$$

with g_{ij} the coupling matrices of the higgs particles to leptons and v the vacuum expectation value.

In the transitory regime as $s \to M^2$ the Majorana particles decouple from the rest of the particles. Also lepton number conserving reactions, like

$$h + N_i \to h + N_i$$

decouple. As $s \to M^2$ the ith particle becomes non-relativistic and interacts very weakly with the other particles and with radiation. They find themselves floating in a thermal bath of massless particles. Space, time and the expansion parameter obey relations of the radiation epoch while at the same time the Majoranas are subjected to an isothermal perturbation, feeling the attraction of gravity toward the spots of higher density. A small fraction of them eventually will produce the lepton asymmetry. Our next task is to calculate the increase of their density.

I shall define the density fluctuations by the equation

$$\varrho = \varrho_0 (1 + \delta_0) + \varrho_M (1 + \delta_M) \tag{3}$$

where ϱ_M is the fluctuation of the Majorana particles and δ_0 the fluctuation of the other particles. In an analogous way I define the uniform velocity by **u** ant the perturbation of the velocity as **v**. The continuity equation is now modified

$$\frac{\partial \delta_M}{\partial t} + \frac{1}{R} \boldsymbol{\nabla}_c \cdot \mathbf{v} = - < r > \delta_M \tag{4}$$

with the gradient operator defined in the co-moving frame and $< r >$ being the averaged decay width. The fluid of particles obeys in addition the Euler equation

$$\frac{\partial \mathbf{v}}{\partial t} + \frac{\dot{R}}{R} \mathbf{v} + \nabla_c \phi = 0 \tag{5}$$

with ϕ the gravitational potential due to the perturbation and $R(t)$ the scale parameter. In this equation I neglected terms $\mathbf{v}\cdot\boldsymbol{\nabla}\mathbf{v}$, being of higher order, and the sound velocity because the Majoranas decouple. Since the vast majority of particles are in the radiation epoch $R(t) \approx t^{1/2}$ and the ratio

$$H = \dot{R}/R = 1/(2t) \tag{6}$$

The potential obeys the equation

$$\nabla_s^2 \phi = 4\pi G(\varrho_0 \delta_0 + \varrho_M \delta_M) \tag{7}$$

with the subscript s denoting stationary coordinates. I followed here the method of [5]. The same result is obtained by following the steps in [6] p. 345.

Combining (4) and (5) in the standard way [5, 6] one obtains

$$\frac{\partial^2 \delta_M}{\partial t^2} + \left(2\frac{\dot{R}}{R} + < r >\right)\frac{\partial \delta_M}{\partial t} = 4\pi G(\varrho_o \delta_0 + \varrho_M \delta_M) - \frac{2\dot{R}}{R} < r > \delta_M. \tag{8}$$

The development of fluctuations depends on the strength of the gravitational potential relative to the decay rate. It is encouraging, however, that the potential depends on both densities. Factorizing out the fluctuation of the Majorana particles

$$4\pi G(\varrho_0 \frac{\delta_0}{\delta_M} + \varrho_M)\delta_M, \tag{9}$$

one sees that the dominant attraction comes from the larger density.

For an Einstein-de Sitter universe the gravitation is approximated by

$$4\pi G \varrho_0 \approx \frac{3}{2}\left(\frac{\dot{R}}{R}\right)^2 = \frac{3}{8t^2} \tag{10}$$

which must be compared to the decay rate. The gravitational term becomes comparable and perhaps larger than the decay term at early times and for high Majorana masses. The size and properties of these effects depend on the solutions of (8) and the parameters of the theory.

In this brief report I described the interactions and decays of heavy Majorana particles in the early universe and in the presence of inhomogenities. I am studying now the magnitudes of the various terms and solutions of (8). The results of this investigation will be presented at a later date.

References

1. Kartavtsev A. (2007) Leptogenesis in the superstring inspired E6 model. Dissertation, University of Dortmund
2. Besak D. (2007) Leptogenesis in two different models of the Universe. Diplom Thesis, University of Dortmund

3. Luty MA (1992) Phys. Rev. D45: 455
4. Pilaftsis A., Underwood T. (2004) Nuclear Physics B692: 303
5. Kolb EW, Turner MS (1990) The Early Universe. Addison Wesley, Redwood City, CA
6. Perkins D (2003) Particle Astrophysics. Oxford University Press, Oxford

Fully Developed Turbulence in Accretion Discs of Binary Stars: Turbulent Viscosity Coefficient and Power Spectrum

A.M. Fridman[1], D.V. Bisikalo[1], A.A. Boyarchuk[1], L. Pustil'nik[2], and Y.M. Torgashin[1]

[1] Institute of Astronomy of the Russian Academy of Sciences, Moscow, Russia
afridman@post.tau.ac.il, bisikalo@inasan.ru
[2] Israel Space Weather and Cosmic Ray Center, Tel Aviv University, Israel
levpust@post.tau.ac.il

Summary. Using 3D numerical simulations we show the existence of a "spiral-vortex" structure in an accretion disc in close binary stars. This structure is not related to the tidal influence of the companion star. It is a density wave containing a one-armed spiral and an anticyclone vortex, centered at the corotation circle.

The latter results were obtained by 2D numerical simulations using main parameters coinciding with those in the 3D numerical simulations. In addition, our 2D numerical simulations show that the first three Fourier harmonics are caused by the "over-reflection" instability and the growth rate of the first harmonic is maximum.

Analytical consideration allows us to outline a scenario of the Kolmogorov turbulence and to estimate the turbulent viscosity coefficient. Using the observed variations of the uneclipsed part of the light curves of cataclysmic binary stars we could estimate the exponent of power turbulent spectrum. In our case, it is different from the Kolmogorov one, because Kolmogorov's hypotheses on homogeneity and isotropy are not fulfilled.

1 Introduction

The efficiency of accretion discs is about[1] $(6 \div 30)\%\ Mc^2$, that appreciably exceeds the amount 0.3% in the case of the nuclear reactions in the stellar cores. So accretion discs are the most effective engine in the Universe.

[1] The efficiency for accretion discs around non-rotating black hole with a "Radius of Innermost Stable Circular Orbit" RISCO $= 6\ GM\,c^{-2}$ is limited by 6%. The corresponding limit of the efficiency of accretion discs around a fast rotating Kerr black hole [1, 2] is 32%. This limitation refers only to radiation efficiency. The apparent efficiency of jets driven by the innermost accretion disc of a highly rotating Kerr black hole [3] can reach 96%.

G. Contopoulos, P.A. Patsis, *Chaos in Astronomy.* Astrophysics and Space Science Proceedings, doi: 10.1007/978-3-540-75826-6,

In astrophysics, accretion is understood as accumulation of matter onto a compact object under action of its gravitation. The object can be, e.g. a normal star, a white dwarf, a neutron star or a black hole. Accretion discs are not rare monsters, but rather prevailing type of objects in the Galaxy, because the majority of stars in our Galaxy are binaries.

The turbulence in accretion discs is necessary for conversion of the energy of the disc rotation into plasma heating, X-ray and gamma-ray emission, jet formation and high-energy cosmic rays acceleration.

Let us consider a binary system with the components of M_1 and M_2 masses, with a separation of A, the angular velocity of the orbital rotation Ω_{orb}, and the period of orbital rotation $P_{orb} = 2\pi/\Omega_{orb}$. Usually in the study of binary systems the Roche approximation is used. It is based on the three assumptions: (a) the gravitational field of the components is equivalent to that of point masses; (b) the orbits of components around the center of mass of the binary system are circular; (c) the proper rotation of the stars is synchronized by tidal forces with orbital rotation.

In our 3D numerical simulations we used the right-hand reference frame corotating with the binary system, with the origin located in the center of the M_1 star, X-axis directed along the line connecting the centers of the stars, Z-axis directed along the axis of rotation (see Fig. 1). The force field in this coordinate system is described by so called Roche potential:

$$\begin{aligned}\Phi = &-\frac{GM_1}{\sqrt{x^2+y^2+z^2}} - \frac{GM_2}{\sqrt{(x-A)^2+y^2+z^2}} - \\ &-\frac{1}{2}\Omega^2_{orb}\left[\left(x - A\frac{M_2}{M_1+M_2}\right)^2 + y^2\right] .\end{aligned} \quad (1)$$

Examples of Roche equipotentials in the equatorial plane, described by this equation are shown in Fig. 1. For large values of $(-\Phi)$ these surfaces are two separated spheroidal lobes, located around the gravitating masses M_1 and

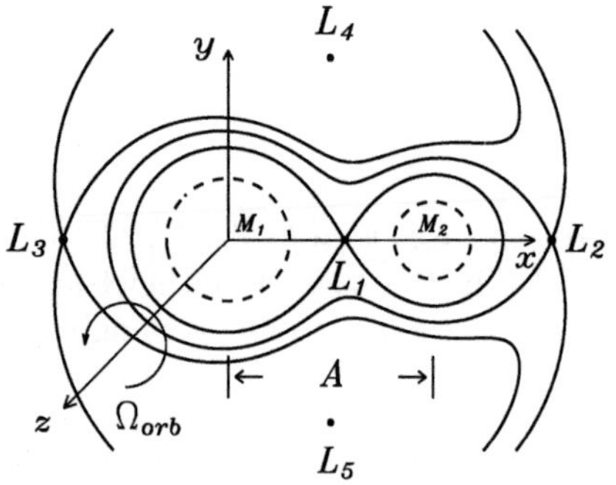

Fig. 1. Roche equipotentials (*dashed and solid lines*) in the equatorial plane for a binary system with the mass ratio $q = M_1/M_2 = 2$. $L_1, L_2, \ldots$ mark the libration points

M_2. When decreasing the value of $(-\Phi)$ the surfaces expand and get in touch with each other in a point, which is known as the first libration point L_1[2].

In the L_1 point, as well as in the four other libration points (Fig. 1) the sum of all forces - two gravitational attractions to masses M_1 and M_2 and centrifugal force, is equal to zero, i.e. $\nabla\Phi = 0$. For binary stars all five libration points are unstable [6] – a probe particle initially located in one of these points will leave it after being infinitesimally disturbed.

The volume restricted by the equipotential surface passing through L_1 is known as Roche lobe. So long as the sizes of stars are smaller than that of the Roche lobe, star's photospheres coincide with spheroidal equipotential surfaces. When one of two stars fills the Roche lobe, the matter of the star begins to flow under action of pressure through the vicinity of the unstable libration point L_1 into the Roche lobe of the neighbor star. If the radius of the latter is sufficiently small, the most part of the inflow material will form an accretion disc. In the close binary stars (CBS) the star surrounded by the disc is called accretor (M_1 in Fig. 1) and the mass losing star is called companion star (M_2 in Fig. 1).

2 Statement of the Problem, Basic Equations, and Numerical Approach

Accretion discs are known as the most dynamical objects in the Universe and their existence is due to the presence of a strong turbulent viscosity being 8-9 orders of magnitude larger than the molecular one. Only such a large value of the viscosity permits the formation of an accretion disc. The search of an instability leading to the turbulization of accretion disc, and thus providing the required rate of the angular momentum loss, is a classical astrophysical problem. In papers [7, 8] authors describe the hydrodynamic shear instability found in the CBS accretion disc, the large scale modes being the most unstable ones. In particular, the azimuthal mode m=1 has a maximum growth rate. It means that a one-armed spiral density wave can arise in the accretion disc.

Here we investigated in details the hydrodynamic shear instability, detected firstly in [7, 8], and came to the conclusion that the instability was the over-reflection one. The one-armed spiral identified in our theoretical calculations (analytically and in numerical simulations) in [7, 8] cannot be directly observed in accretion discs at present time. The accuracy required to observe this feature should be by three orders of magnitude higher than that given by

[2] Note, that in astrophysical literature it is accepted to call L_1 the inner Lagrange point despite it being historically inaccurate, since the first three libration points (L_1, L_2 and L_3) in the restricted three-body problem were discovered by Euler [4]. Five years later Lagrange [5] discovered the last two libration points (L_4, L_5) that in celestial mechanics are refereed to as triangular libration points or sometimes the Lagrange points.

the Hubble Space Telescope. Here we demonstrate the example of one-armed spiral, caused by the over-reflection instability and observed in the "nuclear" disc, situated in the very central part of our Galaxy [9, 10, 12].

Viscosity acts strongly on short wavelength modes and much less on long-wavelength perturbations. Taking into account this fact and being interested mainly in the dynamics of the large scale modes, we performed the 3D simulations of gaseous accretion disc evolution basing on the inviscid compressible hydrodynamic equations (in the reference frame rotating with the binary system, see Fig. 1) [13][3]

$$\frac{\partial \rho}{\partial t} + div\,(\rho \mathbf{V}) = 0, \tag{2}$$

$$\begin{aligned} \frac{\partial \rho \mathbf{V}}{\partial t} + \mathbf{V} \cdot div\,(\rho \mathbf{V}) + \rho\,((\mathbf{V}\nabla)\,\mathbf{V}) + grad\,(p) = \\ = 2\rho\,[\mathbf{V} \times \mathbf{\Omega}_{\rm orb}] - \rho \cdot grad\,(\Phi)\,, \end{aligned} \tag{3}$$

$$\frac{\partial E}{\partial t} + div\,((E + p)\,\mathbf{V}) = 0\,. \tag{4}$$

Here ρ is the volume density; $\mathbf{V} = (V_x, V_y, V_z)$ is the velocity vector; p is the pressure; $E = \rho(\varepsilon + \mathbf{V}^2/2 + \Phi)$ is the energy per unit volume; ε is the specific internal energy; Φ is determined by (1) and does not depend on t. To simulate radiation losses the system (2–4) is closed by an approximately isothermal equation of state for ideal gas [14] with the adiabatic index close to unity (1.01). On the other hand, as it was mentioned above, the very existence of the accretion disc implies the presence of abnormally large (turbulent) viscosity. In the process of the numerical solution of the inviscid hydrodynamic equations, we implicitly introduce a numerical viscosity corresponding to the viscosity coefficient $\nu/(c_s h) \equiv \alpha \sim 0.01 \div 0.1$[4], where c_s and h are the gas sound speed and disc semi-thickness, respectively. In the 3D numerical simulations we used the components mass ratio $q = 1$ (details of the numerical algorithm can be found in [19–24]). At the quasistationary stage, the accretion disc is formed in the system with radius $r_d \sim 0.4A$ and semithickness $h/r_d \sim 0.2 \div 0.3$.

3 Collective Mode m=1

Figure 2 shows the time variations of the density $\bar{\rho}$, averaged over the half-plane XZ (Y=0, X<0), slicing the disc. All curves presented have two common peculiarities: (a) oscillations with practically equal frequencies; (b) the value of

[3] Accretion disc is supposed to be of low mass and hot enough, so the self-gravity of gas in the disc may be ignored, which is typical for CBS.

[4] So the value of our numerical viscosity lies in the middle of the interval $\alpha \sim 0.01 \div 0.1$, which is typical for close binaries, according to the observational data and corresponding theoretical works (see, e.g. [15–18]).

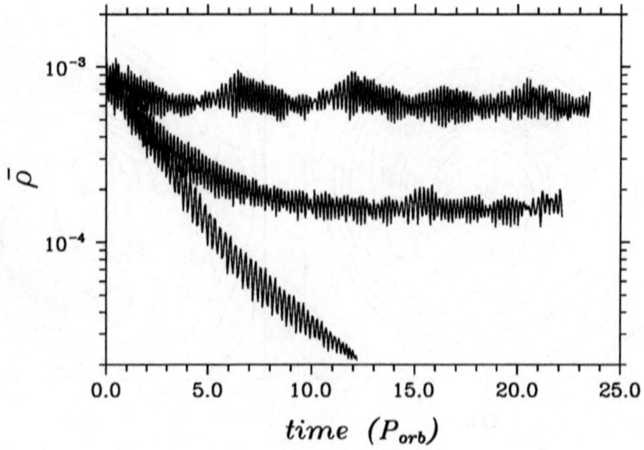

Fig. 2. Time dependence of the density, averaged over a half-plane XZ (Y=0, X<0), after decreasing of the mass transfer rate in L_1 point: twofold decrease (*upper curve*), tenfold decrease (*middle curve*) and decrease by the order 10^5 that corresponds to the practically complete termination of mass transfer (*lower curve*)

$\bar{\rho}$ averaged over the oscillations changes only at the initial transient period and after that remains constant (except the lower curve). The second peculiarity is a result of twofold decrease of mass transfer rate for the upper curve, tenfold for the middle curve and complete termination of the mass transfer for the lower curve [25, 26].

It was found in the papers by Bisikalo et al. [25, 26] that the oscillations on the curves (Fig. 2) can be visually associated with revolution of dense "blob" with angular velocity being approximately 5.5 times larger than Ω_{orb} (see Fig. 3).

In works by Fridman et al. [7, 8] some arguments in favour of the wave nature of the "blob," which is nothing else but a one-armed spiral density wave were presented. The latter is known (from the well developed theory of the galactic spiral density waves) to rotate in rigid-body manner and to be not affected by differential rotation of the disc.

In Fig. 4 (left panel) the Fourier harmonics amplitudes of the non-axisymmetric part $\tilde{\Phi}$ of the potential (1) relative to its axisymmetric part for different radial ranges of the disc are presented. We see that the harmonic m=2 dominates in the Fourier spectrum of the potential $\tilde{\Phi}$ over the whole considered part of the disc. It represents the predominance of the m=2 symmetry in the tidal influence of the companion star. This fact is in accordance with the monotonous growth of the harmonic amplitude from the center to the edge of the disc. Note, that alongside with the harmonic m=2 the Fourier spectrum of the gravitational potential contains also some other large scale harmonics ($m = 1, 3, \dots$).

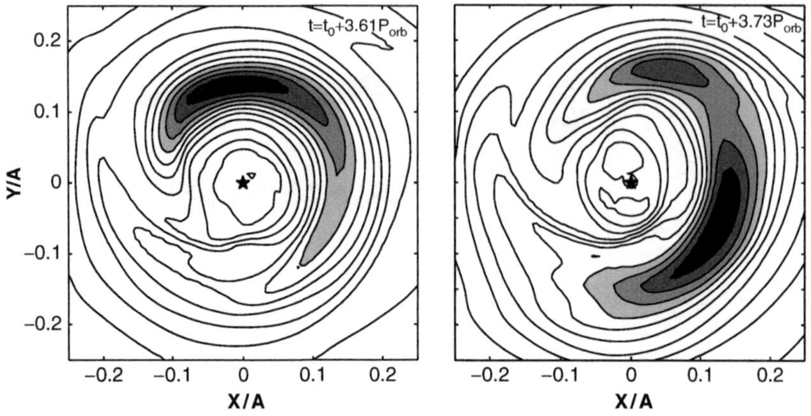

Fig. 3. Density distribution in the equatorial plane for the "blob," caused by the regime of the top curve shown in Fig.2. Results are presented for the moments of time $t_1 = 3.61 \cdot P_{orb}$ and $t_2 = 3.73 \cdot P_{orb}$ after mass transfer termination, so $t_2 - t_1$ approximately equals to the 0.7 of the period of the "blob" revolution in the reference frame, rotating with the binary system. In this coordinate system the rotational period of the "blob" is $\tilde{T}_1 \simeq (1/5.5)P_{orb}$, and external observer in some inertial coordinate system would detect the rotational period $T_1 \simeq (1/6.5)P_{orb}$. The maximum density corresponds to $\rho_{max} \simeq 0.035\rho_{L_1}$. Black asterisk is the accretor [25, 26]

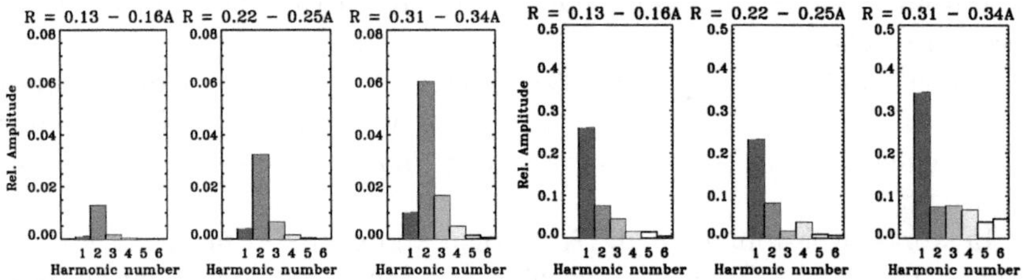

Fig. 4. *Left panel*: The distribution of the non– axisymmetric part of the gravitational potential in the equatorial plane. *Right panel*: The amplitudes of the Fourier harmonics of the non-axisymmetric part of the "refined" surface density of disc relative to its axisymmetric part for different radial ranges of the disc

Figure 4 (right panel) shows that the tidal force does not determine matter distribution in the disc. Indeed, the harmonic m=1 dominates in the Fourier spectrum of the "refined"[5] surface density $\tilde{\sigma}$ everywhere in the accretion disc. It is clear that the domination of the first mode in the spectrum of $\tilde{\sigma}$ is in the agreement with the existence of the "blob" as shown in Fig. 3.

Figure 5 (left panel) presents the phase curves of the maximum density (azimuthal positions of maximum values) of the first Fourier harmonic of

[5] "Refined" functions were calculated by subtracting the averaged over one period distribution from the corresponding field at any moment.

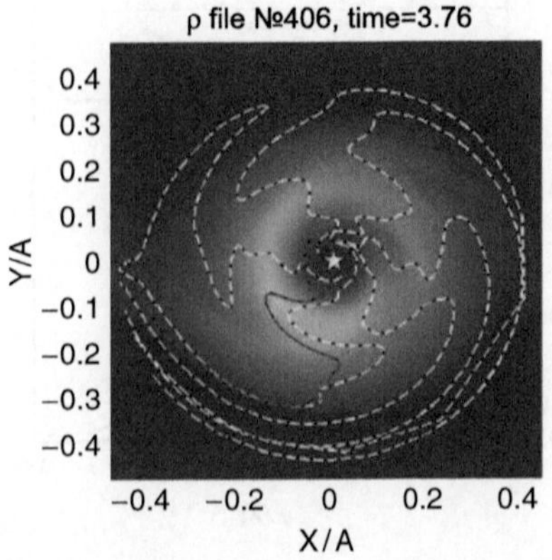

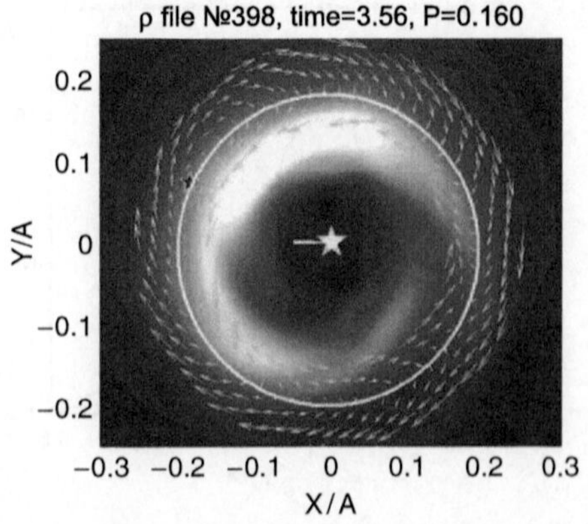

Fig. 5. *Left* – Phase curves of the maximum density of the first azimuthal harmonic of the density profile for different moments of time. Bold black line is the "original" curve. Bold black-white line is the seventh curve which coincides with the original curve if the coefficients of Fourier harmonics do not depend on the time. *Right* – density distribution and velocity field in the equatorial plane for the moment t= $3.56P_{orb}$

"refined" surface density at different radii for different moments of time. We can see that the deformation of the one-armed spiral by differential rotation of the disc is practically absent. It means that the one-armed spiral is a collective mode. In Fig. 5 (right panel) the white star in the center represents the accretor, and the black star – the anticyclone center. Vectors of velocity are shown by white arrows. The short white line near the white star points to the companion star. The center of anticyclone is located in the vicinity of the maximum of the surface density of the "blob" as well as on the corotation circle (solid white curve) that is in accordance with the galactic density wave theory [27]. So, we have proved that "blob" is one-armed density wave.

4 Over-Reflection Instability

In the previous section it was proved that periodical variations on quasi-stationary regime of the accretion disc in 3D numerical simulations are caused by one-armed spiral-vortex structure having rotational period (in a reference frame rotating with binary system) $\tilde{T}_1 \simeq 0.18P_{orb}$ (Figs. 2,3,5), that is not connected with tidal periodicity. In this section we will show that the mode is generated by an over-reflection instability in the disc. To clarify the reason of the origin of this mode and to diminish the tidal forces influence the following steps were made. (a). The results of 3D simulations were recalculated into new reference frame $\{r, \phi, z\}$, that is centered in the accretor and moving with it around the mass center of the system, but has fixed (non-rotating)[6]

[6] Note that in the new coordinate system the period of variations would be $T_1 \simeq 0.15P_{orb}$.

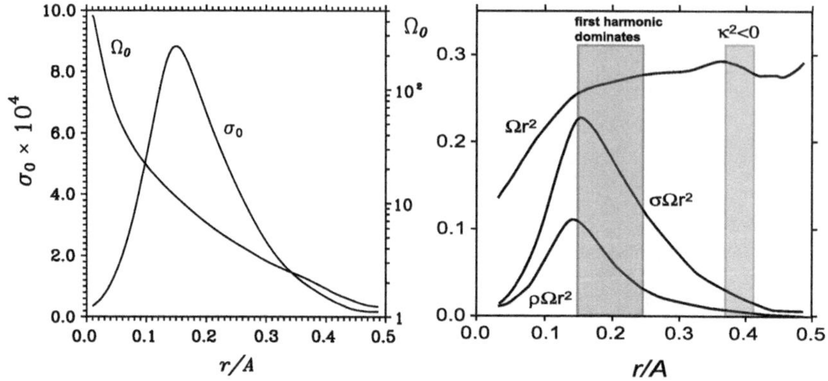

Fig. 6. *Left*: The distribution of the surface density σ_0 and the angular velocity Ω_0 of the isolated 2D model disc in the new reference frame (see text); *Right*: Localization of the specific angular momentum kink in the first harmonic domination region (per unit mass – top, per unit surface – middle, and per unit volume – lower curve, respectively)

axes ($z = 0$ plane in the new and old reference frames being the same). (b). All distributions were integrated over z, and averaged over a period P_{orb} large enough as compared to T_1.

So we analyzed the dynamics of linear perturbations in the prescribed above isolated 2D model disc with steady-state axisymmetric parameters – surface density $\sigma_0(r)$, angular velocity $\Omega_0(r)$, 2D pressure $P_0(r)$ and central potential $\Phi_0(r)$ (see, e.g. Fig. 6, left) [7, 8]. This 2D model accretion disc inherits the main large-scale dynamical properties of the primary 3D accretion disc, but feels no tidal influence. In particular, the distribution of the specific angular momentum in the model disc has an important peculiarity – the (smoothed) kink near the radius where the one-armed spiral structure is located in the primary disc (see Fig. 6, right panel).

To understand the importance of the presence of a kink in a specific (per unit mass) angular momentum[7] radial distribution, let us write the equations describing the dynamics of perturbed state of 2D model accretion disc (in the new reference frame) in the form

$$\frac{\partial \sigma}{\partial t} + \frac{1}{r}\frac{\partial (r\sigma V_r)}{\partial r} + \frac{1}{r}\frac{\partial(\sigma V_\varphi)}{\partial \varphi} = 0, \tag{5a}$$

$$\frac{\partial V_r}{\partial t} + V_r\frac{\partial V_r}{\partial r} + \frac{V_\varphi}{r}\frac{\partial V_r}{\partial \varphi} - \frac{V_\varphi^2}{r} = -\frac{1}{\sigma}\frac{\partial P}{\partial r} - \frac{d\Phi_0}{dr}, \tag{5b}$$

$$\frac{\partial V_\varphi}{\partial t} + V_r\frac{\partial V_\varphi}{\partial r} + \frac{V_\varphi}{r}\frac{\partial V_\varphi}{\partial \varphi} + \frac{V_r V_\varphi}{r} = -\frac{1}{\sigma r}\frac{\partial P}{\partial \varphi}. \tag{5c}$$

[7] Kink in a specific angular momentum (per unit mass) evidently means kink in $\Omega_0(r)$.

Supposing all perturbed quantities have small deviations $\delta\tilde{F}$ from the unperturbed steady-state axisymmetric parameters F_0,

$$F(r,\varphi,t) = F_0(r) + \delta\tilde{F}(r,\varphi,t),$$

where

$$\delta\tilde{F}(r,\varphi,t) = \delta F(r) \cdot \exp\left[i(m\varphi - \omega t)\right],$$

after linearization the system (5), it may be reduced to the ordinary second order differential equation for the perturbed enthalpy δW relative to r [28]:

$$\frac{\chi}{r\sigma_0}\frac{d}{dr}\left[\frac{r\sigma_0}{\chi}\frac{d}{dr}(\delta W)\right] - \frac{\chi}{r\sigma_0\hat{\omega}}\left[\frac{d}{dr}\left(\frac{2m\Omega_0\sigma_0}{\chi}\right)\right]\delta W - \left[\frac{m^2}{r^2} + \frac{\chi}{c_0^2}\right]\delta W = 0. \quad (6)$$

Here the following notations are used:

$$\chi(r,\omega) \equiv \left[\hat{\omega}^2(r) - æ_0^2(r)\right], \; æ_0^2 \equiv 2\Omega_0\left(2\Omega_0 + r\frac{d\Omega_0}{dr}\right),$$
$$\hat{\omega}(r,\omega) \equiv \omega - m\Omega_0(r), \quad c_0^2 \equiv \frac{\partial P_0}{\partial\sigma_0}.$$

In the extreme case of abrupt kink in $\Omega_0(r)$ at radius $r = R_0$, it's derivative would be a θ-function of $\tilde{r} \equiv (r - R_0)$, and the second derivative would be a δ-function of $\tilde{r}$. Just the behaviour the second term in (6) would reveal – in case of abrupt kink in $\Omega_0(r)$ this term would have a δ-function of $\tilde{r}$:

$$\frac{d\chi}{dr} \propto \frac{dæ_0^2}{dr} \propto \frac{d^2\Omega_0}{dr^2} \propto \delta(\tilde{r}).$$

The solution of the (6) in this case would be analogous to the solution of Schroedinger equation with potential energy having a "δ-well" [28], in this case an energy level always exists (Fig. 7a). In case of the real accretion disc, the "δ-well" transforms into the ordinary well (Fig. 7b). Here the "potential well" would have two waves: incident and reflected ones. The latter reflects from the resonance zone (in the vicinity of the corotation circle) with amplitude amplified. This leads to the over-reflection instability, and generation of the one-armed spiral density wave.

To explore spiral waves that would be generated in the model accretion disc by the over-reflection instability, the boundary eigen-value problem for (6) was

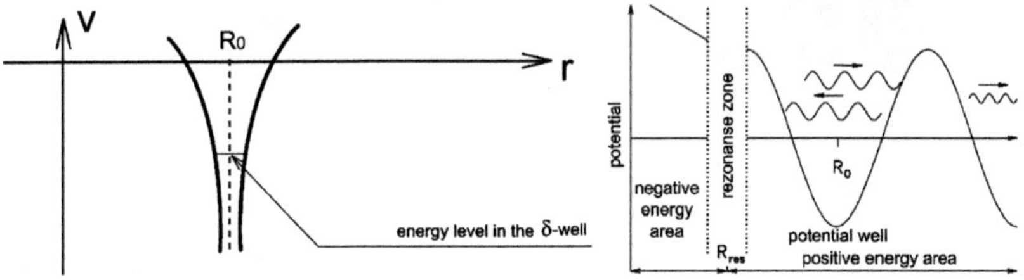

Fig. 7. *Left panel*: Energy level in the "δ -well"; *Right panel*: Ordinary well contains incident and reflected waves

numerically solved. For a set of realistic parameters the eigenvalues – phase angular velocities $\Omega_{ph} \equiv \mathrm{Re}\,(\omega)\,/m$ and growth rates (increments) $\gamma \equiv \mathrm{Im}\,(\omega)$, as far as the main eigen functions – were calculated.

The comparison of the basic results obtained from the linear numerical analysis of the 2D model disc with those from the numerical simulations of the accretion disc in the binary system shows the close correspondence between them [7, 8]. In brief the obtained results are following.

In Fig. 8, left panel, the growth rates γ (in units of Ω_{ph}) for the first three large-scale overreflection unstable modes (m = 1,2,3) are presented. On the right panel of Fig. 8 the main eigen functions for the most unstable one-armed spiral (m=1) mode are shown. The disturbed surface density (only levels above the medium one) is pictured by darkening, the bold black line is the phase curve (azimuth of maximum surface density). The vector field of the disturbed velocities (in the reference frame rotating with the mode) shown by arrows, bold star pointing the center of anti-cyclonic vortex. Dashed circle corresponds to the corotation radius r_c, dash-dot circle – outer Lindblad resonance radius r_{OLR}. Note that (a) the whole spiral-vortex structure rotates with period $T^{(m=1)} \equiv 2\pi/\Omega_{ph}^{(m=1)} \simeq T_1 \simeq (1/6)P_{orb}$ in the same direction (counter-clockwise) as the accretion disc does, so the density spiral is the trailing one; (b) the phase curve of the density wave experiences a $(\pi/2)$-turn in positive direction in the interval $\{r_c, r_{OLR}\}$; (c) the anticyclonic vortex in the perturbed vector velocity field is centered on corotation radius, rotates

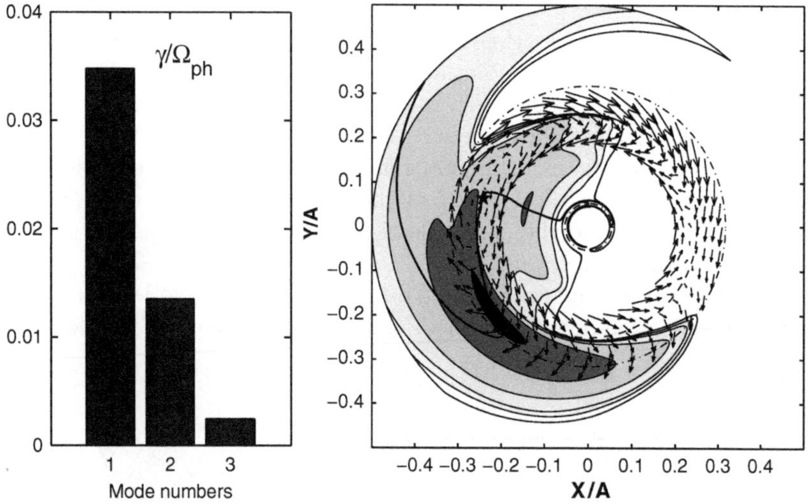

Fig. 8. Results of the numerical investigation of the linearly unstable over-reflection modes in the 2D model accretion disc: *Left* – growth rates for the first three large-scale unstable modes (m = 1,2,3); *Right* – eigen functions for the disturbed surface density (only levels above the medium one are shown) and the disturbed velocities (*arrows*) for the most unstable one-armed spiral mode ($m = 1$)

with the spiral density wave at the same angular velocity $\Omega_{ph}^{(m=1)}$ and plays an important role in disc angular momentum redistribution.

All investigated large-scale over reflection unstable modes demonstrate similar properties, they are located close to their corotation radii. These radii are close to each other, lay in the inner part of the disc, and may be approximated in the following way:

$$r_C^{(m)} \simeq r_C^{(1)} \cdot (1+\beta)^{m-1}\,;\; r_C^{(1)} \simeq 0.25A;\; \beta \simeq 0.024;\; m = 1, 2, 3.$$

The latter results are also close to the value of $r_c^{(1)} \simeq 0.21A$ evaluated from the 3D numerical simulations.

At the same time the rotational frequencies may be approximated as[8]

$$\omega_r^{(m)} \equiv 2\pi \mathrm{f}^{(m)} \equiv m\Omega_{ph}^{(m)} \simeq m\Omega_{ph}^{(1)} \cdot (1+\beta_1)^{-(m-1)}\,;\; \beta_1 \simeq 0.045.$$

So we have some "shifts" of frequencies

$$\Delta \mathrm{f}^{(m)} \equiv \left[\mathrm{f}^{(m)} - m\mathrm{f}^{(1)}\right] / \mathrm{f}^{(m)} \simeq -\beta_1 \cdot (m-1)\,;\; m = 2, 3.$$

These results obtained for the accretion disc in CBS are not surprising for us. The observed rotation velocity of the "nuclear" ionized gaseous disc in the very center of the Galaxy is also revealing a kink near the radius $R_0 \simeq 0.3$ pc, see Fig. 9, left. Such a behaviour is caused by the presence of a massive Black Hole in the center of the disc. Our numerical investigation showed that the one-armed "mini-spiral" observed in the "nuclear" disc can be generated by the over-reflection instability, see Fig. 9, right. It is worth mentioning for

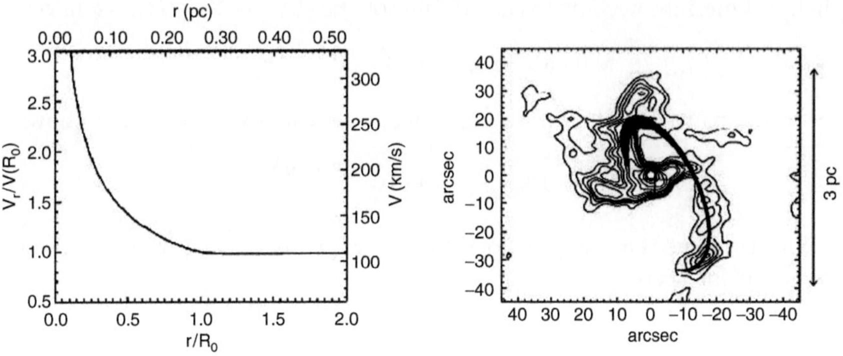

Fig. 9. *Left* – Observed velocity curve of the "nuclear" disc in the center of the Galaxy; *Right* – picture of isophotes, or isodensities (*thin curves*) observed in the disc and forming so-called "mini-spiral," and superimposed calculated one-armed spiral density wave (thick regular structure) generated by the over-reflection instability [10]

[8] It is evident that in the case of purely Keplerian disc we would have $\omega_r^{(m)} \simeq m\Omega_{ph}^{(1)} \cdot (1+\beta)^{-3(m-1)/2}$.

clarity, that (a) the disc is inclined by $\sim 60^o$ with respect to line of sight; (b) only the Northern Arm and Western Arc really belong to the disc. So there is a very good agreement between the observational data and the results of the numerical simulations [10].

5 Estimation of the Turbulent Viscosity Coefficient

In monographs and reviews on plasma physics, the formula for estimation of the turbulent viscosity coefficient ν_{turb} is written in the form:

$$\nu_{turb} \approx \frac{(\gamma_L)_{\max}}{(k_\perp^2)_{\min}}, \tag{7}$$

where $(\gamma_L)_{\max}$ is a maximum growth rate of the linear instability, and $(k_\perp)_{\min}$ is a minimum wave number (see e.g. [29]).

This formula can be easily derived from dimensions analysis. Indeed, the dimension of turbulent viscosity as well as of the molecular one is

$$[\nu_{turb}] = [\nu_{mol}] = \frac{cm^2}{s}.$$

Substituting in the right hand side values of the unstable disturbances: its growth rate γ and the wave vector k ($cm^2 = [k^{-2}]$, $s^{-1} = [\gamma]$), we get the formula (7) for the turbulent viscosity.

The fully developed Kolmogorov turbulence in an accretion disc means, that the maximum scale of the turbulent pulsations is much smaller than the disc semi-thickness h, i.e. $k_\perp h \gg 1$, or in the extreme case $(k_\perp)_{\min} h \geq 1$. Then, taking into account z-equilibrium of the disc, $h \cong c_0/\Omega_0$, we have:

$$(\gamma_L)_{\max} \simeq 0.035 \cdot \Omega_{ph} = 0.035 \cdot \Omega_{cor}, \; (k_\perp)_{\min} \cong h^{-1}.$$

According to the results of the linear instability analysis described above

$$(\gamma_L)_{\max} \simeq 3.5 \cdot 10^{-2} \cdot \left(\frac{c_0}{h}\right).$$

Substituting these expressions in (7), we obtain the turbulent viscosity coefficient for accretion disc in the form

$$\nu_{turb} \approx 3.5 \cdot 10^{-2} \cdot (c_0 h).$$

Then, following [30, 31], we find

$$\alpha \approx 3.5 \cdot 10^{-2}$$

that is in agreement with both the range of a numerical viscosity in our simulations $10^{-2} < \alpha < 10^{-1}$ and the interval given by the corresponding observations and the theoretical works (see, e.g. [15–18]).

6 Spectrum of Turbulence in the Accretion Disc of IP Peg

It is well-known, that according to Kolmogorov scenario of fully developed stationary turbulence the cascade of the energy flux is directed from lower to higher $k_\perp$, herewith the power of the energy flux ε continues to be constant

$$\varepsilon \approx (\delta V)^2 \cdot t^{-1} \approx (\delta V)^3 \cdot \lambda^{-1} = const, \tag{8}$$

where it is assumed that the turbulent pulsations with space size $\sim \lambda$ have pulsation velocity $\sim \delta V$. From (8) we find the Kolmogorov–Obukhov spectrum [33, 34]:

$$(\delta V) \approx (\lambda \varepsilon)^{1/3}, \varepsilon = const. \tag{9}$$

Note, that the condition (8) can be fulfilled only if the medium is homogeneous and isotropic, but the explored above differentially rotating and inhomogeneous accretion disc does not have these properties.

If we put down both arms into the shallow water in a bath and begin to rotate our arms in the same direction, we can see that the two vortices joined in the one - the scale increased unlike the Kolmogorov cascade - it is the case of the Batchelor turbulence. If such a 2D turbulence develops in the disc, than its smallest scale would be greater than the disc semi-thickness: $k_\perp h \ll 1$, or $(k_\perp)_{\max} h \le 1$.

In the earlier works [7, 8] on basis of the preliminary analysis of the observed light curves of the IP Peg close binary system in an active phase [32] we found on the uneclipsed parts of the light curves the apparent intensity variations with the period $T_1 \simeq (1/5)\, T_{orb}$ (here T_{orb} is the orbital period of the binary)[9]. The variations are clearly seen in the ionized HeII emission line 4686 Å - see Fig. 10.

Note that IP Peg is a unique object - a cataclysmic binary system having in active phase the developed accretion disc oriented optimally (nearly edge-on) for studying processes in it. So the radiation emitting from the very hot inner regions of the disc after passing through the exterior regions gives us information about structures in them. It is the ionized HeII high-energy emission line 4686 Å that is radiated from the very inner parts of the accretion disc - the efficient temperature of the corresponding HeII excited energy level is approximately $\simeq$ 51Ev.

According to the results described in Sect. 4 the one-armed spiral density wave can be generated by the over-reflection instability in the accretion disc. This density structure is localized in the medium radius range $r \sim (0.15 \div 0.30)\, A$, and it's period of rotation is close to T_1[10]. That is why

[9] Of course the tidal variations with the large amplitude A_{tdl} and the orbital period $T_{tdl} \simeq T_{orb}/2$ are also presented.

[10] In IP Peg the components mass ratio $q = M_1/M_2 = 2$.

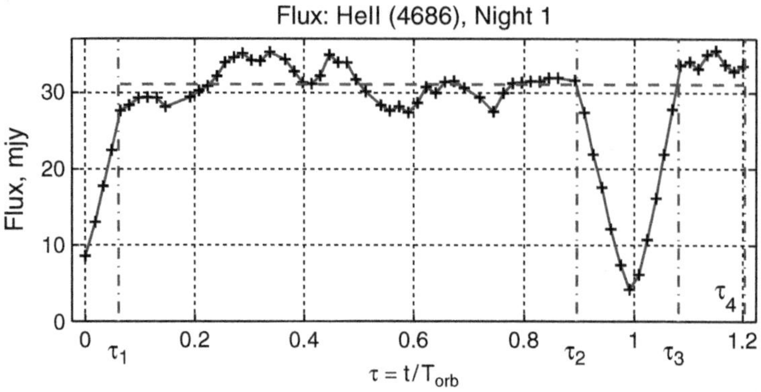

Fig. 10. The observed light curve for emission line of ionized helium HeII 4686 Å of the IP Peg (see [32], Fig. 5, lower panel, the first night of observations). The vertical dashed lines correspond to the uneclipsed regions limits, the horizontal dashed line denotes the medium intensity level (after eclipsed regions were removed)

we believe that the variations of the intensity in the HeII emission line 4686 Å [32] are caused by the partial modulation of radiation when it passes through the rotating density wave structure. The maximum relative amplitude of the intensity variations in this emission line is of the order of ∼ 10%, so the optical depth of the gas of the accretion disc exterior regions is much less then unity. Consequently the intensity decay is proportional to the density on the light path.

Let us represent the density in the middle plane of the accretion disc at some medium radius and fixed azimuth in the form

$$\rho(t) = \langle\rho\rangle + \Delta\rho(t)\,;\; \Delta\rho(t) = \delta\rho_{tdl}(t) + \sum_{m=1}^{m_{\max}} \delta\rho^{(m)}(t)\,.$$

Here we suppose that the density variations $\Delta\rho(t)$ around some medium level $\langle\rho\rangle$ are caused by the superposition of the tidal harmonic $\delta\rho_{tdl}(t)$ and the main over-reflection eigen modes – spiral density waves with azimuthal mode numbers m ($m_{\max} \leq (3 \div 4)$). Then we can in a similar way approximate the observed radiation intensity of the uneclipsed interval of the IP Peg in the HeII line as the superposition of the medium level $\langle I\rangle$ and some pulsations $\Delta I(t)$, i.e.

$$I(t) = \langle I\rangle + \Delta I(t)\,;\; \Delta I(t) = \delta I_{tdl}(t) + \sum_{m=1}^{m_{\max}} \delta I^{(m)}(t)\,.$$

Here $\delta I_{tdl}(t)$ – the tidal harmonic with the frequency $\mathrm{f}_{tdl} \simeq 2\mathrm{f}_{orb}$, $\mathrm{f}_{orb} \equiv 1/T_{orb}$, and the amplitude A_{tdl}. The other pulsations supposed to correspond to the main over-reflection eigen modes and have a simple harmonic form

$$\delta I^{(m)}(t) \cong A_m \cdot \cos\left[2\pi \mathrm{f}_m (t - t_{0m})\right].$$

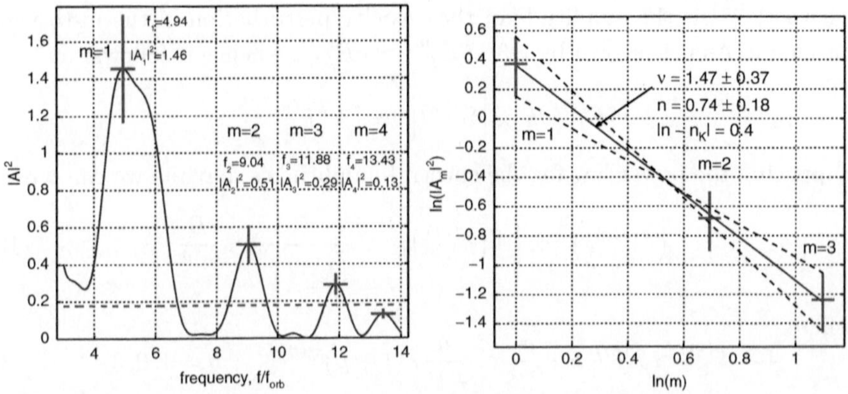

Fig. 11. *Left*: The Fourier power spectrum for the uneclipsed variability of the emission from IP Peg in the line HeII 4686 (night 1), see Fig. 10, the crosses denote maximum errors of the spectrum calculations, the horizontal dashed line shows the observational noise level; *Right*: The linear interpolated logarithm of the power spectrum main maxima (using the Least Mean Square method), the crosses correspond to the errors denoted on the left panel, the dashed lines show maximum variance of the linear approximation in the error limits

Therefore the amplitude A_m of the m-th intensity pulsation harmonic would be proportional to the amplitude $\left|\delta\rho^{(m)}\right|$ of the corresponding eigen mode, $A_m \equiv A\left(k_\perp^{(m)}\right) \propto \left|\delta\rho^{(m)}\right|$. The calculated frequency Fourier power spectrum of the pulsations $\Delta I\,(t)$ shows the four main peaks besides the tidal harmonic, see Fig. 11, left. The first peak at the frequency $\mathrm{f}_1 \simeq 4.94$ is supposed to correspond to the $m = 1$ over-reflection mode. Therefore, according to our results of the numerical investigation of the linearly unstable over-reflection modes in the 2D model accretion disc, the two subsequent spectrum peaks would correspond to the modes $m = 2$ and $m = 3$.[11]

The value of the specific thermal (elastic) energy of the m-th density wave harmonic may be estimated as (see, e.g. [13])

$$\delta E_T^{(m)} \cong \frac{c_0^2}{2\left<\rho\right>}\left(\delta\rho^{(m)}\right)^2 \propto \left(\delta\rho^{(m)}\right)^2 .$$

The value of the specific kinetic energy in the same wave is [13]

$$\delta E_K^{(m)} \cong \frac{\left<\rho\right>}{2}\left(\delta V^{(m)}\right)^2 \propto \left(\delta V^{(m)}\right)^2 ,$$

[11] The "shifts" of the corresponding peak frequencies in the spectrum on Fig. 11 (left) are estimated as $\Delta\mathrm{f}^{(2)} \simeq -0.09$, and $\Delta\mathrm{f}^{(3)} \simeq -0.25$, that is in a qualitative accordance with our model results described in Sect. 4.

where $\delta V^{(m)}$ is the amplitude of the velocity perturbation in the m-th wave harmonic. As it is shown by [13], $\delta E_T^{(m)} \simeq \delta E_K^{(m)}$, whence it follows:

$$\delta E_T^{(m)} dk \propto \left|\delta\rho^{(m)}\right|^2 \propto \left|A\left(k_\perp^{(m)}\right)\right|^2 .$$

Supposing the power law for the spectral amplitude's maxima we can write

$$\left(\delta V^{(m)}\right)^2 \propto \left|A\left(k_\perp^{(m)}\right)\right|^2 \simeq \frac{B}{\left(k_\perp^{(m)}\right)^\nu} \simeq \frac{\tilde{B}}{m^\nu}, \tag{10}$$

where

$$k_\perp^{(m)} \simeq k_\varphi^{(m)} \simeq \frac{m}{r_C^{(m)}} \simeq \frac{m}{r_C^{(1)}} \cdot (1+\beta)^{-(m-1)} ; m = 1, 2, 3$$

and the corotation radii $r_C^{(m)}$ of the large-scale over-reflection modes believed to be close to each other (according to the Sect. 4, we can approximate the β value as $\beta \sim 0.05 \ll 1$ here).

Taking the logarithm of (10) we can rewrite

$$\ln\left\{|A_m|^2\right\} \simeq \tilde{B} - \nu \ln(m) .$$

As we see from the Fig. 11, right, the three peaks corresponding to the modes $m = 1, 2, 3$ fit this linear dependence sufficiently well. The slope of this linear dependence (calculated by the Mean Least Square method) gives us the power exponent n

$$\left(\delta V^{(m)}\right) \propto \left(k_\perp^{(m)}\right)^{-n}, n \equiv \nu/2 \simeq 0.74 \pm 0.18$$

Here the estimated value of the errors corresponds to the uncertainties of the calculated (by the discrete Fourier transform) power spectrum (these errors are denoted by the crosses on the Fig. 11, left). The value $n \simeq 0.74 \simeq 3/4$ of the turbulent power spectrum exponent calculated for the IP Peg accretion disc differs significantly from the Kolmogorov-Obukhov $n_K = 1/3$ exponent for the fully developed turbulence (see the formulae (9)) by $\Delta n \simeq 0.4$, that exceeds by approximately two times the estimated possible errors corresponding to the observations and the spectrum calculations. This principal difference occurs due to the fact that in any accretion disc the conditions of the Kolmogorov hypotheses of the homogeneity and isotropy are not fulfilled.

7 Conclusions

The complex analysis of large-scale processes in the accretion disc of a close binary system, involving 3D numerical simulations, the numerical solution of the linear eigen-value problem for the 2D model disc, and the analysis of the Fourier power spectrum for the IP Peg accretion disc radiation intensity pulsations $\Delta I(t)$, shows the following:

1. From the Fourier power spectrum for the IP Peg radiation intensity pulsations we can see the global tidal mode caused by the presence of the companion star and located in the exterior regions of the disc.
2. The other power spectrum harmonics are supposed to correspond to the main large-scale eigen modes caused in the IP Peg accretion disc by the development of the over-reflection instability.
3. The latter, first of all, results in the dominant growth of the largest scale mode $m = 1$.
4. The calculation of the growth rate of the dominating mode $m = 1$ gave the possibility to evaluate the turbulent viscosity coefficient as well as the α-coefficient. The value of the latter is in agrement with the values presented in many observational and theoretical papers.
5. The rotational frequencies of the large-scale over-reflection eigen modes (m=1,2,3) found by the linear numerical analysis in the 2D model accretion disc, are corresponded well to the peaks on the Fourier spectrum for the IP Peg radiation intensity pulsations.
6. On basis of the principal results of the linearly unstable over-reflection modes numerical investigation in the 2D model accretion disc, the power-law dependence was found for the IP Peg Fourier spectrum amplitudes peaks upon the $k_{\perp}^{(m)}$ wave numbers.

Resuming our studies of power spectra of the fully developed turbulence in the medium of different astrophysical discs we can conclude that:

$V_\lambda \propto \lambda^{1/3}$ – $\rho_0 = const$, $\Omega_0 \equiv 0$ – theory: [33, 34].
$V_\lambda \propto \lambda^{1/2}$ – Galactic disc, $\rho_0 \neq const$, $\Omega_0 = const \neq 0$ – observations: [16]; theory: [11].
$V_\lambda \propto \lambda^{3/4}$ – accretion disc, $\rho_0 \neq const$, $\Omega_0 = \Omega_0\,(r) \neq const$ – from theory and observations: [35].

The relative error of the last result (the power exponent) is estimated to be ~20%. The errors may appear mainly during the power spectrum calculations (on the reasons of unknown phases of the harmonics in the observed signal), and the presence of the tidal harmonic with the large amplitude.

Acknowledgments

The work was supported by the Russian Foundation for Basic Research (projects no. 06-02-16097, 05-02-17874a), the Scientific Schools Program (grants no. NSh-4820.2006.2, NSh-7629.2006.2, NSh-900.2008.2), and the basic research programs "Origin and evolution of stars and galaxies" and "Extensive objects in the Universe" of the RAS.

References

1. I.D. Novikov, K.S. Thorne: Astrophysics of black holes, In: *Black Holes. Les Astres Occlus*, C. DeWitt and B.S. DeWitt (eds.) (Gordon & Breach, New York, 1973), p. 343.
2. D.N. Page, K.S. Thorne: Astrophys. J. **191**, 499 (1974).
3. I. Dutan, P. Biermann: The efficiency of using accretion power of Kerr black holes, In: *Neutrinos and Explosive Events in the Universe*, M.M. Shapiro et al. (eds.) (Springer, Netherlands, 2005), p. 175.
4. L. Euler: Novi Comm. Acad. Sci. Petrop. **11**, 144 (1767).
5. J.L. Lagrange: Euvres **6**, 272 (1772).
6. V. Szebehely: *Theory of Orbits: the restricted Problem of three Bodies*, (Academic, New York, 1967).
7. A.M. Fridman, A.A. Boyarchuk, D.V. Bisikalo, O.A. Kuznetsov, O.V. Khoruzhii, Yu.M. Torgashin, A.A. Kilpio: Phys. Lett. A **317**, 181 (2003).
8. A.M. Fridman, O.V. Khoruzhii: Space Sci. Rev. **1-3**, 1 (2003).
9. A.M. Fridman, O.V. Khoruzhii, V.V. Lyakhovich, L. Ozernoi, L. Blitz: In: *Proc. Unsolved Problems of the Milky Way*, L. Blitz and P. Tauben (eds.) (Kluwer, Dordecht, 1996), p. 241.
10. A.M. Fridman, O.V. Khoruzhii, V.V. Lyakhovich, L. Ozernoi, O.K. Sil'chenko L. Blitz: In: *The Galactic Center*, R. Gredel (ed.) (Astron. Soc. of the Pacific Conf. Series, vol. 102, 1996), p. 335.
11. V.V. Dolotin, A.M. Fridman: Sov. Phys. JETP **72** (1), 1 (1991).
12. A.M. Fridman, O.V. Khoruzhii, V.V. Lyakhovich, L. Ozernoi, L. Blitz: In: *Barred Galaxies and Circumnuclear Activity, Proc. of the NOBEL SYMPOSIUM 98 held at Stockholm Observatory, Salltsjobaden, Sweden, 30 November – 3 December 1995*, A.Sandqvist and P.O.Lindblad (eds.) (Springer, Berlin, 1996). Also *Lecture Notes in Physics*, vol. 474, p. 193.
13. L.D. Landau, E.M. Lifshitz: *Fluid Mechanics*, (Pergamon Press, 1984).
14. L.D.Landau, E.M. Lifshitz: *Statistical Physics*, (Volume 5 of Theoretical Physics, Butterworth-Heinemann, 1980).
15. D. Lynden – Bell, G.E. Pringle: Mon. Not. Roy. Astron. Soc. **158**, 603 (1974).
16. E. Meyer – Hofmeister, H. Ritter: In: *The Realm of Interacting Binaries*, J.Sahade, G.E.McCluskey, Jr., Y.Kondo (eds.) (Kluwer, Dordrecht, 1993) p. 143.
17. J. Smak: Acta Astron. **49**, 391 (1999).
18. F.H. Shu, D. Galli, S. Lisano, A.F. Glassgjld, P.H. Diamond: Astrophys. J. **665**, 535 (2007).
19. D.V. Bisikalo, A.A. Boyarchuk, V.M. Chechetkin, O.A. Kuznetsov, D. Molteni: Monthly Notices Roy. Astron. Soc., **300**, 39 (1998).
20. D.V. Bisikalo, P. Harmanec, A.A. Boyarchuk, O.A. Kuznetsov, P. Hadrava: Astron. Astrophys. **353**, 1009, (2000).

21. D.V. Bisikalo, A.A. Boyarchuk, P.V. Kaigorodov, O.A. Kuznetsov, T. Matsuda: Astron. Rep. **48**, 449, (2004)
22. D.V. Bisikalo: Astrophys. Space Sci. **296**, 391, (2005).
23. A.Yu. Sytov, P.V. Kaigorodov, D.V. Bisikalo, A.A. Boyarchuk, O.A. Kuznetsov: Astron. Rep. **51**, 836, (2007).
24. A.A. Boyarchuk, D.V. Bisikalo, O.A. Kuznetsov, V.M. Chechetkin: *Mass transfer in close binary stars*, (Taylor and Francis, London, 2002).
25. D.V. Bisikalo, A.A. Boyarchuk, A.A. Kilpio, O.A. Kuznetsov, V.M. Chechetkin: Astron. Rep. **45**, 611, (2001).
26. D.V. Bisikalo, A.A. Boyarchuk, A.A. Kilpio, O.A. Kuznetsov, V.M. Chechetkin: Astron. Rep. **45**, 676, (2001).
27. A.M. Fridman: Physics Uspekhi. **50**, 121 (2007).
28. V.V. Lyakhovich, A.M. Fridman, O.V. Khoruzhii: In *Unstable processes in Space*, A.G.Masevich (ed.) (Kosmosinform, Moscow, 1997) p.194 (in Russian).
29. B.B. Kadomtsev: In: *Voprosi Teorii Plazmy Vol. 4*, M.A.M. Leontovich (ed.) (Atomizdat, Moscow, 1964), p. 188/ (Reviews in Plasma Physics, Consultants Bureau, New York, 1971).
30. N.I. Shakura: Sov. Astron. **16**, 756 (1972).
31. N.I. Shakura, R.A. Sunyaev: Astron. Astrophys. **24**, 337 (1973).
32. L. Morales-Rueda, T.R. Marsh, I. Billington: Mon. Not. R. Astron. Soc. **313**, 454 (2000).
33. A.N. Kolmogorov: Doklady Akademii Nauk USSR. **30**, 4, 299 (1941).
34. A.M. Obukhov: Doklady Akademii Nauk USSR. **32**, 1, 22 (1941).
35. A.M. Fridman, D.V. Bisikalo, A.A. Boyarchuk, Yu.M. Torgashin, L. Pustilnik: accepted for publication in Astron. Rep. (2008).

Solar and Stellar Active Regions

A Cosmic Laboratory for the study of Complexity*

Loukas Vlahos

Department of Physics, University of Thessaloniki, 54124 Thessaloniki, Greece
vlahos@astro.auth.gr

Summary. Solar active regions are driven dissipative dynamical systems. The turbulent convection zone forces new magnetic flux tubes to rise above the photosphere and shuffles the magnetic fields which are already above the photosphere. The driven 3D active region responds to the driver with the formation of Thin Current Sheets in all scales and releases impulsively energy, when special thresholds are met, on the form of nano-, micro-, flares and large scale coronal mass ejections. It has been documented that active regions form self similar structures with area Probability Distribution Functions (PDF's) following power laws and with fractal dimensions ranging from 1.2–1.7. The energy release on the other hand follows a specific energy distribution law $f(E_T) \sim E_T^{-a}$, where $a \sim 1.6$–1.8 and E_T is the total energy released. A possible explanation for the statistical properties of the magnetograms and the energy release by the active region is that the magnetic field formation follows rules analogous to percolating models, and the 3D magnetic fields above the photosphere reach a Self Organized Critical (SOC) state. The implications of these findings on the acceleration of energetic particles during impulsive phenomena will briefly be outlined.

1 Solar Active Regions as Driven Non Linear Systems

The most energetic phenomena above the solar surface are associated with "active regions (AR)." The 3D AR is a theater of intense activity of various (spatiotemporal) scales. The 3D AR has a visual boundary at the photosphere (although its physical boundary, as we will see in the next section, is inside the turbulent convection zone), and is subject to external forcing caused by the flux emergence from the solar interior and by the shuffling motions at the photosphere.

Our main goal in this article is to show that a variety of well known solar phenomena (e.g. coronal heating, flares, CME's, particle acceleration etc) can be understood in a unified manner by considering the solar active regions as externally (sub-photospherically) driven non linear systems. There

* In the memory of Prof. Nikos Voglis

G. Contopoulos, P.A. Patsis, *Chaos in Astronomy.* Astrophysics and Space Science Proceedings, doi: 10.1007/978-3-540-75826-6,

are many well known statistical observations which suggest that ARs are far from equilibrium: (1) The magnetic structures at the solar photosphere establish a fractal form and have power law size distributions, (2) the explosive phenomena (i.e flares) follow a very stable power law frequency distribution (3) the high energy particles, accelerated during solar flares, establish before leaving the accelerator a power law energy distribution. The main question addressed in this review then is: How an active region achieves all these statistical regularities? We argue that the formation of AR's follows laws analogous to well known percolation models and simultaneously drives the extrapolated 3D magnetic structures above the photosphere, leading the entire 3D structure to a Self Organized Critical (SOC) state. The coupling of two well known mechanisms of complex systems (percolation (as the driver) and self organized criticality for the energy dissipation) are behind all the observed regularities recorded on the data.

2 Active Region Formation: A Percolating Driver?

We can learn a lot about the sub-photospheric activity by "reading" carefully the magnetograms. Both full disk and more detail magnetograms around specific AR are extremely useful tools (see Fig. 1). Two of their striking properties are found in the Probability Distribution Function (PDF) of their sizes and their fractal properties.

Numerous observational studies have investigated the statistical properties of active regions, using full-disc magnetograms. These studies have examined among other parameters the size distribution of active regions, and their fractal dimension: *The size distribution function* of the newly formed active regions exhibits a well defined power law with index ≈ -1.94 and active regions cover only a small fraction of the solar surface (around $\sim 8\%$) [10]. The *fractal dimension* of the active regions has been studied using high-resolution magnetograms by [6], and more recently by [26]. These authors found, using not always the same method, a fractal dimension D_F in the range $1.2 < D_F < 1.7$. Fractal dimensions in solar magnetic fields are typically

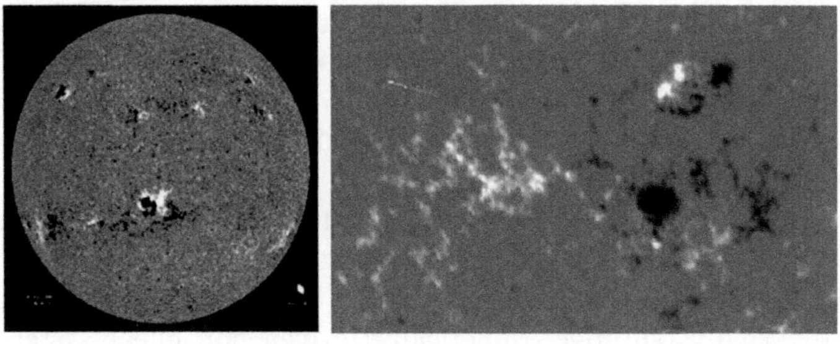

Fig. 1. (**a**) Sunspots in a full disk magnetogram, (**b**) Magnetogram around an active region

calculated using the box-counting technique. The values of the fractal dimension depends on whether the structures themselves or just their boundaries are box counted. The analysis has been pursued even further using the concepts of multi-fractality. It is well known that an AR includes multiple types of structures such as different classes of sunspots, plages, emerging flux subregions, etc. The physics behind the formation and evolution of each of these structures is not believed to be the same, so the impact and the final outcome of convection zone turbulence in each of these configurations should not be the same. Numerous other tools have been used to uncover aspects of the complex behavior "mapped" by the convection zone on the photospheric boundary, e.g. generalized correlation dimension, structure formations, wavelet power spectrum [16].

Theoretical studies on the formation of AR can be divided in two main categories: (1) The evolution of one or two slender and isolated flux tubes (see recent articles on this topic [2, 13]), (2) using standard percolation techniques. We stress the second method here since we place special emphasis on the complexity of AR.

A percolation model was proposed to simulate the formation and evolution of active regions at the photosphere in [37, 44]. In this model, the evolution of the magnetograms is followed by reducing all the complicated convection zone dynamics into three dimensionless parameters. The emergence and evolution of magnetic flux on the solar surface in the 2-D cellular automaton (CA) is probabilistic and based on the competition between two "fighting" tendencies: **stimulated** or **spontaneous** emergence of new magnetic flux, and the disappearance of flux due to **diffusion** (i.e. dilution below observable limits), together with random **motion** of the flux tubes on the solar surface (this processes mimics the shuffling). This percolation model explains the observed size distribution of active regions and their fractal characteristics [26, 40]. It was later used for the reconstruction of 3-D active regions using the force free approximation and many of the observational details reported in [41] were reproduced [11].

3 How Active Regions Evolve by Dissipating Energy: Are AR in a Self Organized Critical State?

3.1 3-D Extrapolation of Magnetic Field Lines and the Formation of Unstable Current Sheets

The energy needed to power solar flares is provided by photospheric and subphotospheric motions and is stored in non-potential coronal magnetic fields. Since the magnetic Reynolds number is very large in the solar corona, MHD theory states that magnetic energy can only be released in localized regions where the magnetic field forms small scale structures and steep gradients, i.e. in thin current sheets (TCS).

Numerous articles (see recent reviews [8, 23]) are devoted to the analysis of magnetic topologies which can host TCSs. The main trend of current research in this area is to find ways to realistically reconstruct the 3-D magnetic field topology in the corona based on the available magnetograms and large-scale plasma motions at the photosphere. A realistic magnetic field generates many "poles and sources" [23] and naturally has a relatively large number of TCSs. We feel that detailed representations inside the 3D AR of the TCS's are mathematically appealing only for relatively simple magnetic topologies (dipoles, quadrupoles, symmetric magnetic arcades [3]). When such simple topologies are broken in the photosphere, for example due to large-scale sub-Alfvénic photospheric motions or the emergence of new magnetic flux that disturbs the corona, such tools may be less useful. All these constraints restrict our ability to reconstruct fully the dynamically evolving magnetic field of an active region (and it is not clear that an exact reconstruction will ever be possible).

Many of the widely used magnetograms measure only the line of sight component of the magnetic field. The component of the magnetic field vertical to the surface matches the measured magnetic field only at the center of the disk and becomes increasingly questionable as the limb is approached. Extrapolating the measured magnetic field is relatively simple if we assume that the magnetic field is in force-free equilibrium:

$$\nabla \times \mathbf{B} = \alpha(\mathbf{x})\mathbf{B} \tag{1}$$

where the function $\alpha(\mathbf{x})$ is arbitrary except for the requirement $\mathbf{B}\cdot\nabla\alpha(\mathbf{x}) = 0$, in order to preserve $\nabla \cdot \mathbf{B} = 0$. Equation (1) is non-linear since both $\alpha(\mathbf{x})$ and $\mathbf{B}(\mathbf{x})$ are unknown. We can simplify the analysis of (1) when α=constant. The solution is easier still when $\alpha = 0$, which is equivalent to assuming the coronal fields to contain no currents (potential field), hence no free energy, and thus they are uninteresting. A variety of techniques have been developed for the reconstruction of the magnetic field lines above the photosphere and the search for TCSs [23]. It is beyond the scope of this article to discuss these techniques in detail. For instructive purposes, we use the simplest method available, a linear force free extrapolation, and search for "sharp" magnetic discontinuities in the extrapolated magnetic fields as in Vlahos and Georgoulis [41]: They use an observed active-region vector magnetogram and then: (a) resolve the intrinsic azimuthal ambiguity of 180° [14], and (b) find the best-fit value α_{AR} of the force-free parameter for the entire active region, by minimizing the difference between the extrapolated and the ambiguity-resolved observed horizontal field (the "minimum residual" method of [17]). They perform a linear force-free extrapolation [1] to determine the three-dimensional magnetic field in the active region (see Fig. 2). Although it is known that magnetic fields at the photosphere are not force-free [15], they argue that a linear force-free approximation is suitable for the statistical purposes of their study.

Two different selection criteria were used in order to identify potentially unstable locations (identified as the afore-mentioned TCSs) [41]. These are (a) the Parker angle, and (b) the total magnetic field gradient. The angular

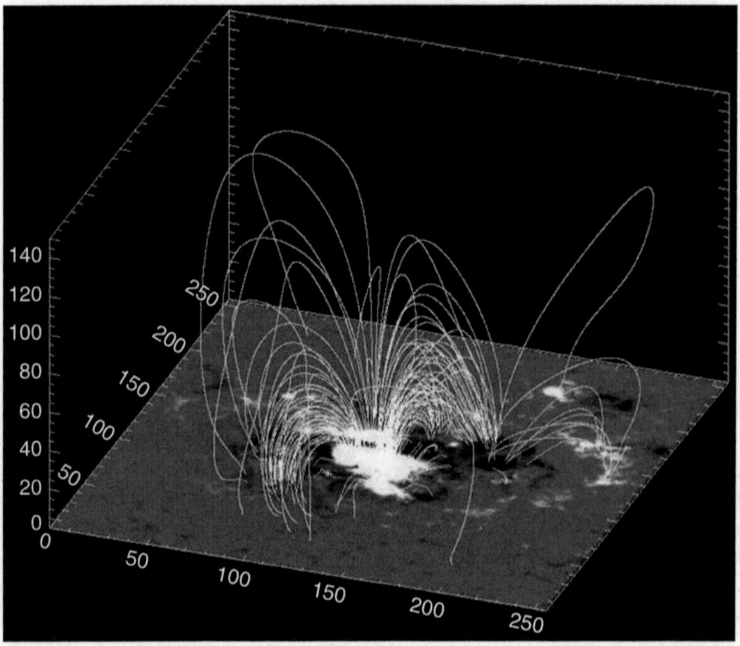

Fig. 2. Force free extrapolation of magnetograms and reconstruction of the magnetic fields inside an AR

difference $\Delta\psi$ between two adjacent magnetic field vectors, $\mathbf{B_1}$ and $\mathbf{B_2}$, is given by $\Delta\psi = \cos^{-1}[\mathbf{B_1} \cdot \mathbf{B_2}/(B_1 B_2)]$. Assuming a cubic grid, they estimated six different angles at any given location, one for each closest neighbors. The location is considered potentially unstable if at least one $\Delta\psi_i > \Delta\psi_c$, where $i \equiv \{1,6\}$ and $\Delta\psi_c = 14^o$. The critical value $\Delta\psi_c$ is the Parker angle which, if exceeded locally, favors tangential discontinuity formation and the triggering of fast reconnection [33, 34]. In addition, the total magnetic field gradient between two adjacent locations with magnetic field strengths B_1 and B_2 is given by $|B_1 - B_2|/B_1$. Six different gradients were calculated at any given location. If at least one $G_i > G_c$, where $i \equiv \{1,6\}$ and $G_c = 0.2$ (an arbitrary choice), then the location is considered potentially unstable. When a TCS obeys one of the criteria listed above, it will be transformed to an Unstable Current Sheet (UCS). Potentially unstable volumes are formed by the merging of adjacent selected locations of dissipation. These volumes are given by $V = N\lambda^2\delta h$, where N is the number of adjacent locations, λ is the pixel size of the magnetogram and δh is the height step of the force-free extrapolation. The free magnetic energy E in any volume V is given by

$$E = \frac{\lambda^2 \delta h}{2\mu_0} \sum_{l=1}^{N} (\mathbf{B_{ff}}_l - \mathbf{B_p}_l)^2 \tag{2}$$

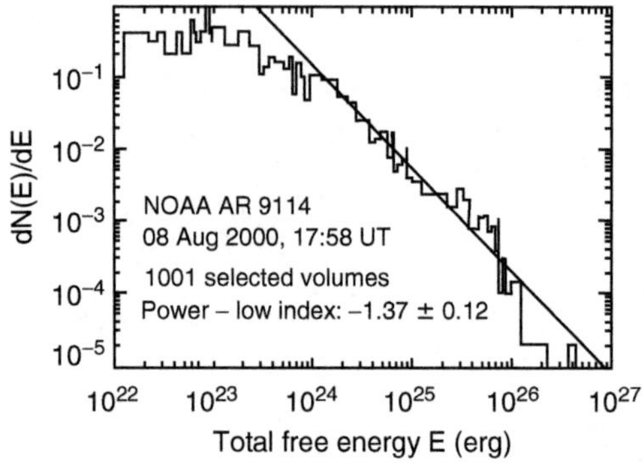

Fig. 3. Typical distribution function of the total free energy in the selected volume, on using a critical angle 14° [41].

where $\mathbf{B}_{\mathbf{ff}\,l}$ and $\mathbf{B}_{\mathbf{p}\,l}$ are the linear force-free and the potential fields at location l , respectively,,. The assumption used is that any deviation from a potential configuration implies a non-zero freen magnetic energy which is likely to be released if certain conditions are met. UCS are created naturally in active regions even during their formation and the free energy available in these unstable volumes follows a **power law distribution** with a well defined exponent (Fig.n 3). Vlahos and Georgoulis concluded that active regions store energy in many unstable locations, forming UCS of all sizes (i.e. the UCS have a self-similar structure). The UCS are fragmented and distributed inside the global 3-D structure. Viewing the flare in the context of the UCS scenario presented above, we can expect, depending of the size distribution and the scales of the UCS, to have flares of all sizes.

3.2 A Cellular Automata Model for the Energy Release in the Solar Corona

Coronal energy release observed at various wavelengths shows impulsive behaviour with events from flares to bright points exhibiting intermittency in time and space. Intense X-ray flare emission typically lasts several minutes to tens of minutes, and only a few flares are recorded in an active region that typically lives several days to several weeks. The flaring volume is small compared to the volume of an active region, regardless of the flare size. Intermittency is the dynamical response of a turbulent system when the triggering of the system is the result of a critical threshold for the instability [7]. In a turbulent system one also expects **self organization**, i.e. the reduction of the numerous physical parameters (degrees of freedom) present in the system to a small number of significant degrees of freedom that regulates the system's response to external forcing [28]. This is the reason for the success of concepts such as Self Organized Criticality (SOC) [4, 5] in explaining the statistical behavior of flares. Cellular Automata (CA) models typically employ one variable (the magnetic field, vector potential, etc) and study its evolution subject to external perturbations. When a critical threshold is exceeded (when the TCS becomes an UCS), parts of the configuration are unstable, and will be restructured to re-establish stability. The rearrangement may cause instabilities in adjacent locations, so the relaxation of the system may proceed as an avalanche-type process. In SOC flare models [24, 25, 39] each elementary relaxation is viewed as a single magnetic reconnection event, so magnetic reconnection is explicitly assumed to occur with respect to a critical threshold.

In solar MHD an UCS disrupts either when its width becomes smaller than a critical value [35], or when the magnetic field vector forms tangential discontinuities exceeding a certain angle [32], or when magnetic field gradients are steep enough to trigger restructuring [36]. We noticed that a critical threshold is involved in all cases: the first process points to the turbulent evolution in the magnetic field configuration and the onset of anomalous resistivity, while the latter two imply magnetic discontinuities caused either by the orientation of the magnetic field vector or by changes of the magnetic field strength. Magnetic field gradients and discontinuities imply electric currents via Ampére's law, so a critical magnetic shear or gradient implies a critical electric current accumulated in the current sheet which in turn leads to the onset of anomalous resistivity [31, 33].

One way of modeling the appearance, disappearance, and spatial organization of UCS inside a large-scale topology is with the use of the Extended Cellular Automaton (X-CA) model [18–20]. Figure 4 illustrates some basic features of the X-CA model. The X-CA model has as its core a cellular automaton model of the sand-pile type and is run in the state of Self-Organized Criticality (SOC). It is extended to be fully consistent with MHD: the primary grid variable is the vector-potential, and the magnetic field and the current are calculated by means of interpolation as derivatives of the vector potential in the usual sense of MHD, guaranteeing $\nabla \cdot \mathbf{B} = 0$ and $\mathbf{J} = (1/\mu_0)\nabla \times \mathbf{B}$ everywhere in the simulated 3-D volume. The electric field is defined as $\mathbf{E} = \eta \mathbf{J}$, with η the diffusivity. The latter usually is negligibly small, but if a threshold in the current is locally reached ($|\mathbf{J}| > J_{cr}$), then current-driven instabilities are assumed to occur, η becomes anomalous in turn, and the resistive electric field locally increases drastically. These localized regions of intense electric fields are the UCS in the X-CA model.

The X-CA model yields distributions of total energy and peak flux which are compatible with the observations. The UCSs in the X-CA form a set which is highly fragmented in space and time: the individual UCS are small scale regions, varying in size, and are short-lived. They do not form in their ensemble a simple large-scale structure, but form a fractal set with fractal dimension roughly $D_F = 1.8$ [42]. The individual UCS also do not usually split into smaller UCS, but they trigger new UCSs in their neighborhood, so that different chains of UCS travel through the active region, triggering new side-chains of UCS on their way. It is obvious that the rules of this simulation do not include the fragmentation of the UCS, in many ways through the results coincide with the MHD simulations [12].

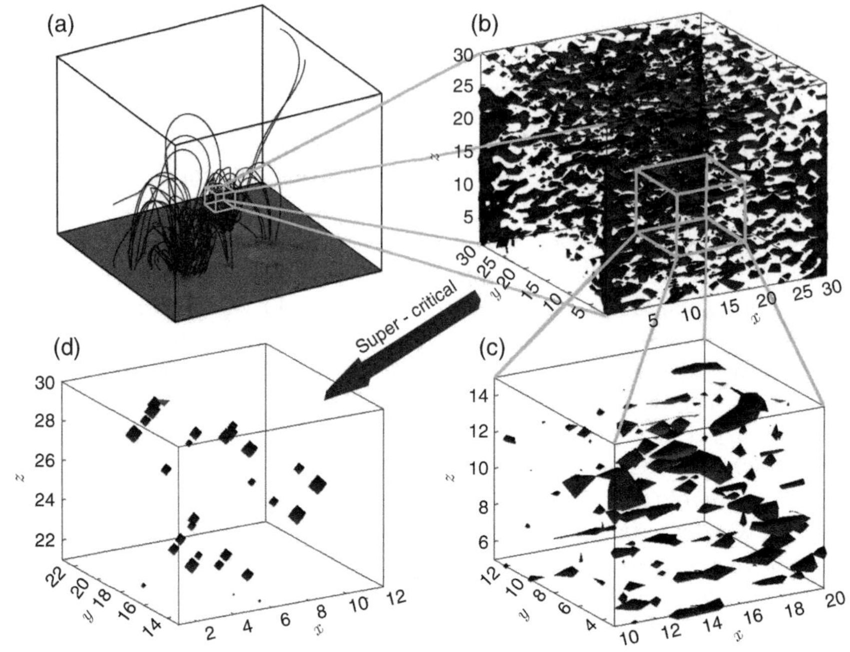

Fig. 4. (**a**) Simulated magnetogram of a photospheric active region and force-free magnetic field-lines, extrapolated into the corona (**b**) Sub-critical current iso-surfaces in space, as yielded by the X-CA model, which models a sub-volume of a coronal active region. (**c**) The same as (**b**), but zoomed. (**d**) Temporal snap-shot of the X-CA model during a flare, showing the spatial distribution of the UCS (super-critical current iso-surface) inside the complex active region [42]

4 Active Regions as Multi-Scale Physics Laboratories

So far we have discussed very briefly: (1) the formation of an AR as it is mapped in the magnetogram, (2) the use of the magnetogram as non-linearly evolving driver for the 3D AR, (3) the reconstruction of the 3D AR using simple techniques and the search for Thin Current Sheets (TCS) where energy may be dissipated. The critical transition of a TCS to a rapidly reconnecting structure (UCS) is essential for the 3D AR to reach a SOC state. The TCS formed inside the AR extend from the large scales (10^{10} cm) which are very unstable and rapidly fragment down to a few meters (on the order of the ion gyro radius) where the fast reconnection ignites. We have already discovered on all these levels enormous complexity. The main question now is: Do the UCS remain stable and dissipate magnetic energy or are they fragmented even further?

Onofri et al. [29] studied the nonlinear evolution of current sheets using the 3-D incompressible and dissipative MHD equations in a slab geometry. The nonlinear evolution of the system is characterized by the formation of

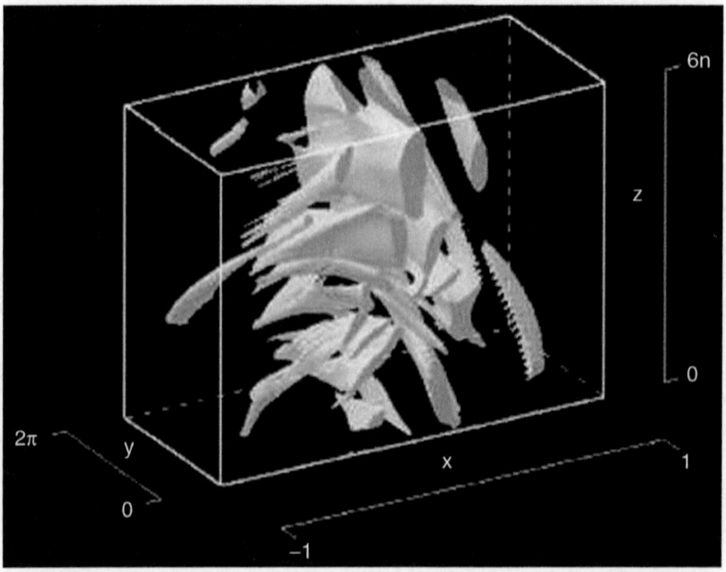

Fig. 5. Current isosurfaces showing the formation of current filaments, [29]

small scale structures, especially in the lateral regions of the computational domain, and coalescence of magnetic islands in the center.

This behavior is reflected in the 3-D structure of the current (see Fig. 5), which shows that the initial equilibrium is destroyed by the formation of current filaments, with a prevalence of small scale features. The final stage of these simulations is a turbulent state, characterised by many spatial scales, with small structures produced by a cascade with wavelengths decreasing with increasing distance from the current sheet. In contrast, inverse energy transfer leads to the coalescence of magnetic islands producing the growth of two-dimensional modes. The energy spectrum approximates a power law with slope close to two at the end of the simulation [30]. Similar results have been reported by many authors using several approximations [9, 22, 27, 38]. It is also interesting to note that similar results are reported from magnetic fluctuations in the Earth's magnetotail [45].

We have now created current structures on all scales and the next question is how particles will respond to the electric fields developed at all these scales due to the presence of enhanced resistivity at the small scales? The problem of particle acceleration is beyond the domain of MHD theory or even the two-fluid description of the plasmas. Only kinetic theory can follow the evolution of the particle distribution inside a fractally distributed electric field environment. Particle (anomalous) diffusion and the build up of non-thermal velocity distributions in localized structures distributed inside the 3D AR was the subject of many recent articles [42, 43].

5 Summary

In this review we have attempted to show that many well known solar phenomena, treated separately in many recent reviews (e.g. coronal heating, flares, CME's and particle acceleration) are symptoms of the formation and evolution of ARs. The complexity of the magnetograms, the formation of millions of TCS which, after passing a threshold become UCS at sporadic places inside the AR, and the further fragmentation of all UCS practically to all scales, play an important role in the formation of high energy particles. Therefore in 3D ARs all scales are active from the very large scales (thousands of Kilometers, treated by MHD) down to meters (treated by kinetic equations). This multi-scale and complex environment still maintains many interesting statistical regularities which are manifested in the numerous statistical laws recorded from the data so far.

Acknowledgement

I would like to thank my colleagues Drs. H. Isliker, M. Georgoulis and Mr T. Fragos for many useful conversations.

References

1. C.E. Alissandrakis: Astron. Astrophys., 100, 197 (1981)
2. V. Archontis, A.W. Hood, C. Brady: Astron. Astrophys., **466**, 367 (2007)
3. G. Aulanier, E. Pariat, P. Demoulin: Astron. Astrophys., **444**, 961 (2005)
4. P. Bak, C. Tang, K. Wiesenfeld: 1987, Phys. Rev. Lett., **59**, 381 (1987)
5. P. Bak: *How Nature Works*, New York: Springer Verlag (1996)
6. A.C. Balke,C.J. Schrijver, C. Zwaan, T.D. Tarbell: Solar Phys., **143**, 215, (1993)
7. P.A. Cassak, M.A. Shay, J.F. Drake: Phys. Rev. Lett., **95**, 235002 (2005)
8. P. Démoulin: Adv. Space Res., **39**, 1367 (2007)
9. J.F. Drake, M. Swisdak, K.M. Schoeffler, B.N. Rogers, S. Kobayashi: J. Geophys. Res., **33**, L13105 (2006)
10. K. L. Harvey and C. Zwaan: Solar Phys., **148**, 85 (1993)
11. T. Fragos, M. Rantziou, L. Vlahos: Astron. Astrophys., (2003)
12. K. Galsgaard: in SOLMAG: Magnetic Coupling of the Solar Atmosphere, ESA SP-505 (2002)
13. K. Galsgaard, V. Archontis, F. Moreno-Insertis, A.W. Hood: Astrophys. J., **666**, 516 (2007)
14. M. K. Georgoulis, B.J. LaBonte, T.R. Metcalf: Astrophys. J., **602**, 447 (2004)
15. M.K. Georgoulis and B.J. LaBonte: Astrophys. J., **615**, 1029 (2004)
16. M.K. Georgoulis: Solar Phys., **228**, 5 (2005)

17. K.D. Leka and A. Skumanich: Solar Phys., **188**, 3 (1999)
18. H. Isliker, A. Anastasiadis, D. Vassiliadis, L. Vlahos: Astron. Astrophys., **363**, 1134 (1998)
19. H. Isliker, A. Anastasiadis, L. Vlahos: Astron. Astrophys., **363**, 1134 (2000)
20. H. Isliker, A. Anastasiadis, L. Vlahos: Astron. Astrophys., **377**, 1068 (2001)
21. H. Isliker and L. Vlahos: Phys. Rev. E, **67**, 026413 (2003)
22. T.N. Larosa and R.L. Moore: Astrophys. J., **418**, 912 (1993) K.D. Leka and A. Skumanich: Solar Phys., **188**, 3 (1999)
23. D.W. Longcope: Living Rev. Solar Phys. **2**, 7 (2005)
24. E.T. Lu and R.J. Hamilton: Astrophys. J. Lett., **380**, L89. (1991)
25. E.T. Lu, R.J. Hamilton, J.M. McTiernan, K.R. Bromund: Astrophys. J., **412**, 841 (1993)
26. N. Meunier: Astrophys. J., **515**, 801 (1999)
27. L.J. Milano, P. Dmitruk, C.H. Mandrini, D.O. Gómez, P. Demoulin: Astrophys. J., **521**, 889 (1999)
28. G. Nicolis and I. Prigogin: *Exploring Complexity: An introduction*, New York: Freeman Co. (1989)
29. M. Onofri, L. Primavera, F. Malara, P. Veltri, Phys. Plasmas, **11**, 4837
30. M. Onofri, H. Isliker, L. Vlahos: Phys. Rev. Lett., **96**, 151102 (2006)
31. K. Papadopoulos: Rev. Geophys. Space Phys., **15**, 113 (1977)
32. E.N. Parker: Astrophys. J. **174**, 642 (1972)
33. E.N. Parker: Astrophys. J., **264**, 642 (1983)
34. E.N. Parker: Astrophys. J. **330**, 474 (1988)
35. H.E. Petschek: in *The physics of solar flares* (ed.: Hess, W. N.), Proc. AAS-NASA Symposium, SP-50 p. 425 (1964)
36. E.R. Priest, G. Hornig, D.I. Pontin: J. Geophys. Res., **108**, 1285 (2003)
37. P.E. Seiden and D.G. Wentzel: Astrophys. J., **460**, 522 (1996)
38. K. Shibata and S. Tanima: Earth Planets Space, **53**, 473 (2001)
39. L. Vlahos, M. Georgoulis, R. Kluiving, P. Paschos: Astron. Astrophys., **299**, 897 (1995)
40. L. Vlahos, T. Fragos, H. Isliker, M. Gergoulis: Astrophys. J. Lett.,, **575**, L87. (2002)
41. L. Vlahos, M. Georgoulis: Astrophys. J. Lett., **603**, L61 (2004)
42. L. Vlahos, H. Isliker, F. Lepreti: Astrophys. J., **608**, 540 (2005)
43. Ł. Vlahos, S. Krucker, P. Cargill: in "Turbulence in Space Plasmas" eds L. Vlahos and P. Cargill, (Lecture Notes in Physics, Springer Verlag), in press (2008)
44. D.G. Wentzel and P.E. Seiden: Astrophys. J., **390**, 280 (1992)
45. L. M. Zelenyi, A.V. Milanov, G. Zimabrdo: in *Multiscale magnetic structures in the daistant tail*, ed. A. Nishida, D.N. Baker and S.N. Cowley (American Geophysical Union, Washington DC, p. 37 (1998)

Chaos and Self-Organization in Solar Flares: A Critical Analysis of the Present Approach

L. Pustil'nik

Israel Space Weather and Cosmic Ray Center, Tel Aviv University, Israel
levpust@post.tau.ac.il

Summary. We review the present approach to the origin of solar flares with a critical analysis of different aspects of this problem:

1. In the pre-flare equilibrium state, the observations show numerous ultra fine structures of force free magnetic fields playing the dominant role in active regions. These structures, compose a stochastic ensemble of magnetic ropes at dynamical equilibrium with a fractal dimension ("current magnetic jungle")
2. The loss of pre flare equilibrium is possibly triggered by a catastrophic transition from a previous dispersed state of the currents threads to singular states where the dominant part of the coronal current is concentrated into very thin sheets and strings. There, anomalous magnetic dissipation and reconnection takes place.
3. During the flare energy release, there is the problem of "survival" of the flare's current sheet caused by its strong (threshold-like) dependence to the current density and to the overheating and "splitting" of the current sheet region.

We show that the percolation of the current in the random resistor's network, produces a feed-back of the resistors to the current. This approach is able to explain naturally the main observed features of solar flares. Properties as the "threshold-like" start of a flare energy release, the power law frequency-amplitude spectra of the flare burst, the power law energy spectra of accelerated particles, precursors of the main flare event, are manifestations of a self-organization and phase transition in a percolated system.

1 Introduction

1.1 Historical Introduction

The first mention of the role of primary "chaos" in the Universe formation is the first sentence in the Bible: "In the beginning God created the heaven and the earth (Gen 1:1), And the earth was without form, and void (Gen 1:2)." The definition "without form, and void," translated from the Hebrew

G. Contopoulos, P.A. Patsis, *Chaos in Astronomy*. Astrophysics and Space Science Proceedings, doi: 10.1007/978-3-540-75826-6,

form "tou va vou" is the simplification of a much more complex notion. In the more extensive variant it may be translated from Hebrew as: and the earth ("Universe") was the "absolutely disordered nothing." This definition of the "great chaos of the nothing" may be considered as primordial "grand-chaos" or "chaos-0." All states of chaotic phenomena, as well as ordered and self organized processes, dealing with the energy and entropy flux in open systems (like clusters of galaxies, galaxies, planetary systems, near planetary rings, regular structures of convective zones inside stars, and the quasi-regular cycle of the magnetic activity in our Sun and red dwarfs), are a weak "echo" of the primordial chaos.

All the complex phenomena encountered on the Sun can be described as manifestations of chaos and order generated at three levels:

1. **CHAOS I:** Formation of the **convective zone with developed turbulence** and **dynamo process** (solar cycle, differential rotation, global circulation, granulation, super-granulation,. . .). The reason of this manifestation is the disruption of the primary energy flux from the solar core (generated by thermonuclear reactions) into random convective motions by thermal convective instability.
2. **CHAOS II:** The rise of the random magnetic threads into the solar atmosphere from the magnetic ropes generated in the convective zone forms the dynamic current-magnetic jungle above the active regions. These features are the cause of the coronal heating, the formation of the high structured solar atmosphere and solar wind, the mega and nano-flares with a power law frequency-amplitude spectrum. The reason of this manifestation is the disruption of the primary energy flux of magnetic fields from the convective zone to corona by complex MHD instabilities (described in the very stimulating lecture of Professor Vlahos during the present conference).
3. **CHAOS III:** Solar flares could be described as a current percolation phenomenon through a network of chaotic resistors, with a threshold dependence of the local resistance as a function of the current intensity. There is a dynamic equilibrium in the resistors network controlled by a feed-back scheme of resistance-current-resistance. The result of this chaos-III process is a flare energy release with effective particles acceleration up to very high energy and with the formation of a power law energetic spectrum. The reason of this manifestation, is the disruption of the energy flux of a regular current through turbulent current sheets resulting from plasma instabilities in this extreme current-magnetic configuration.

The general feature for all levels of chaotic processes in solar activity is the disruption of the initial flux of energy and entropy by instabilities in threshold state of the medium and next randomization and self-organization of the disrupted flux. We will remind shortly after about the main properties of these complex phenomena in states Chaos I and Chaos II and will discuss in details processes and manifestations of the Chaos II in solar flares.

1.2 Chaos I: Disruption of the Primary Flux of Energy into Numerous Convection Cells: primary Chaos in Solar Convective Zone

As we said before, the primary source of the Chaos I is the flux of energy from solar center generated by energy release in thermonuclear reactions in the solar core. Another factor is the low opacity of the solar plasma which prevents the heat transfer by thermal conduction and initiates conditions for thermo-convection instability. As a result, an extensive turbulent zone is formed in the upper part of the solar radius. A threshold condition in order to start a convective process is the generation of an over-adiabatic temperature gradient. The resulting manifestations caused by the Chaos I processes, observed on the Sun (see Fig. 1), are granulation, super-granulation, meso-granulation, giant cells, meridian circulation and differential rotation.

The movies of the images above are in: www.suntrek.org/images/movies/pb_gran_movie1.mpg and in sohowww.nascom.nasa.gov/hotshots/2003_01_02/supergranulation.mpg correspondingly. The next scales (mezo- and giant cell turbulence) are seen only in brightness or in magnetic field and are responsible, first of all, for global circulation processes.

We would like to note here that the spectrum of the velocity oscillation generated by convective chaos and observed as turbulence on the solar surface (see Fig. 2) is very far from the simple power law $v_l \propto l^n$ with slope n=1/3, expected for the case of full developed Kolmogorov turbulence in the isotropic and uniform media (this situation is like the same one in the turbulent accretion discs, described in the previous report of Prof. Fridman and colleagues in this conference). Convective chaos and turbulence are a natural consequence

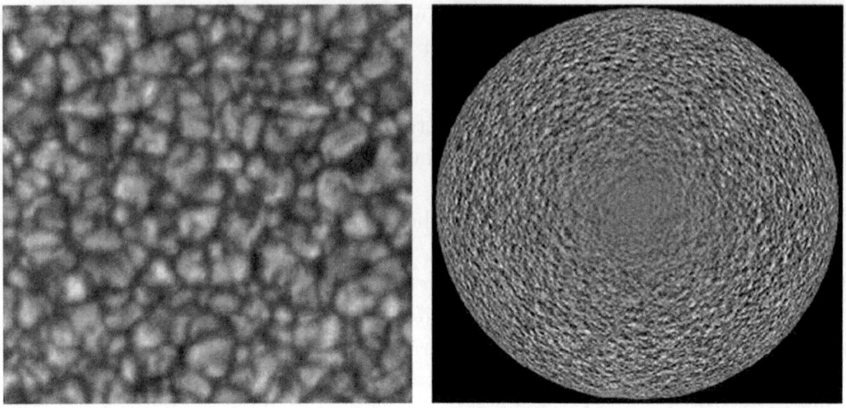

Fig. 1. Examples of observational manifestations of convective motions in the small scales (*left panel*: granulation seen in intensity, recorded from the Swedish Vacuum Solar Telescope) and in intermediate scale (supergranulation seen in Doppler shifts, recorded with SOHO/MDI)

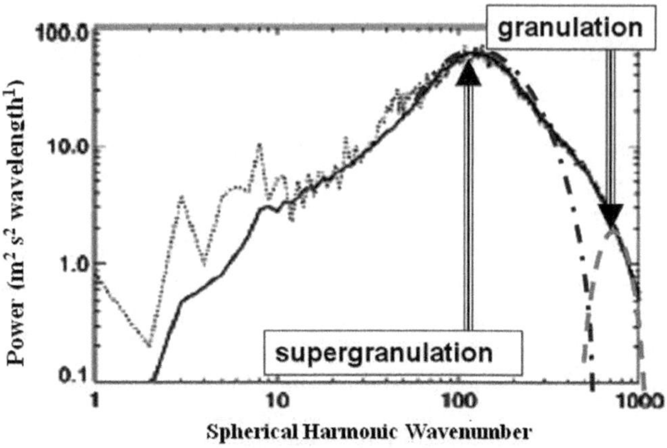

Fig. 2. Comparison between the observed photospheric convection spectrum (*solid line*) and that for the simulation (*dotted line*) created from granule and supergranule components [1]. The peak at wave numbers near 110 is due to supergranulation and the peak at 500–1,000 is due to granulation

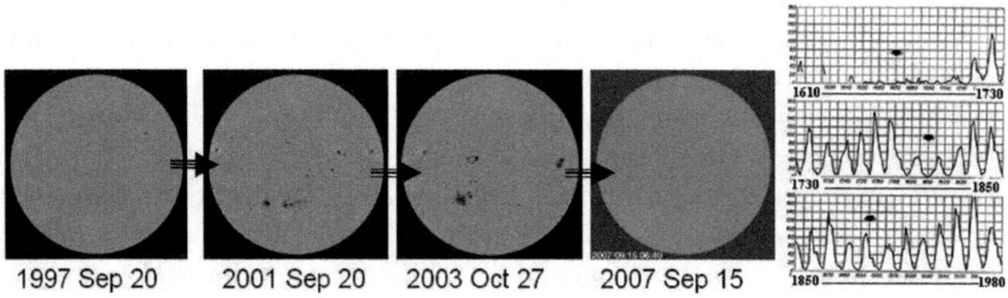

Fig. 3. *Left* - Example of the variation of the sunspots number and location during the last solar cycle, *Right* - sunspots number variation during the last 400 years with Maunder minimum and Dalton Minimum

of the very inhomogeneous and anisotropic state of the solar convective zone, of its strong radial gradients and of the existence of the fast differential solar rotation.

Another typical manifestation of the Chaos I phenomena is the generation of the magnetic field by the dynamo process resulting from the combination of gyrotropic turbulence in the convective solar zone (with strong radial gradient and Coriolis effect) and the strong differential character of the solar rotation which is a function of the solar radius and the solar latitude. The generation of the magnetic field has a cyclic character both in the global field direction, latitude localization of the field and its amplitude. The most known manifestation are: quasi-periodic 11-year cycle of the solar activity (Fig. 3.), Maunder butterfly of the sunspots location, periodic reverse of global magnetic polarity with 22-year mean period [2].

The specific non-linear connection between the solar convection and the dynamo process, the generation of magnetic fields and the suppression of turbulent motions, as well as the change of state of the period and the amplitude between two predominant equilibriums, may lead to the appearance of a "strange attractor." This effect is detected as regular long time periods of disappearances of the observed sunspot activity in Fig. 3. like the Maunder minimum (1645–1715), the Spoerer Minimum (1416–1534) and the Dalton Minimum (1790–1820).

1.3 Chaos II: The Disruption of the Primary Regular Magnetic Ropes into Numerous Interacting Current-Magnetic Threads during their Emergence into the Solar Atmosphere

The interest to this group of phenomena was initiated by the unexpected discovery of permanent very fine structures of the current- magnetic configuration, observed both on the photospheric level and in the coronal plasma (see Fig. 4.)

Observations of the photospheric magnetic fields show, that the observed magnetic flux is concentrated in very thin threads, forming magnetic ropes with almost no magnetic flux between them (fractal dimension). These very compact regions of magnetic flux emerge from clusters with fractal dimension on a wide range of scales.

Observations on the coronal level demonstrate that the coronal plasma-magnetic structure consist of numerous (tens-hundreds) thin arc-like threads with constant cross sections, where the magnetic tubes must have a constant 136 magnetic field, (about 100–300 G). The magnetic configuration of the individual threads may be described as almost force-free $rot\overrightarrow{H} = \alpha\overrightarrow{H}$, but the complexity of the numerous threads can not be described by a force-free

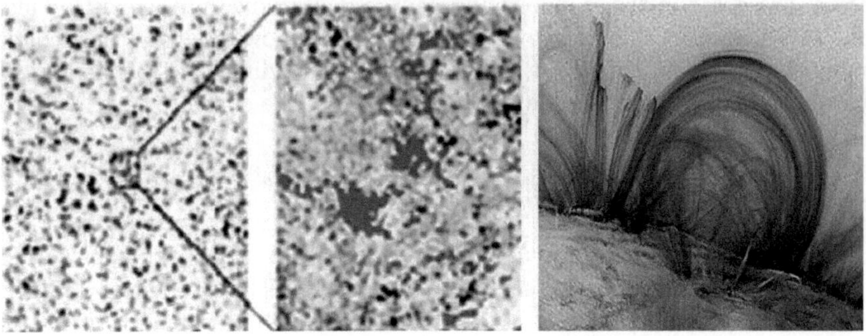

Fig. 4. *Left* - fine structure of the photospheric magnetic field with evident conservation of the fractal dimension of the magnetic flux points [3]; *Right* - fine structure of the coronal plasma-magnetic loops observed by TRACE. These loops with the constant cross section, consist of magnetic ropes which are linked to the photosphere at their footpoints (apod.nasa.gov/apod/ap000928.html)

equilibrium configuration as a whole. These threads are in permanent strong interaction one with another and this state may be described as dynamical equilibrium of the system of strong interacting elements.

The above interpretation of the observations is in contradiction with the standard description of the coronal features as quasi static and quasi potential magnetic structures including small disturbances caused by twists and distortions of the magnetic field. As we may deduce from observations, the state of the corona is a large scale dynamic equilibrium between the plasma, the currents and the magnetic fields which are in a random interaction. This is not a static and force free equilibrium. Anomalous dissipation in numerous current concentrations on the boundaries between individual current-magnetic elements is a source of nano-flares, coronal heating and background particles acceleration. This group of questions of the Chaos II processes is studied successfully by Prof. L. Vlahos and his team and reflected very vividly in his stimulating lecture.

2 Solar Flares as Manifestation of Chaos III: "Disruption of the Turbulent Current Sheet of the Flare into Random Resistors Network"'

2.1 Current Description of the Flare Energy Release

The present approach to solar flares concentrate its attention on the question of generation of current sheets on the boundary between different magnetic fluxes, interacting during their evolution. In the frame of this approach, a current sheet is the final state of the preflare evolution, it is the necessary element of the global magnetic equilibrium and it is the engine of the flare energy release. The question of the equilibrium and stability of the current sheet, presented in the standard approach, is not studied in the existing models. It is suggested that the current sheet will somehow dissipate the incoming magnetic energy, without a detailed analysis of the difficulties of this process. Nevertheless, the very high level of instability of the current sheet made this scheme an object of criticism.

I would like to formulate here several positions about the nature of the solar flares, essential for us in future considerations:

1. A solar flare is the result of loss of equilibrium of the pre-flare current-magnetic configuration above an active region. This pre-flare equilibrium has a dynamical character (it includes numerous current threads with strong interactions between neighborhood elements. In some cases it may be described as a gas of strong interacting current threads).
2. The loss of equilibrium and the transition to the "flare state" is caused by spontaneous or induced current accumulations with scales from about 100,000 km down to 1–10 km (a factor of 10,000) forming a "thin turbulent

current sheet." The mechanism of this self-concentration is unknown at present time.

3. Possible external agents (candidates for the trigger mechanisms), causing the current self concentration into a thin "solar flare" current sheet are:
   (a) New Emerging Magnetic Fluxes (EMF), penetrating into the photosphere and atmosphere from the convective zone (and observed for all strong flares 1–3 days before the flare oneset)
   (b) Flute instability of the main plasma magnetic elements, existing in the stationary solar atmosphere before flares and activated (possibly by new EMF) immediately before a flare:
      – Quiescent Prominences, activated as "disappeared filaments" during 0.5–1 h before >85% of the strong Hα-flares (Hα class>2)
      – Stationary coronal loops, observed in X-ray and radio emission in the pre-flare corona, present strong oscillations, with periods of about 10 mins, immediately before strong proton flares.

The fundamental element of solar flares, the main engine of the flare energy release, is a "thin turbulent current sheet." The use of this exotic notion is not an extreme perversion of the theorist's mind, but a forced reaction on the necessity to explain the observed conversion of the bulk of the magnetic energy, in the volume above active regions, into anomalous heating and particles' acceleration during extremely short times (seconds or tens of seconds).

This necessity may be illustrated on the base of the Joule law of the power P of magnetic energy dissipation by current density **j** in the medium with electrical conductivity $\mathbf{P} \propto \frac{j^2}{\sigma}$

For standard conditions of the solar atmosphere with a magnetic structure size of about $(10-100)*10^3$km and a Coulomb conductivity in the corona with temperature of about 10^6 K, the estimated power of the magnetic dissipation is too low, so that the expected duration of the flare event would be more than million years. Evidently, for solving of this contradiction we must suggest:

1. or very strong current densities **j** in the current sheet with correspondingly very thin thickness of the sheet $a_* = (c/4\pi)(\Delta\mathbf{H}/\mathbf{j})$
2. or a very low conductivity $\sigma{=}\sigma_* \ll \sigma_{Coulomb}$

Fortunately, these two forced suggestions are equivalent, because when the current density is larger than a critical value $j_{cr} = neu_{cr}$ the plasma in the current sheet will automatically switch into a turbulent state with very low conductivity σ_*

The next question about a possible estimation of the turbulent conductivity σ_* needs a quantitative description of the plasma turbulence in the current sheet with estimation of the spectrum and amplitude of plasma turbulent pulsation of electric fields and effective frequency of scattering of current's electrons in these fields ν_{eff}. The resulting anomalous conductivity [4] will be $\sigma_* = \frac{\omega_{0e}^2}{4\pi\nu_{eff}}$ where $\omega_{0e}^2 = \frac{4\pi n_e e^2}{m_e}$ is the electronic Langmuir frequency. Most popular models with ion-sound wave turbulence estimated effective frequency of scattering of current's electrons $\nu_{eff(i-s)} = A \cdot \omega_{0i}$, where A is a constant, less or about 0.1-0.01 and proportional to the plasma turbulence level, plasma wave anisotropy and another inner parameters of plasma turbulence in current sheet. For other case of plasma waves, dominating in the current sheets (ion cyclotron waves, for example, considered in part of models) we will have the same estimation after replacing the ion Langmuir frequency with the ion Larmor frequency.

The estimation of the expected power of the energy release in this "turbulent currents sheet in the solar corona with density $n = 10^{8-10} cm^{-3}$, magnetic fields about 100–300 G and current sheet with sizes $10^5 * 10^{10} * 10^{10}$ cm is about $P_* = 10^{28-29}$ erg s^{-1} and integrated energy of flares is about $W = 10^{32}$erg in very good agreement with observed properties of the strong flares.

2.2 The Problem of the Instabilities of the Turbulent Current Sheet

The main problem in the described picture is the evident instability of the turbulent current sheet. It may be illustrated by an analogy. The concentration of the current, dispersed in the coronal volume with size 100,000 km in a thin current sheet with size 1–10 km is similar to the concentration of a weak breeze in the conference hall with about 10 m size into a narrow sheet with thickness about 1–10 mm and corresponding increasing of wind velocity in this layer from the initial 1 m s^{-1} up to the super-sound wind of a hurricane. Evidently, even if ones succeed with the primary concentration, in the next moment this configuration will be disrupted by fast instabilities of this extreme configuration.

The same situation must take place in a turbulent current sheet. We list here the main instabilities and processes that must lead to local disruption of the turbulent current sheet:

1. **Drift-dissipative instabilities** [5] including instabilities like the tearing-mode, swiping mode ($\nabla\sigma \neq 0$), … . The result of these instabilities is the fast disruption of the primary thin current sheet into numerous linear currents – magnetic islands (in 3-d -it will lead to the formation of the numerous current threads and interlacing between them). The characteristic time of these instabilities is:

$$\tau_{tearing} \approx \tau_A^s \tau_d^{1-s}, \quad (\tau_A = d/V_A;\ \tau_d = d^2/(c^2/4\pi\sigma))$$

   where $s \approx 1/2 - 2/3$ depending on the mode of the instability.
2. **Fast "pinch" like instabilities of the linear currents** in the disrupted current sheet – "sausage", "kink", "twist" instabilities with Alfven time of disruption of current threads:

$$\tau_{pinch} \approx \tau_A \approx d/V_A$$

3. **Accumulation of linear currents in numerous bundles** with $u = j/ne > u_{cr}$ will lead to the next generation of plasma turbulence in selected regions-"domains" with fast anomalous current dissipation in the points of a current rupture (like the explosion in the electric lamp, caused by glower disruption)
4. **From the other side, fast anomalous overheating of turbulent domains up to** $c_{Si} > u_{cr} = j/ne$ will suppress plasma waves by very fast Landau damping of the excited waves in the heated plasma. This "overheating" paradox is the fundamental problem for supporting a turbulent state in a plasma with currents. Possible solutions are: abnormal fast cooling of the heated electrons or consideration of the alternative variants of the turbulent heating with formation of the non-Maxwell tail of the not numerous high energy electrons with some "cold" Maxwell-like distribution of the bulk of background electrons.
5. **"Splitting" of boundary of turbulent current sheet** caused by the threshold like dependence of conductivity $\sigma(\mathbf{j})$ on the current value **j.** On the boundary ($j = j_{cr}$) the continuity condition for the transversal electric field component ($rot(\vec{E}) = 0$) leads to the discontinuity of the current with amplitude $[\mathrm{J}] \cong J_{cr} = neu_{cr}$. Fast diffusion of the field and current through this boundary leads to the splitting of the current on the boundary which will then produce a new discontinuity and then a new splitting. This process causes the current sheet to be in a non-stationary state helping the formation of numerous splitted currents.

The final result of all these processes will be the fast disruption of the thin turbulent current sheet into a random network of numerous domains characterized by "normal" or "turbulent" "resistor-elements" (Fig. 5). This

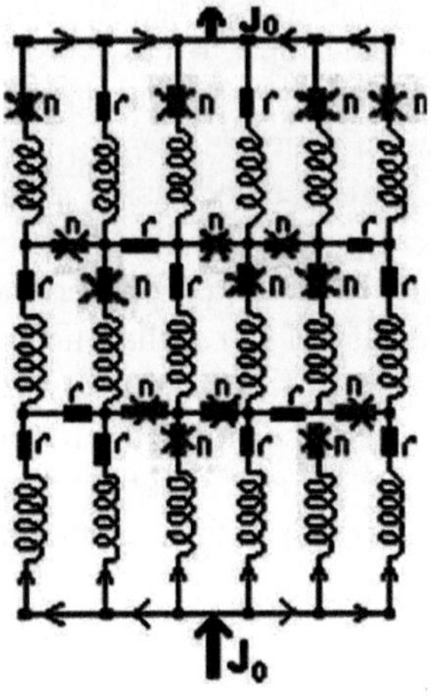

Fig. 5. Scheme of percolated resistor's network with "good"-"r" and "turbulent"-"n" resistors and global current J_0

effect will force current percolation [6] through this resistor's network under the condition of global current conservation (which is the natural consequence of the high inductance of the corona around the current sheet). The strong dependence of the resistors' conductivity as a function of the current densities flowing through them on one hand and the permanent redistribution of the local currents j_{ik} for cases when some domains switch from normal to turbulent state when $j_{ik} > j_{cr}$, will change the equilibrium of this system. The resulting change in the system's equilibrium will induce a strong feed-back on the dynamical equilibrium of the system and a permanent change of the local state of the individual elements and the inner currents redistributions.

Therefore we must consider the flare process as a dynamic equilibrium in percolated random network of resistors with saltatory resistance R_{ik}, depending on the local current level J_{ik} as:

$$\begin{cases} R_{ik} = R_0 \text{ for } \mathrm{j_{ik}} \leq \mathrm{j_{cr}}; \\ R_{\mathrm{ik}} = R_t \approx 10^6 \mathrm{R_0} \text{ for } \mathrm{j_{ik}} > \mathrm{j}_{\mathrm{cr}} \end{cases}$$

with Kirchhoff's laws in resistors network

$$\begin{cases} j_i = \sum_j j_{ij} = I_0 \neq f(\mathrm{sec}\,tion_i); \\ \Delta \mathrm{V_j} = \sum_i j_{ij} R_{ij} \neq f(\mathrm{circuit}_j) \end{cases}$$

2.3 Flare as Current Percolation Through Random Resistor's Network

The percolation through the random network of conducting elements (or clusters of elements) is a well known field of investigation and has numerous applications, describes very different phenomena, including this process of propagation [7, 8].

1. Self-conductors can be described using percolation through a mixture of conductors and insulators: critical parameters – density of conductive elements (or bonds between them)
2. Superconductors can be described using percolation through a metal-ceramic mixture (critical parameter like to the previous one)
3. Forest fire can be described using percolation of the fire from tree to tree (critical parameters: density of trees (distance between them) and life time of the fired tree)
4. Epidemic and pandemic percolation (like to the previous one)
5. Gossip percolation in pre-radio-TV-Internet epoch (like to the previous cases, but with using as bonds information channels from person to person/(s))
6. And lots of other situations (polymer's chain, neuron net, ...).

The main instrument for studying similar system is the numerical simulation, taking into account both the character of inter-elements connection, the dimension of the connection space and the fractal dimension of the clusters formed by conducted elements. This process was studied in experiments (in superconductive ceramic samples [9] and in conductive graphite paper with random holes [10] , in numerical simulations [11] and in theoretical models [12].

All these systems have some universal properties:

(a) A phase transition type of behavior on the threshold of percolation with drastic change of the global properties of the system as a whole (transition "conductor"–"insulator", "normal conductor"–"superconductor", pandemic, forest fire, hurricane , ...);
(b) On the threshold of phase transition, the excitation of large amplitude disturbances take place. These high amplitude abnormal disturbances are precursors, preceding global phase transition;
(c) The self-organization of clusters of elements in the network with ensemble properties of fractal type and power-law statistics (probabilities, length ways, life time, amplitude,...) of the global output.

The existence of the feed-back of elements on the current (depending on level of percolated flux) adds to the behavior of the network additional very interesting non-linear effects, especially near the boundary of transition.

For the concrete case of the current percolation through a network of random resistors, a very essential property for us is the threshold dependence of the global net conductivity Σ_{net} on the relative density of "bad" elements p_c (caused for our case by the global current J_0):

$$\Sigma_{net} \propto (J - J_{cr})^{-\alpha}$$

which must lead to a flare-like increasing of the global resistance of the resistors network (Fig. 6) ($\Sigma_{net} \Longrightarrow \infty$) on the boundary of the phase transition $J = J_{cr}$.

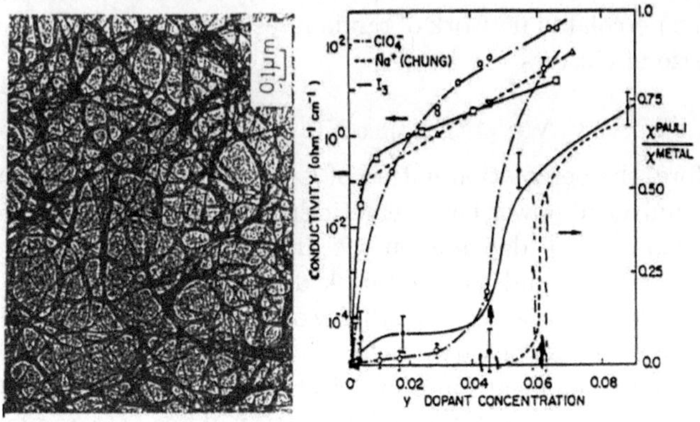

Fig. 6. Experimental net conductivity dependence on the density of conducted dopants in percolated polymer's network. These graphs illustrate "phase transition" effect of the catastrophic decreasing of the global conductivity in percolated network on the threshold of percolation

Another fundamental property of the percolated random network is the power law dependence $N(x) \propto x^{-n}$of the main statistical properties of the manifestations in this network from the statistic of clusters of network elements (size, volume, length) and up to statistics of the energy dissipation in this clusters of resistors. This fundamental property allows to explain the universal "power law" character of the amplitude-frequency spectrum $N(W) \propto W^{-m}$for solar flares and similar phenomena on the Sun and in the flare red dwarfs of UV Cety type.

Another application of the percolation approach to solar flares is the possibility of the natural explanation of the observed universal "power law" like the energy spectrum of the particle, accelerated during the flare energy release (solar cosmic rays). For this aim we must take into account, that the turbulent domain with $\mathbf{j} > j_{cr} = nec_{Si}$ and with abnormal conductivity

$$\sigma_* = \omega_{0e}^2/(4\pi\nu_{eff}) = 10^{2 \div 3}\Omega_{0i}(\nu_{eff}/0.1 \cdot \Omega_{0i})$$

is an "electrostatic turbulent double layer" with a giant electric field:

$$E = j_{cr}/\sigma_* = (ne \cdot c_{si})/(10^{2 \div 3}\Omega_{0i})$$

which lead to the acceleration of charged particle in this place. The effective energy, that a particle obtains in the current sheet during its propagation (random walk) through clusters of these turbulent domains-"electrostatic double layers" will be proportional to the number of turbulent elements in the way along the regular electric field (z-direction) or the length of the cluster of double electric layers, which the particle crosses $\varepsilon \approx eE_* l_z$ with maximal value

$$\varepsilon_{\max} \approx eE_* L = (1 \div 10) Gev \cdot (n_8^{1/2} T_7^{1/2} L_9)$$

From another point of view the probability or number of clusters with size $l > l_z$ in the percolated network of random resistors has power law dependence from the size of clusters

$$N(\varepsilon = eE_* l_z) \propto N(l_z) \propto l_z^{-k} \propto \varepsilon^{-k}$$

Therefore, the percolation nature of flare's current propagation leads naturally to a universal power law energetic spectrum of the accelerated particles where the exponent k depends on the dimension of the process in the turbulent current sheet and the feed-back dependence in chain "resistance$\Longrightarrow$ current$\Longrightarrow$resistance$\Longrightarrow$...." In other words, the observed power law spectrums of solar cosmic rays (particles accelerated in the percolated current sheet, included random mixture of "electrostatic double layers") is a direct sequence of power law statistics of clusters in percolated random resistor's network near percolated threshold.

Hereby the "percolation" approach to current percolation through random (stochastic) network of "normal" and "turbulent" domains-resistors in the flare's current sheet may explain:

1. The fundamental "flare" nature of solar flares (sudden start of flare energy release) as "phase transition" in percolated systems
2. The universal power law like statistical properties of the flare events, observed both in solar flares and in their inner fine time structure. If on the large range of energetic scales this power law like frequency spectrum is a manifestation of the percolation of magnetic structures from underphotosphere layers (lecture of Prof. Vlahos in the conference), lower amplitude variations on the inner flare events is a manifestation of current percolation through the fragmented turbulent current sheet.
3. Universal power law character of the energy spectrum of high energy particles (solar cosmic rays), accelerated in the flare's turbulent current sheet during propagation through turbulent elements-"electrostatic turbulent double layers"

As a conclusion to this lecture I would like to formulate several final remarks:

1. The main observational properties of solar flares may be understood naturally as a manifestation of self-organized chaos in a percolated current sheet, consisting of a resistors network with strong threshold like dependence of the resistance of local elements on the local current level and strong feed back of the current distribution on change of local resistance in network elements.
2. If you observe something universal and regular – let search for chaos as a reason.
3. Solar flares are no exception.

Acknowledgement

I would like to thank Organizing Committee of the conference and Prof. P. Patsis for kindly invitation to participate in the conference and for creation of the friendly and creative atmosphere during conference itself. I would like to thank Prof. A.Fridman, stimulated my participation in the conference. I am very thankful to referee for his great work on improvement of my article.

References

1. P. E. Williams, D. Hathaway, M. Cuntz, The Astrophysical Journal, **662**, L135 (2007)
2. Ya. B. Zeldovich, A. Ruzmaikin, D. Sokoloff:*Magnetic Fields in Astrophysics*, Gordon and Breach, NewYork (1983)
3. J. O. Stenflo, Nature, **430**, 304, (2004).

4. R. Z.Sagdeev, A. A. Galeev, *Nonlinear Plasma Theory.* Benjamin, New York, (1969).
5. F. F. Cap, *Handbook on plasma instabilities*, Academic, New York, (1978)
6. L. Pustil'nik, Astrophysics and Space Science, **264**(1–4), (1998)
7. D. Stauffer, A. Aharony, *Introduction to Percolation Theory*, Taylor& Francis Group, London, (1999)
8. J. Feder, *Fractals*, Plenum Press, New York, (1988)
9. N. F. Vedernikov, K. M. Mukimov, G. P. Sigal, B. Yu. Sokolov (Superconductivity) Sverhprovodimost: fizika,ximiya,technika(russian), **7**, 316, (1994)
10. M. E. Levinshtein, M. S. Shur, A. L. Efros, Soviet Physics-JETP, **42**, 1120, (1976)
11. S. Kirkpatrik, Reviews of Modern Physics, **45**, 574, (1973).
12. S. Render, Annals Israel Physics Society, **5**, 447 , (1983)

Charged Particles' Acceleration through Reconnecting Current Sheets in Solar Flares

C. Gontikakis[1], C. Efthymiopoulos[1], and A. Anastasiadis[2]

[1] Research Center for Astronomy, Academy of Athens, Soranou Efessiou 4, GR-11527, Athens, Greece
cgontik@academyofathens.gr, cefthim@academyofathens.gr

[2] National Observatory of Athens, Institute for Space Applications and Remote Sensing, GR-15236, Penteli, Greece
anastasi@space.noa.gr

Summary. This study focuses on charged particles' acceleration in Reconnecting Current Sheets (RCS) taking place during solar flares. A Harris type topology was used to describe the electric and magnetic field of reconnecting current sheets. The use of a Hamiltonian formalism provided analytical laws describing the motion of the particles. Numerical simulations of a thermal distribution of particles interacting with single or multiple current sheets are presented. The final kinetic energy distributions after an interaction with a single current sheet present short energy ranges. When thermal particles interact with multiple current sheets a broad kinetic energy distribution is created. However, it is shown that there is an upper bound in the amount of kinetic energy that the particles can gain.

1 Introduction

The magnetic reconnection taking place in current sheets releases the energy stored in the magnetic field, in various forms, but primarily as the energy of charged particles accelerated, e.g., by DC electric fields in the reconnecting region. Several facts indicate that magnetic reconnection is responsible for the energy release during solar flares [6]. The study of particles' orbits inside a current sheet can provide important information on the physics of the acceleration process. Simple models of magnetic and electric fields inside current sheets gave the first estimation of the kinetic energy gain for the Earth magnetotail [5] and for solar flares [4]. In a series of papers ([1],[3],[2]) we studied the physics of particles acceleration using a Harris type reconnection topology. Several results were derived using a Hamiltonian formalism.

G. Contopoulos, P.A. Patsis, *Chaos in Astronomy.* Astrophysics and Space Science Proceedings, doi: 10.1007/978-3-540-75826-6,

2 Analytical Formulation of the Problem

The Harris type topology represents the inner part of a current sheet where dissipation of the magnetic energy takes place. The electric and magnetic field components are

$$E = (0, 0, E), \qquad B = (-y/a, \xi_\perp, \xi_\parallel)B_0 \qquad \text{for } |y| \leq a. \tag{1}$$

In (1), $\xi_\perp = B_\perp/B_0$ and $\xi_\parallel = B_\parallel/B_0$. We set B_0=100 Gauss as a reference value of the main magnetic field component. A particle is considered as being inside the current sheet when $|y| \leq a,\ a = 1$. In that case the particle is subject to acceleration. The equations of motion are

$$\ddot{x} = \xi_\parallel \dot{y} - \xi_\perp \dot{z}, \qquad \ddot{y} = -\xi_\parallel \dot{x} - y\dot{z}, \qquad \ddot{z} = \epsilon + \xi_\perp \dot{x} + y\dot{y}, \tag{2}$$

where, ϵ is a scaled value of the electric field. The equations of motions can be cast in a Hamiltonian form which, using the translation symmetry with respect to the x-axis, is reduced in a 2 d.o.f. problem, namely

$$H = \frac{1}{2}p_y^2 + \frac{1}{2}(c_4 + \frac{1}{2}y^2)^2 + \frac{1}{2}(I_2 - \xi_\perp z + \xi_\parallel y)^2 - \epsilon z \tag{3}$$

where p_y, and $c_4 = \dot{z} - y^2/2$ are the canonical momenta and $I_2 = \dot{x} - \xi_\parallel y + \xi_\perp z$ is a second integral of motion [1].

3 Numerical Results

Poincaré surfaces of section

The Poincaré sections shown in Fig. 1 yield an overview of the types of orbits encountered in the phase space of the problem under study. The orbits are

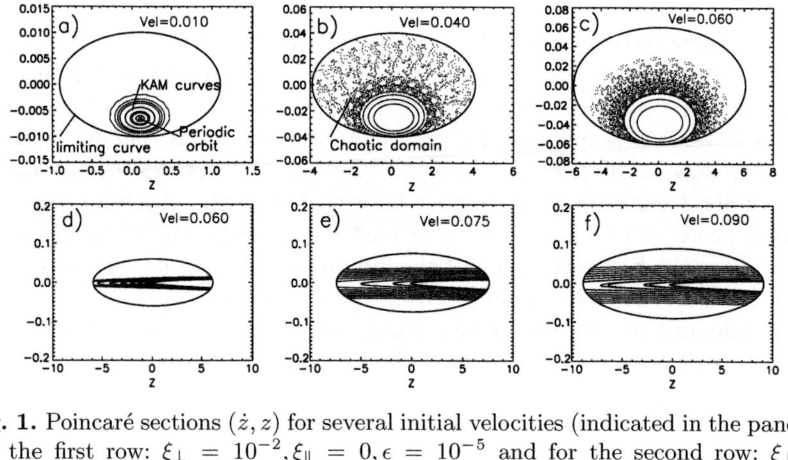

Fig. 1. Poincaré sections $(\dot{z}, z)$ for several initial velocities (indicated in the panels). For the first row: $\xi_\perp = 10^{-2}, \xi_\parallel = 0, \epsilon = 10^{-5}$ and for the second row: $\xi_\perp = 10^{-2}, \xi_\parallel = 1, \epsilon = 10^{-5}$

traced on the plane $(\dot{z}, z)$ whenever they cross the y=0 plane with a positive velocity. In all cases the particles' maximum velocity is restricted by a limiting curve derived by the condition of energy conservation. In the first row ($\xi_{\|} = 0$), the phase space shows KAM curves and a domain of chaotic motion. An adiabatic invariant exists on the KAM curves, in addition to the constant energy E and second integral I_2. In this case the regular orbits describe an oscillation of the particles as if the RCSs were acting like in a magnetic trap. When $\xi_{\|} = 1$, the KAM curves (Fig. 1) are sections of ellipses, which dominate the phase space.

Kinetic energy distribution of the accelerated particles

We now present results from numerical simulations of the interaction of 30,000 particles, initially at thermal equilibrium (10^6 K, a typical coronal temperature), with a current sheet. Particles are injected from both sides and from the centre of the RCS ($y_{in} = -1, 0, 1$). Figure 2 shows the kinetic energy distributions for electrons (panels a,b) and protons (panels d,e) after such an interaction. The accelerated particles yield distributions in a very narrow

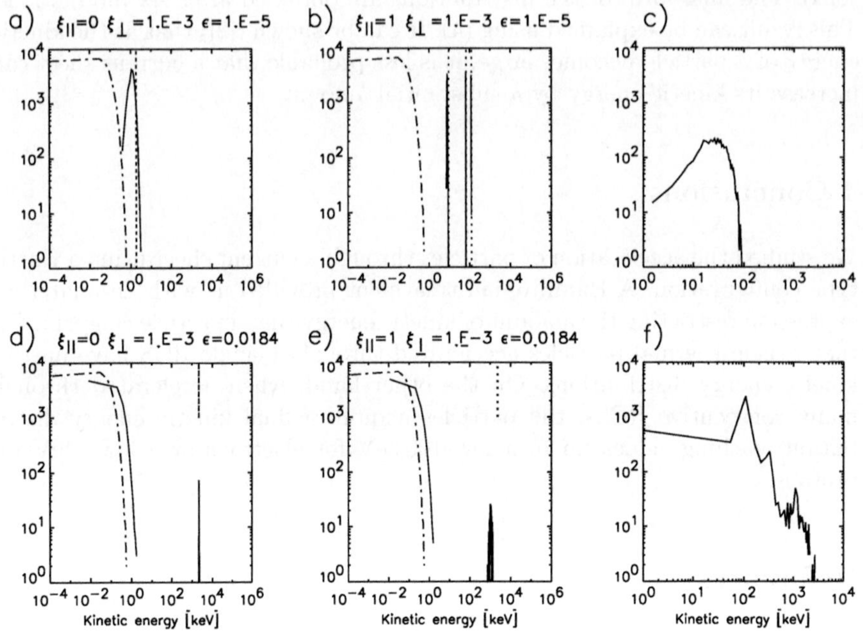

Fig. 2. Thermal distribution of particles interacting with single (panels a,b,d,e) or multiple (c,f) RCSs. The first row refers to electrons and the second row to protons. The RCS parameters for a single particle-RCS interaction are shown above each panel. The dot-dashed lines show the initial thermal distributions. The vertical lines show the analytic approximation (4). Sixty to seventy percent of protons are not accelerated by the current sheet

range of energies. This can be explained as follows: Using the Hamiltonian (see (3)) with an initial kinetic energy E_0, one can find the following relation for the acceleration length along the Z-axis

$$z_{min,max} = \frac{1}{\xi_\perp^2}\left(\xi_\perp I_2 + \xi_\parallel \xi_\perp y_{out} + \epsilon \pm \sqrt{2\xi_\perp I_2 \epsilon + 2\xi_\parallel \xi_\perp y_{out}\epsilon + \epsilon^2 + 2\xi_\perp^2 E_0}\right) \tag{4}$$

$\Delta E_{min,max} = \epsilon z_{min,max}$ are the constraints on the kinetic energy that a particle can gain. The three peaks when $\xi_\parallel \neq 0$ (Fig. 2b) correspond to the three different injection points along the y-axis (such peaks also exist in the case of Fig. 2e but they are hardly visible on the scale of the figure). The $z_{min,max}$ function depends on the injection points in (4) through $I_2(y_0)$. In our next numerical experiment, we studied the consecutive events of acceleration of particles with an initially thermal distribution interacting with 10 different current sheets. This experiment simulates the acceleration process in a solar flare where a large number of reconnecting current sheets are present. The parameters for each current sheet are selected randomly. The experiment shows (Fig. 2c,f) that the kinetic energy distribution is converging to a final form with maximum kinetic energy of 100 keV (for electrons) and 1 MeV (for protons). The final form of the distributions are obtained after six interactions. This result can be explained using (4). It can be shown ([2]) that as the kinetic energy of a particle becomes large, it is less probable that a current sheet can increase its kinetic energy by a substantial amount.

4 Conclusions

We studied the acceleration of particles through a current sheet using a Harris type configuration. A Hamiltonian treatment provided us with an analytical expression restricting the amount of kinetic energy that a particle can gain. For that reason thermal particles accelerated through a single RCS have narrow kinetic energy distributions. On the other hand, when accelerated through many consecutive RCSs, the particles acquire a final kinetic energy distribution reaching values up to a few 100 keV for electrons or a few MeV for protons.

References

1. Efthymiopoulos C., Gontikakis C., Anastasiadis A., *Astron. Astrophys.*, **443**, 663 (2005)
2. Gontikakis C., Anastasiadis A., Efthymiopoulos C., Mon. Not. R. Astr. Soc., **378**, 1019 (2007)
3. Gontikakis C., Efthymiopoulos C., Anastasiadis A., Mon. Not. R. Astr. Soc., **368**, 293 (2006)

4. Litvinenko Y.E., Astrophys. J., **462**, 997 (1996)
5. Speiser T.W., *J. Geophys. Res.*, **70**, 4219 (1965)
6. Priest E., & Forbes T., *Magnetic Reconnection*, 1st edn (Cambridge University Press, Cambridge, 2000)

The Perturbed Photometric-Magnetic Dynamical Model for the Sunspot Evolution

G. Livadiotis and X. Moussas

Department of Astrophysics Astronomy and Mechanics, National University of Athens, Panepistimiopolis, GR15784 Zografos, Athens, Greece
glivad@phys.uoa.gr, xmousas@phys.uoa.gr

We present the perturbed version of the Photometric-Magnetic Dynamical (PhMD) model for describing the sunspot evolution. This involves fluctuating the magnetic flux inflow into the sunspot, by a harmonic perturbation with frequency equal to that of the solar rotation. The nonlinear dynamical system is studied either for the conservative or for the dissipative case.

1 Introduction

The PhMD model, introduced in [1], describes the sunspot evolution within the framework of a 2-dimensional system of ordinary differential equations of the first order with respect to time, having the form

$$\dot{A} = -a_1(A-A_G)+a_2(B^2-B_G^2), \quad \dot{B} = b_{11}A+b_{12}-b_{13}\sqrt{A}+b_2(B-B_G) \quad (1)$$

where $A(t)$, $B(t)$, are the photospheric area and the average magnetic strength of the sunspots, respectively, while $A_G \leq A(t)$, $B_G \leq B(t)$, $\forall t$, are relevant characteristic, small scales. The term $a_2(B^2-B_G^2)$ comes from the fact that the magnetic flux fragments which coalesce forming the flux tube of the sunspot [2], erupt from below the photosphere due to the magnetic buoyancy which is proportional to B^2 [3]. The terms $b_{11}A + b_{12}$ and $-b_{13}\sqrt{A}$ represent the magnetic flux inflow and outflow, respectively, (combining the well-known "collar flow" [2]), which together with the negative $-a_1(A-A_G)$ and positive $b_2(B-B_G)$ feedback, help to stabilize the sunspot [2], [4] - [6].

Furthermore, by considering a proportionality relation between $A(t)$, $B(t)$, and A_G, B_G, respectively, we define the dimensionless variables $x(t) \equiv A(t)/A_G$, $y(t) \equiv B(t)/B_G$, rewriting the system (1) as follows

$$\dot{x} = a_1[-x+q(y^2-1)+1], \quad \dot{y} = b_2[p_1x+m-p_2\sqrt{x}+y] \quad (2)$$

where the new set of dimensionless parameters q, p_1, m, p_2 is expressed in terms of the old parameters involved in (1). The fixed value of $q = 310$ expresses the universality of the sunspots creation mechanism. Fixed values

G. Contopoulos, P.A. Patsis, *Chaos in Astronomy*. Astrophysics and Space Science Proceedings, doi: 10.1007/978-3-540-75826-6,

have been also found to be characterizing both the inflow parameters, namely $p_1 = 0.00185$, $m = 4.9$, while the values of the outflow parameter p_2 may differ for different sunspots, constituting however, an equidistribution [1], [7].

As well as any 2-dimensional nonlinear dynamical system, (2) is integrable [8], but only its conservative version, namely when $b_2 = a_1$, can suitably describe the sunspot evolution [1]. Throughout in this analysis, we deal with the non-integrable version of the perturbed PhMD model, where the inflow parameter m is given by $m(t) = m_0 + \delta \cos(\Omega t)$, where $m_0 = 4.9$ (the unperturbed m value), $\Omega = 15 \deg d^{-1}$ (cyclic frequency of the solar rotation at the equator, with d corresponding to days). In particular, the following Sect. 2 deals with the conservative version of the model, while in Sect. 3 the dissipative version is studied.

2 The Conservative Perturbed PhMD Model

We identify A_G, B_G, as the characteristic scales of the granules surrounding the sunspot. Indeed, once the magnetic microstructures emerging the photosphere are accumulated in one main magnetic core, this has to be arranged between the granules. In other words, the coalescence of the various magnetic flux fragments into the flux tube of the sunspot, is a procedure initially taking place into the enclosed intermediate space between adjacent granules. Hence, this is the given space being disposed to the sunspot initial area A_0.

If the granules have a low density arrangement, the initial area is quite larger than the characteristic area of the granules, i.e. $A_0 > A_G$. Then, the system shall be spontaneously driven to a higher density arrangement. In the meanwhile, the enclosed area is compressed, being opposed to the emergence of the magnetic elements from below the photosphere, leading thus, to a resistance $a_1(A_0 - A_G) > 0$ on the area initial growth. On the other hand, in a high density granules arrangement, we have $A_0 < A_G$, and the system shall be spontaneously driven to a lower density arrangement, with the enclosed area being expanded, giving advance to the sunspot creation, that is expressed by a "negative" resistance $a_1(A_0 - A_G) < 0$ on the area initial growth. The case of neither high nor low density granules arrangement can be thought as an equilibrium, expressed by the initial conditions $\dot{A} \approx 0$, $A_0 \approx A_G$, leading to $B_0 \approx B_G$, hence, $x_0 \approx 1$, $y_0 \approx 1$. In [1], [7], it was discussed that these are the most appropriate initial conditions for describing the sunspot evolution. In general, however, we may have $x_0 \geq 1$, $y_0 \geq 1$ (any other initial conditions do not yield to a sunspot of finite dimensions).

We consider the conservative case of the perturbed model, that is for $b_2 = a_1$. The Hamiltonian function is time-dependent and it does not constitute a constrain factor (an integral) [1]. Then, chaos emerges for any perturbation amplitude $\delta > 0$. In Fig. 1a,b, the stroboscopic maps [1] of the pairs (x_n, y_n) and $(x_n, \dot{x}_n)$ are, respectively, depicted for $n = 1, \cdots, N = 10^3$, and $\delta = 0.2$.

In Fig. 2 we plot the first statistical moment ν_ϕ of the distribution function of the twist angle, defined in [9], for $x = 1, 360$.

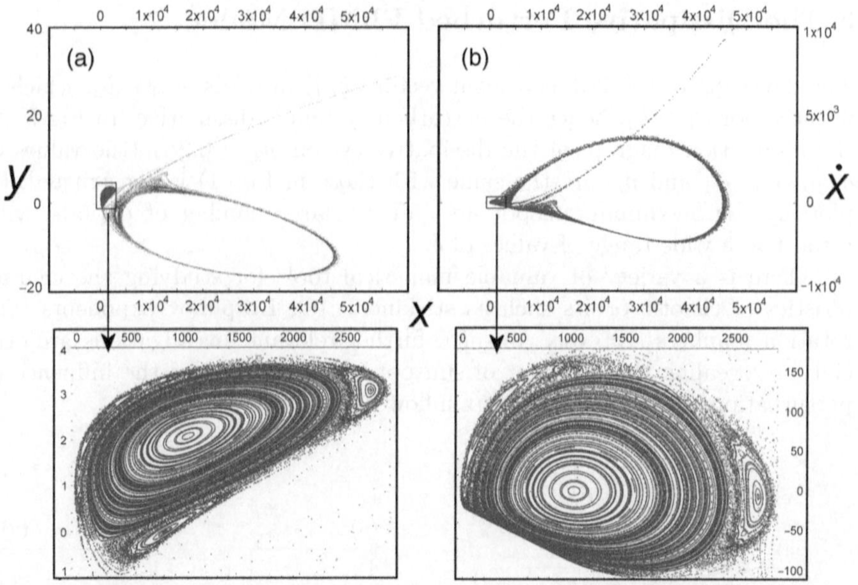

Fig. 1. The stroboscopic maps of $N = 10^3$ pairs (x_n, y_n) (a) and $(x_n, \dot{x}_n)$ (b) depicted for $\delta = 0.2$, $a_1 = 0.068\ \mathrm{d}^{-1}$, $p_2 = 0.276$ and 10^3 different initial conditions. The insets magnify the region of central stable point and its surrounding curves. (The values of the rest of the parameters are given in the text.)

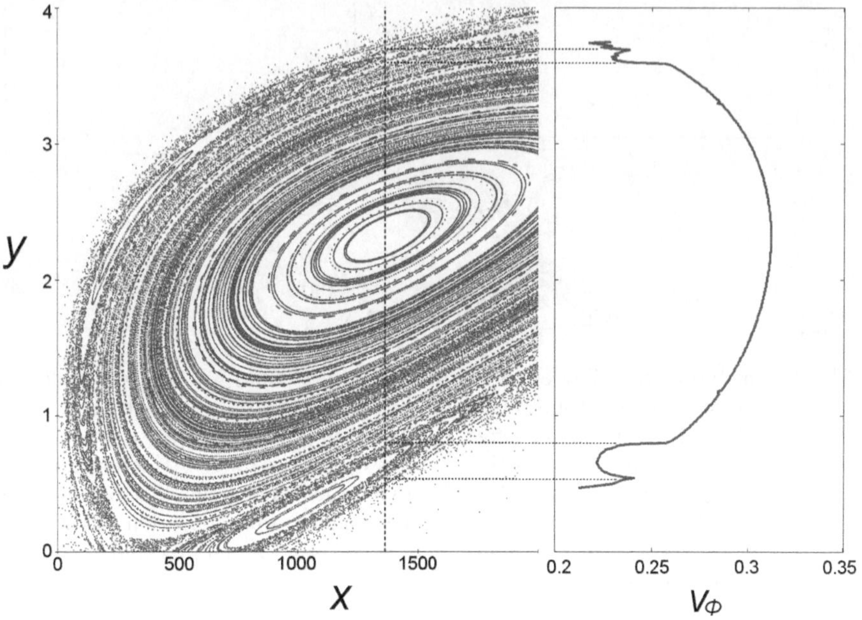

Fig. 2. The twist angle statistical moment ν_ϕ depicted for $x = 1,360$ ($p_2 = 0.263$). The "U" shape dips, corresponding to high multiplicity islands, are shown

3 The Dissipative Perturbed PhMD Model

The deviation of the 2-dimensional vector $(\dot{x}, \dot{y})$ in (2) is $-a_1 + b_2$, which is negative for $a_1 > b_2$, hence the perturbed system is dissipative. In Fig. 3a,b, the bifurcation diagram of the dissipative system $b_2 = 0.2\alpha_1$ (the values of q, p_1, m_0, α_1 and p_2, are the same with those in Fig. 1) is constructed, by plotting the maximum sunspot area after a large number of periods, with respect to a wide range of values of δ.

There is a variety of valuable numerical tools for studying the characteristics of chaotic orbits such as stickiness [10], Lyapunov exponents [11], rotation numbers [12], etc. As such, further relevant investigations are crucial for revealing the behavior of sunspots evolution, under the influence of perturbations of the magnetic flux inflow parameter.

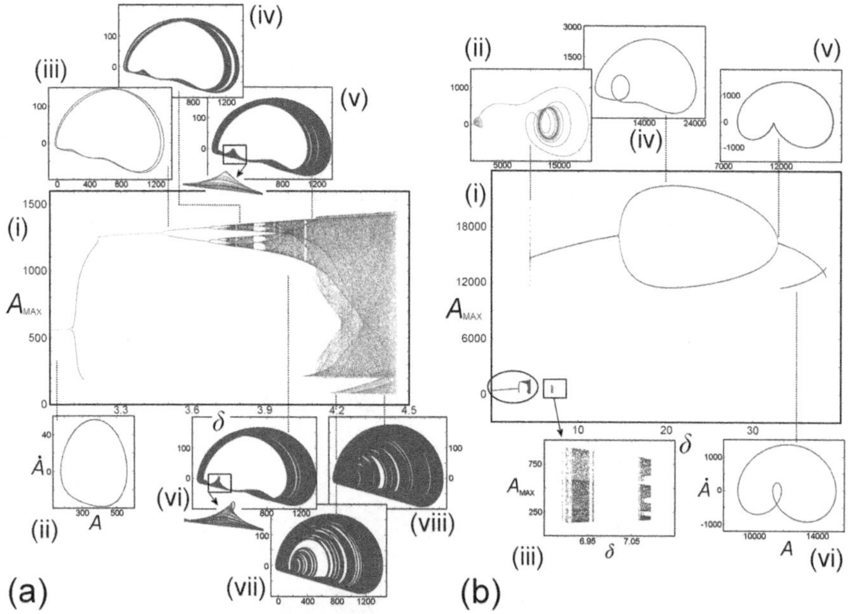

Fig. 3. The bifurcation diagram of the dissipative system $b_2 = 0.2\alpha_1$ constructed by plotting the maximum sunspot areas for the initial conditions $x_0 = 1$, $y_0 = 1$, and $A_0 = 0.71$ MSH (average granule area, calculated in [1]). The insets (orbits in $(A, \dot{A})$ phase space) illustrate the transition of limit cycle to strange attractor, except of (ii) showing the whole orbit and (iii) magnifying the frame detail. The oval detail in (**b**) is the whole diagram depicted in (**a**). The sudden orbit transition to the larger area values is attained for $\delta \simeq 4.45$. (The values of q, p_1, m_0, α_1 and p_2, are the same with those in Fig. 1.)

References

1. G. Livadiotis, X. Moussas: Physica A **379**, 436 (2007)
2. S.K. Solanki: Astron. Astrophys. Rev. **11**, 153 (2003)
3. M.G. Kivelson, C.T. Russell: *Introduction to Space Physics*, (Cambridge University Press, 1995) pp 73–74
4. T.L. Duvall Jr., S. DSilva, S.M. Jefferies et al: Nature **379**, 235 (1996)
5. N.E. Hurlburt, A.M. Rucklidge: Mon. Not. R. Astron. Soc. **314**, 793 (2000)
6. A.G. Kosovichev: Astron. Nachr. **323**, 186 (2002)
7. G. Livadiotis, X. Moussas: Upper limit of the total magnetic flux in an Active Region. In: *Proc. of Symposium "Solar Extreme Events"*, (in prep. for 2008)
8. G. Contopoulos: *Order and Chaos in Dynamical Astronomy*, (Springer, Berlin, Heidelberg, New York, 2004) pp 17,20,48
9. N. Voglis, C. Efthimiopoulos: J. Phys. A **31**, 2913 (1998)
10. G. Contopoulos: Astron. J. **76**, 147 (1971)
11. G. Livadiotis: Adv. Comp. Syst. **8**, 15 (2005)
12. G. Livadiotis, N. Voglis: J. Phys. A **39**, 15231 (2006)

The Dynamics of Non-Symmetrically Collapsing Stars

G.S. Bisnovatyi-Kogan[1,2,3] and O.Y. Tsupko[1,3]

[1] Space Research Institute of Russian Academy of Science, Profsoyuznaya 84/32, Moscow 117997, Russia
gkogan@iki.rssi.ru, tsupko@iki.rssi.ru
[2] Joint Institute for Nuclear Research, Dubna, Russia
[3] Moscow Engineering Physics Institute, Moscow, Russia

1 Introduction

The dynamical stability of spherical stars is determined by an average adiabatic power $\gamma = \frac{\partial \ln P}{\partial \ln \rho}|_S$ [1, 2]. The collapse of a spherical star can only be stopped by its equation of state becoming stiff. Without such stiffening a spherical star in the newtonian theory would collapse into a point with infinite density.

Here we show that deviations from the spherical symmetry in a non-rotating star lead to dynamic stabilization, and non-spherical star without dissipative processes will never reach a singularity in the newtonian gravity. Therefore collapse to singularity is connected with a secular type of instability, even without rotation.

We calculate the dynamical behavior of a non-spherical, non-rotating star after its loss of a linear stability, and investigate nonlinear stages of contraction. We use an approximate system of dynamical equations, describing 3 degrees of freedom of a uniform self-gravitating compressible ellipsoidal body [3, 4]. We obtain that the development of instability leads to formation a regularly or chaotically oscillating body, in which dynamical motion prevents formation of the singularity. At the end we discuss qualitatively effects of general relativity in a non-spherical collapse of a non-rotating body.

2 Equations of Motion and Numerical Results

Let us consider a 3-axis ellipsoid with semi-axes $a \neq b \neq c$ and uniform density ρ. The mass m of the uniform ellipsoid is written as (V is the volume of the ellipsoid) $m = \rho V = \frac{4\pi}{3}\rho abc$. Let us assume a linear dependence of velocities on coordinates $v_x = \dot{a}x/a$, $v_y = \dot{b}y/b$, $v_z = \dot{c}z/c$. The gravitational energy of the uniform ellipsoid is defined as:

G. Contopoulos, P.A. Patsis, *Chaos in Astronomy*. Astrophysics and Space Science Proceedings, doi: 10.1007/978-3-540-75826-6,

$$U_g = -\frac{3Gm^2}{10} \int\limits_0^\infty \frac{du}{\sqrt{(a^2+u)(b^2+u)(c^2+u)}}. \tag{1}$$

The equation of state $P = K\rho^\gamma$ is considered here, with $\gamma = 4/3$. For $\gamma = 4/3$, the thermal energy $E_{th} \sim V^{-1/3} \sim (abc)^{-1/3}$, and the value $\varepsilon = E_{th}(abc)^{1/3} = 3(3m/4\pi)^{1/3}K$ remains constant in time. A Lagrange function yielding the equations of motion of the three semiaxes of the ellipsoid reads

$$L = U_{kin} - U_{pot}, \quad U_{pot} = U_g + E_{th}, \quad E_{th} = \frac{\varepsilon}{(abc)^{1/3}}, \tag{2}$$

$$U_{kin} = \frac{1}{2}\rho \int\limits_V (v_x^2 + v_y^2 + v_z^2)\, dV = \frac{m}{10}(\dot{a}^2 + \dot{b}^2 + \dot{c}^2). \tag{3}$$

To obtain a numerical solution of equations we write them in non-dimensional variables. We solve here numerically the equations of motion for a spheroid with $a = b \neq c$, which are written as

$$\ddot{a} = \frac{3}{2a^2(1-k^2)}\left[k - \frac{\arccos k}{\sqrt{1-k^2}}\right] + \frac{1}{a}\frac{\varepsilon}{(a^2c)^{1/3}}, \tag{4}$$

$$\ddot{c} = -\frac{3}{a^2(1-k^2)}\left[1 - \frac{k \arccos k}{\sqrt{1-k^2}}\right] + \frac{1}{c}\frac{\varepsilon}{(a^2c)^{1/3}} \tag{5}$$

for the oblate spheroid with $k = c/a < 1$,

$$\ddot{a} = -\frac{3}{2a^2(k^2-1)}\left[k - \frac{\cosh^{-1} k}{\sqrt{k^2-1}}\right] + \frac{1}{a}\frac{\varepsilon}{(a^2c)^{1/3}}, \tag{6}$$

$$\ddot{c} = \frac{3}{a^2(k^2-1)}\left[1 - \frac{k \cosh^{-1} k}{\sqrt{k^2-1}}\right] + \frac{1}{c}\frac{\varepsilon}{(a^2c)^{1/3}} \tag{7}$$

for the prolate spheroid with $k = c/a > 1$, and

$$\ddot{a} = -\frac{1-\varepsilon}{a^2} \tag{8}$$

for the sphere. For the given mass there is only one equilibrium value $\varepsilon_{eq} = 1$ at which the spherical star has zero total energy, and it may have an arbitrary radius.

In these variables the non-dimensional Hamiltonian (or non-dimensional total energy) is written as

$$H = \frac{\dot{a}^2}{5} + \frac{\dot{c}^2}{10} - \frac{3}{5a}\frac{\arccos k}{\sqrt{1-k^2}} + \frac{3}{5}\frac{\varepsilon}{(a^2c)^{1/3}}, \quad \text{(oblate)}$$

$$H = \frac{\dot{a}^2}{5} + \frac{\dot{c}^2}{10} - \frac{3}{5a}\frac{\cosh^{-1} k}{\sqrt{k^2-1}} + \frac{3}{5}\frac{\varepsilon}{(a^2c)^{1/3}}, \quad \text{(prolate)}$$

$$H = \frac{3\dot{a}^2}{10} - \frac{3}{5a}(1-\varepsilon). \quad \text{(sphere)} \tag{9}$$

Solution of the system of equations (4)–(7) was performed for initial conditions at $t = 0$: $\dot{c}_0 = 0$, different values of initial a_0, $\dot{a}_0$, k_0, and different values of the constant parameter ε. Evidently, at $k_0 = 1$, $\dot{a}_0 = 0$, $\varepsilon < 1$ we have the spherical collapse to singularity. The most interesting result was obtained at $k_0 \neq 1$, and all other cases with deviations from spherical symmetry. It is clear, that at $\varepsilon = 0$ a weak singularity is reached during formation of a pancake with infinite surface density, and finite gravitational force. At $\varepsilon > 0$ no singularity was reached. At $\varepsilon \geq 1$ the total energy of spheroid $H > 0$, determining the full disruption. At $H < 0$ the oscillatory regime is established at any value of $\varepsilon < 1$, and may be represented either by regular periodic oscillations, or by chaotic behavior.

3 The Poincaré Section

To investigate the regular or chaotic dynamics we use the method of Poincaré section [5] and obtain the Poincaré map for different values of the total energy H. A spheroid with semi-axes $a = b \neq c$ has two degrees of freedom. Therefore in this case the phase space is four-dimensional: a, $\dot{a}$, c, $\dot{c}$. If we choose a value of the Hamiltonian H_0, we fix a three-dimensional energy surface $H(a, \dot{a}, c, \dot{c}) = H_0$. During the integration of the equations of motion which preserve the constant H, we fix moments t_i, when $\dot{c} = 0$. At each moment t_i we put a dot on the plane $(a, \dot{a})$. These points are the intersection points of the trajectories on the three-dimensional energy surface with a two-dimensional plane $\dot{c} = 0$, called the Poincaré section.

For each fixed combination of ε, H we get the Poincaré map. The condition $\dot{c} = 0$ holds at both the local minima or local maxima of c with respect to t. The Poincaré maps for these two cases must be plotted separately. The regular oscillations are represented by closed curves on the Poincaré map, while the chaotic orbits fill stochastically domains of a finite area (see Fig. 1).

The variables a and $\dot{a}$ cannot occupy the whole plane $(a, \dot{a}) : \ 0 < a < \infty$, $-\infty < \dot{a} < +\infty$. We obtain analytically a curve bounding the area of the values a and $\dot{a}$. This limiting curve is shown by a bold curve in Fig. 1.

4 Conclusions

The main result following from our calculations is the indication of a degenerate nature of the formation of a singularity in unstable newtonian self-gravitating gaseous bodies. Only pure spherical models can collapse to singularity, but any kind of nonsphericity leads to nonlinear stabilization of the collapse by a dynamical motion, and the formation of a regularly or chaotically oscillating body. This conclusion is valid for all unstable equations of

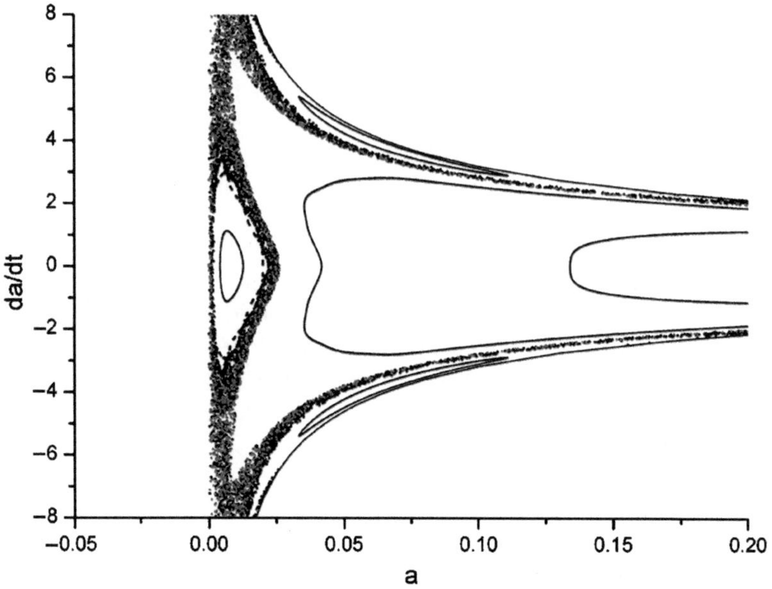

Fig. 1. The part of Poincaré map for three regular and two chaotic trajectories in case of $H = -1/5$, $\varepsilon = 2/3$. Points of minimum of c. The bounding curve is painted by heavy line

state, namely, for adiabatic with $\gamma < 4/3$. In addition to the case with $\gamma = 4/3$, we have calculated the dynamics of the model with $\gamma = 6/5$, and have obtained similar results. Note that, for $\gamma = 4/3$, the area of the chaotic domain on the Poincaré map increases gradually, as the entropy ϵ, or the total energy H decreases.

In the frame of a general relativity dynamic stabilization against collapse by nonlinear nonspherical oscillations cannot be universal. When the size of the body approaches gravitational radius no stabilization is possible at any γ. Nevertheless, the nonlinear stabilization may happen at larger radii, so after damping of the oscillations the star would collapse to the black hole. Due to development of nonspherical oscillations there is a possibility for emission of gravitational waves during the collapse of nonrotating stars with the intensity similar to rotating bodies, or even larger.

Account of general relativity will introduce a new non-dimensional parameter, which can be written as $p_g = \frac{2Gm_0}{c^2 a_0}$. The fate of the gravitating body will depend on the value of this parameter, and we may expect a direct relativistic collapse to a black hole at increasing p_g, approaching unity. It is known, that a nonrotating black hole is characterized only by its mass [1]. In absence of other dissipative processes, the excess of energy, connected with a nonspherical motion should be emitted by gravitational waves during a formation of the black hole.

Acknowledgments

We are grateful to A.I. Neishtadt for the valuable advices in constructing the Poincaré map.

This work was partially supported by RFBR grants 08-02-00491 and 08-02-90106, the RAN Programme "Formation and evolution of stars and galaxies" and the Grant for Leading Scientific Schools NSh-2977.2008.2.

Work of O. Yu. Tsupko was also partially supported by the Dynasty Foundation.

References

1. Ya.B. Zeldovich, I.D. Novikov: *Relativistic Astrophysics*, (Nauka, Moscow, 1967 (in Russian))
2. G.S. Bisnovatyi-Kogan: *Physical problems in the theory of stellar evolution*, (Nauka, Moscow, 1989 (in Russian)). (English translation: *Stellar Physics*, Vol. 1,2, (Springer, 2001))
3. G.S. Bisnovatyi-Kogan: MNRAS **347**, 163 (2004)
4. G.S. Bisnovatyi-Kogan, O.Yu. Tsupko: MNRAS **364**, 833 (2005)
5. A.J. Lichtenberg, M.A. Lieberman: *Regular and Stochastic motion*, (Springer, New York, 1983)

Cosmic Order out of Primordial Chaos: A Tribute to Nikos Voglis

B. Jones and R. van de Weygaert

Kapteyn Astronomical Institute, University of Groningen, P.O. Box 800, 9700 AV Groningen, The Netherlands
bernard@astrag.demon.co.uk, weygaert@astro.rug.nl

Summary. Nikos Voglis had many astronomical interests, among them was the question of the origin of galactic angular momentum. In this short tribute we review how this subject has changed since the 1970s and how it has now become evident that gravitational tidal forces have not only caused galaxies to rotate, but have also acted to shape the very cosmic structure in which those galaxies are found. We present recent evidence for this based on data analysis techniques that provide objective catalogues of clusters, filaments and voids.

1 Some Early History

It was in the 1970s that Nikos Voglis first came to visit Cambridge, England, to attend a conference and to discuss a problem that was to remain a key area of personal interest for many years to come: the origin of galaxy angular momentum. It was during this period that Nikos teamed up with Phil Palmer to create a long lasting and productive collaboration.

The fundamental notion that angular momentum is conserved leads one to wonder how galaxies could acquire their angular momentum if they started out with none. This puzzle was perhaps one of the main driving forces behind the idea that cosmic structure was born out of some primordial turbulence. However, by the early 1970s the cosmic turbulence theory was falling into disfavour owing to a number of inherent problems (see [19] for a detailed review of this issue).

The alternative, and now well entrenched, theory was the gravitational instability theory in which structure grew through the driving force of gravitation acting on primordial density perturbations. The question of the origin of angular momentum had to be addressed and would be central to the success or failure of that theory. Peebles [32] provided the seminal paper on this, proposing that tidal torques would be adequate to provide the solution. However, this was for many years mired in controversy.

Tidal torques had been suggested as a source for the origin of angular momentum since the late 1940s when Hoyle [17] invoked the tidal stresses exerted

G. Contopoulos, P.A. Patsis, *Chaos in Astronomy.* Astrophysics and Space Science Proceedings, doi: 10.1007/978-3-540-75826-6,

by a cluster on a galaxy as the driving force of galaxy rotation. Although the idea as expounded was not specific to any cosmology, there can be little doubt that Hoyle had his Steady State cosmology in mind. The Peebles [32] version of this process specifically invoked the tidal stresses between two neighbouring protogalaxies, but it was not without controversy. There were perhaps three sources for the ensuing debate:

- Is the tidal force sufficient to generate the required angular momentum?
- Are tidal torques between proto-galaxies alone responsible for the origin of galactic angular momentum?
- How does the zero vorticity shear flow generated by the tidal torques get converted into circular motion?

Oort [29] and Harrison [15] had both argued that the interaction between low-amplitude primordial perturbations would be inadequate to drive the rotation: they saw the positive density fluctuations as being "shielded" by a surrounding negative density region which would diminish the tidal forces. This doubt was a major driving force behind "alternative" scenarios for galaxy formation. The last of these was a more subtle problem since, to some, even if tidal forces managed to generate adequate shear flows, the production of rotational motion would nonetheless require some violation of the Kelvin circulation theorem. Although the situation was clarified by Jones [19] it was not until the exploitation of N-Body cosmological simulations that the issue was considered to have been resolved.

It was into this controversy that Nikos stepped, asking precisely these questions. A considerable body of his later work (much of it with Phil Palmer, see for example [30]) was devoted to addressing these issues at various levels. Since these days our understanding of the tidal generation of galaxy rotation has expanded impressively, mostly as a result of ever more sophisticated and large N-body simulation (e.g. [3, 5, 12, 20, 38, 45]). What remains is Nikos' urge for a deeper insight, beyond simulation, into the physical intricacies of the problem.

2 Angular Momentum Generation: The Tidal Mechanism

In order to appreciate these problems it is helpful to look at a simplified version of the tidal model as proposed by Peebles. Consider two neighbouring, similar sized, protogalaxies **A** and **B** (Fig. 1). We can view the tidal forces exerted on **B** by **A** from either the reference frame of the mass center of **A** or from the reference frame of the mass center of **B** itself. These forces are depicted by arrows in the figure: note that relative to the mass center of **B** the tidal forces act so as to stretch **B** out in the direction of **A**.

To a first approximation, the force gradient acting on **B** can be expressed in terms of the potential field $\phi(\mathbf{x})$ in which **B** is situated:

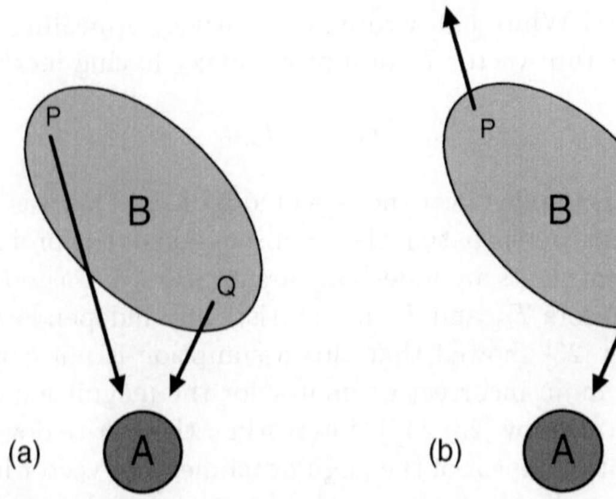

Fig. 1. An extended object, **B**, acted on by the gravitational field of a nearby object, **A**. (**a**) depicts the forces as seen from the point of view of the forcing object: both points P and Q fall towards the A, albeit at different rates. (**b**) depicts the forces as seen from the point of view of the mass center of **B** where both P and Q recede from the mass center

$$T_{ij} = \frac{\partial F_i}{\partial x_j} = \frac{\partial^2 \phi}{\partial x_i \partial x_j} - \frac{1}{3}\delta_{ij}\nabla^2\phi \tag{1}$$

where the potential field is determined from the fluctuating component of the density field via the Poisson equation[1]. The flow of material is thus a shear flow determined by the principal directions and magnitudes of inertia tensor of the blob **B**. Viewed as a fluid flow this is undeniably a shear flow with zero vorticity as demanded by the Kelvin circulation theorem[2].

So how does the vorticity that is evident in galaxy rotation arise? The answer is twofold. Shocks will develop in the gas flow and stars will form: the Kelvin Theorem holds only for nondissipative flows. Then, a "gas" of stars does not obey the Kelvin Theorem since it is not a fluid (though there is a six-dimensional phase space analogue for a stellar "gas").

The magnitude and direction of angular momentum vector is related to the inertia tensor, I_{ij}, of the torqued object and the driving tidal forces described by the tensor T_{ij} of equation (1). In 1984, based on simple low-order

---

[1] The Poisson equation determines only the trace of the symmetric tensor $\frac{\partial^2 \phi}{\partial x_\alpha \partial x_\beta}$. The interesting exercise for the reader is to contemplate what determines the other five components?

[2] This raises the technically interesting question as to whether a body with zero angular momentum can rotate: most undergraduates following a classical dynamics course with a section on rigid bodies would unequivocally answer "no." The situation is beautifully discussed in Feynman's famous "Lectures in Modern Physics" [13].

perturbation theory, White [49] wrote an intuitively appealing expression for the angular momentum vector L_i of a protogalaxy having inertia tensor I_{mk}:

$$L_i \propto \epsilon_{ijk} T_{jm} I_{mk}, \tag{2}$$

where summation is implied over the repeated indices. This was later taken up by [9] in a high-order perturbation theory discussion of the problem. However, there is in these treatments an underlying assumption, discussed but dismissed by [9], that the tensors T_{ij} and I_{ij} are statistically independent. Subsequent numerical work by [23] showed that this assumption is not correct and that ignoring it results in an incorrect estimator for the magnitude of the spin.

The approach taken by [23, 24] is interesting: they write down an equation for the autocorrelation tensor of the angular momentum vector in a given tidal field, averaging over all orientations and magnitudes of the inertia tensor. On the basis of equation (2) one would expect this tensor autocorrelation function to be given by

$$\langle L_i L_j | \mathbf{T} \rangle \propto \epsilon_{ipq} \epsilon_{jrs} T_{pm} T_{rn} \langle I_{mq} I_{ns} \rangle \tag{3}$$

where the notation $\langle L_i L_j | \mathbf{T} \rangle$ is used to emphasise that T_{ij} is regarded as a given value and is not a random variable. The argument then goes that the isotropy of underlying density distribution allows us to replace the statistical quantity $\langle I_{mq} I_{ns} \rangle$ by a sum of Kronecker deltas leaving only

$$\langle L_i L_j | \mathbf{T} \rangle \propto \frac{1}{3}\delta_{ij} + (\frac{1}{3}\delta_{ij} - T_{ik}T_{kj}) \tag{4}$$

It is then asserted that if the moment of inertia and tidal shear tensors were uncorrelated, we would have only the first term on the right hand side, $\frac{1}{3}\delta_{ij}$: the angular momentum vector would be isotropically distributed relative to the tidal tensor.

In fact, in the primordial density field and the early linear phase of structure formation there is a significant correlation between the shape of density fluctuations and the tidal force field [7, 46]. Part of the correlation is due to the anisotropic shape of density peaks and the internal tidal gravitational force field that goes along with it [18]. The most significant factor is that of intrinsic spatial correlations in the primordial density field. It is these intrinsic correlations between shape and tidal field that are at the heart of our understanding of the Cosmic Web, as has been recognized by the Cosmic Web theory of [8]. The subsequent nonlinear evolution may strongly augment these correlations (see e.g. Fig. 2), although small-scale highly nonlinear interactions also lead to a substantial loss of the alignments: clusters are still strongly aligned, while galaxies seem less so.

Recognizing that the inertia and tidal tensors may not be mutually independent, [23, 24] write

$$\langle L_i L_j | \mathbf{T} \rangle \propto \frac{1}{3}\delta_{ij} + c(\frac{1}{3}\delta_{ij} - T_{ik}T_{kj}) \tag{5}$$

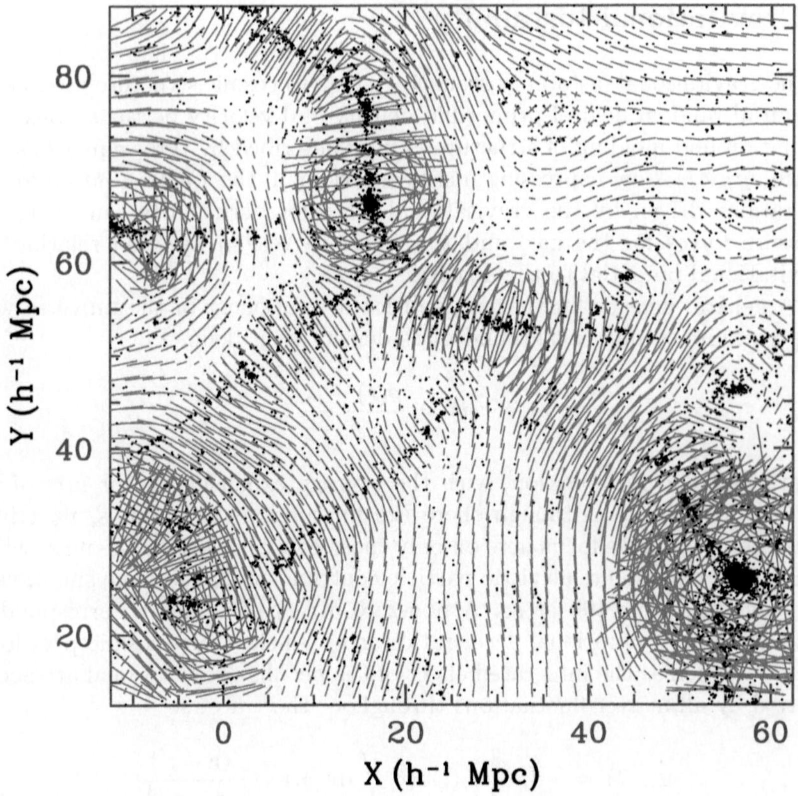

Fig. 2. The relation between the *cosmic web*, the clusters at the nodes in this network and the corresponding compressional tidal field pattern. It shows the matter distribution at the present cosmic epoch, along with the (compressional component) tidal field bars in a slice through a simulation box containing a realization of cosmic structure formed in an open, $\Omega_o = 0.3$, Universe for a CDM structure formation scenario (scale: $R_G = 2h^{-1}$Mpc). The frame shows structure in a $5h^{-1}$Mpc thin central slice, on which the related tidal bar configuration is superimposed. The matter distribution, displaying a pronounced weblike geometry, is clearly intimately linked with a characteristic coherent compressional tidal bar pattern. From: van de Weygaert 2002

where $c = 0$ for randomly distributed angular momentum vectors. The case of mutually independent tidal and inertia tensors is described by $c = 1$ (see (4)). They finally introduce a different parameter $a = 3c/5$ and write

$$\langle L_i L_j | \mathbf{T} \rangle \propto \frac{1+a}{3}\delta_{ij} - aT_{ik}T_{kj} \qquad (6)$$

which forms the basis of much current research in this field. The value derived from recent study of the Millennium simulations by [25] is $a \approx 0.1$.

3 Gravitational Instability

In the gravitational instability scenario (e.g. [33]) cosmic structure grows from an initial random field of primordial density and velocity perturbations. The formation and molding of structure is fully described by three equations, the *continuity equation*, expressing mass conservation, the *Euler equation* for accelerations driven by the gravitational force for dark matter and gas, and pressure forces for the gas, and the *Poisson–Newton equation* relating the gravitational potential to the density.

A general density fluctuation field for a component of the universe with respect to its cosmic background mass density ρ_{u} is defined by

$$\delta(\mathbf{r},t) \;=\; \frac{\rho(\mathbf{r})-\rho_{\mathrm{u}}}{\rho_{\mathrm{u}}}\,. \tag{7}$$

Here $\mathbf{r}$ is comoving position, with the average expansion factor $a(t)$ of the universe taken out. Although there are fluctuations in photons, neutrinos, dark energy, etc., we focus here on only those contributions to the mass which can cluster once the relativistic particle contribution has become small, valid for redshifts below 100 or so. A non-zero $\delta(\mathbf{r},t)$ generates a corresponding total peculiar gravitational acceleration $\mathbf{g}(\mathbf{r})$ which at any cosmic position $\mathbf{r}$ can be written as the integrated effect of the peculiar gravitational attraction exerted by all matter fluctuations throughout the Universe:

$$\mathbf{g}(\mathbf{r},t) \;=\; -4\pi G\bar{\rho}_m(t)a(t)\int \mathrm{d}\mathbf{r}'\,\delta(\mathbf{r}',t)\,\frac{(\mathbf{r}-\mathbf{r}')}{|\mathbf{r}-\mathbf{r}'|^3}\,. \tag{8}$$

Here $\bar{\rho}_m(t)$ is the mean density of the mass in the universe that can cluster (dark matter and baryons). The cosmological density parameter $\Omega_m(t)$ is defined by ρ_{u}, via the relation $\Omega_m H^2 = (8\pi G/3)\bar{\rho_m}$, in terms of the Hubble parameter H. The relation between the density field and gravitational potential Φ is established through the Poisson–Newton equation:

$$\nabla^2\Phi = 4\pi G\bar{\rho}_m(t)a(t)^2\;\delta(\mathbf{r},t). \tag{9}$$

The peculiar gravitational acceleration is related to $\Phi(\mathbf{r},t)$ through $\mathbf{g} = -\nabla\Phi/a$ and drives peculiar motions.

In slightly overdense regions around density excesses, the excess gravitational attraction slows down the expansion relative to the mean, while underdense regions expand more rapidly. The underdense regions around density minima expand relative to the background, forming deep voids. Once the gravitational clustering process has progressed beyond the initial linear growth phase we see the emergence of complex patterns and structures in the density field.

Large N-body simulations all reveal a few "universal" characteristics of the (mildly) nonlinear cosmic matter distribution: its hierarchical nature, the

Fig. 3. The hierarchical Cosmic Web: over a wide range of spatial and mass scales structures and features are embedded within structures of a larger effective dimension and a lower density. Image courtesy of V. Springel & Virgo consortium, also see Springel et al. 2005. Reproduced with permission of Nature

anisotropic and weblike spatial geometry of the spatial mass distribution and the presence of huge underdense voids. These basic elements of the Cosmic Web [8, 48] exist at all redshifts, but differ in scale.

Figure 3, from the state-of-the-art "Millennium simulation", illustrates this complexity in great detail over a substantial range of scales. The figure zooms in on the dark matter distribution at five levels of spatial resolution and shows the formation of a filamentary network connecting to a central cluster. This network establishes transport channels along which matter will flow into the cluster. The hierarchical nature of the structure is clearly visible. The dark matter distribution is far from homogeneous: a myriad of tiny dense clumps indicate the presence of dark halos in which galaxies, or groups of galaxies, will have formed.

Within the context of gravitational instability, it is the gravitational tidal forces that establish the relationship between some of the most prominent manifestations of the structure formation process. It is this intimate link between the Cosmic Web, the mutual alignment between cosmic structures and the rotation of galaxies to which we wish to draw attention in this short contribution.

4 Tidal Shear

When describing the dynamical evolution of a region in the density field it is useful to distinguish between large scale "background" fluctuations $\delta_{\rm b}$ and small-scale fluctuations $\delta_{\rm f}$. Here, we are primarily interested in the influence of the smooth large-scale field. Its scale R_b should be chosen such that it remains (largely) linear, i.e. the r.m.s. density fluctuation amplitude $\sigma_\rho(R_b,t) \lesssim 1$.

To a good approximation the smoother background gravitational force $\mathbf{g}_{\rm b}(\mathbf{x})$ (Eqn. 8) in and around the mass element includes three components (apart from rotational aspects). The *bulk force* $\mathbf{g}_{\rm b}(\mathbf{x}_{pk})$ is responsible for the acceleration of the mass element as a whole. Its divergence $(\nabla \cdot \mathbf{g}_{\rm b})$ encapsulates the collapse of the overdensity while the tidal tensor T_{ij} quantifies its deformation,

$$g_{\rm b,i}(\mathbf{x}) = g_{\rm b,i}(\mathbf{x}_{pk}) + a \sum_{j=1}^{3} \left\{ \frac{1}{3a} (\nabla \cdot \mathbf{g}_{\rm b})(\mathbf{x}_{pk})\, \delta_{\rm ij} - T_{ij} \right\} (x_j - x_{pk,\rm j})\,. \quad (10)$$

The tidal shear force acting over the mass element is represented by the (traceless) tidal tensor T_{ij},

$$T_{\rm ij} \equiv -\frac{1}{2a} \left\{ \frac{\partial g_{{\rm b},i}}{\partial x_i} + \frac{\partial g_{{\rm b},j}}{\partial x_j} \right\} + \frac{1}{3a} (\nabla \cdot \mathbf{g}_{\rm b})\, \delta_{\rm ij} \quad (11)$$

in which the trace of the collapsing mass element, proportional to its overdensity δ, dictates its contraction (or expansion). For a cosmological matter distribution the close connection between local force field and global matter distribution follows from the expression of the tidal tensor in terms of the generating cosmic matter density fluctuation distribution $\delta(\mathbf{r})$ [46]:

$$\begin{aligned} T_{ij}(\mathbf{r}) &= \frac{3\Omega H^2}{8\pi} \int {\rm d}\mathbf{r}'\, \delta(\mathbf{r}') \left\{ \frac{3(r_i' - r_i)(r_j' - r_j) - |\mathbf{r}' - \mathbf{r}|^2\, \delta_{ij}}{|\mathbf{r}' - \mathbf{r}|^5} \right\} \\ &\quad -\frac{1}{2} \Omega H^2\, \delta(\mathbf{r},t)\, \delta_{ij}. \end{aligned}$$

The tidal shear tensor has been the source of intense study by the gravitational lensing community since it is now possible to map the distribution of large scale cosmic shear using weak lensing data. See for example [16, 27].

5 The Cosmic Web

Perhaps the most prominent manifestation of the tidal shear forces is that of the distinct *weblike geometry* of the cosmic matter distribution, marked by highly elongated filamentary, flattened planar structures and dense compact clusters surrounding large near-empty void regions (see Fig. 3). The recognition of the *Cosmic Web* as a key aspect in the emergence of structure in the Universe came with early analytical studies and approximations concerning the emergence of structure out of a nearly featureless primordial Universe. In this respect the Zel'dovich formalism [50] played a seminal role. It led to the view of structure formation in which planar pancakes form first, draining into filaments which in turn drain into clusters, with the entirety forming a cellular network of sheets.

The Megaparsec scale tidal shear forces are the main agent for the contraction of matter into the sheets and filaments which trace out the cosmic web. The anisotropic contraction of patches of matter depends sensitively on the signature of the tidal shear tensor eigenvalues. With two positive eigenvalues and one negative, $(-++)$, we will see strong collapse along two directions. Dependent on the overall overdensity, along the third axis collapse will be slow or not take place at all. Likewise, a sheetlike membrane will be the product of a $(--+)$ signature, while a $(+++)$ signature inescapably leads to the full collapse of a density peak into a dense cluster.

For a proper understanding of the Cosmic Web we need to invoke two important observations stemming from intrinsic correlations in the primordial stochastic cosmic density field. When restricting ourselves to overdense regions in a Gaussian density field we find that mildly overdense regions do mostly correspond to filamentary $(-++)$ tidal signatures [37]. This explains the prominence of filamentary structures in the cosmic Megaparsec matter distribution, as opposed to a more sheetlike appearance predicted by the Zeld'ovich theory. The same considerations lead to the finding that the highest density regions are mainly confined to density peaks and their immediate surroundings.

The second, most crucial, observation [8] is the intrinsic link between filaments and cluster peaks. Compact highly dense massive cluster peaks are the main source of the Megaparsec tidal force field: filaments should be seen as tidal bridges between cluster peaks. This may be directly understood by realizing that a $(-++)$ tidal shear configuration implies a quadrupolar density distribution (12). This means that an evolving filament tends to be accompanied by two massive cluster patches at its tip. These overdense protoclusters are the source of the specified shear, explaining the canonical *cluster-filament-cluster* configuration so prominently recognizable in the observed Cosmic Web.

6 The Cosmic Web and Galaxy Rotation: MMF Analysis

With the cosmic web as a direct manifestation of the large scale tidal field we may wonder whether we can detect a connection with the angular momentum of galaxies or galaxy halos. In sect. 2 we have discussed how tidal torques generate the rotation of galaxies. Given the common tidal origin we would expect a significant correlation between the angular momentum of halos and the filaments or sheets in which they are embedded. It was [23] who pointed out that this link should be visible in alignment of the spin axis of the halos with the inducing tidal tensor, and by implication the large scale environment in which they lie.

In order to investigate this relationship it is necessary to isolate filamentary features in the cosmic matter distribution. A systematic morphological analysis of the cosmic web has proven to be a far from trivial problem, though there have recently been some significant advances. Perhaps the most rigorous program, with a particular emphasis on the description and analysis of filaments, is that of the *skeleton* analysis of density fields by [28, 41]. Another strategy has been followed by [14] who identify clusters, filaments, walls and voids in the matter distribution on the basis of the tidal field tensor $\partial^2\phi/\partial x_i \partial x_j$, determined from the density distribution filtered on a scale of $\approx 5h^{-1}$Mpc.

The one method that explicitly takes into account the hierarchical nature of the mass distribution when analyzing the weblike geometries is the Multiscale Morphology Filter (MMF), introduced by [2]. The MMF dissects the cosmic web on the basis of the multiscale analysis of the Hessian of the density field. It starts by translating an N-body particle distribution or a spatial galaxy distribution into a DTFE density field (see [47]). This guarantees a morphologically unbiased and optimized density field retaining all features visible in a discrete galaxy or particle distribution. The DTFE field is filtered over a range of scales. By means of morphology filter operations defined on the basis of the Hessian of the filtered density fields the MMF successively selects the regions which have a bloblike (cluster) morphology, a filamentary morphology and a planar morphology, at the scale at which the morphological signal is optimal. By means of a percolation criterion the physically significant filaments are selected. Following a sequence of blob, filament and wall filtering finally produces a map of the different morphological features in the particle distribution.

With the help of the MMF we have managed to find the relationship of shape (inertia tensor) and spin-axis of halos in filaments and walls and their environment. On average, the long axis of filament halos is directed along the axis of the filament; wall halos tend to have their longest axis in the plane of the wall. At the present cosmic epoch the effect is stronger for massive halos. Interestingly, the trend appears to change in time: low mass halos tended to be more strongly aligned but as time proceeds local nonlinear interactions affect the low mass halos to such an extent that the situation has reversed.

The orientation of the rotation axis provides a more puzzling picture (Fig. 4). The rotation axis of low mass halos tends to be directed along the filament's axis while that of massive halos appears to align in the perpendicular direction. In walls there does not seem to exist such a bias: the rotation-axis of both massive and light haloes tends to lie in the plane of the wall. At earlier

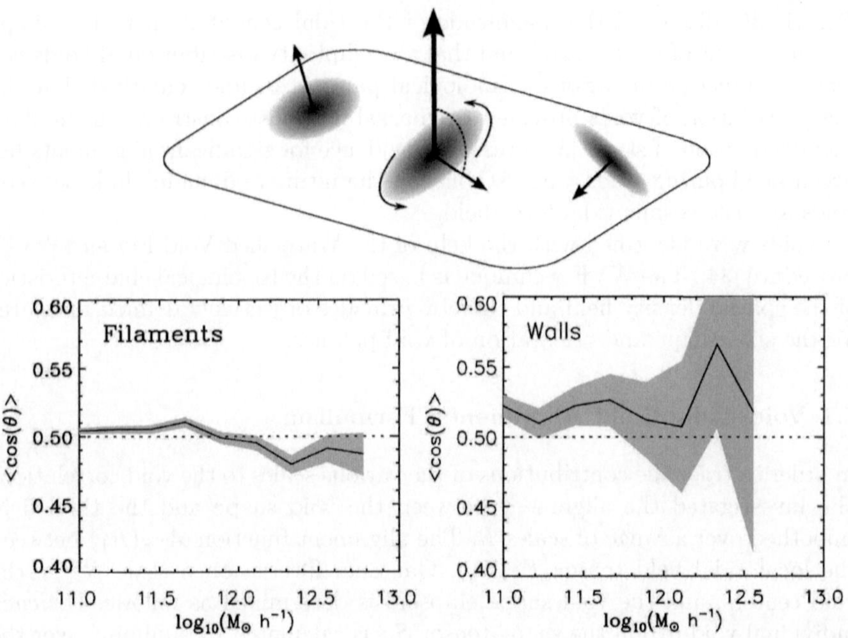

Fig. 4. Average alignment angle $\cos\theta$ between the halo spin direction and the orientation of the host structure as a function of halo mass, for filaments (left) and walls (right) in a ΛCDM N-body simulation. Filaments and walls were identified using the MMF technique. The dotted line indicates a uniform distribution of halo orientations. The shaded area corresponds to the standard deviation of 1,000 random realisations with the same number of galaxies as the halo sample and is wider in the case of walls due to the lower number of haloes in walls. From [1]

cosmic epochs the trend in filaments was entirely different: low mass halo spins were more strongly aligned as large scale tidal fields were more effective in directing them. During the subsequent evolution in high-density areas, marked by strongly local nonlinear interactions with neighbouring galaxies, the alignment of the low mass objects weakens and ultimately disappears.

7 Tidal Fields and Void Alignment

A major manifestation of large scale tidal influences is that of the alignment of shape and angular momentum of objects (see [8, 10]). The alignment of the orientations of galaxy haloes, galaxy spins and clusters with larger scale structures such as clusters, filaments and superclusters has been the subject of numerous studies (see e.g. [1, 4, 6, 7, 21, 26, 36, 39, 43]).

Voids are a dominant component of the Cosmic Web (see [40, 44], occupying most of the volume of space. Recent analytical and numerical work

[22, 31, 35] discussed the magnitude of the tidal contribution to the shape and alignment of voids. [22] found that the ellipticity distribution of voids is a sensitive function of various cosmological parameters and remarked that the shape evolution of voids provides a remarkably robust constraint on the dark energy equation of state. [35] presented evidence for significant alignments between neigbouring voids, and established the intimate dynamic link between voids and the cosmic tidal force field.

Voids were identified with the help of the Watershed Void Finder (WVF) procedure [34]. The WVF technique is based on the topological characteristics of the spatial density field and thereby provides objectively defined measures for the size, shape and orientation of void patches.

7.1 Void-Tidal Field Alignments: Formalism

In order to trace the contributions of the various scales to the void correlations [35] investigated the alignment between the void shape and the tidal field smoothed over a range of scales R. The alignment function $\mathcal{A}_{TS}(R_1)$ between the local tidal field tensor $T_{ij}(R_1)$, Gaussian filtered on a scale R_1 at the void centers, and the void shape ellipsoid is determined as follows. For each individual void region the shape-tensor $\mathcal{S}_{ij}$ is calculated by summing over the N volume elements k located within the void,

$$\begin{aligned} \mathcal{S}_{ij} &= -\sum_k x_{ki} x_{kj} \qquad &\text{(offdiagonal)} \\ \mathcal{S}_{ii} &= \sum_k \left(\mathbf{x}_k^2 - x_{ki}^2\right) \qquad &\text{(diagonal)}\,, \end{aligned} \tag{12}$$

where $\mathbf{x}_k$ is the position of the k-th volume element within the void, with respect to the (volume-weighted) void center $\bar{\mathbf{r}}_v$, i.e. $\mathbf{x}_k = \mathbf{r}_k - \bar{\mathbf{r}}_v$. The shape tensor $\mathcal{S}_{ij}$ is related to the inertia tensor $\mathcal{I}_{ij}$. However, it differs in assigning equal weight to each volume element within the void region. Instead of biasing the measure towards the mass concentrations near the edge of voids, the shape tensor $\mathcal{S}_{ij}$ yields a truer reflection of the void's interior shape.

The smoothing of the tidal field is done in Fourier space using a Gaussian window function $\hat{W}^*(\mathbf{k}; R)$:

$$T_{ij}(\mathbf{r}; R) = \frac{3}{2}\Omega H^2 \int \frac{d\mathbf{k}}{(2\pi)^3} \left(\frac{k_i k_j}{k^2} - \frac{1}{3}\delta_{ij}\right) \hat{W}^*(\mathbf{k}; R)\, \hat{\delta}(\mathbf{k})\, e^{-i\mathbf{k}\cdot\mathbf{r}}$$

Here, $\hat{\delta}(\mathbf{k})$ is the Fourier amplitude of the relative density fluctuation field at wavenember $\mathbf{k}$.

Given the void shape $\mathcal{S}_{ij}$ and the tidal tensor T_{ij}, for every void the function $\Gamma_{TS}(m, R_1)$ at the void centers is determined:

$$\varGamma_{TS}(m;R_1) = -\frac{\sum_{i,j}\tilde{S}_{m,ij}\,T_{ij}(\mathbf{r}_m;R_1)}{\tilde{S}_m\,T(\mathbf{r}_m;R_1)} \tag{13}$$

where $T(\mathbf{r}_m;R_1)$ is the norm of the tidal tensor $T_{ij}(\mathbf{r_m})$ filtered on a scale R_1 and $\tilde{S}_m$ is the norm of the shape tensor of void m.

The void-tidal alignment $\mathcal{A}_{TS}(R_1)$ at a scale R is then the ensemble average

$$\mathcal{A}_{TS}(R_1) = \langle\,\varGamma_{TS}(R_1)\,\rangle\,. \tag{14}$$

which we determine simply by averaging $\varGamma_{TS}(m,R_1)$ over the complete sample of voids.

7.2 Void-Tidal Field Alignments: Results

A visual impression of the strong relation between the void's shape and orientation and the tidal field is presented in the lefthand panel of Fig. 5 (from [35]). The tidal field configuration is depicted by means of (red-coloured) tidal bars. These bars represent the compressional component of the tidal force field in the slice plane, and have a size proportional to its strength and are directed along the corresponding tidal axis. The bars are superimposed on the pattern of black solid watershed void boundaries, whose orientation is emphasized by means of a bar directed along the projection of their main axis.

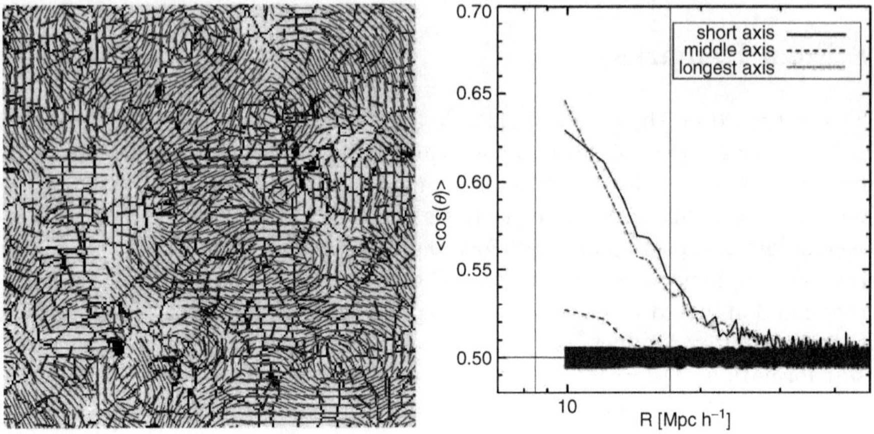

Fig. 5. *Left*: on the landscape with WVF void boundaries the tidal field compressional component is represented by tidal bars (*red*), representing the direction and strength of the tidal field. Also depicted are the void shape bars (*blue*). *Right*: the dotted line shows $\mathcal{C}_{TS}$, the alignment between the compressional direction of the tidal field and the shortest shape axis. For comparison the short axis alignment is also superimposed. From Platen et al. 2008

The compressional tidal forces tend to be directed perpendicular to the main axis of the void. This is most clearly in regions where the forces are strongest and most coherent. In the vicinity of great clusters the voids point towards these mass concentrations, stretched by the cluster tides. The voids that line up along filamentary structures, marked by coherent tidal forces along their ridge, are mostly oriented along the filament axis and perpendicular to the local tidal compression in these region. The alignment of small voids along the diagonal running from the upper left to the bottom right is particularly striking.

A direct quantitative impression of the alignment between the void shape and tidal field, may be obtained from the righthand panel of Fig. 5. The figure shows $\mathcal{C}_{TS}$ (dotted line), the alignment between the compressional direction of the tidal field and the shortest shape axis. It indicates that the tidal field is instrumental in aligning the voids. To further quantify and trace the tidal origin of the alignment one can investigate the local shape-tide alignment function $\mathcal{A}_{TS}$ (14) vs. the smoothing radius R_1.

This analysis reveals that the alignment remains strong over the whole range of smoothing radii out to $R_1 \approx 20 - 30h^{-1}$Mpc and peaks at a scale of $R_1 \approx 6h^{-1}$Mpc. This scale is very close to the average void size, and also close to the scale of nonlinearity. This is not a coincidence: the identifiable voids probe the linear-nonlinear transition scale. The remarkably strong alignment signal at large radii than $R_1 > 20h^{-1}$Mpc (where $\mathcal{A}_{TS} \approx 0.3$), can only be understood if large scale tidal forces play a substantial role in aligning the voids.

8 Final Remarks

The last word on the origin of galactic angular momentum has not been said yet. It is now a part of our cosmological paradigm that the global tidal fields from the irregular matter distribution on all scales is the driving force, but the details of how this works have yet to be explored. That is neither particularly demanding nor particularly difficult, it is simply not trendy: there are other problems of more pressing interest. The transition from shear dominated to rotation dominated motion is hardly explored and will undoubtedly be one of the principal by-products of cosmological simulations with gas dynamics and star formation.

The role of tidal fields has been found to be more profound than the mere transfer of angular momentum to proto-objects. The cosmic tidal fields evidently shape the entire distribution and dynamics of galaxies: they shape what has become known as the "cosmic web". Although we see angular momentum generation in cosmological N-Body simulations it is not clear that the simulations do much more than tell us what happened: galaxy haloes in N-Body models have acquired spin by virtue of tidal interaction. We draw comfort from the fact that the models give the desired result.

Nikos Voglis' approach was somewhat deeper: he wanted to understand things at a mechanistic level rather than simply to simulate them and observe the result. In that he stands in the finest tradition of the last of the great Hellenistic scientists, Hipparchus of Nicaea, who studied motion of bodies under gravity. Perhaps we should continue in the spirit of Nikos' work by trying to understand things rather than simply simulate them.

Nikos was a good friend, a fine scientist and certainly one of the kindest people one could ever meet. It was less than one year ago when we met for the last time at the *bernard60* conference in Valencia. We were of course delighted to see him and we shall cherish that brief time together.

Acknowledgments

We thank Panos Patsis and George Contopoulos for the opportunity of delivering this tribute to our late friend. We are grateful to Volker Springel for allowing us to use Fig. 3. We particularly wish to acknowledge Miguel Aragón-Calvo and Erwin Platen for allowing us to use their scientific results: their contributions and discussions have been essential for our understanding of the subject.

References

1. Aragón-Calvo M.A., Jones B.J.T., van de Weygaert R., van der Hulst J.M.: Astrophys. J. **655**, L5 (2007)
2. Aragón-Calvo M.A., Jones B.J.T., van de Weygaert R., van der Hulst J.M.: Astron. Astrophys. **474**, 315 (2007)
3. Barnes J., Efstathiou G.: Astrophys. J. **319**, 575 (1987)
4. Basilakos S., Plionis M., Yepes G., Gottlöber S., Turchaninov V.: Mon. Not. Roy. Astron. Soc. **365**, 539 (2006)
5. Bett P., Eke V., Frenk C.S., Jenkins A., Helly J., Navarro J.: Mon. Not. Roy. Astron. Soc. **376**, 215 (2007)
6. Binggeli B.: Astron. Astrophys. **107**, 338 (1982)
7. Bond J.R.: in *Nearly Normal Galaxies*, ed. S. Faber (Springer, New York), p. 388 (1987)
8. Bond J. R., Kofman L., Pogosyan D.: Nature **380**, 603 (1996)
9. Catelan, P., and Theuns, T.: Mon. Not. Roy. Astron. Soc. **282**, 436 (1999)
10. Desjacques V.: arXiv0707.4670 (2007)
11. Doroshkevich A.G.: Astrophysics **6**, 320 (1970)
12. Efstathiou G., Jones B. J. T.: Mon. Not. Roy. Astron. Soc. **186**, 133 (1979)
13. Feynman, R.P.: *Feynman Lectures On Physics*, (Addison Wesley Longman) (1970)
14. Hahn O., Porciani C., Carollo M., Dekel A.: Mon. Not. R. Astron. Soc. **375**, 489 (2007)

15. Harrison, E.R.: Mon. Not. Roy. Astron. Soc. **154**, 167 (1971)
16. Hirata C.M., Seljak U.: Phys. Rev. D, 70, 063526 (2004)
17. Hoyle F.: in *Problems of Cosmical Aerodynamics*, ed. J. M. Burgers and H. C. van de Hulst (Dayton: Central Air Documents Office), 195 (1949)
18. Icke V.: Astron. Astrophys. **27**, 1 (1973)
19. Jones B.J.T.: Rev. Mod. Phys. **48**, 107 (1976)
20. Jones B. J. T., Efstathiou G.: Mon. Not. Roy. Astron. Soc. **189**, 27 (1979)
21. Lee J., Evrard A.E.: Astrophys. J. **657**, 30 (2007)
22. Lee J., Park D.: arXiv0704.0881 (2007)
23. Lee J., Pen U.: Astrophys. J. **532**, L5 (2000)
24. Lee J., Pen U.: Astrophys. J. **555**, 106 (2001)
25. Lee J., Pen U.: arXiv0707.1690 (2007)
26. Lee J., Springel V., Pen U-L., Lemson G.: arXiv0709.1106 (2007)
27. Massey R., Rhodes J., Ellis R., Scoville N., Leauthaud A., Finoguenov A., Capak P., Bacon D., Aussel H., Kneib J.-P., Koekemoer A., McCracken H., Mobasher B., Pires S., Refregier A., Sasaki S., Starck J.-L., Taniguchi Y., Taylor A., Taylor J.: Nature **445**, 286 (2007)
28. Novikov D., Colombi S., Doré O.: Mon. Not. R. Astron. Soc. **366**, 1201 (2006)
29. Oort J.H.: Astron. Astrophys. **7**, 381 (1970)
30. Palmer P.L., Voglis N.: Mon. Not. Roy. Astron. Soc. **205**, 543 (1983)
31. Park D. and Lee J.: Astrophys. J. **665**, 96 (2007)
32. Peebles P. J. E.: Astrophys. J. **155**, 393 (1969)
33. Peebles P. J. E.: *The large-scale structure of the universe*, (Princeton University Press) (1980)
34. Platen E., van de Weygaert R., Jones B.J.T.: Mon. Not. Roy. Astron. Soc. **380**, 551 (2007)
35. Platen E., van de Weygaert R., Jones B.J.T.: Mon. Not. R. Astron. Soc., in press (2008)
36. Plionis M., Basilakos S.: Mon. Not. Roy. Astron. Soc. **329**, L47 (2002)
37. Pogosyan D.Yu.m Bond J.R., Kofman L., Wadsley J.: Cosmic Web: Origin and Observables. In: *Wide Field Surveys in Cosmology, 14th IAP meeting), eds. S. Colombi, Y. Mellier, (Editions Frontieres*, p. 61 (1998)
38. Porciani C., Dekel A., Hoffman Y.: Mon. Not. Roy. Astron. Soc. **332**, 325 (2002)
39. Rhee G., van Haarlem M., Katgert P.: Astron. J. **103**, 6 (1991)
40. Romano-Díaz E., van de Weygaert R.: Mon. Not. Roy. Astron. Soc. **382**, 2 (2007)
41. Sousbie T., Pichon C., Colombi S., Novikov D., Pogosyan D.: astroph/07073123 (2007)
42. Springel V., White S.D.M., Jenkins A., Frenk C.S., Yoshida N., Gao L., Navarro J., Thacker R., Croton D., Helly J., Peacock J.A., Cole S., Thomas P., Couchman H., Evrard A., Colberg J.M., Pearce F.: Nature **435**, 629 (2005)
43. Trujillo I., Carretero C., Patiri S. G.: Astrophys. J. **640**, L111 (2006)

44. Tully R.B., Shaya E.J., Karachentsev I.D., Courtois H., Kocevski D.D., Rizzi L., Peel A.: arXiv:0705.4139 (2007)
45. van den Bosch F. C., Abel T., Croft R. A. C., Hernquist L., White S. D. M.: Astrophys. J. **576**, 21 (2002)
46. van de Weygaert R., Bertschinger E.: Mon. Not. R. Astron. Soc. **281**, 84 (1996)
47. van de Weygaert R., Schaap W.E: The Cosmic Web: Geometric Analysis. In: *Data Analysis in Cosmology*, lectures summerschool Valencia 2004, ed. by V. Martínez, E. Saar, E. Martínez-González, M. Pons-Borderia (Springer-Verlag), 129pp. (2008)
48. van de Weygaert R., Bond J.R.: Clusters and the Theory of the Cosmic Web. In: *A Pan-Chromatic View of Clusters of Galaxies and the LSS*, eds. M. Plionis, D. Hughes, O. López-Cruz (Springer 2008)
49. White S. D. M.: Astrophys. J. **286**, 38 (1984)
50. Zel'dovich Ya. B.: Astron. Astrophys. **5**, 84 (1970)

Cosmological Inflation: A Personal Perspective

D. Kazanas

Astrophysics Science Division, Code 663, NASA/GSFC Greenbelt, MD 20771, USA
Demos.Kazanas@nasa.gov

Summary. We present a brief review of Cosmological Inflation from the personal perspective of the author who almost 30 years ago proposed a way of resolving the problem of Cosmological Horizon by employing certain notions and developments from the field of High Energy Physics. Along with a brief introduction of the Horizon and Flatness problems of standard cosmology, this lecture concentrates on personal reminiscing of the notions and ideas that prevailed and influenced the author's thinking at the time. The lecture then touches upon some more recent developments related to the subject and concludes with some personal views concerning the direction that the cosmology field has taken in the past couple of decades and certain speculations some notions that may indicate future directions of research.

1 Introduction

The development of General Relativity and the possibility it offers to probe the issues of the overall geometry, topology and evolution of the Universe as a whole it is certainly one of the great achievements of human spirit and captured my own imagination when I first came across an article by G. Chasapis on the "Universe" in the Greek encyclopedia "Helios." Since then, the field of Cosmology has been for me an avocation of sorts, honed in time, as I pursued studies in physics first at the University of Thessaloniki, through courses offered by professors G. Contopoulos and S. Persides and then through a large number of discussions with my late roommate and fellow graduate student at the University of Chicago B. Xanthopoulos as well as from interactions with my late thesis advisor D. N. Schramm. Since this is a brief personal account and not a review, I would like to apologize in advance to many for the absence of a large number of important references and contributions to the subject.

Following the original cosmological models of Einstein, de Sitter, Lemaitre, Friedman and others and the discovery of the expansion of the Universe by Hubble, the next development came through the realization (Gamow, Alpher) that the present expansion of the Universe implies that at an earlier stage it should have been sufficiently hot for nuclear reactions to take place. This then,

G. Contopoulos, P.A. Patsis, *Chaos in Astronomy.* Astrophysics and Space Science Proceedings, doi: 10.1007/978-3-540-75826-6,

supported by the discovery of the Cosmic Microwave Background (CMB) radiation, led to the development of Big Bang Nucleosynthesis (BBN) by Wagoner, Fowler & Hoyle [1] that still serves as a ruler against which all cosmological models have to be measured.

In the mid to late 70s, the subject of Cosmology was much less prominent than today, at least from the perspective of a graduate student, even one that specialized in astrophysics. The primary Cosmology text was Weinberg's book [2], wherein one could find the fundamentals of General Relativity and its application to relativistic objects, i.e. neutron stars and black holes, as well as the Universe itself. Its exposition of Cosmology provided, in addition to the general cosmological models, also the details of the thermal evolution of the universe and some of the open outstanding issues of standard cosmology namely the entropy (number of photons) per baryon $1/\eta$ in the cosmological fluid, and the issue of horizons.

The issue of the high value of $1/\eta$ ($\simeq 10^9$), compared to that found in a typical star ($\eta \simeq 1$), was given a prominent position both in [2] and also in Weinberg's, then new, more popular book "The First Three Minutes"[3]. Particular emphasis was given at the difficulty of producing such a large value for $1/\eta$ through dissipative processes given that the homogeneity and isotropy of the Universe that allows only for the effects bulk viscosity. However, as argued by the author, even this process could not add much more than a photon per baryon to the value of $1/\eta$.

2 The Cosmological Problems

At this point I would like to make a brief digression to outline the dynamics of the Universe and formulate the Cosmological problems of Horizon and Flatness. The root of both these problems lies in the fact that, in the system of units in which $h = c = 1$, the gravitational constant G has dimension of $(\text{mass})^{-2}$, the so-called Planck mass; this is the mass of particles for which the Schwarzschild and Compton lengths are equal, i.e. $2GM_P/c^2 = h/M_Pc$, or $G = hc/2M_P^2$ or $M_P = (hc/G)^{1/2} \simeq 10^{-5}$ gr. To this mass scale one can assign equivalent length, time and temperature scales of corresponding values $l_P \simeq 10^{-33}$ cm, $t_P \simeq l_P/c \simeq 10^{-43}$ s and $T_P \simeq 10^{32}$ K.

It is most amazing that the dynamics of the Universe as determined by the equations of General Relativity can be derived from purely Newtonian considerations. The facts that allow a Newtonian treatment of cosmology are that: (1) the Universe is homogeneous and isotropic, so any point can serve as the origin of a spherically symmetric coordinate system and (2) the property of the Newtonian potential that for a spherically symmetric matter distribution, the dynamics of the matter within a volume of radius a is determined only by the matter interior to a. Therefore, for a homogeneous and isotropic distribution, such as that of the Universe, one can choose the radius a arbitrarily and study the dynamics this sphere, all matter exterior to a being irrelevant. The

Hubble law indicating that velocities are proportional to the distance, then, guarantees that shells of different radii expand homologously and do not run onto each other.

One can, hence, write the equations of motions of a sphere of arbitrary radius a simply using the total energy integral, E, namely

$$\frac{1}{2}\dot{a}^2 - \frac{GM}{a} = \frac{1}{2}\dot{a}^2 - \frac{4\pi G\rho}{3}a^2 = E \quad \text{or} \quad H^2 = \frac{\dot{a}^2}{a^2} = \frac{2E}{a^2} + \frac{8\pi G\rho}{3} \tag{1}$$

It is instructive to compare this equation the corresponding Einstein equation for a homogeneous and isotropic Universe of spacial curvature $k = 1,\ 0, -1$ corresponding to a closed, flat or open Universe:

$$\frac{\dot{a}^2}{a^2} + \frac{k}{a^2} = \frac{8\pi G\rho}{3} \tag{2}$$

The role of the energy is played by the spatial curvature, $-k$, indicating that in a closed Universe ($k > 0$, $E < 0$) the radius of the sphere reaches a maximum while in flat and open universes it can reach infinity.

The solution of this equation requires an assumption about the variation of the density with time (or with a); this can be obtained from the conservation of energy, which reads $\rho \propto a^{-3}$ for pressureless matter and $\rho \propto a^{-4}$ for radiation while for temperature implies $T \propto 1/a$.

The only difference between the Newtonian and Einstein version of Cosmology becomes apparent only by differentiating (1) or (2) taking into account the relation between ρ and the pressure P from local energy conservation (Eq. (8) below) to obtain the corresponding force equation

$$\frac{\ddot{a}}{a} = -\frac{4\pi G}{3}(3P + \rho)\ . \tag{3}$$

This equation incorporates the contribution of pressure to the gravitational force, as it should, since pressure is energy density and all energy gravitates. The presence of this term, significant in the radiation era, has been verified by comparing the outcome of BBN to observation[5].

2.1 The Horizon Problem

The finite age of the Universe $t_U \simeq 1.4 \times 10^{10}$ yr $\simeq 5 \times 10^{17}$ s, along with the finite speed of light indicate that light signals since the creation of the Universe have traveled a distance $R_H \simeq ct_U \simeq 10^{28}$ cm. One can now estimate the size of R_H at the time its age was t_P and its temperature T_P, by scaling R_H by the ratio of the CMB temperatures at the two epochs, namely $R_P \simeq R_H(3\text{K}/10^{32}\ \text{K}) \simeq 10^{-3}$ cm. This size is 30 orders of magnitude larger than the horizon size at that time $ct_P \simeq l_P$, indicating that the Universe at that time comprised $\sim 10^{90}$ causally disconnected regions, all of which must have had approximately the same temperature since the Cosmic Microwave

Background (CMB) appears to be quite uniform across the observed Universe. This constitutes the Horizon problem. More formally, the size of the horizon must take into account the fact that the photon signal co-moves with the expanding Universe and it is thus given by

$$S_H = a(t) \int_0^t \frac{cdt}{a(t)} \tag{4}$$

One can see that for a power-law expansion rate $a(t) \propto t^p$ with $p < 1$ (as is the case for a radiation ($p = 1/2$) or matter ($p = 2/3$) dominated Universe) the horizon size is just a multiple of ct_U. However, for $p \geq 1$ the integral is dominated by the lower limit and the horizon diverges at $t \to 0$.

2.2 The Flatness Problem

For a given value of the ratio $H \equiv \dot{a}/a$, (1) defines a characteristic value of the density $\rho_c = 3H^2/8\pi G$, i.e. the density for which the explosion energy E is equal to zero, and use it to define the ratio of the density to the critical one as $\Omega = \rho/\rho_c$. We can now divide (1) by $\dot{a}^2$ to obtain

$$1 - \Omega = \frac{2E}{\dot{a}^2} = -\frac{k}{(Ha)^2} \tag{5}$$

Applying the above relation at two different values of a and the corresponding values of Ω we obtain

$$\Omega_1 - 1 = \frac{\dot{a}_0^2}{\dot{a}_1^2}(\Omega_0 - 1) \tag{6}$$

One can now see that if the present value $|\Omega_0 - 1| \simeq \mathcal{O}(1)$, then, given than in standard cosmology $a(t) \simeq K\,t^{1/2}$, $\dot{a}_0^2/\dot{a}_1^2 \simeq t_1/t_0$; since $t_0 \simeq 10^{17}$ s, at an earlier epoch with $t_1 \ll t_0$, $\Omega_1 \to 1$. If, in particular we set $t_1 \sim t_P \sim 10^{-43}$ s, $t_1/t_0 \simeq 10^{-60}$, i.e. under Standard Cosmology, at the Planck time, the radiation density was equal to the critical density to within 1 part in 10^{60}!

3 Phase Transitions, Baryogenesis

The focus placed by Weinberg on the value of η helped galvanized a couple of fellow graduate students including myself to take an independent look at this parameter in search for mechanisms that could account for its value. As far as I can now recall, our first attempt was to use Weinberg's prescription of bulk viscosity[4] but dare to consider its application to much higher temperatures and include much more massive particles than had been considered till then. However, we soon realized that no matter what the temperature and the particle masses, this process could add but a small number of photons per baryon in the cosmological fluid.

In search of other entropy producing processes I stumbled upon the idea of phase transitions and the entropy associated with the latent heat. Being aware that quarks were confined into baryons by a potential that grows (linearly) with distance, I considered that if this transition could be somehow delayed during the expansion of the Universe to densities lower than nuclear, the linear quark interaction could produce extremely large values of entropy *from the vacuum!* Because I considered such a situation rather contrived and poorly constrained, I suggested (in a publication[6] that received just a single citation[7]) that, even though there are overall no free quarks, it is possible that within a horizon volume there may be an excess of color, which would now interact via the quark linear potential with a similar color excess in an adjacent horizon volume. Assuming that the local color excess to be purely statistical, i.e. proportional to the square root of the particles within a given horizon volume, then one can calculate the amount of entropy produced as the universe expands. However, under these conditions the entropy thus released does not contribute significantly to $1/\eta$. Despite this fact, I was impressed by the possibility of energy production from the vacuum and thought it could have potentially significant consequences.

The issue of the value of $1/\eta$ was resolved in 1978 in an altogether different and far more subtle way (e.g. [8, 9] and others): The production of a large number of photons per baryon was supplanted by the production of a small excess of baryons over antibaryons in an originally symmetric cosmic fluid; this entailed invoking processes that violated baryon conservation, the CP symmetry and thermodynamic equilibrium. These processes were apparently possible within the context of Grand Unified Theories, i.e. theories that unified the strong with the weak and electromagnetic interactions at energies $\sim 10^{15}$ GeV.

While the issue of the photon to baryon ratio $1/\eta$ was resolved in principle as above, the issue of entropy production from the vacuum was still extremely appealing to me and my thought was that perhaps this could help resolve the remaining open cosmological problem, that of the Horizon.

4 Resolving the Horizon Problem

At the end of 1978 I got my PhD, left Chicago and spent the following year (1979) in the Greek military. Upon my discharge I returned to the US having been offered an NRC fellowship at GSFC by Floyd Stecker. On my way back to the US I spent a few days at Nordita in Denmark, where K. Sato had been also a visitor. He was very much interested in phase transitions in the early universe and we did discuss some of the issues of the quark–baryon one outlined above with one of his comments being that he was interested "in a different type of phase transition."

This last comment caught my attention enough to launch a (not so thorough, as it turned out) search for this different type of phase transition; the search produced only one relevant paper [10], which however involved the quark–baryon transition I was already aware of. At the same time, my interest in the horizon problem was rekindled by a paper by Brown & Stecker[11], which considered the interesting possibility of a matter - antimatter domain Universe produced by a phase transition-like violation of the CP symmetry with the order parameter taking randomly values of either -1 or $+1$ within each domain. The Horizon Problem is at the very heart of this proposal because the size of these domains is limited by the Horizon size at temperatures $\sim 10^{15}$ GeV, at which the baryon asymmetry is formed. In one of the references of [11] I found then a citation to [12] who discussed very much the same problem. The authors of [12] showed that because of the discrete nature of the CP–symmetry the corresponding phase transition produced a network of walls separating the two phases and that the wall network corresponds to a perfect fluid with equation of state $P = -2\rho/3$; this then leads to an expansion rate for the Universe $a(t) \propto t^2$ which, as discussed above can lead to domain sizes sufficiently large to avoid contradiction with observations on the existence of antimatter in space. This provided a resolution of sorts of the Horizon Problem, except for the fact that the resulting Universe would be very inhomogeneous due to the presence of these walls, in contradiction with observation.

At this point, I noticed a paper [15] discussing phase transitions within the context of Spontaneous Symmetry Breaking (SSB), a subject that had been extensively treated by [13, 14]. These are not unlike those discussed in [12] but the broken symmetries are not necessarily discrete and hence they do not have to lead to inhomogeneities. The work of [13] was very instructive: It showed that the energy stored in the self-interacting Higgs field ϕ, a fundamental ingredient of SSB, acts as a perfect fluid with an equation of state $P_v = -\rho_v$ with the vacuum expectation value of ϕ and the energy density ρ_v having the temperature dependence given in Fig. 1: (*i*) For $T > T_c$, $\langle\phi\rangle = 0$ and its energy density is $\rho_v \simeq a_{bb}T_c^4 = \text{constant} < \rho_r = a_{bb}T^4$ (ρ_r is the radiation energy density, a_{bb} is the black body constant). (*ii*) For $T < T_c$, $\langle\phi\rangle \neq 0$ and the energy density $\rho_v = \epsilon_v = a_{bb}T^4$ decreases with the temperature T but remains comparable to that of radiation ρ_r. This phase transition does not involve the confinement of quarks and does not suffer from the problems with that discussed earlier.

The effects of such a phase transition on the evolution of the Universe can be easily studied by considering that the total pressure and energy density consist of the sum of radiation and the vacuum, i.e. $P = P_r + P_v$ and $\rho = \rho_r + \rho_v$, with each obeying its own equation of state, i.e. $P_r = \rho_r/3$ and $P_v = -\rho_v$. The solution to Einstein's equation

$$\left(\frac{\dot{a}}{a}\right)^2 + \frac{k}{a^2} = \frac{8\pi G}{3}(\rho_r + \epsilon_v) \tag{7}$$

requires also the knowledge of variation of ρ, ϵ_v with time or with a. The relation between a and T is given by the first law of thermodynamics [16]

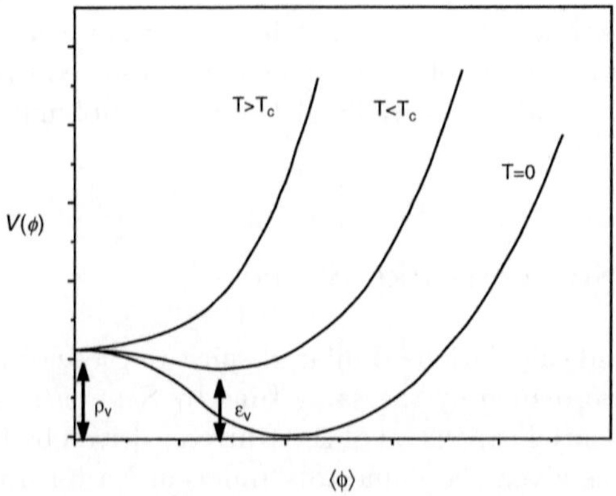

Fig. 1. The temperature dependent Higgs potential. For $T > T_c$, $\langle\phi\rangle = 0$ and the vacuum energy density ρ_v is constant but insignificant. For $T < T_c$ $\langle\phi\rangle \neq 0$ and the vacuum energy density ϵ_v depends on T, being zero for $T \to 0$

$$a^3 \frac{d\rho}{da} = -3(P+\rho)a^2 \quad \text{or} \quad a^3 \frac{d}{da}(\rho_r + \epsilon_v) = -4\rho_r a^2 \tag{8}$$

after taking into consideration the corresponding equations of state. This last equation then leads to the following relations between a and T or ϵ and T:

$$a \propto \frac{1}{T^2} \quad \text{or} \quad T \propto \frac{1}{a^{1/2}} \quad \text{and} \quad \epsilon_v,\ \rho_r \propto \frac{1}{a^2} \tag{9}$$

The presence of a vacuum component makes therefore a great deal of difference in the cooling of the universe: As long as $\epsilon_v \propto T^4$, the universe has to expand by twice as many decades to cool by the same factor as under adiabatic conditions. During this period the vacuum energy is never dominant but it is comparable to that of radiation and keeps feeding into it as ϵ_v slowly decreases. It was argued in [16] that this behavior should terminate at some point, else it would over-dilute the baryon/photon ratio.

With the relation between a and T (9) it is easy to compute the evolution of a (2) to obtain $a \propto t$, indicating that the horizon diverges logarithmically for $t \to 0$. However, this divergence is very mild and it is unlikely that it can resolve the Horizon Problem. Motivated by the work of [13, 14] and prompted by the referee of the paper I considered also the case $\epsilon_v \propto T^2$ for $T < T_c$. The slower decrease in the vacuum energy density then gave a very different relation between a and T, namely[16]

$$\frac{a}{a_c} = \frac{T_c}{T} exp\left[\frac{1}{4}\left(\frac{T_c^2}{T^2} - 1\right)\right] \quad \text{for } T < T_c \tag{10}$$

where a_c is the value of a when the temperature drops to the critical one T_c. This expression leads to a much slower decrease of T with a, which, when substituted into (2) yields an exponential expansion $a \propto exp[t^{1/3}]$, which can expand the Horizon size to values much larger than R_H, thereby resolving the Horizon Problem in a robust way. One can also see that an exponential expansion quickly renders the RHS of (5) $\ll 1$, resolving also the Flatness Problem.

5 "Nothing Succeeds like Success"

Considerations and calculations similar in spirit to those discussed above were worked out at approximately the same time by Sato [17] and Guth [18]; the early stage exponential expansion of the Universe driven by the energy density of the vacuum was given the name[18] "Inflation," a term resonant with the state of the US economy at the time, which has been since adopted universally, despite the subsequent change in the state of the US economy. The evolution of the Universe as described in [17, 18] proceeds through the formation of bubbles with $\rho_v = 0$ surrounded by exponentially expanding space of $\rho_v \neq 0$; the hope was that eventually the $\rho_v = 0$ regions would occupy the entire volume of the universe, which in the mean time had inflated enough to resolve the Horizon and Flatness problems. The problem was that, due to a secondary minimum of $V(\phi)$ at $\phi = 0$, the transition rate to $\rho_v = 0$ was too slow to complete the transition. This shortcoming was overcome in the "New Inflation" [19] where the Universe was considered to "slowly roll" down on a potential similar to that corresponding to $T = 0$ in Fig. 1, with the expansion dominated by a roughly constant ϵ_v and with the present horizon constituting a small patch of the expanding universe with $\Omega = 1$ with extremely high accuracy.

However, the most important feature of the "New Inflation" is that it affords a process that can produce the fluctuations necessary for the formation of cosmological structure: During the "slow-roll" period of the evolution of the Universe the geometry of space is that of de Sitter space with a cosmological horizon at a constant coordinate distance. Quantum fluctuations of the field ϕ created with constant amplitude $\delta\phi$ decrease until they cross the de Sitter horizon; then, as they are stretched by the expansion of the Universe to super-horizon scales, their amplitude freezes to a constant value; this is so because the metric perturbation at a given length scale is proportional to $\delta\phi$ while the latter $\delta\phi \propto V_{,\phi}/V$ is almost constant during the "slow roll" phase. After the end of the phase transition, the Universe resumes its conventional expansion; as the horizon size increases the fluctuations come within the horizon at roughly the constant amplitude they had when exiting the de Sitter horizon to produce the Harrison–Zeldovich spectrum of cosmological perturbations. These have subsequently left their imprint as fluctuations on the CMB temperature which were recently measured by both the COBE and WMAP [20] missions confirming the general predictions of the inflationary scenario.

A most interesting feature of the above process is that the amplitude of perturbations depends on the shape of the potential $V(\phi)$ near $\phi = 0$ and the energy scale of inflation. Furthermore, small deviations of the fluctuation spectrum from the precise Harrison–Zeldovich form, can also give an estimate of the number of e–foldings of inflation which was found (for the simplest models) to be of order of 60–70[20], suggesting an expansion by a factor of roughly 10^{30}, the minimum required to reconcile the disparity between the size of the universe and the Planck length at $t = t_P$ discussed in Sect. 2.

While the issue of the horizon size or the flatness of the universe are resolved in an appealing way by the inflationary scenario, these issues provide little additional quantitative evidence in support of its fundamental premises. However, the production, amplitude and spectrum of the resulting matter fluctuations and their imprint on the CMB, the result of the quantum fluctuations of the field ϕ, provides a much more rigorous instrument of scrutiny of the above ideas. The interested reader can find of all these in the modern literature (e.g. the monograph by Mukhanov[21]). While there has been at least one (sound in my opinion) objection against the entire "Inflationary" edifice[22], in the absence of a competing framework that can account for the CMB observations, this has gained little traction; it is then not unreasonable to conclude that in science as in business "nothing succeeds like success."

6 Conclusions and Speculations

It is fair to say that the ideas of Cosmological Inflation provided the impetus and the mathematical tools for tracing (with great success) the evolution of the Universe at an era impossible to imagine 30 years ago. In my personal view, a great deal of the appeal of this scenario lies in its simplicity: The mathematics of the original inflation proposal are almost trivial, while the complexity of even the theory of fluctuations is moderate.

That being said, again in my personal view, the characterization of Inflation as a "scenario" rather than a "theory" is also not unfair. To begin with, while in its original versions the scalar field employed to resolve the cosmological puzzles was considered to be the Higgs field, for reasons unknown to me, this association was dropped in favor of an altogether independent scalar field ϕ (the inflaton), unencumbered by such an association (perhaps because of the constraints it imposed on the models). Furthermore, the all important self-interaction potential $V(\phi)$ of this field remains (again in my personal, poorly informed view) in the realm of phenomenology. Despite these objections, the concordance of its predictions with the CMB data has set the bar for future alternative, competing schemes.

On the other hand, the confidence in the "slow-roll" aspects of the theory led to the notion of eternal inflation, i.e. that, though unlikely, sufficiently large fluctuations, can produce volumes of the universe that inflate much faster than the parent domains which in their turn also self-reproduce and so on (see e.g. [23] and references therein). Each such region becomes, then, a universe of its own, with (presumably) different values of the inflation field ϕ and possibly different value of the physical constants. With the apparent proliferation of "universes," the question is whether our present domain of the universe is special. To the best of my understanding, the prevailing view is that we live in the domain with the proper parameters to foster life, thereby enunciating a truly cosmic version of the Copernican view. So while the success of the inflationary scenario rests with its consistency with the CMB observations, some of its other (more far reaching) consequences are outside the domain of the observable. The question that arises, then, is whether one should accept all these implications as true or consider the inflationary scenario as an *ansatz* that simply provides a framework within which one can work out and fit the CMB fluctuations. The answers to these questions lie possibly in future more accurate observations or alternative theoretical developments.

To provide an example of a theory that addresses coincidences with fine tuning akin in precision to that of the standard cosmology, I will refer to the locally scale invariant theory of gravity considered in [25, 26]. This theory is defined by the unique action

$$I_W = -\alpha \int d^4x(-g)^{1/2} C^{\alpha\beta\gamma\delta} C_{\alpha\beta\gamma\delta} \tag{11}$$

whose static spherically symmetric geometry with a charge Q reads

$$g_{00} = 1/g_{rr} = 1 - 3\beta\gamma - \beta(2 - 3\beta\gamma)/r - Q^2/(8\alpha\gamma r) + \gamma r - kr^2 \tag{12}$$

where β, γ, k are integration constants. One should note first that in this theory charge modifies geometry the same way as mass, possibly evading the problems that the Q^2/r^2 term of the Einstein gravity solution entails! For $\gamma = 0$ this metric is that of Schwarzschild–de Sitter. However the linear term (analogous to the quark potential) is totally novel and being asymptotically non-flat, it is reasonable to associate γ with the inverse Hubble length R_H. The presence of this term suggests deviations of order 1 from the Newtonian potential at distances such that $2M/r \simeq r/R_H$ or $r^2 \simeq 2MR_H$ i.e. at a radius that is the geometric mean of the Schwarzschild and the Hubble radius. It is interesting to note that classes of virialized of objects (including the Universe for which $2M \simeq R_H$) lie on this line (Fig. 2a). This is relevant to inflation because the mass-radius relation associated with galaxies and their clusters supposedly originates in the inflationary perturbations. One could suggest that these systematics (known as Tully–Fisher and Larson relations in galaxies and star forming regions respectively) are due to the non-linear dynamics of clustering. However, extrapolation of this relation by 60(!) orders of magnitude to the mass of the electron (Fig. 2b), yields for the radius the classical electron radius!!! So, it is not only the size of the Universe at the Planck time that presents us with a fine tuning problem. There are numerical relations equally astounding but very little understood even outside the standard gravity and cosmology, apparently related to the metric of (12).

As discussed in Sect. 2 the origin of the Horizon and Flatness problems has its origin in the presence of a scale in the gravitational Lagrangian. Actually, Inflation proposes that the universe at creation had a size equal to the

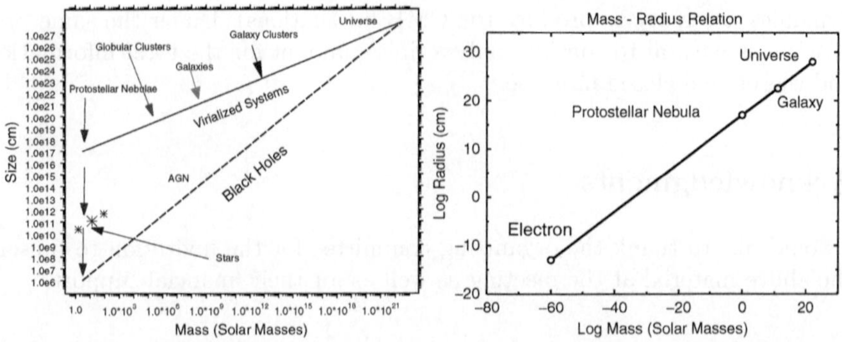

Fig. 2. *Left*: The Mass-Radius relation $r^2 = 2MR_H$ (solid) along with black hole line $r = 2M$; arrows point to the regions occupied by classes of the objects noted. *Right*: The same relation extrapolated to the mass of the electron.

Planck length and mass equal to the Planck mass, gaining mass as it inflated. As described above, inflation as of today does not predict the real size of the universe, since we cannot estimate how long this period lasts and, even worse (depending on one's view), it allows for the possibility of a huge number of disjoint domains (Multiverse). At this section of speculations I would like to venture to a totally different point of view which at present provides only hints on directions to be followed in the future. It involves the notion of information, which, as it has been suggested, may lie at the root of all physics [24]. To be sure, it is easy to see that the Special Theory of Relativity rests on and can be formulated on the condition of a finite, maximum information transfer speed, namely c. Pursuing a similar line of thought, I would suggest Quantum Mechanics as the framework for imposing a finite, maximum information density, namely h. Moving on, the role of Gravity, then, is to provide the free energy necessary to process the available information; as such, it also provides a sense for the direction of time, in fact gives rise to time itself. What about the total amount of information? A (pre-Inflation) universe of size equal to l_P and mass M_P contains only one bit of information. In my view such a universe is rather uninteresting and likely to remain virtual. This immediately raises the issue of whether any amount of information can be converted from virtual to real (while these considerations border the metaphysical, so are those pondering the existence of other universes totally inaccessible to us). Perhaps this is possible only for sufficiently large number of bits (10^{90}; one of the reasons gravity is so weak?) and a very specific geometrical arrangement is needed (as discussed in [22]) for their conversion from virtual to real presumably by the influence of gravity [22] (assuming one bit per horizon size at $t = t_P$, the observed universe contained at the Planck time 10^{90} bits; the value of α in the action of (11), needed to make the term $Q^2/\alpha\gamma$ in (12) equal to the electron gravitational radius is $\sim 10^{60}$). Such a proposal would resolve in a different way the issues of Horizon and Flatness (but it would also need to provide the dynamics necessary to produce the CMB fluctuations). Under the same proposal, it is natural to consider only a finite amount for the total information and therefore a closed universe.

Acknowledgments

I would like to thank the organizing committee for the invitation to present the above material at the meeting as well as for their financial support.

References

1. R.V. Wagoner, W.A. Fowler & F. Hoyle: ApJ, **148**, 3 (1967)
2. S. Weinberg, "Gravitation and Cosmology", Wiley, New York (1972)
3. S. Weinberg "The First Three Minutes", Basic Books, New York (1977)
4. S. Weinberg: ApJ, **167**, 175 (1971)
5. S. Rappaport et al.: Phys. Rev. D, **77**, 023515 (2008)
6. D. Kazanas, *Proc. of the Seminar on Proton Stability*, ed. D. Cline, Univ. of Wisconcin Madison (1978)
7. A. D. Dolgov & Y. B. Zel'dovich: Rev. Mod. Phys., **53**, 1 (1981)
8. M. Yoshimura: Phys. Rev. Lett., **41**, 381, (1978)
9. S. Dimopoulos & L. Susskind: Phys. Rev., **D18**, 4500 (1978)
10. G. Lasher: Phys. Rev. Lett., **42**, 164, (1979)
11. R. W. Brown & F. W. Stecker: Phys. Rev. Lett., **43**, 315 (1979)
12. Y. B. Zel'dovich, Y. I. Kobzarev & B. L. Okun: Sov. Phys.–JEPT, **40**, 1 (1975)
13. D. A. Kirzhnits & A. D. Linde: Ann. Phys., **101**, 195 (1976)
14. A. D. Linde: Rep. Prog. Phys., **42**, 389 (1979)
15. A. Zee: Phys. Rev. Lett., **44**, 703 (1980)
16. D. Kazanas: ApJ, **241**, L59 (1980)
17. K. Sato: Phys. Lett. B, **99**, 66 (1981)
18. A. Guth: Phys. Rev. D, **23**, 347 (1981)
19. A. Albrecht & P. J. Steinhardt: Phys. Rev. Lett., **48**, 1220 (1982)
20. D. Spergel et al.: ApJS, **170**, 377 (2007)
21. V. F. Mukhanov: "Physical Foundations of Cosmology", Cambridge University Press, Cambridge (2005)
22. R. D. Penrose: "*The Large the Small and the Human Mind*", Cambridge University Press, Cambridge (1997)
23. A. Linde: hep-th/0211048 (2002)
24. S. Lloyd: Nature, **406**, 1047 (2000)
25. P. D. Mannheim & D. Kazanas: ApJ, **342**, 635 (1989)
26. P. D. Mannheim & D. Kazanas: Phys. Rev. D, **44**, 417 (1991)

Astrophysics and Space Science Proceedings

Diffuse Matter from Star Forming Regions to Active Galaxies, edited by T.W. Hartquist, J.M. Pittard, S.A.E.G. Falle
Hardbound ISBN 978 1-4020-5424-2, December 2006

Solar, Steller and Galactic Connections between Particle Physics and Astrophysics, edited by A. Carramiñana, F.S. Guzmán, and T. Matos
Hardbound ISBN 978 1-4020-5574-4, December 2006

Island Universes, Structure and Evolution of Disk Galaxies, by R.S. de Jong
Hardbound ISBN 978-1-4020-5572-0, December 2006

Mapping the Galaxy and Nearby Galaxies, edited by K. Wada, F. Combes
Hardbound ISBN 978-0-387-72767-7, September 2007

African Cultural Astronomy, Current Archaeoastronomy and Ethnoastronomy research in Africa, edited by J.C. Holbrook, R.T. Medupe, J.O. Urama
Hardbound ISBN 978-1-4020-6638-2, December 2007

Galaxies in the Local Volume, edited by B.S. Koribalski, H. Jerjen
Hardbound ISBN 978-1-4020-6932-1, June 2008

Chaos in Astronomy, Conference 2007, edited by G. Contopoulos, P.A. Patsis
Hardbound ISBN 978-3-540-75825-9, August 2008

For further information about this book series we refer to the following web site: www.springer.com

GPSR Compliance

The European Union's (EU) General Product Safety Regulation (GPSR) is a set of rules that requires consumer products to be safe and our obligations to ensure this.

If you have any concerns about our products, you can contact us on ProductSafety@springernature.com

In case Publisher is established outside the EU, the EU authorized representative is:

Springer Nature Customer Service Center GmbH
Europaplatz 3
69115 Heidelberg, Germany

Batch number: 10755310

Printed by Printforce, the Netherlands